Protein Expression Technologies

Current Status and Future Trends

Edited by

François Baneyx
Department of Chemical Engineering,
University of Washington, Seattle, WA 98195, USA

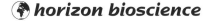

Copyright © 2004
Horizon Bioscience
32 Hewitts Lane
Wymondham
Norfolk NR18 0JA
U.K.

www.horizonbioscience.com

British Library Cataloguing-in-Publication Data

A catalogue record for this book is available from the British Library

ISBN: 0-9545232-5-3

Printed and bound in Great Britain

Contents

Contributors

François Baneyx*
Department of Chemical Engineering
Box 351750
University of Washington
Seattle
WA 98195
USA
baneyx@u.washington.edu

Michael Betenbaugh*
Department of Chemical and
Biomolecular Engineering
Johns Hopkins University
Baltimore
Maryland 21218
USA
beten@jhu.edu

Sierd Bron
Department of Molecular Genetics
Rijksuniversiteit Groningen
Kerklaan 30
9751 NN Haren
The Netherlands

Yu-chan Chao
Institute of Molecular Biology
Academia Sinica
Nankang
Taipei 115
Taiwan

Donald A. Comfort
Department of Chemical Engineering
North Carolina State University
Raleigh
NC 27695-7615
USA

Kelly A. Fitzgerald
Department of Chemical Engineering
Box 352180
University of Washington
Seattle
WA 98195
USA

Amy M. Grunden*
Department of Microbiology
North Carolina State University
Raleigh
NC 27695-7615
USA
amy_grunden@ncsu.edu

Siger Holsappel
Department of Molecular Genetics
Rijksuniversiteit Groningen
Kerklaan 30
9751 NN Haren
The Netherlands

John T. A. Hsu
Division of Biotechnology and
Pharmaceutical Research
National Health Research Institutes
Taipei
Taiwan

Sun Ok Hwang
Department of Biological Sciences
Korea Advanced Institute of Science
and Technology
373-1 Kusong-Dong
Yusong-Gu
Daejon 305-701
Korea

Sanjay Jain
Process R & D
Shire Biologics Inc
Northborough
MA 01532
USA

Jay D. Keasling*
Department of Chemical Engineering
University of California
Berkeley
CA 94720
USA
keasling@socrates.berkeley.edu

Aaron S. Kelley
Genencor International, Inc.
925 Page Mill Road
Palo Alto
CA 94304
USA

Robert M. Kelly*
Department of Chemical Engineering
North Carolina State University
Raleigh
NC 27695-7615
USA
rmkelly@eos.ncsu.edu

Gyun Min Lee*
Department of Biological Sciences
Korea Advanced Institute of Science
and Technology
373-1 Kusong-Dong
Yusong-Gu
Daejon 305-701
Korea
gmlee@mail.kaist.ac.kr

Gary J. Lesnicki
Fermentation Pilot Plant
University of British Columbia
Vancouver
B.C. V6T 1Z3
Canada

Mary E. Lidstrom*
Department of Chemical Engineering
Box 352180
University of Washington
Seattle
WA 98195
USA
lidstrom@u.washington.edu

Erica L. Malotky
Department of Microbiology
North Carolina State University
Raleigh
NC 27695-7615
USA

Vincent J.J. Martin
Department of Chemical Engineering
University of California
Berkeley
CA 94720
USA

Hendrik J. Meerman*
Genencor International, Inc.
925 Page Mill Road
Palo Alto
CA 94304
USA
hmeerman@genencor.com

Rob Meima*
DSM Food Specialties
R&D Genetics
Wateringseweg 1
2600 MA Delft
The Netherlands
Rob.Meima@dsm.com

Mirna Mujacic
Department of Chemical Engineering
Box 351750 University of Washington
Seattle
WA 98195
USA

Ronald T. Niebauer
Department of Chemical Engineering
University of Delaware
Newark
DE 19711
USA

John F. Nomellini
Department of Microbiology and
Immunology
University of British Columbia
Vancouver
British Columbia V6T 1Z3
Canada

Anne Skaja Robinson*
Department of Chemical Engineering
University of Delaware
Newark
DE 19711
USA
robinson@che.udel.edu

John Smit*
Department of Microbiology and
Immunology
University of British Columbia
Vancouver
British Columbia V6T 1Z3
Canada
jsmit@interchange.ubc.ca

Christina D. Smolke
Department of Chemical Engineering
University of California
Berkeley
CA 94720
USA

Michael C. Toporowski
Department of Microbiology and
Immunology
University of British Columbia
Vancouver
British Columbia V6T 1Z3
Canada

Jan Maarten van Dijl
Department of Pharmaceutical Biology
Rijksuniversiteit Groningen
A. Deusinglaan 1
9713 AV Groningen
The Netherlands

Karthik Viswanathan
Department of Chemical and
Biomolecular Engineering
Johns Hopkins University
Baltimore
Maryland 21218
USA

Michael Ward
Genencor International, Inc.
925 Page Mill Road
Palo Alto
CA 94304
USA

Bruce L. Zamost*
BioProcess Development
ZymoGenetics Inc. Seattle Washington
98102
USA
zamostb@zgi.com

*Corresponding author

Books of Related Interest

Full details of all these books at: **www.horizonbioscience.com**

Preface

With a growing demand for purified proteins for biochemical studies, structural genomics endeavors and clinical trials, the field of protein expression has been very active in recent years. Genetic modification of host organisms has been successfully implemented to improve the yield and quality of recombinant proteins and to reduce production costs. Yet, no single organism is without disadvantage and many options are available. Usually, the choice of an expression system is driven by a number of considerations. These include whether or not the desired gene product requires post-translational modifications (*e.g.*, the formation of complex patterns of disulfide bonds or human-like glycosylation), whether it should be produced in a soluble or insoluble form or secreted, whether fusion tags are acceptable, and what yields will be required. Many other factors affect this equation including proteolytic stability of the target protein, acceptable ratio of active to inactive product, as well as budget and time available for production. Often, however, the choice of an expression system is based upon in-house expertise and familiarity with a subset of expression systems.

The objective of this book is to showcase the desirable features and point out the limitations of current and emerging expression technologies under a single cover. All chapters were designed to provide both beginning and expert readers with the necessary information to make an educated choice on the expression technology that is best suited for their needs. Following a thorough review of extremophile biology, Chapter 1 sets the stage for the rest of the volume by highlighting the difficulties associated with the production of extremophilic proteins and introducing the various expression systems that have been used for this purpose. The next two chapters deal with *Escherichia coli* expression, one of the oldest and most extensively engineered expression systems. Chapter 2 focuses on recent advances in the understanding of protein folding, degradation and export in *E. coli* and the practical implications of these findings. The complementary Chapter 3 describes metabolic engineering tools and approaches to facilitate the production of complex proteins and fine chemicals in *E. coli*. *Bacillus* species genetics and expression is the focus of Chapter 4, where particular emphasis is placed on protein secretion. Chapter 5 addresses potential pitfalls and solutions in *Saccharomyces cerevisiae* expression and describes emerging technologies involving this organism. The genetic tools and fermentation strategies available for protein production in the methylotrophic yeasts *Pichia pastoris* and *Pichia methanolica* are the subject of Chapter 6. Chapter 7 deals with filamentous fungi expression with a focus on secretion and fermentation. Insect cell expression and recent development involving the baculovirus system are reviewed in Chapter 8, while Chapter 9 showcases the basic features and strategies for the construction of Chinese Hamster Ovary (CHO) cell lines for large-scale protein production. The last two chapters deal with emerging expression systems: Chapter 10 assesses the current state of methylotrophic bacteria expression with an emphasis on *Methylobacterium extorquens* AM1, while Chapter 11 describes the usefulness of *Caulobacter crescentus* for heterologous protein production and cell surface display.

It is my hope that this volume will prove a valuable reference on current and emerging expression technologies and will instill a desire to consider alternative organisms for heterologous protein expression. In closing, I would like to thank all authors for their good humor, hard work and quality contributions.

François Baneyx
Department of Chemical Engineering
University of Washington
Seattle, WA

December 12, 2003

1

Expression of Extremophilic Proteins

Amy M. Grunden, Donald A. Comfort,
Erica L. Malotky, and Robert M. Kelly

Abstract

Proteins from extremophilic microorganisms, those inhabiting niches which are characterized by biologically-extreme temperatures, ionic strengths, pressures, and/or pHs, have been examined for their scientific merit and biotechnological potential. Difficulties encountered in obtaining the native forms of these proteins have fostered efforts to express recombinant versions in foreign hosts and, in some cases, in the extremophile itself. While proteins from a variety of extremophiles have been successfully produced in recombinant forms, there are significant problems that can be encountered that go beyond those normally found with recombinant protein production. Overcoming these problems is key to future scientific efforts as well as to use of extremophilic proteins in technological applications.

Extremophile Biology

Extremophiles are organisms that thrive in biologically extreme environments. The study of these organisms has engendered a great deal of interest in the physical and geochemical factors that limit life, as well as in ways to exploit enzymes from these environments for technologically significant applications (Horikoshi *et al.*, 1998; Madigan *et al.*, 1997; Seckbach, 2000). The boundaries of the biosphere include physical extremes, such as temperature, radiation, or pressure, as well as geochemical extremes, that include salinity, pH, and redox potential. The focus here is on recombinant protein expression associated with those organisms living in the extremes of temperature, salinity and pH - thermophiles, hyperthermophiles, psychrophiles, halophiles, acidophiles, and alkaliphiles. As can be seen from Table 1, a variety of extremophiles have served as sources of recombinant proteins. It should be noted that the disproportionate number of thermophilic and hyperthermophilic recombinant proteins produced, compared to other extremophilic sources, reflects the considerable amount of genome sequence information for these organisms in addition to particular biotechnological interest in thermostable and

Table 1. Comparison of numbers of recombinant proteins produced from various extremophiles[a]

Type of Extremophile	Number of Organisms That Are Sources of Recombinant Protein	Number of Different Recombinant Proteins Produced
Thermophiles/ Hyperthermophiles	34	177
Psychrophiles	11	13
Halophiles	7	14
Acidophiles	3	4
Alkaliphiles	8	8

[a]The numbers listed are based on referenced papers from the Pubmed database.

thermoactive enzymes (Adams *et al.*, 1995; Adams *et al.*, 1998). Interest in other types of extremophiles as recombinant proteins sources will likely increase in coming years.

Thermophiles and Hyperthermophiles

Thermophiles, organisms with optimal growth temperatures ranging from 45 to 80 °C, have been isolated from a number of environments on Earth including geothermal springs, solfataric fields, sunlight-heated soils, and compost piles (Madigan *et al.*, 1999). Hyperthermophiles, organisms with optimal growth temperatures >80 °C, are routinely isolated from natural environments, such as deep-sea and shallow thermal sediments, hydrothermal vent systems, geothermal springs and solfataras, and from hot industrial environments, such as the outflow of geothermal power plants and deep-earth drilling samples (Huber *et al.*, 2000). To date, a significant number of isolates have successfully been propagated from these thermal environments in pure laboratory culture, including members of both the Bacteria and Archaea and representing a broad spectrum of growth physiologies (methanogens, fermentative anaerobes, sulfur-reducers, sulfate reducers, and microaerophiles) (Huber *et al.*, 2001).

From study of these organisms, a number of molecular adaptations to thermophily have been identified which involve thermal stabilization of major cell components - proteins (Adams *et al.*, 1995; Ladenstein *et al.*, 1998), nucleic acids (Grogan, 1998; Marguet *et al.*, 2001), and lipids (van de Vossenberg *et al.*, 1998). The levels of thermal stabilization intrinsic to proteins from hyperthermophiles can be exceptional with some enzymes remaining active for extended periods of times at temperatures up to 140 °C (Adams *et al.*, 1994). Although exact attributes governing protein thermostabilization have not been completely defined, structural studies of thermostable enzymes suggest that a number of noncovalent features are important. These include: high numbers of hydrophobic residues in the protein cores, reduced surface-to-volume ratios, decreased glycine contents, high numbers of surface ionic interactions, and shortened loose N- and C-terminal regions, all of which contribute to overall thermostability to various extents in specific proteins (Adams *et al.*, 1995; Ladenstein *et al.*, 1998). The reduction of N- and C-terminal regions and the loss of destabilizing loops is evident in the reduced length of hyperthermophilic protein relative to mesophilic proteins, 268 ± 38 residues and 310 ± 16 residues, respectively (Chakravarty *et al.*, 2000; Thompson *et al.*, 1999). The general trend of increased surface ionic interactions is accomplished by the replacement of uncharged polar residues (Gln, Asn, Ser, Thr, and Cys) with charged residues (Arg, Lys, His, Asp, and Glu) (Chakravarty *et al.*, 2000). The amino acid composition can be important in some cases with hyperthermophilic proteins typically having only have the number of thermally labile glutamine residues compared to mesophiles (Vieille *et al.*, 2001).

Factors contributing to the thermal stabilization of DNA in thermophiles and hyperthermophiles have also been identified and include high intracellular levels of K^+ (Marguet *et al.*, 2001), reverse DNA gyrase (Forterre *et al.*, 1996; Lopez-Garcia, 1998), and histones and other DNA-binding proteins coating the chromosomal DNA (Pereira *et al.*, 1998; Sandman *et al.*, 2001). Potassium is present at greater than 1 M concentrations in some hyperthermophiles and has been shown to prevent depurination of DNA at high temperatures (Marguet *et al.*, 2001). The accumulation of high levels of K^+, however, does not appear to be a universal DNA thermostabilization adaptation, as many thermophiles and hyperthermophiles do not possess high intracellular K^+ levels. Positive supercoiling of DNA, on the other hand, does appear to be a universal thermal adaptation. All hyperthermophilic bacteria and archaea analyzed to date contain reverse DNA gyrase, which is responsible for imposing positive supercoils in chromosomal DNA (Lopez-Garcia, 1998); this is in contrast to non-hyperthermophilic organisms that typically have negatively supercoiled DNA. Positive supercoiling is thought to stabilize DNA against thermal effects by producing a higher linking number in the coiled DNA compared to negative supercoiling. Furthermore, several hyperthermophilic methanogens are known to contain histone-like proteins, which closely resemble eukaryotic histones both in function and structure (Sandman *et al.*, 2001). Archaeal histones wind chromosomal DNA into compact nucleosome-like structures that help maintain DNA in a double-stranded form at high temperature (Pereira *et al.*, 1997; Soares *et al.*, 1998). To date, archaeal histones have only been identified in members of the Euryarchaeota branch of Archaea, whereas members of the Crenarchaeota appear to contain other types of small DNA binding proteins, such as Sac7d from *Sulfolobus acidocaldarius*, which function in a manner similar to true histones (Robinson *et al.*, 1998).

Many thermophiles and hyperthermophiles have been characterized physiologically, and some have recently been the focus of genome sequencing efforts (Table 2). These sequences have substantially aided in the identification of new enzymes of both scientific and technological interest. They have also provided valuable gene sequence information required for cloning and recombinant expression of extremophile proteins in more commonly used mesophilic hosts, such as *Escherichia coli*. As can be seen from Table 2, the majority of available genome sequences come from thermophilic and hyperthermophilic organisms (19 genomes, compared to 12 genomes for all of the other extremophiles), which reflects the high degree of interest in these organisms from many perspectives including evolution and biotechnology.

Psychrotolerant Organisms and Psychrophiles

The majority of the Earth's biosphere (~70%) is permanently cold and contains a diverse set of organisms, including members of the Bacteria, Archaea and Eukarya (Cavicchioli, 2000). Most of the microbes isolated from these cold environs are either psychrotolerant organisms (can grow at temperatures near the freezing point of water but grow optimally at temperatures >20 °C) or psychrophiles (optimal growth at temperatures <15 °C) (Cavicchioli *et al.*, 2002). The cold environments psychrophiles have been isolated from include polar and alpine regions, polar and deep waters, shallow subterranean regions, the upper atmosphere, refrigerated appliances and in or on animals and plants inhabiting cold regions (Cavicchioli *et al.*, 2000; Russell *et al.*, 1998). From preliminary findings, it has been suggested that there are unexpectedly large numbers of diverse psychrotolerant and psychrophilic bacteria that exist in cold surface waters and Arctic

Table 2. Extremophiles that have had their genomes sequenced[a] or are in the process of being sequenced

Type of Extremophile	Organism	Domain	Optimal Growth Condition	Organization Responsible for Sequence Generation	References
Thermophile	*Bacillus stearothermophilus*	Bacteria	65°C	University of Oklahoma	
	Geobacillus kaustophilus	Bacteria	60°C	Japan Marine Science and Technology Center	(Smith *et al.*, 1997)
	Methanobacterium thermoautotrophicum	Archaea	65°C	Genome Therapeutics and Ohio State University	(Bao *et al.*, 2002)
	Thermoanaerobacter tencongensis	Bacteria	75°C	Beijing Genomics Institute	
	Thermus thermophilus	Bacteria	65°C	Goettingen Genomics Laboratory	
	Thermoplasma volcanium	Archaea	60°C	AIST	(Kawashima *et al.*, 2000)
Thermophilic acidophile	*Thermoplasma acidophilum*	Archaea	59°C, pH 2	Max-Planck Institute for Biochemistry	(Ruepp *et al.*, 2000)
Hyperthermophile	*Aeropyrum pernix*	Archaea	95°C	NITE	(Kawarabayasi *et al.*, 1999)
	Aquifex aeolicus	Bacteria	85°C	Diversa	(Deckert *et al.*, 1998)
	Archaeoglobus fulgidus	Archaea	87°C	TIGR	(Klenk *et al.*, 1997)
	Methanococcus jannaschii	Archaea	85°C	TIGR	(Bult *et al.*, 1996)
	Methanopyrus kandleri	Archaea	98°C	Fidelity Systems, Inc.	(Slesarev *et al.*, 2002)
	Pyrobaculum aerophilum	Archaea	100°C	Caltech and UCLA	(Fitz-Gibbon *et al.*, 2002)
	Pyrococcus abyssi	Archaea	103°C	GENOSCOPE	
	Pyrococcus furiosus	Archaea	100°C	Center of Marine Biotechnology and University of Utah	(Robb *et al.*, 2001)
	Pyrococcus horikoshii	Archaea	100°C	NITE	(Kawarabayasi *et al.*, 1998)
	Thermotoga maritima	Bacteria	80°C	TIGR	(Nelson *et al.*, 1999)
Hyperthermophilic acidophile	*Sulfolobus solfataricus*	Archaea	80°C, pH 2-4	Canadian & European Consortium	(She *et al.*, 2001)
	Sulfolobus tokodaii	Archaea	80°C, pH 2-4	NITE	(Kawarabayasi *et al.*, 2001)
Psychrophile	*Colwelia psychrerythraea*	Bacteria	15°C	TIGR	
	Desulfotalea psychrophila	Bacteria	18°C	REGX-Project	
	Methanococcoides burtonii	Archaea	23°C	JGI	
	Methanogenium frigidum	Archaea	15°C	UNSW	
Halophile	*Haloarcula marismortui*	Archaea	3.5 M NaCl	UMBI and Institute for Systems Biology	
	Halobacterium sp. NRC-1	Archaea	4.5 M NaCl	Halobacterium genome consortium	(Ng *et al.*, 2000)
	Halobacterium salinarium	Archaea	4.2 M NaCl	Max-Plank Institute for Biochemistry	
	Haloferax volcanii	Archaea	3.0 M NaCl	University of Scranton	
Acidophile	*Acidothiobacillus ferrooxidans*	Bacteria	pH 2	TIGR	
	Ferroplasma acidarmanus	Archaea	pH 0	TIGR and Univ. of Washington	
Alkaliphile	*Bacillus halodurans*	Bacteria	pH 10	Japan Marine Science and Technology Center	(Takami *et al.*, 2000)
	Oceanobacillus iheyensis	Bacteria	pH 9.5	Japan Marine Science and Technology Center	(Takami *et al.*, 2002)

[a]References are listed for those genomes that have been completed.

marine surface sediments (Brown *et al.*, 2001; Deming, 2002; Ravenschlag *et al.*, 2001), whereas Archaea, especially members of the Crenarchaeota, predominate in deep cold waters (Karner *et al.*, 2001).

Although relatively little is currently known about the true diversity and molecular adaptations of psychrotolerant and psychrophilic organisms compared to that known about the diversity and molecular adaptations of thermophiles and hyperthermophiles, research activity in these areas has been increasing. For instance, four sequencing projects have recently commenced for psychrotolerant/psychrophilic organisms that, when finished, will undoubtedly provide insight into molecular adaptations possessed by these organisms and information about potentially useful enzymes (See Table 2).

Given that the cold temperatures completely permeate these cells, it is known that significant molecular adaptations must occur for cold-adapted organisms, including cell membranes, transport systems, nucleic acids, and proteins. Studies of cold-adapted organisms have identified cellular responses to counteract reduced enzyme reaction rates and solute uptake, decreased membrane fluidity, inhibitory nucleic acid structures and the formation of intracellular crystalline ice (Cavicchioli *et al.*, 2002). For molecular adaptation of proteins from psychrophiles, it has become clear that overall structural strategies are employed to improve catalytic efficiency of the enzyme at low temperature either by increasing the specific activity (k_{cat}) or the affinity of substrates (K_m), or both (Zecchinon *et al.*, 2001). The improved low-temperature catalytic activity of psychrophilic enzymes is likely achieved through an increase in flexibility, either of a selected region or the entire protein structure. This provides the enhanced capacity to undergo catalytically-required conformational changes at low temperatures (Fields *et al.*, 1998; Zavodszky *et al.*, 1998). Increased flexibility in psychrophilic enzymes is generally achieved by a combination of structural features, some of which are juxtaposed to those found in thermophiles. These can include: a decrease in protein core hydrophobicity; reduced surface ionic and electrostatic interactions; increased charge of surface residue that promote surface solvent interaction; increased numbers and longer surface loops;

substitutions of glycines for prolines in loops; decreased arginine/lysine ratios; and fewer aromatic residue interactions (Gerday *et al.*, 2000; Gianese *et al.*, 2001; Lonhienne *et al.*, 2000). It has also been shown that this higher degree of protein flexibility directly correlates to increased thermolability of psychrophilic enzymes (Zecchinon *et al.*, 2001). Thermolability of psychrophilic enzymes is an obvious concern for the recombinant expression of cold-adapted enzymes.

Halophiles

Microorganisms that require the presence of salt for growth are referred to as halophiles and can be found in all three domains of life, Bacteria, Archaea, and Eukarya. Various types of halophiles can be distinguished based on the organism's salt tolerance/requirement as follows: slight halophiles grow optimally at 3% (w/v) NaCl, moderate halophiles grow optimally at 3-15% (w/v) NaCl, and extreme halophiles grow optimally at 25% (w/v) NaCl (Kushner, 1978). Many marine microbes are considered slight halophiles, as seawater is typically 3% (w/v) NaCl. Bacteria belonging to the genus *Halomonas* can adapt to salt over a broad concentration range and archaeal Halobacteriales are unable to grow below 15-20% (w/v) NaCl and are therefore extreme halophiles (Madigan *et al.*, 1999). Halophilic organisms can be isolated from hypersaline lakes such as the Dead Sea, the Great Salt Lake, saltern evaporation ponds, salted foods, and saline soils (Madigan *et al.*, 1999).

There are two predominant strategies employed by halophilic organisms to maintain osmotic balance in their high salt environments. The first strategy that is used by members of the Halobacteriales involves the intracellular accumulation of high concentrations of inorganic salts. Typically, these organisms accumulate K^+ as the primary cation and Cl^- as the primary anion, through a combination of a Na^+/H^+ antiporter and chloride transporters (Eisenberg *et al.*, 1992; Oren, 1999). The second strategy commonly used by halophiles, including most halophilic Bacteria, eukaryotic algae and fungi, and methanogenic Archaea, involves maintaining low intracellular concentrations of salt by manufacturing high levels of other intracellular solutes (compatible solutes) to establish osmotic balance (Galinski, 1995). Common compatible solutes produced by halophiles include glycerol, glycine betaine, and ectoine (Galinski, 1995). These compatible solutes, which in addition to regulating osmotic balance in halophilic cells, also have been shown to stabilize proteins by preventing unfolding induced by heating, freezing, or drying (Galinski, 1995).

In the case of halophiles that maintain osmotic balance by accumulating high intracellular concentrations of counter ions, fundamental adaptations to structural and enzymatic proteins occur in order to retain function (Dennis *et al.*, 1997; Lanyi, 1974; Madern *et al.*, 2000). Recent study of the halophilic glutamate dehydrogenase from *Halobacterium salinarum* has revealed that protein stability under high salt concentrations is maintained by having a high number of acidic amino acids relative to basic amino acids and a decreased surface hydrophobic character (Britton *et al.*, 1998). A thermodynamic 'solvation-stabilization hypothesis' has been proposed to explain the stabilization of halophilic proteins in terms of protein-solvent interactions (Ebel *et al.*, 1999). Based on this hypothesis, stability of folded protein structure requires a network of hydrated ions associated with acidic residues at the surface of the protein. High salt is needed to maintain weak hydrophobic interactions due to the low hydrophobic nature of halophilic enzymes. A second hypothesis proposes that halophilic proteins have evolved to harness the high ionic concentrations of their environments to balance their stability and solubility (Richard *et al.*, 2000). In fact, the salt-stabilization mechanism is not specific to halophiles - they have just evolved to take advantage of these higher salt concentrations (Wright *et al.*, 2002). Studies with halophilic enzymes, especially those isolated from extreme halophiles, have demonstrated that NaCl concentrations of at least 1-2 M are required to maintain protein integrity and that protein denaturation rapidly occurs as salt concentrations are reduced (Eisenberg *et al.*, 1992).

Recently the biotechnological potential of halophilic organisms has received more attention with respect to developing industrial application related to halophilic enzymes, compatible solutes, and various halophilic biopolymers, such as biosurfactants, lectins, and bioplastics (Margesin *et al.*, 2001). This increased interest in industrial application of halophiles and the continued desire to understand molecular adaptations to halophily has prompted the inception of a number of sequencing projects devoted to halophilic organisms (see Table 2), as well as the development of expression systems for the production of halophilic proteins.

Acidophiles

Acidophiles are those organisms that grow optimally in the pH range of 0 – 5.5 and have representatives in all three domains of life. Well-characterized eukaryotic acidophiles include the red alga *Cyanidium caldarium* (optimum growth pH 2-3) (Doemel *et al.*, 1971), the green alga *Dunaliella acidophila* (optimum growth pH 1), and various fungi including *Cephalosporium* sp. and *Trichosporon cerebriae* (Schleper *et al.*, 1995). Archaeal acidophiles include the aerobic heterotrophs *Picrophilus oshimae* and *Ferroplasma acidarmanus* (optimal growth at 0.7 and 0 pH, respectively) (Edwards *et al.*, 2000; Schleper *et al.*, 1995), while the best-characterized bacterial acidophile is *Acidiothiobacillus ferrooxidans* (Ingledew, 1982). In nature, acidophiles can be isolated from acidic soils, solfataras, acidic mine drainages, and acidic industrial outflows (Rothschild *et al.*, 2001).

Studies have shown that acidophiles have adapted to their low-pH environments primarily by maintaining the pH of their cytoplasm at near neutral pH. The ability to maintain a neutral intracellular pH obviates the need for acidophilic organisms to have developed specific internal molecular adaptations to low pH (Rothschild *et al.*, 2001). Maintenance of near-neutral internal pHs have been shown to result from high levels of positive cell surface charges, high internal buffer capacity, and the overexpression of H^+ export proteins (Pick, 1999). The genome sequences of two acidophiles, *Acidothiobacillus ferrooxidans* and *Ferroplasma acidarmanus*, have recently been initiated in an effort to more fully understand the physiology of acidophiles and the molecular adaptations to acidophily (see Table 2).

Industrial interest in acidophiles has been spurred by discoveries that the sulfur oxidation ability of certain acidophiles can be exploited to remove sulfur from coal and rubber and that acidophiles can be used to improve metal recovery from mine outflows (Demirjian *et al.*, 2001). Efforts have also recently been initiated to identify and isolate acidophilic sugar degradation enzymes, such as amylases and glucose isomerases, which could potentially be used in acidic steps of high fructose sugar conversion processes (Demirjian *et al.*, 2001).

Alkaliphiles

Alkaliphiles are organisms capable of growing optimally at a pH range of 8.5 – 11.5; alkaliphilic representatives of the Bacteria, Archaea, and Eukarya have been isolated (Rothschild *et al.*, 2001). Alkaliphiles are routinely isolated from natural habitats, such as alkaline soils, soda lakes, soda deserts, and alkaline hot springs existing in volcanic areas. They can also be found in alkaline environments produced as a result of industrial activity, including cement manufacture, mining, paper and pulp manufacture and food-processing effluents (Javor, 1989). The greatest diversity of alkaliphilic organisms is arguably associated with soda lakes, as numerous bacterial, archaeal, and eukaryotic organisms have been isolated from such environments (Jones *et al.*, 1998).

As with acidophiles, alkaliphiles tend to cope with their environmental pH extreme by maintaining near neutral intracellular pH, again alleviating the necessity of evolving molecular adaptations to extreme pH conditions for internal structures (Rothschild *et al.*, 2001). Maintenance of internal neutral pH in alkaliphiles involves active import of H^+, the production of negatively charged cell-wall polymers, and the use of unusual membrane complexes for bioenergetics (Krulwich *et al.*, 1998).

Historically, alkaliphiles and their enzymes have been studied with an eye towards identifying and characterizing enzymes with industrial utility. To date, a number of alkaliphilic cellulose-degrading enzymes, starch-degrading enzymes, lipases, esterases, and proteases have been isolated (Demirjian *et al.*, 2001; Takami *et al.*, 2000). Interest in alkaliphilic enzymes triggered recent genome sequencing efforts for two alkaliphiles, *Bacillus halodurans* and *Oceanobacillus iheyensis* (See Table 2). Furthermore, the completed sequence and subsequent genome analysis of *Bacillus halodurans* has led to the identification of a wealth of potential candidate enzymes of industrial interest which are currently being studied (Takami *et al.*, 2000).

Genetic Elements and Shuttle/Expression Systems from Extremophiles

Genetic Elements from Thermophiles and Hyperthermophiles

In the past few years, a number of genetic elements have been isolated from a variety of thermophilic and hyperthermophilic archaea and bacteria, including plasmids, viruses, and insertion sequences (Noll *et al.*, 1997; Sowers *et al.*, 1999; Zillig *et al.*, 1998). The focus here is on identification of plasmids in extremophiles, since a number of these naturally isolated plasmids are potentially good candidates for, or have been engineered for use as, shuttle vectors and/or expression vectors.

As more thermophilic and hyperthermophilic organisms are being isolated and characterized, many endogenous plasmids are also being discovered. These endogenous plasmids range in size from 0.8 to 41 kb (Table 3) and most appear to be cryptic, since no phenotypic trait has been successfully linked to the plasmid in the majority of cases. In hyperthermophilic archaea, a number of plasmids have been identified and sequenced. Most of these plasmids have been isolated from either species of the aerobic, acidophilic Crenarchaeote, *Sulfolobus*, or from the strictly anaerobic, heterotrophic Euryarchaeote, *Pyrococcus*. In fact, the distribution of plasmids in the Thermococcales (*Pyrococcus* is a member of the order Thermococcales) has been investigated, and it was determined that out of 57 strains, 11 harbored plasmids which ranged in size from 3-24 kb. This indicates that the plasmids are diverse in nature and widely distributed among the

Table 3. Naturally occurring plasmids from extremophiles[a]

Plasmid Name	Host Organism	Plasmid Size (kb)	Plasmid Copy Number	Sequence ID	Special Features	References
Thermophiles/Hyperthermophiles						
pDL10	Acidianus ambivalens	7.6	high	AJ225333	Amplified in S-reducing growth mode; Involved in H_2S autotrophy	(Kletzin et al., 1999)
pGS5	Archaeoglobus profundus	2.8	high	N.D.	Negatively supercoiled	(Lopez-Garcia et al., 2000)
pGT5	Pyrococcus abyssi	3.4	35	U49503	Used in construction of shuttle vector	(Erauso et al., 1996)
pRT1	Pyrococcus sp. JT1	3.4	>30	AF393813	Good candidate for shuttle vector	(Ward et al., 2002)
pHE7	Sulfolobus islandicus	7.5	15	AJ294536		(Zillig et al., 1996)
pRN1	Sulfolobus islandicus	5.5	20	NC001771	Can also replicate in S.solfataricus	(Keeling et al., 1998)
pRN2	Sulfolobus islandicus	6.9	35	NC002101	Good candidate for shuttle vector	(Keeling et al., 1998)
pSSVx	Sulfolobus islandicus	5.7	high	AJ243537	Derived from recombination of pRN-like plasmid and SSV1 virus	(Arnold et al., 1999)
pSTH3	Sulfolobus neozealandicus	35.0	high	N.D.		(Zillig et al., 1998)
pSTH4	Sulfolobus neozealandicus	4.7	high	N.D.		(Zillig et al., 1998)
pTAV4	Sulfolobus neozealandicus	6.2	high	N.D.		(Zillig et al., 1998)
pTIK4	Sulfolobus neozealandicus	14.3	high	N.D.	May confer ability of host strain to outcompete other strains in culture	(Zillig et al., 1998)
pWHI1	Sulfolobus neozealandicus	6.8	low	N.D.		(Zillig et al., 1998)
pWHI2	Sulfolobus neozealandicus	15.4	high	N.D.		(Zillig et al., 1998)
pIT3	Sulfolobus solfataricus	4.9	~10	N.D.		(Zillig et al., 1998)
pMC24	Thermotoga maritima	0.85	high	Not released	Has 2 point mutations that distinguish it from the replication ORF in pRQ7	(Akimkina et al., 1999)
pRQ7	Thermotoga maritima	0.85	200	L19928	Negatively supercoiled, has 1 ORF	(Harriott et al., 1994)
Psychrophiles						
pFL1	Flavobacterium sp. KP1	2.3	16	AB007196	Has 2 ORFs with no known homologs; good candidate for shuttle vector	(Ashiuchi et al., 1999)
pMtBL	Pseudomonas haloplanktis	4.1	20-50	AJ224742	Has no detectable transcriptional activity	(Tutino et al., 2001)
pTADw	Psychrobacter sp. TA144	1.3	N.D.	AJ224744	Has novel replicon	(Tutino et al., 2000)
pTAUp	Psychrobacter sp. TA144	1.9	N.D.	AJ224743		(Tutino et al., 2000)

Table 3. Naturally occurring plasmids from extremophiles (continued)

Plasmid Name	Host Organism	Plasmid Size Number (kb)	Plasmid Copy ID	Sequence	Special Features	References
Halophiles						
pCM1	*Chromobacter marismortui*	17.5	N.D.	X86092	Narrow host range	(Mellado *et al.*, 1995)
pMH1	*Deleya halophila*	11.5	N.D.	N.D.	Stable in *E. coli*; has Kn^r, Tet^r, Neo^r	(Fernandez-Castillo *et al.*, 1992)
pGRB1	*Halobacterium sp.* GRB	1.8	180	XS2610	Used in bop expression in *H. halobium*	(Krebs *et al.*, 1991)
pHSB1	*Halobacterium sp.* SBS	1.7				(Hackett *et al.*, 1989)
pHV2	*Halobacterium volcanii*	6.4	6	NC001375		(Charlebois *et al.*, 1987)
pHK2	*Haloferax sp.* Aa2.2	10.5	N.D.	L29110	Used for construction of pMDS shuttle vectors	(Holmes *et al.*, 1995)
pHE1	*Halomonas elongata*	4.2	N.D.	AJ243135		(Vargas *et al.*, 1995)
pPL1	*Marinococcus halophiles*	3.9	N.D.	NC002095		(Louis *et al.*, 1997)
Acidophiles						
pTC-F14	*Acidothiobacillus caldus*	14.0	12-16	AF325537	Has protic plasmid stabilization; can replicate in *E. coli*	(Gardner *et al.*, 2001)
pTAV1	*Parococcus versutus*	107.0	low	N.D.		(Bartosik *et al.*, 1997)
pTF5	*Thiobacillus ferrooxidans*	19.8	N.D.	U73041	Codes for electron transport system	(Dominy *et al.*, 1998)
pTF191	*Thiobacillus ferrooxidans*	9.8	N.D.	N.D.		(Chakravarty *et al.*, 1995)
pTF-FC2	*Thiobacillus ferrooxidans*	12.2	N.D.	M64981	Cannot replicate in *E. coli*	(Rawlings *et al.*, 1994)
pTik12	*Thiobacillus intermedius*	65.0	low	LB6865		(English *et al.*, 1995)
pT3.2I	*Thiobacillus T3.2*	15.4	N.D.	AJ007958	Has Insertion Sequence elements	(Aparicio *et al.*, 2000)
Alkaliphiles[b]						
pAH1	Alkaliphilic halomonad	5.9	N.D.	N.D.	Has several unique restriction sites	(Fish *et al.*, 1999)
pAH2	Alkaliphilic halomonad	5.3	N.D.	N.D.	Has several unique restriction sites	(Fish *et al.*, 1999)
pAH3	Alkaliphilic halomonad	15.0	N.D.	N.D.		(Fish *et al.*, 1999)
pAH4	Alkaliphilic halomonad	33.0	N.D.	N.D.		(Fish *et al.*, 1999)
pAH5	Alkaliphilic halomonad	20.5	N.D.	N.D.		(Fish *et al.*, 1999)

Table 3. Naturally occurring plasmids from extremophiles (continued)

Plasmid Name	Host Organism	Plasmid Size Number (kb)	Plasmid Copy ID	Sequence	Special Features	References
Conjugative Plasmids						
pARN3/2	Sulfolobus sp.	26.1	low	N.D.		(Prangishvili et al., 1998)
pARN4/2	Sulfolobus sp.	26.5	low	N.D.		(Prangishvili et al., 1998)
pHEV14/5	Sulfolobus sp.	36.5	low	N.D.		(Prangishvili et al., 1998)
pING1	Sulfolobus sp.	24.5	low	AF23340	Has 38 ORFs	(Stedman et al., 2000)
pING4	Sulfolobus sp.	24.8	low	AF23340	Derived from transposition insertion in pING1	(Stedman et al., 2000)
pING6	Sulfolobus sp.	26.6	low	AF23340	Derived from transposition insertion in pING4	(Stedman et al., 2000)
pNOB8	Sulfolobus sp.	41.2	low	AJ010405	Can be maintained in other strains	(She et al., 1998)
pNOB8-33	Sulfolobus sp.	33.0	high	AJ010405	Occurs when pNOB8 is transferred to Sulfolobus strains that cannot maintain it	(She et al., 1998)

[a]The abbreviations used in the tables are as follows: Kn^r, kanamycin resistance; Neo^r, neomycin resistance; Tet^r, tetracycline resistance.
[b]The alkaliphilic halomonads are polyextremophiles exhibiting optimal growth at both high salt concentrations and high pH.

Thermococcales (Benbouzid-Rollet *et al.*, 1997). Of the Thermococcales plasmids, only the small, cryptic plasmids pGT5 from *Pyrococcus abyssi* (Erauso *et al.*, 1996) and pRT1 from *Pyrococcus* sp. JT1 (Ward *et al.*, 2002) have been characterized in any detail. Both plasmids are 3.4 kb, have two open reading frames (ORFs) each, and apparently replicate using a rolling-circle mechanism. The larger of the two open reading frames encodes the Rep protein required for plasmid replication, while the function of the smaller ORF is unknown, but may participate in copy number control. Plasmid pGT5 has served as the parent plasmid for construction of a shuttle vector for use in *Pyrococcus* strains, *Sulfolobus solfataricus* and *Escherichia coli* (Aravalli *et al.*, 1997). Work is currently underway to convert plasmid RT1 into a shuttle/expression vector for use in *Pyrococcus* (Ward *et al.*, 2002).

The diversity and distribution of plasmids from hyperthermophiles have best been described thus far for members of the genus *Sulfolobus*. To date, more than 25 plasmids and 30 novel viruses have been identified from *Sulfolobus* strains (Zillig *et al.*, 1998). These plasmids have been grouped into conjugative and nonconjugative plasmids, based on their modes of transmission. The nonconjugative plasmids range in size from 4.7 to 35 kb and for the most part seem to be cryptic (Zillig *et al.*, 1998). Four of the isolated and sequenced *Sulfolobus* plasmids (pRN1, pRN2, pHE7, and pSSVx) appear to be closely related and are grouped together as the pRN family of plasmids (Arnold *et al.*, 1999). These plasmids share significant homology in three ORFs as well as two non-coding regions. Two of the three ORFs appear to be involved in plasmid replication and its control with one ORF encoding the Rep protein and the other encoding a CopG homolog that has been implicated in plasmid copy number control (Lipps *et al.*, 2001). Plasmid pSSVx is a somewhat unique member of this pRN family of plasmids, as it appears to be a cross between a plasmid and a virus (Arnold *et al.*, 1999). Plasmid pSSVx was isolated from *Sulfolobus islandicus* strain REY15/4, which also harbors the novel virus SSV2. Plasmid pSSVx could be transferred to *Sulfolobus solfataricus* but only when the cells were also carrying the virus SSV2. Various shuttle/expression vectors have been developed for use in *Sulfolobus* using the autonomously replicating sequence from the virus

SSV1, based in part on studies with plasmid pSSVx (Cannio *et al.*, 1998; Contursi *et al.*, 2003; Stedman *et al.*, 1999).

A fairly large number of conjugative plasmids have also been isolated from various *Sulfolobus* strains. In fact a recent study revealed that 11 conjugative plasmids ranging in size from 24 – 41 kb were identified in 300 *Sulfolobus* isolates (Prangishvili *et al.*, 1998). The conjugative plasmids were shown to transfer readily from donor to recipient cells through direct contact, and without the aid of pili (Prangishvili *et al.*, 1998). The copy number of the plasmids was determined to be low in the host cell but increased to ~35 copies in the recipient cell; this plasmid amplification ultimately resulted in growth inhibition of the recipient cells (Prangishvili *et al.*, 1998; Schleper *et al.*, 1995). Significant genetic variation and rearrangement of the *Sulfolobus* conjugative plasmids has also been observed. For instance, conjugative plasmids pING4 and pING6 are transpositional derivatives of parent conjugative plasmid pING1 (Stedman *et al.*, 2000). The conjugative plasmid pNOB8 routinely sustains a 10 kb deletion when transferred to other *Sulfolobus* strains, thereby generating plasmid pNOB8-33 (She *et al.*, 1998). The relatively large size, seeming genetic instability, and growth inhibition of recipient strains has currently limited the possible use of the conjugative plasmids as shuttle/expression vectors.

To date, only two plasmids, pRQ7 and pMC24, have been isolated and characterized from hyperthermophilic bacteria (Akimkina *et al.*, 1999; Yu *et al.*, 1997). Both plasmids had been isolated from *Thermotoga* strains and are only 846 bp. The plasmids have been sequenced and were shown to code for a single Rep protein, with the Rep ORF in plasmid pMC24 containing two mutations distinguishing it from the Rep ORF from plasmid pRQ7. Further study of the distribution pattern of these plasmids in the *Thermotogales* suggests that they are very conserved and widely distributed, as highly related plasmids have been isolated from *Thermotoga* strains from distantly located hydrothermal vent systems (Akimkina *et al.*, 1999).

Genetic Elements from Psychrophiles

Until recently, information regarding the identification and characterization of genetic elements in psychrophiles has been fairly limited (Dahlberg *et al.*, 1997; Sobecky *et al.*, 1997). New studies, however, have reported the discoveries of plasmids from the psychrophilic bacteria *Psychrobacter* sp. strain TA144, *Pseudoalteromonas haloplanktis* strain TAC 125, and *Flavobacterium* sp. KP1 (Ashiuchi *et al.*, 1999; Tutino *et al.*, 2000; Tutino *et al.*, 2001). Two plasmids, pTAUp and pTADw, were isolated from the Gram-negative psychrophile, *Psychrobacter* sp. strain TA144. Plasmid pTAUp (1.9 kb) was sequenced and shown to encode a Rep protein and a second ORF of unknown function. The sequence for plasmid pTADw (1.3 kb) revealed the presence of a novel replication gene, which bears no homology to any previously known Rep protein. Analysis of the Rep gene of plasmid pTAUp suggests that the plasmid replicates using the rolling circle method (Tutino *et al.*, 2000). Transfer of the Rep protein from plasmid pTAUp into the *E. coli* compatible plasmid pUC18 enabled the successful maintenance of this plasmid (pUC-ORIT/rep) in both *E. coli* and psychrophilic bacteria (Tutino *et al.*, 2000).

Another novel plasmid, pMtBL, was recently isolated from the psychrophilic bacterium *Pseudoalteromonas haloplanktis* TAC 125 (Tutino *et al.*, 2001). Plasmid pMtBL (4.1 kb) appears to contain a cryptic replicon with no discernable transcriptional activity, however, the plasmid autonomous replication sequence demonstrated a broad host range when subcloned into a mesophilic plasmid (Clone Q). Furthermore, introduction of the cold-adapted α-amylase gene into plasmid Clone Q allowed for the overexpression of the enzyme in *Pseudoalteromonas haloplanktis* (Tutino *et al.*, 2001).

The cryptic plasmid pFL1 has been identified from the psychrophilic bacterium *Flavobacterium* sp. KP1 (Ashiuchi *et al.*, 1999). The 2.3 kb plasmid was determined to bear little homology to other known plasmids of similar size and contains two ORFs. One of them is similar to replication proteins from *Bacteroides fragilis* plasmid pBI143 and *Zymomonas mobilis* plasmid pZM2. The other ORF has similarity to recombination genes from Gram positive plasmids, such as *Staphylococcus aureus* plasmid pT181 (Ashiuchi *et al.*, 1999). Plasmid pFL1 seems to be a good candidate for use as a psychrophilic-based shuttle/expression vector since it is maintained stably in high copy in psychrophilic bacteria and also has unique *Nde*I and *Hind*III restriction sites for use in subcloning.

Genetic Elements from Halophiles

Until the early 1990s, researchers had obtained little information about genetic systems in halophilic organisms. Since that time, however, a number of endogenous plasmids have been identified from various moderate and extreme halophiles. Subsequently, several of these plasmids have been adapted for use as shuttle/expression vectors (Sowers *et al.*, 1999). One such plasmid is pCM1, a 17.5 kb plasmid isolated from the moderate halophile *Chromohalobacter marismortui*, that has a 1.6 bp region coding for the plasmid replicon (Mellado *et al.*, 1995). Initial studies of the plasmid indicated that it likely had utility as the basis for a halophilic shuttle/expression vector, and it subsequently has served as the parent plasmid for construction of the shuttle vector pEE3 (Mellado *et al.*, 1995). Another plasmid, pMH1 (11.5 kb) was isolated from the moderately halophilic bacteria *Deleya halophila*, *Halomonas halmophila*, *Halomonas elongata* and *Vibrio costicola*. It could be maintained in *E. coli*, carries resistance markers for the antibiotics kanamycin, tetracycline, and neomycin, and has unique restriction sites for *Eco*RI, *Eco*RV and *Cla*I (Fernandez-Castillo *et al.*, 1992). All of these properties would seemingly ensure that plasmid pMH1 would be an ideal shuttle/expression vector candidate, however, it has been shown to be difficult to isolate and thus far can only be detected by transformation into *E. coli* cells (Mellado *et al.*, 1995). Additional endogenous plasmids isolated from halophilic sources, including pHSB1 from *Halobacterium halobium* strain SB3 (Hackett *et al.*, 1989), pHE1 from *Halomonas elongata* (Vargas *et al.*, 1995), and pHK2 from *Haloferax* sp. Aa2.2 (Holmes *et al.*, 1995), have also all been identified as amenable to shuttle/expression vector conversion.

Genetic Elements from Acidophiles

A number of native plasmids have been isolated from acidophiles, with the majority coming from various *Thiobacillus ferrooxidans* strains (Aparicio *et al.*, 2000). As indicated by Table 3, many of these plasmids are fairly large in size ranging from 9.8 to 107 kb. Although a majority of these plasmids have been characterized in terms of their gene sequences, methods of replication, and involvement in host cellular metabolism (Aparicio *et al.*, 2000; Dominy *et al.*, 1998; English *et al.*, 1995; Gardner *et al.*, 2001; Rawlings *et al.*, 1994), shuttle/ expression vectors based on these characterized plasmids for use in acidophiles have yet to be developed. The delay in generation of shuttle/ expression vectors for acidophiles may be due in part to the relatively large size of the native acidophile plasmids, which poses difficulties for vector construction.

Genetic Elements from Alkaliphiles

Similar to the case of the dearth of shuttle/expression vectors for acidophiles, there is very little known regarding native plasmids isolated from alkaliphiles. In fact the only reported instance of native plasmids isolated from a true alkaliphile is a 110 kb endogenous plasmid from the alkaliphile *Bacillus firmus* which contains cadmium resistance and transposition genes (Gronstad *et al.*, 1998). A number of plasmids, however, have been described from alkaliphilic Halomonads isolated from soda lakes in the Kenyan Tanzanian Rift Valley; two of which were characterized (pAH1, 5.9 kb and pAH2, 5.3 kb) (Fish *et al.*, 1999). Both plasmids contain unique restriction sites that could be used in the construction of shuttle/expression vectors for alkaliphiles.

Shuttle/Expression Systems developed for Use in Extremophiles

As indicated in Table 4 and from discussion of the various genetic elements from extremophiles, a number of shuttle/expression vectors have been developed using existing endogenous plasmids, especially in the cases of thermophiles/hyperthermophiles and halophiles. At this point, there are now several vectors that have been constructed that can be maintained stably in both *E. coli* and thermophilic/ hyperthermophilic microbes; a subset of these have also been successfully used for the expression of cloned genes. One of the first shuttle vectors created for use in *Pyrococcus* sp., plasmid pAG21, was generated using the native *Pyrococcus* plasmid pGT5 and the origin of replication (Ori), ampicillin resistance gene, and multiple cloning site (MCS) from the *E. coli*-compatible plasmid pUC19 (Aravalli *et al.*, 1997). Plasmid pAG21 also contained the alcohol dehydrogenase (*adh*) gene from *S. solfataricus* for maintenance by butanol selection in *Pyrococcus* (Aravalli *et al.*, 1997). Although initial studies reported that plasmid pAG21 could be maintained in *Pyrococcus* cells, subsequent work indicated that butanol selection failed to support stable maintenance of the plasmid in *Pyrococcus* after multiple generations. Recently, however, a second shuttle vector, plasmid pYS2, has been constructed for use in *Pyrococcus* that again uses native plasmid pGT5, but this time also includes the *pyrE* gene for selection of uracil prototrophy in *Pyrococcus* uracil auxotrophs (Lucas *et al.*, 2002). This new shuttle vector has been shown to be stably maintained over many generations in *Pyrococcus* cells and will likely be used in the future for the overexpression of genes in *Pyrococcus*. The recently developed pEXADH vector, which contains the *Sulfolobus adh* promoter and aspartate aminotransferases terminator, has already successfully been used to drive the overexpression of the *Bacillus stearothermophilus* alcohol dehydrogenase gene in *S. solfataricus* (Contursi *et al.*, 2003). The pEXADH construct was derived from the shuttle vector pEXSc, which has a hygromycin resistance gene for plasmid maintenance in *Sulfolobus* and a SSV1 virus autonomously replicating sequence (ARS) for plasmid replication in *Sulfolobus* (Cannio *et al.*, 1998). In another recent example of successful thermophilic/hyperthermophilic expression systems, heterologous hydrolytic enzymes were produced using a shuttle/expression vector (pRUKM1) developed for maintenance in the thermophile *Thermoanaerobacterium* sp. (Mai *et al.*, 2000).

Table 4. Shuttle vectors developed for use in extremophiles[a]

Vector	Vector Host	Genetic Markers	Vector Size (kb)	Vector Copy Number	Unique RE Sites	Special Features	References
Thermophiles/ Hyperthermophiles							
pYS2	*Pyrococcus abyssi*	Ura$^+$/Ap^r	6.3	high	7	Based on pGT5 and pLitmus38	(Lucas et al., 2002)
pAG21	*Pyrococcus furiosus*	Adh/Ap^r	6.5	mid/high	11	Derived from pGT5 and pUC19	(Aravalli et al., 1997)
pKMS48	*Sulfolobus solfataricus*	Ap^r	18.5	20–40		Cloned genes can be controlled by UV induction	(Stedman et al., 1999)
pEXADH	*Sulfolobus sp.* *Sulfolobus solfataricus*	Hph/Ap^r	7.7	high	7	pEXSc derivative with *Sulfolobus* *Adh* promoter to drive expression	(Contursi et al., 2003)
pEXSc	*Sulfolobus solfataricus*	Hph/Ap^r	6.4	high	7	Uses SSV1 virus autonomously replicating sequence	(Cannio et al., 1998)
pRUKM1	*Thermoanaerobacterium*	Cm^r/Kn^r	7.2	high	19	Uses RP9 vector from *Bacillus Stearothermophilus*	(Mai et al., 2000)
pINV	*Thermus thermophilus*	Ap^r	not listed	1	5	Can be used to produce targeted gene disruptions	(Tamakoshi et al., 1997)
pUC19-pYK189	*Thermus thermophilus*	Ap^r/Kn^r	11.7	low		Based on *Thermus* pTT8 plasmid	(Wayne et al., 1997)
Psychrophiles							
pFF	*Pseudoalteromonas haloplanktis*	Ap^r	4.0	high	9	Has cold-adapted *aspC* promoter and RBS	(Tutino et al., 2002)
Halophiles							
pEE3	*Deleya halophila*	Tp^r/Ap^r	6.6	high	6	Derived from pCM1	(Mellado et al., 1995)
pHRZH	*Halobacterium salinarium*	Ani^r/Thi^r/Ap^r	16.7	high		Has multiple selection markers	(Mankin et al., 1992)
pMPK62	*Halobacterium salinarium*	Mv^r/Ap^r	10.2	high		Has *bop* gene	(Krebs et al., 1993)
pNG168	*Halobacterium salinarium*	Mv^r/Ap^r	8.9	high	8	Has blue/white selection	(DasSarma, 1995)
pUBP2	*Halobacterium salinarium*	Mv^r/Ap^r	12.3	high	3		(Cline et al., 1992)

Table 4. Shuttle vectors developed for use in extremophiles (continued)

Vector	Vector Host	Genetic Markers	Vector Size (kb)	Vector Copy Number	Unique RE Sites	Special Features	References
Halophiles							
pGB70	*Haloferax volcanii*	Ap^r	3.3	high	5	Can produce gene disruptions using *pyrE* gene	(Bitan-Banin et al., 2003)
pKAR112	*Haloferax volcanii*	Mv^r/Ap^r	10.5	high	5	Derived from pWL222 and has *P. furiosus gdh* promoter	(Schreier et al., 1999)
pMDS20	*Haloferax volcanii*	Nb^r/Ap^r	10.0	high	8		(Holmes et al., 1994)
pMLH3	*Haloferax volcanii*	Mv^r/Nb^r/Ap^r	11.3	high	8	Has multiple selection markers	(Holmes et al., 1994)
pSD1-R1/6	*Haloferax volcanii*	Nb^r		high	4	Uses synthetic pR16 promoter	(Danner et al., 1996)
pWL102	*Haloferax volcanii*	Mv^r/Ap^r	10.5	high	7		(Lam et al., 1989)
pWL104	*Haloferax volcanii*	Mv^r/Ap^r	8.7	high	7		(Lam et al., 1992)
pWL222	*Haloferax volcanii*	Mv^r/Ap^r	10.5	high	5	Has *Saccharomyces* tRNApro reporter	(Palmer et al., 1994)
pEES	*Halomonas elongata*	Tp^r/Ap^r	7.0	high	6	Has blue/white selection	(Mellado et al., 1995)
Acidophiles							
pAH101	*Acidophilum facilus*	Ap^r	8.8	high			(Inagaki et al., 1993)

aThe abbreviations used in the tables are as follows: Adh, alcohol resistance (alcohol dehydrogenase); Ani^r, anisomycin resistance; Ap^r, ampicillin resistance; Cm^r, chloramphenicol resistance; gdh, glutamte dehydrogenase; Hph^r, hygromycin resistance; Kn^r, kanamycin resistance; Mv^r, mevinolin resistance; Nb^r, novobiocin resistance; Thi^r, thiostrpton resistance; Tp, trimethoprim; ura$^+$, uracil prototrophy.

In the case of psychrophilic shuttle/expression vectors, Tutino *et al.* constructed an expression vector (pFF) for use in the psychrophile *Pseudoalteromonas haloplanktis* strain TAB23, based on plasmid Clone Q derived from native plasmid pMtBL (Tutino *et al.*, 2002). The pFF expression plasmid contained the transcription and translation control elements from the cold-adapted aspartate aminotransferases (*aspC*) gene from *P. haloplanktis* strain TAC125. These elements were used to drive expression of the cloned *P. haloplanktis* α-amylase gene in the pFF construct (Tutino *et al.*, 2002). The present intense biotechnological interest in cold-adapted proteins, coupled with the recent findings of numerous endogenous plasmids from psychrophiles, will undoubtedly drive the rapid development of more shuttle/ expression systems for use in psychrophiles.

One of the first examples of the use of an endogenous halophilic plasmid for expression of halophilic proteins involved the bacteriorhodopsin (bop) gene that was cloned into the *Halobacterium halobium* strain GRB1 plasmid pGRB1 and recombinantly expressed in the bop⁻ strain R1 of *Halobacterium halobium* (Krebs *et al.*, 1991). A more recent example of heterologous gene expression in a halophile showed that mRNA was successfully expressed in the halophile *Haloferax volcanii* from the glutamate dehydrogenase promoter (gdh) of *P. furiosus* that had been cloned into the shuttle vector pKAR112 (Schreier *et al.*, 1999). Heterologous expression of eukaryotic G protein-coupled receptors has also been undertaken in *Haloferax volcanii* using the pSD1-R1/6 shuttle vector as the host-cloning vector (Danner *et al.*, 1996; Patenge *et al.*, 1999). In this case, the rat G protein-coupled receptors were cloned as fusions with the dihydrofolate reductase gene from *H. volcanii* and expression of the fusions were driven with the strong, synthetic haloarchaeal promoter pRr16. Although high-level expression of the eukaryotic genes was ultimately prevented by extensive proteolysis in the host, the fact that eukaryotic G protein-coupled receptors could be expressed at all in the halophilic expression system is impressive given the difficulty inherent to heterologous G protein-coupled receptor expression (Patenge *et al.*, 1999).

For shuttle/expression vectors developed for use in acidophiles, little has been done thus far as previously discussed. There is one reported example, however, of a shuttle vector (pAH101) constructed for use in the acidophile *Acidophilum facilus* (Inagaki *et al.*, 1993). As biotechnological interest in the production of acidophilic proteins increases, it is expected that the generation of suitable shuttle/expression vectors will also occur. This is especially important because the directed evolution of protein to confer acid stability is challenging.

Recombinant Expression Systems for Extremophilic Proteins

To meet the needs for scientific and technological evaluation of extremophilic enzymes, production strategies are needed to generate sufficient amounts of material for characterization. Unfortunately, the characteristics that make these organisms so intriguing exacerbate production of the native forms of their proteins. Difficulties in cultivating the organisms on large scale, low basal level expression of the enzymes of interest in the native host, and lack of suitable expression systems in the particular extremophile, have prevented high-level production and purification of the enzymes from the source extremophile (Adams *et al.*, 1995; Adams *et al.*, 1998; Cowan, 1992). As a result, when possible, extremophilic proteins are currently produced using various recombinant foreign expression systems described in more details in other chapters of this book. Aside from enabling the high-level production of extremophile proteins, expression systems in use today generally allow for the tagging/modification of the recombinant proteins for the purpose of simplifying downstream protein purification procedures.

Early on both in the discovery and isolation of extremophiles and in the development of recombinant expression technologies, the idea of relying on recombinant expression for the high-level production of active extremophile enzymes was considered to be rife with difficulty. In fact it was initially thought that some of the molecular adaptations intrinsic to extremophile proteins and differences in codon usage or

GC content of the extremophile genes compared to the host would pose insurmountable obstacles for successful recombinant production of extremophile proteins in traditional non-extremophile expression hosts such as *E. coli*, yeast, and *Bacillus subtilis*. However, with the continual improvement and development of new expression systems, as well as greater understanding of the nature of molecular adaptations of extremophile proteins, the successful recombinant expression of extremophile proteins has been demonstrated for a number of different proteins from a variety of different extremophiles. This ability to recombinantly express proteins from thermophiles and hyperthermophiles, as well as proteins from the other extremophiles, is clearly illustrated in Tables 5 and 6, which contain examples of proteins grouped under different functional categories. It is important to note that these functional categories are derived from those established as a result of whole genome sequencing of organisms and that there are extremophile protein representatives in every category.

Analysis of the information presented in Tables 5 and 6 provides some insight concerning the current trends in recombinant extremophile protein expression. For instance, the majority of proteins that have been successfully recombinantly expressed, dating back to the early 1990s, come from thermophilic or hyperthermophilic sources (152 thermophile/hyperthermophile proteins compared to 39 total for all other extremophile proteins). This seeming disparity in recombinant protein production depending on the type of extremophile is likely driven by the significant biotechnological interest in, and the availability of, many characterized thermophile/hyperthermophile strains and not due to any inherent difficulty in recombinant expression of the other types of extremophile proteins. Another observation is that the vast majority of the recombinant proteins are produced using *E. coli* as the expression host. The recombinant protein data presented in Tables 5 and 6 indicate that just over 90% of all of the recombinant extremophile proteins are produced using *E. coli* expression hosts.

Escherichia coli-based Expression Systems

The Gram negative bacterium, *Escherichia coli*, has traditionally served as the expression host of choice for recombinant protein production mainly because of its ability to grow quickly to high cell density in large scale on relatively inexpensive media, its well-characterized physiology/genetics, and its amenability to genetic manipulation (Baneyx, 1999). And, since the beginning of the development of *E. coli*-based expression systems in the early 1980s, a number of different vector and *E. coli* host combinations have been created (Makrides, 1996). According to the information listed in Tables 5 and 6, only a handful of the available *E. coli* expression systems are routinely employed for recombinant extremophile protein expression.

The most commonly used of the *E. coli* systems is the T7 promoter-based expression vectors (Studier *et al*., 1986; Tabor *et al*., 1985), with ~60% of the recombinant extremophile proteins being produced with this system. The second most often used *E. coli* expression system for extremophile protein production is *lac* promoter-based expression vectors (used ~20% of the time, See Tables 5 and 6). A synthetic derivative of the *lac* promoter, the *tac* promoter, has also been used for recombinant extremophile protein production. The *tac* promoter has the –35 promoter region of the *trp* promoter fused to the –10 region of the *lac* promoter (Brosius *et al*., 1985). Other *E. coli* expression systems that have been used successfully for recombinant extremophile protein production include the arabinose-inducible, arabinose promoter-based pBAD vectors (Invitrogen Corp. Carlsbad, CA) (Taylor *et al*., 1992), the IPTG-inducible, T5 phage-*lac* operator promoter-based pQE vectors (Qiagen, Inc. Valencia, CA) (Bujard *et al*., 1987), and the nalidixic acid-inducible *recA* promoter-based pREC7 vectors.

To this point, it appears that the extremophilic recombinant proteins retain their native characteristics. Less than 10% of all extremophilic enzymes produced in *E. coli* have been reported to have stability, catalytic, or structural properties significantly different from the same enzymes purified from the native hosts. Although this is an encouraging finding, it is most likely an underestimate since failed recombinant

Table 5. Types of recombinantly expressed thermophile/hyperthermophile proteins and the expressions systems used[a]

Type of Protein	Source Organism	Expression System Used	Reference
Regulators			
DrrA response regulator	*Thermotoga maritima*	*E. coli* – T7 promoter	(Lee *et al.*, 1996)
Leucine responsive regulators	*Methanococcus jannaschii*	*E. coli* – pET system (T7 promoter)	(Ouhammouch *et al.*, 2001)
Protein tyrosine/serine phosphatase	*Thermococcus kodakaraensis*	*E. coli* – pET system (T7 promoter)	(Jeon *et al.*, 2002)
Transporters			
Trehalose/maltose ABC transporter	*Thermococcus litoralis*	*E. coli* – T7 promoter	(Greller *et al.*, 2001)
DNA Replication, Transcription, Translation			
DNA ligase	*Methanobacterium thermoautotrophicum*	*E. coli* – pET system (T7 promoter)	(Sriskanda *et al.*, 2000)
	Thermus thermophilus	*E. coli* – Lambda phage promoter	(Lauer *et al.*, 1991)
DNA polymerase	*Thermus aquaticus*	*E. coli* – pET system (T7 promoter)	(Dabrowski *et al.*, 1998)
	Thermus thermophilus	*E. coli* – pET system (T7 promoter)	(Dabrowski *et al.*, 1998)
	Pyrococcus furiosus	*E. coli* - T7 promoter	(Goldman *et al.*, 1998)
Endonuclease III	*Pyrobaculum aerophilum*	*E. coli* – pQE30 (T5 promoter)	(Yang *et al.*, 2001)
Endonuclease IV	*Thermotoga maritima*	*E. coli* – pET system (T7 promoter)	(Haas *et al.*, 1999)
Flap endonuclease	*Pyrococcus horikoshii*	*E. coli* – pET system (T7 promoter)	(Matsui *et al.*, 1999)
Histone	*Methanothermus fervidus*	*E. coli* – *tac* promoter	(Decanniere *et al.*, 2000)
Intein-encoded PfuI, PfuII endonucleases	*Pyrococcus furiosus*	*E. coli* – T7 promoter	(Komori *et al.*, 1999)
O6-methylguanosine-DNA methyltransferase	*Thermococcus kodakaraensis*	*E. coli* – *lac* promoter	(Leclere *et al.*, 1998)
ORF56, plasmid copy control	*Sulfolobus islandicus* (A)	*E. coli* – pET system (T7 promoter)	(Lipps *et al.*, 2001)
Replication factor C	*Pyrococcus furiosus*	*E. coli* –T7 promoter	(Cann *et al.*, 2001)
	Sulfolobus solfataricus (A)	*E. coli* – pET system (T7 promoter)	(Pisani *et al.*, 2000)
Reverse gyrase	*Methanopyrus kandleri*	*E. coli* – pET system (T7 promoter)	(Krah *et al.*, 1997)
RNA polymerase	*Thermus aquaticus*	*E. coli* – pET system (T7 promoter)	(Minakhin *et al.*, 2001)
RNase HII	*Thermococcus kodakaraensis*	*E. coli* – T7 promoter	(Haruki *et al.*, 1998)
S7 ribosomal protein	*Thermus thermophilus*	*E. coli* – T5 promoter	(Karginov *et al.*, 1995)
S14 ribosomal protein	*Thermus thermophilus*	*E. coli* – pET system (T7 promoter)	(Tsiboli *et al.*, 1998)
TBP-interacting protein	*Thermococcus kodakaraensis*	*E. coli* – pET system (T7 promoter)	(Matsuda *et al.*, 1999)
Topoisomerase IV	*Sulfolobus shibatae* (A)	*E. coli* – pET system (T7 promoter)	(Buhler *et al.*, 1998)
Uracil-DNA glycosylase	*Pyrobaculum aerophilum*	*E. coli* – pQE30 (T5 promoter)	(Yang *et al.*, 2002)
	Thermotoga maritima	*E. coli* – pET system (T7 promoter)	(Sandigursky *et al.*, 2001)

Type of Protein	Source Organism	Expression System Used	Reference
Energy Metabolism			
ATP sulfurylase	*Archaeoglobus fulgidus*	*E. coli* – pET system (T7 promoter)	(Sperling *et al.*, 1998)
Cytochrome P450	*Sulfolobus solfataricus* (A)	*E. coli* – *lac* promoter	(McLean *et al.*, 1998)
F420-dependent methylene-5,6,7,8-tetrahydromethanopteran dehydrogenase	*Methanobacterium thermoautotrophicum*	*E. coli* promoter	(Mukhopadhyay *et al.*, 1995)
Formylmethanofuran:tetrahydro-methanopterin formyl transferase	*Methanopyrus kandleri*	*E. coli* – pET system (T7 promoter)	(Shima *et al.*, 1995)
Hydrogenase, γ subunit	*Thermotoga maritima*	*E. coli* – pET system (T7 promoter)	(Verhagen *et al.*, 2001)
8-hydroxy-5-deazaflavin reducing hydrogenase	*Methanobacterium thermoautotrophicum*	*E. coli* – *lac* promoter	(Alex *et al.*, 1990)
P-type ATPase	*Methanococcus jannaschii*	*Saccharomyces* – *PMA1* promoter	(Morsomme *et al.*, 2002)
	Archaeoglobus fulgidus	*E. coli* – pBAD (arabinose promoter)	(Mandal *et al.*, 2002)
Pyrophosphatase	*Bacillus stearothermophilus*	*E. coli* – *lac* promoter	(Satoh *et al.*, 1999)
	Thermus thermophilus	*E. coli* – pET system (T7 promoter)	(Satoh *et al.*, 1998)
Central Intermediary Metabolism			
Acetyl-CoA Synthetase	*Archeaoglobus fulgidus*	*E. coli* – pET system (T7 promoter)	(Musfeldt *et al.*, 2002)
	Methanococcus jannaschii	*E. coli* – pET system (T7 promoter)	(Musfeldt *et al.*, 2002)
	Pyrococcus furiosus	*E. coli* – pET system (T7 promoter)	(Musfeldt *et al.*, 1999)
ADP-dependent glucokinase	*Methanococcus jannaschii*	*E. coli* – pET system (T7 promoter)	(Sakuraba *et al.*, 2002)
Alcohol dehydrogenase	*Bacillus stearothermophilus*	*E. coli* – *tac* promoter	(Fiorentino *et al.*, 1998)
	Thermococcus hydrothermalis	*E. coli* – pET system (T7 promoter)	(Antoine *et al.*, 1999)
ATP-dependent glucokinase	*Aeropyrum pernix*	*E. coli* – pBAD (arabinose promoter)	(Hansen *et al.*, 2002)
ATP-dependent 6-phospho-fructokinase	*Aeropyrum pernix*	*E. coli* – pBAD (arabinose promoter)	(Hansen *et al.*, 2001)
Carbonic anhydrase	*Methanobacterium thermoautotrophicum*	*E. coli* – pET system (T7 promoter)	(Smith *et al.*, 1999)
Citrate synthase	*Pyrococcus furiosus*	*E. coli* – *recA* promoter	(Arnott *et al.*, 2000)
Fructose-1,6,-bisphosphate aldolase	*Pyrococcus furiosus*	*E. coli* – pET system (T7 promoter)	(Siebers *et al.*, 2001)
Glyceraldehyde-3-phophate dehydrogenase	*Solfolobus solfataricus* (A)	*E. coli* – pET system (T7 promoter)	(Isupov *et al.*, 1999)
	Thermoproteus tenax	*E. coli* – *lac* promoter	(Brunner *et al.*, 1998)
	Thermotoga maritima	*E. coli* – *lac* promoter	(Tomschy *et al.*, 1993)
2-Keto-3-deoxygluconate aldolas	*Sulfolobus solfataricus* (A)	*E. coli* – *recA* promoter	(Buchanan *et al.*, 1999)

Central Intermediary Metabolism (continued)

Type of Protein	Source Organism	Expression System Used	Reference
Glycerol kinase	*Thermococcus kodakaraensis*	*E. coli* – pET system (T7 promoter)	(Koga *et al.*, 1998)
	Thermus aquaticus	*E. coli* – *tac* promoter	(Huang *et al.*, 1998)
Glyoxylate reductase	*Thermococcus litoralis*	*E. coli* – *lac* promoter	(Ohshima *et al.*, 2001)
Indolepyruvate:Fd oxidoreductase	*Thermococcus kodakaraensis*	*E. coli* – pET system (T7 promoter)	(Siddiqui *et al.*, 1998)
Isocitrate dehydrogenase	*Thermus aquaticus*	*E. coli* – *lac* promoter	(Miyazaki, 1996)
	Thermus thermophilus	*E. coli* – *tac* -promoter	(Miyazaki *et al.*, 1994)
Isopropyl malate dehydrogenase	*Thermus thermophilus*	*E. coli* – *lac* promoter	(Numata *et al.*, 1995)
Lactate dehydrogenase	*Archaeoglobus fulgidus*	*E. coli* – pMal (maltose-binding protein)	(Reed *et al.*, 1999)
	Methanococcus jannaschii	*E. coli* – pET system (T7 promoter)	(Lee *et al.*, 2000)
	Thermotoga maritima	*E. coli* – pET system (T7 promoter)	(Dams *et al.*, 1996)
Phosphoglycerate kinase	*Thermotoga maritima*	*E. coli* – T7 promoter	(Grattinger *et al.*, 1998)
Pyruvate:ferredoxin oxidoreductase	*Pyrococcus furiosus*	*E. coli* – pET system (T7 promoter)	(Menon *et al.*, 1998)
Pyruvate kinase	*Thermoproteus tenax*	*E. coli* – *lac* promoter	(Schramm *et al.*, 2000)
Ribulose bisphosphate carboxylase	*Thermococcus kodakaraensis*	*E. coli* – pET system (T7 promoter)	(Maeda *et al.*, 1999)

Carbon Compound Catabolism

Type of Protein	Source Organism	Expression System Used	Reference
α-Amylase	*Alicyclobacillus acidocaldarius* (A)	*E. coli* – T7 promoter	(Schwermann *et al.*, 1994)
	Pyrococcus furiosus	*E. coli* – *lac* promoter	(Laderman *et al.*, 1993)
	Pyrococcus woesei	*Halomonas elongata*	(Frillingos *et al.*, 2000)
	Sulfolobus solfataricus (A)	*Candida utilise* – GAP promoter	(Miura *et al.*, 1999)
	Thermococcus hydrothermalis	*E. coli* – *lac* promoter	(Leveque *et al.*, 2000)
	Thermotoga maritima	*E. coli* – *lac* promoter	(Liebl *et al.*, 1997)
Amylopullulanase	*Pyrococcus furiosus*	*E. coli* – *lac* promoter	(Dong *et al.*, 1997)
	Desulfurococcus mucosus	*Bacillus subtilis*- α-amylase promoter	(Duffner *et al.*, 2000)
Arabinose isomerase	*Thermotoga neapolitana*	*E. coli* – pET system (T7 promoter)	(Kim *et al.*, 2002)
Cellulase	*Rhodothermus marinus*	*E. coli* – pET system (T7 promoter)	(Halldorsdottir *et al.*, 1998)
	Thermotoga maritima	*E. coli* – *lac* promoter	(Liebl *et al.*, 1996)
Chitinase	*Thermococcus kodakaraensis*	*E. coli* – pET system (T7 promoter)	(Tanaka *et al.*, 1999)
Endogluconase	*Aquifex aeolicus*	*E. coli* – *lac* promoter	(Kim *et al.*, 2000)
	Pyrococcus furiosus	*E. coli* – pQE30 (T5 promoter)	(Bauer *et al.*, 1999)
	Pyrococcus horikoshii	*E. coli* – pET system (T7 promoter)	(Ando *et al.*, 2002)
Endo-(1,4)-β-mannanase	*Rhodothermus marinus*	*E. coli* – pET system (T7 promoter)	(Politz *et al.*, 2000)
	Bacillus stearothermophilus	*E. coli* – *lac* promoter	(Fridjonsson *et al.*, 1999)
α-Galactosidase	*Thermotoga maritima*	*E. coli* – pET system (T7 promoter)	(Liebl *et al.*, 1998)

Type of Protein	Source Organism	Expression System Used	Reference
Carbon Compound Catabolism (continued)			
β-Galactosidase	Sulfolobus solfataricus (A)	Saccharomyces – G2 hybrid promoter	(Moracci et al., 1992)
β-Glucosidase	Pyrococcus furiosus	E. coli – pET system (T7 promoter)	(Bauer et al., 1998)
		Saccharomyces cerevisiae- Gal1-10	(Smith et al., 2002)
Glucan glucohydrolase	Thermotoga neapolitana	E. coli – pBAD (arabinose promoter)	(Yernool et al., 2000)
4-α-glucotransferase	Thermococcus litoralis	E. coli – lac promoter	(Jeon et al., 1997)
α-glucosidase	Sulfolobus solfataricus (A)	Schistosoma japonicum (GST-fusion)	(Rolfsmeier et al., 1998)
β-glycosidase	Sulfolobus solfataricus (A)	E. coli – pET system (T7 promoter)	(Moracci et al., 1996)
Laminarinase	Rhodothermus marinus	Trichoderma reeseii (pcbh promoter)	(Krah et al., 1998)
	Dictyoglomus thermophilum	E. coli – pET system (T7 promoter)	(Te'o et al., 2000)
	Rhodothermus marinus	Kluyveromyces lactis – lac promoter	(Karlsson et al., 1998)
	Thermotoga strain FjSS3B.1		(Walsh et al., 1998)
Xylanase	Thermotoga maritima	E. coli – lac promoter (GST-fusion)	(Wassenberg et al., 1997)
Amino Acid Biosynthesis and Metabolism			
Acylamino acid-releasing enzyme	Pyrococcus horikoshii	E. coli – pET system (T7 promoter)	(Ishikawa et al., 1998)
Alanine aminotransferases	Pyrococcus furiosus	E. coli – T7 promoter	(Ward et al., 2000)
Aminoacylase	Pyrococcus furiosus	E. coli – T7 promoter	(Story et al., 2001)
Aminopeptidase	Pyrococcus horikoshii	E. coli – pET system (T7 promoter)	(Ando et al., 1999)
Anthranilate synthase	Sulfolobus solfataricus (A)	E. coli – tac- promoter	(Tutino et al., 1997)
	Thermococcus kodakaraensis	E. coli – pET system (T7 promoter)	(Tang et al., 2001)
Aqualysin I	Thermus aquaticus	E. coli – T7	(Sakamoto et al., 1995)
Aromatic aminotransferases	Pyrococcus horikoshii	E. coli – pET system (T7 promoter)	(Matsui et al., 2000)
Asparaginase	Archaeoglobus fulgidus	E. coli – pBAD (arabinose promoter)	(Li et al., 2002)
Aspartate transaminase	Bacillus stearothermophilus	E. coli – lac promoter	(Bartsch et al., 1996)
Carboxypeptidase	Pyrococcus horikoshii	E. coli – pET system (T7 promoter)	(Ishikawa et al., 2001)
Glutamate dehydrogenase	Aeropyrum pernix	E. coli – lac promoter	(Bhuiya et al., 2000)
	Pyrococcus furiosus	E. coli – pET system (T7 promoter)	(Diruggiero et al., 1995)
	Thermotoga maritima	E. coli – lac promoter	(Lebbink et al., 1999)
Glutamate synthase	Thermococcus kodakaraensis	E. coli – T7 promoter	(Jongsareejit et al., 1997)
Glutamine synthetase	Thermococcus kodakaraensis	E. coli – Tet promoter	(Adul Rahman et al., 1997)
Indoleglycerol phosphate synthase	Thermotoga maritima	E. coli – T7 promoter	(Merz et al., 1999)
L-Isoaspartyl methyltransferase	Pyrococcus furiosus	E. coli – Tac promoter	(Ichikawa et al., 1998)
Lon protease	Thermococcus kodakaraensis	E. coli – pET system (T7 promoter)	(Fukui et al., 2002)

Type of Protein	Source Organism	Expression System Used	Reference
Amino Acid Biosynthesis and Metabolism (continued)			
Methionyl aminopeptidase	*Pyrococcus furiosus*	*E. coli* – pET system (T7 promoter)	(Meng *et al.*, 2002)
Ornithine decarboxylase	*Thermus thermophilus*	*E. coli* – pBR322	(Pantazaki *et al.*, 1999)
Phosphoribosyl anthranilate isomerase	*Thermotoga maritima*	*E. coli* – T7 promoter	(Sterner *et al.*, 1996)
Prolidase	*Pyrococcus furiosus*	*E. coli* – pET system (T7 promoter)	(Ghosh *et al.*, 1998)
Protease PfpI	*Pyrococcus furiosus*	*E. coli* – T7 promoter	(Halio *et al.*, 1996)
20S Proteasome	*Methanococcus jannaschii*	*E. coli* – pET system (T7 promoter)	(Frankenberg *et al.*, 2001)
Pyrrolidone carboxyl peptidase	*Pyrococcus furiosus*	*E. coli* – *lac* promoter	(Tsunasawa *et al.*, 1998)
	Thermococcus litoralis	*E. coli* – *tac* promoter	(Singleton *et al.*, 2000)
S-adenosylhomocysteine hydrolase	*Sulfolobus solfataricus* (A)	*E. coli* – *tac* promoter	(Porcelli *et al.*, 2000)
Trytophan synthase (TrpA, TrpB)	*Pyrococcus furiosus*	*E. coli* – *lac* promoter	(Ishida *et al.*, 2002)
Nucleotide Biosynthesis			
Aspartate transcarbamoylase	*Methanococcus jannaschii*	*E. coli* – T7 promoter	(Hack *et al.*, 2000)
	Thermotoga maritima	*E. coli* – *lac* promoter	(Chen *et al.*, 1998)
Carbamoyl phosphate synthetase	*Aquifex aeolicus*	*E. coli* – T7 promoter	(Ahuja *et al.*, 2001)
	Pyrococcus abyssi	*E. coli* —T7 promoter	(Purcarea *et al.*, 2001)
	Pyrococcus furiosus	*E. coli* – pET system (T7 promoter)	(Uriarte *et al.*, 1999)
Dihydrofolate reductase	*Thermotoga maritima*	*E. coli* – *tac* promoter	(Dams *et al.*, 1998)
Inositol monophosphatase	*Thermotoga maritima*	*E. coli* – pET system (T7 promoter)	(Chen *et al.*, 1999)
Nicotinamide mononucleotide adenylyltransferase	*Methanococcus jannaschii*	*E. coli* – T7 promoter	(Raffaelli *et al.*, 1997)
	Sulfolobus solfataricus (A)	*E. coli* – T7 promoter	(Raffaelli *et al.*, 1997)
Nucleoside diphosphate kinase	*Methanococcus jannaschii*	*E. coli* – pET system (T7 promoter)	(Min *et al.*, 2000)
Ornithine carbamoyltransferase	*Pyrococcus furiosus*	*E. coli* – *tac* promoter *Saccharomyces* – pYeF1	(Legrain *et al.*, 1997)
Thymidine kinase	*Rhodothermus marinus*	*E. coli* – pET system (T7 promoter)	(Blondal *et al.*, 1999)
Fatty Acid and Phospholipid Metabolism			
3-Deoxy-D-manno-octulosonic acid 8-phosphate synthase	*Aquifex aeolicus*	*E. coli* – T7 promoter	(Duewel *et al.*, 1999)
Farnesylgeranyl diphosphate synthase	*Aeropyrum pernix*	*E. coli* – T7 promoter, GST fusion	(Tachibana *et al.*, 2000)

Type of Protein	Source Organism	Expression System Used	Reference
Fatty Acid and Phospholipid Metabolism (continued)			
Geranylgeranyl diphosphate synthase	*Thermus thermophilus*	*E. coli* – GST fusion	(Ohto *et al.*, 1999)
Lipase	*Archaeoglobus fulgidus*	*E. coli* – pET system (T7 promoter)	(Manco *et al.*, 2000)
Housekeeping and Stress Response			
Catalase-peroxidase	*Archaeoglobus fulgidus*	*E. coli* – pET system (T7 promoter)	(Kengen *et al.*, 2001)
Cold shock protein	*Thermotoga maritima*	*E. coli* – pET system (T7 promoter)	(Welker *et al.*, 1999)
DnaJ, DnaK, GrpE	*Thermus thermophilus*	*E. coli* – pET system (T7 promoter)	(Osipiuk *et al.*, 1997)
GroEL, GroES	*Bacillus stearothermophilus*	*E. coli* – *lac* promoter	(Schon *et al.*, 1995)
Heat shock protein	*Methanococcus jannaschii*	*E. coli* – pET system (T7 promoter)	(Kim *et al.*, 1998)
NAD(P)H:Rubredoxin oxidoreductase	*Pyrococcus furiosus*	*E. coli* – T7 promoter, intein fusion	(Grunden *et al.*, 2003)
Peroxiredoxin	*Thermus aquaticus*	*E. coli* – pET system (T7 promoter)	(Logan *et al.*, 2000)
Rubredoxin	*Pyrococcus furiosus*	*E. coli* – pET system (T7 promoter)	(Eidsness *et al.*, 1997)
Superoxide reductase	*Archaeoglobus fulgidus*	*E. coli* – T7 promoter	(Abreu *et al.*, 2001)
	Pyrococcus furiosus	*E. coli* – pET system (T7 promoter)	(Clay *et al.*, 2002)
Thermosome	*Methanopyrus kandleri*	*E. coli* – pET system (T7 promoter)	(Minuth *et al.*, 1999)
UvrA	*Thermus thermophilus*	*E. coli* – *lac* promoter	(Yamamoto *et al.*, 1996)

[a](A) indicates that the organism is an acidophile as well

Table 6. Types of recombinantly expressed proteins from psychrophilic, halophilic, acidophilic, and alkaliphic organisms[a]

Type of Protein	Source Organism	Expression System Used	Reference
Regulators			
GvpE (regulator of *gvpA* gene)	*Halobacterium salinarum* (H)	*Haloferax volcanii* – pJAS35 vector	(Plosser *et al.*, 2002)
Serine/tyrosine phosphatase I	*Schewanella sp.* (P)	*E. coli* – *lac* promoter, GST fusion, 25°C	(Murakawa *et al.*, 2002)
Transporters			
SecE (secretion apparatus)	*Acidianus ambivalens* (A)	*E. coli* – pET system	(Moll *et al.*, 1997)
TeaA (ectoine transporter)	*Halomonas elongata* (H)	*E. coli* – pET system (T7 promoter)	(Tetsch *et al.*, 2002)
DNA Replication, Transcription, Translation			
DNA polymerase	*Cenarchaeum symbiosum* (P)	*E. coli* – pET system (T7 promoter), 13°C	(Schleper *et al.*, 1997)
		E. coli – Tac promoter, CBP-intein fusion	
Uracil-DNA glycosylase	Strain BMTU3346 (P)	*E. coli* – pQE30 (T5 promoter)	(Jaeger *et al.*, 2000)
Energy Metabolism			
Alkaline phosphatase	Antarctic strain TAB5 (P)	*E. coli* – T7 promoter	(Rina *et al.*, 2000)
Diphosphate kinase	*Halobacterium salinarum* (H)	*E. coli* – T7 promoter	(Ishibashi *et al.*, 2001)
Central Intermediary Metabolism			
Citrate synthase	Strain DS2-3R (P)	*E. coli* – *recA*- promoter	(Gerike *et al.*, 1997)
Glucose dehydrogenase	*Haloferax mediterranei* (H)	*E. coli* – pET system (T7 promoter)	(Pire *et al.*, 2001)
Isocitrate dehydrogenase	*Haloferax volcanii* (H)	*E. coli* – pET system (T7 promoter)	(Camacho *et al.*, 2002)
	Schewanella sp. (P)	*E. coli* – *tac* promoter	(Murakawa *et al.*, 2002)
Malate dehydrogenase	*Photobacterium sp.* strain SS9 (P)	*E. coli* – *lac* Promoter	(Welch *et al.*, 1997)
	Haloarcula marmortui (H)	*E. coli* – pET system	(Cendrin *et al.*, 1993)
Triose phosphate isomerase	*Pseudoalteromonas haloplanktis* (P)	*E. coli* – T7 promoter, 22°C	(Rentier-Delrue *et al.*, 1993)
Carbon Compound Catabolism			
α-Amylase	*Alteromonas haloplanktis* (P)	*E. coli* – T7 promoter	(Feller *et al.*, 1998)
		Pseudoalteromonas haloplanktis TAC125	(Tutino *et al.*, 2002)
		Pseudoalteromonas haloplanktis TAD	
	Halothermothrix orenii (H)	*E. coli* – *tac* promoter	(Mijts *et al.*, 2002)
	Bacillus sp. strain KSM-1378 (AL)	*Bacillus subtilis* – vector pHSP64	(Ikawa *et al.*, 1998)

Type of Protein	Source Organism	Expression System Used	Reference
Carbon Compound Catabolism (continued)			
α-Amylase-pullulunase	Bacillus sp. XAL601 (AL)	E. coli – lac promoter	(Lee et al., 1994)
β-Galactosidase	Carnobacterium piscicola AB (P)	E. coli – lac- promoter, 18 °C	(Coombs et al., 1999)
Chitiobiase	Arthrobacter sp. TAD20 (P)	E. coli – pET system (T7 promoter)	(Lonhienne et al., 2001)
Endo-1,4-β-glucanase	Bacillus sp. KSM-S237 (AL)	Bacillus subtilis – pHY300PLK shuttle vector	(Hakamada et al., 2000)
Pectate lyase	Bacillus sp. P-4-N (AL)	Bacillus subtilis – pHY300PLK shuttle vector	(Kobayashi et al., 2000)
Xylanase	Bacillus sp. NCIM59 (AL)	E. coli – lac promoter	(Kulkarni et al., 1999)
Amino Acid Biosynthesis and Metabolism			
Alanine racemace	Bacillus psychrosaccharolyticus (P)	E. coli – pYOK3 vector	(Okubo et al., 1999)
Alkaline metalloprotease	Pseudomonas sp. TACII18 (P)	E. coli – lac promoter	(Chessa et al., 2000)
Alkaline serine protease	Bacillus sp. AH-101 (AL)	Bacillus subtilis	(Takami et al., 1992)
Aspartate aminotransferases	Pseudoalteromonas haloplanktis (P)	E. coli – Tac promoter	(Birolo et al., 2000)
Halolysin R4 (serine protease)	Haloferax mediterranei (H)	Haloferax volcanii – pMD30	(Kamekura et al., 1996)
Peptidyl-prolyl cis-trans isomerase	Halobacterium cutirubrum (H)	E. coli – GST fusion	(Iida et al., 2000)
20 S proteosome	Haloferax volcanii (H)	E. coli – pET system (T7 promoter)	(Kaczowka et al., 2003)
Nucleotide Biosynthesis			
Dihydrofolate reductase	Haloferax volcanii (H)	E. coli – pET system (T7 promoter)	(Blecher et al., 1993)
Seryl-tRNA synthetase	Haloarcula marismortui (H)	E. coli – T7 promoter	(Taupin et al., 1997)
Fatty Acid and Phospholipid Metabolism			
Cardiolipin synthase	Bacillus firmus (AL)	E. coli – pET system (T7 promoter)	(Guo et al., 1998)
Dihydrolipoamide dehydrogenase	Haloferax volcanii (H)	H. volcanii – rRNA promoter	(Jolley et al., 1996)
Housekeeping and Stress Response			
GroEL	Bacillus sp. C-125 (AL)	E. coli – lac promoter	(Xu et al., 1996)

[a]The abbreviations used in the tables are as follows: A, acidophile; AL, alkaliphile; H, halophile; P, pscyhrophile.

protein expressions, either at the level of protein production or failure to produce enzymatically active protein, rarely are recorded in the literature. As a result the recombinant expression field often has to rely on anecdotal discussions of problems rather than use cited sources, which tends to diminish the reliability of the information and slow real experimental progress.

A structural genomics project with objectives to recombinantly express, purify and structurally characterize by X-ray crystallographic analysis all of the genes of the hyperthermophilic archaeon *Pyrococcus furiosus* provides a benchmark to evaluate the success rate for recombinant extremophile protein production (URL: www.secsg.org/pfu.htm). At this point in the project, 1900 of the 2200 *P. furiosus* ORFs have been cloned into pET expression vectors. One-liter scale expressions have been attempted for 411 of the 1900 clones so far. *E. coli* strain BL21(λDE3), harboring the rare codon plasmid pRIL (Stratagene, La Jolla, CA), was used as the expression host. The expression cultures were grown in LB medium at 37 °C and expression was induced using 400 μM IPTG. Of the 411 different *P. furiosus* clones, 258 yielded expressed protein and of the 258 expressed protein products, 175 were soluble and have been purified. These initial protein expressions reveal that ~63% of the *P. furiosus* genes could be expressed using the indicated conditions and that ~68% of the expressed protein was produced in soluble form.

Alternate Expression Systems Used for Recombinant Extremophile Protein Production

Yeast Expression Systems

Although the majority of recombinant extremophile proteins have been produced using *E. coli*-based expression systems, yeast-based expression systems are beginning to emerge as a popular alternative. There are a number of reasons why using yeast expression systems may prove advantageous for recombinant extremophile protein production. First, many extremophiles belong to the Archaea and it is known that archaea have some eukaryotic-like characteristics, such as having histone-like proteins and having eukaryotic-like transcription and translation initiation (Bell *et al.*, 1998). Therefore, production of archaeal extremophile proteins may benefit from expression in the eukaryotic yeast host system. Further, an often used yeast expression host, *Saccharomyces cerevisiae*, is a Generally-Regarded-As-Safe (GRAS) organism that can easily grow to high cell densities using relatively cheap culture media (Smith *et al.*, 2002). In addition, yeast expression systems can generally be used to produce secreted recombinant proteins, which can simplify the downstream protein purification process significantly.

An example of the successful expression of an extremophile protein in a yeast-based system includes the production of *Sulfolobus solfataricus* β-galactosidase (LacS) in *S. cerevisiae* using the *Gal1-10* promoter (Moracci *et al.*, 1992). In this case, a thermostable β-galactosidase, which had optimal activity at >90 °C was produced. Recombinant production of *S. solfataricus* β-galactosidase in *S. cerevisiae* was later performed at an industrial scale (100 L fermentation). The result was a 56-fold increase in protein production of the yeast-expressed protein compared to that obtained in *S. solfataricus* and enabled purification of the protein using a simple two-step purification procedure (Morana *et al.*, 1995).

A more recent example of the use of *S. cerevisiae* as an expression vector for recombinant extremophile production involved the expression of *Methanococcus jannaschii* P-type ATPase (Morsomme *et al.*, 2002). The recombinant expression of P-type ATPases is generally considered difficult since the protein is membrane-embedded. The use of a suitable *S. cerevisiae* expression strain (one that lacks both chromosomally-encoded host P-type ATPases, that had an active yeast ATPase supplied in *trans* under the control of the *Gal1-10* promoter, and had an expression vector bearing the PMA1 (P-type ATPase promoter), however, allowed for the successful production of *M. jannaschii* ATPase. The use of this strain/vector system or derivatives of it may prove useful for the recombinant production of other integral-membrane extremophile proteins.

S. cerevisiae has also been used to recombinantly express the β-glucosidase from *P. furiosus* (Smith *et al.*, 2002). In this case, the β-glucosidase gene was expressed using the yeast *Gal1-10* promoter, and ~10 mg/L of secreted recombinant protein was obtained. Analysis of the recombinant β-glucosidase expression has identified a bottleneck in the overall production of secreted protein in the yeast. The majority of the protein failed to fold properly into monomeric subunits in the endoplasmic reticulum (ER) and, therefore, was not subsequently trafficked to the yeast secretory pathway (Smith *et al.*, 2002). It was determined that increasing the expression temperature from 30 °C to 37 °C improved cell secretion levels by 440% suggesting that higher expression temperatures are required for the proper folding of the *P. furiosus* β-glucosidase subunits. Identification of the problem with proper folding recombinant protein in the yeast cell ER should ultimately aid in the development of improved expression methodologies and yeast expression strains.

Other yeast expression strains have also been used for the recombinant expression of extremophile proteins. For example, xylanase from the hyperthermophilic bacterium *Thermotoga* strain FjSS3B.1 was expressed in yeast strain *Kluyveromyces lactis*, which is considered an ideal strain choice for high-level secretion of protein (Walsh *et al.*, 1998). For this expression, the *K. lactis* secretion signal from the killer toxin was fused in frame with *Thermotoga xynA* and the resulting recombinant xylanase protein constituted over 95% of the excreted protein. Furthermore, the food yeast *Candida utilis* was used to express *S. solfataricus* α-amylase (Miura *et al.*, 1999). Previous recombinant expression of *S. solfataricus* α-amylase failed to produce high levels of protein. As such, the α-amylase gene was engineered to contain codons preferentially found in the highly expressed *C. utilis* glyceraldehyde-3-phosphatedehydrogenase (GAP) gene. This optimized *S. solfataricus* α-amylase gene was expressed in *C. utilis* under the control of the GAP promoter and the recombinant α-amylase protein constituted 50% of the soluble cell protein, yielding 12.3 g/L of α-amylase (Miura *et al.*, 1999).

Bacillus Expression Systems

Another alternative expression host that has recently been used more frequently, especially for the production of proteins from alkaliphiles, is *Bacillus subtilis*. For example, α-amylase from the alkaliphilic *Bacillus* sp. KSM-1378 was recombinantly expressed at levels of 1 g/L using the pHSP64 vector, which was designed expressly for the hyperexcretion of recombinant protein in *B. subtilis* (Ikawa *et al.*, 1998). Alkaline pectate lyase from the alkaliphilic *Bacillus* strain P-4-N was excreted at very low levels in the native *Bacillus* strain and was, therefore, expressed in *B. subtilis* using the *Bacillus* shuttle vector pHY300PLK (Ishiwa *et al.*, 1986). The recombinant expression of the alkaliphilic pectate lyase in *B. subtilis* resulted in expression levels of 220 mg/L (Kobayashi *et al.*, 2000).

B. subtilis has also been used recently to express proteins from thermophiles. In this case, thermoactive pullulanase from *Desulfurococcus mucosus* was initially expressed in *E. coli*, however it resulted in such low expression that the recombinant protein could not be purified. The *D. mucosus* gene was then cloned into the *B. subtilis* expression vector pJA803 and placed under the control of the *Bacillus* maltogenic α-amylase (P*amyM*) promoter (Duffner *et al.*, 2000). Expression of *D. mucosus* pullulanase in *B. subtilis* yielded 15 mg/L of recombinant protein. The ability of the pullulanase protein to be overexpressed in *B. subtilis* but not *E. coli* likely is the result of the codon bias of the *D. mucosus* pullulanase gene. The pullulanase gene reflects the codon biases present in *B. subtilis* better than that of *E. coli* since it contains nine arginine, isoleucine, and leucine codons within the first 22 amino acids that are considered rare in *E. coli* (Duffner *et al.*, 2000). With the increasing identification of new alkaliphiles and potentially useful alkaliphilic enzymes, it is anticipated that the recent development of new *Bacillus* expression systems will continue to occur.

Extremophile Expression Systems

It is obvious from discussion of recombinant extremophile protein expression in the standard non-extreme expression hosts, such as *E. coli, S. cerevisiae,* and *B. subtilis,* that none of these systems are necessarily ideal for the overexpression of recombinant extremophile proteins. These non-extreme expression hosts are unable to mimic or withstand the extreme conditions under which the native proteins are ideally produced. Thus, as established from previous discussion of extremophile shuttle/expression vectors, the need for extremophile expression systems is evident, however, the field is still in its infancy. Progress is being made, however, as there are now successful examples of expression systems for every type of extremophile in the literature (See Table 4).

Factors Influencing the Success of Recombinant Extremophile Protein Expression

Codon Usage

It has been well documented that all living organisms exhibit a nonrandom usage of synonymous codons (Gouy *et al.*, 1982; Ikemura, 1985). Analysis of codon usage patterns in *E. coli* has resulted in the following observations: (1) bias exists for one or two codons for almost all degenerate codon families, (2) some codons are preferentially used by all the organism's genes regardless of the relative abundance of the protein, (3) highly expressed genes tend to have a greater degree of codon bias compared to poorly expressed genes, and (4) the frequency of the use of synonymous codons generally correlates to the abundance of available cognate tRNAs (de Boer *et al.*, 1986). Codon usage has been identified as a significant factor that influences the efficiency of recombinant protein expression (Kane, 1995). It has been determined that the limited availability of cognate charged tRNAs can promote translational stalling and an increased chance of reading frame shifts. This causes premature translational termination and the production of truncated protein products (Kurland *et al.*, 1996).

Table 7. Codons used by *E. coli* at a frequency of <1% compared to codon frequencies of representative extremophiles[a]

Codon	Encoded Amino Acid	Frequency of Particular Codon Per 1000 Codons						
		E. coli	*M. jannaschii*	*P. furiosus*	*S. solfataricus*	*T. maritima*	*P. haloplanktis*	*H. volcanii*
AGG	Arg	1.2	9.9	**21.1**	**17.4**	**15.8**	0.6	0.8
AGA	Arg	2.1	**27.5**	**27.8**	**25.6**	**29.7**	4.3	0.9
CGA	Arg	3.6	0.4	1.2	1.3	3.0	2.7	4.8
CUA	Leu	3.9	8.5	**17.8**	**19.1**	2.6	**12.8**	0.7
AUA	Ile	4.4	**45.4**	**40.6**	**49.3**	**27.9**	**12.3**	1.1
UGU	Cys	5.2	8.7	3.7	4.2	4.2	7.3	1.4
CGG	Arg	5.4	0.1	0.7	0.5	1.8	1.3	**10.6**
CCC	Pro	5.5	1.3	9.2	5.4	**10.6**	2.8	**18.2**
UGC	Cys	6.5	4.1	2.2	2.0	2.9	5.1	2.5
CCU	Pro	7.0	8.5	**10.6**	**12.5**	9.9	**12.3**	0.5
ACA	Thr	7.1	**19.7**	**15.6**	**13.8**	**12.9**	**23.1**	1.5
UCA	Ser	7.2	**14.5**	**10.2**	**15.3**	7.1	**16.3**	0.5
GGA	Gly	8.0	**35.5**	**37.0**	**25.8**	**32.5**	8.5	5.3
CCA	Pro	8.4	**22.2**	**19.9**	**16.3**	9.6	**15.6**	1.0
UCU	Ser	8.5	**10.6**	9.1	**15.3**	**12.2**	**12.3**	1.6
UCC	Ser	8.6	2.8	6.3	7.5	**12.1**	1.3	**17.7**
AGU	Ser	8.8	**11.1**	**10.2**	**16.5**	7.8	**19.1**	1.6
UCG	Ser	8.9	0.8	2.5	4.8	7.8	8.0	**21.9**
CAC	His	9.7	4.0	8.1	4.7	**10.6**	6.9	**16.8**

[a]The data presented were derived from the Codon Usage Database accessed at www.kazusa.or.jp/codon/. The reported codon frequencies were based on the following numbers of genes per organisms: 4,291, *Escherichia coli*; 1,773, *Methanococcus jannaschii*; 2,064 *Pyrococcus furiosus*; 3,594, *Sulfolobus solfataricus*; 1,960 *Thermotoga maritima*; 18, *Pseudoalteromonas haloplanktis*; 71, *Haloferax volcanii*. Codons which are used by the various organisms at frequencies >1% are indicated in bold type.

In the case of recombinant expression of extremophile proteins in *E. coli*, there have been a number of reported instances where incompatible expression host and gene source codon usage has resulted in inefficient protein expression. The prevalence of codon usage problems for expression of recombinant extremophile proteins can be readily explained by observing that a number of rarely used codons in *E. coli* are used at much greater frequency in various extremophiles (see Table 7). For instance, in *E. coli*, the arginine codons AGG and AGA, leucine codon CUA, and isoleucine codon AUA are used at frequencies of 0.12, 0.21, 0.39, and 0.44%, respectively, whereas in the hyperthermophilic archaeon *P. furiosus* these same codons are used at frequencies of 2.1, 2.8, 1.8, and 4.1%, respectively.

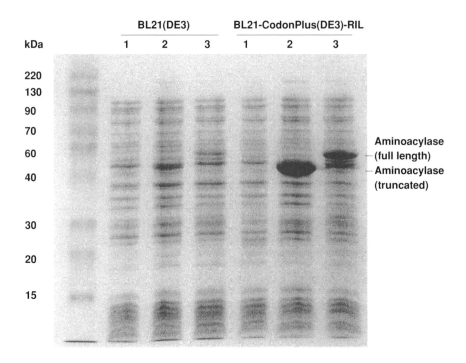

Figure 1. Expression of *Pyrococcus furiosus* aminoacylase in *E. coli* strain BL21(DE3) in the presence and absence of the rare tRNA expression plasmid pRIL. Lane 1, BL21(DE3) with empty vector only; Lane 2, BL21(DE3) with truncated-aminoacylase expression vector; Lane 3, BL21(DE3) with full-length aminoacylase expression vector. In the *P. furiosus* aminoacylase gene there are two possible translation start sites, one which yields a 382 amino acid product (truncated aminoacylase), and an upstream translation start site that produces a 440 amino acid aminoacylase (full-length aminoacylase). Messenger RNA analysis indicates that transcripts for both products are produced in *P. furiosus*.

The recombinant expression of *P. furiosus* aminoacylase in *E. coli* is one example of how differences in codon usage between the expression host and the recombinant gene source results in poor protein production (Story *et al.*, 2001). In this case, the *P. furiosus* aminoacylase gene contains 8 AGG (2.1% codon usage frequency in the gene), 15 AGA (3.9 %), 8 CUA (2.1%), and 13 AUA (3.4%) codons, and eight of these codons are located in the first 50 amino acid positions. T7- RNA polymerase-driven expression of *P. furiosus* aminoacylase in *E. coli* strain BL21(λDE3) resulted in very poor expression (Figure 1). When the expression was conducted with the host strain also harboring a plasmid expressing the rare tRNAs for the codons AGG, AGA, CUA, and AUA (plasmid pRIL; Stratagene, La Jolla, CA), overexpression of *P. furiosus* aminoacylase was successfully attained.

In a second example, expression of a 66.7 kDa translation initiation factor IF2 (*Afu*IF5B) from the sulfate-reducing hyperthermophilic archaeon *Archaeoglobus fulgidus* was unsuccessful in *E. coli* strain BL21(λDE3) without coexpression of rare codon tRNAs (Sorensen *et al.*, 2003). Sequence analysis of *Afu*IF5B revealed that the gene contained 6.2% AGA/AGU, 0.7% CUA, and 1.2% AUA codons. Recombinant expression of *Afu*IF5B did occur when the rare tRNA containing plasmids pRIG [codes for *argU* (AGA/AGG), *ileX* (AUA), *glyT* (GGA), Baca and Hol, 2000], pCODON$^+$ [codes for *argU* (AGA/ AGG), *ileX* (AUA), *leuW* (CUA), Stratagene], or pLysSR.A.R.E. [codes for *argU* (AGA/AGG), *ileX* (AUA), *leuW* (CUA), *glyT* (GGA), *proL* (CCC), Novagen] were also transformed into the expression strain (Sorensen *et al.*, 2003). However, the expression of *Afu*IF5B when conducted in the presence of these rare tRNA expression plasmids gave rise to the overexpression of both the full-length 66.7 kDa protein as well as ~40 kDa truncated product. From analysis of the ~40 kDa expression product, both proteolytic degradation and secondary translation initiation were ruled out as possible causes for accumulation of this truncated protein. It was determined that the presence of rare AAG codon doublets was in fact the responsible factor. Mutation of the consecutive AAGs to AAAs and expression of this substituted *Afu*IF5B gene in the presence of the pLysSR.A.R.E. plasmid enabled the overproduction of the 66.7 kDa protein and completely eliminated the unwanted accumulation of the ~40 kDa product (Sorensen *et al.*, 2003). Thus, this work highlights how use of a combination of both selected substitution of rare codons for more frequently used codons and coexpression of rare tRNAs can be employed to overexpress full-length recombinant protein.

Rare codon usage was also implicated as the primary reason for poor expression of a thermoactive pullulanase from *Desulfurococcus mucosus* in *E. coli* (Duffner *et al.*, 2000). Sequence analysis of the *D. mucosus* gene revealed the presence of nine rare arginine, leucine, and isoleucine codons in the first 22 coding amino acids. In this case, incompatible codon usage was somewhat ameliorated by expression of the *D. mucosus* pullulanase in *Bacillus subtilis* rather than *E. coli*. Codon usage analysis of *Bacillus subtilis* indicates that the codons for arginine (AGA/AGG) and isoleucine that are particularly rare in *E. coli* are less rare in *Bacillus* (AGA, 1.1%, AGA 0.4%, AUA, 0.9% frequency). As a result of having a more compatible codon usage pattern for expression in *Bacillus subtilis*, successful overexpression of the *D. mucosus* pullulanase was observed (15 mg/L of recombinant protein). Thus, this illustrates how appropriate matching of codon usage patterns of both the expression host and source organism genes can help ensure successful recombinant expression of genes. Furthermore, the fact that problems of codon usage can so dramatically affect the ability to overexpress recombinant protein has certainly been a factor leading to increased interest in developing effective extremophile expression systems, thereby eliminating incompatible codon usage as a potential expression bottleneck.

Recombinant Expression of AT and GC-Rich Genes

From the analysis of a number of the completed genomes of hyperthermophilic archaea such as *Methanococcus jannaschii, Pyrococcus abyssi, P. furiosus, P. horikoshii, Sulfolobus solfataricus,* and *Sulfolobus tokodaii* (Bult *et al.*, 1996; Kawarabayasi *et al.*, 2001; Kawarabayasi *et al.*, 1998; Robb *et al.*, 2001; She *et al.*, 2001), significant portions of AT-rich DNA present in coding regions have been identified. It has been found that AT-rich genes from archaeal sources generally express poorly in *E. coli* (Ishida *et al.*, 2002). A method was recently developed, however, for the improvement of recombinant AT-rich hyperthermophile gene expression in *E. coli* (Ishida *et al.*, 2002). This method was used to express the AT-rich *P. furiosus trpA* and *trpB* genes, coding for α and β subunits of tryptophan synthase, respectively. For this procedure, a 36-bp leader ORF, containing both a ribosomal binding site (RBS) for translation of the *P. furiosus* gene and terminator, was introduced in front of both the *trpA* and *trpB* genes. The 36-bp-ORF-*trpA* and 36-bp-ORF-*trpB*-genes were placed under the control of the *E. coli lac* promoter for expression in *E. coli* strain JM109. Using this method, overexpression of both *P. furiosus* TrpA and TrpB was achieved, whereas no expression was observed for the *trpB* gene without the 36-bp leader ORF (Ishida *et al.*, 2002).

In discussing the results of their study, Ishida *et al.* point out that most researchers attribute the poor expression of AT-rich genes in *E. coli* to a tendency for rare *E. coli* codons such as AGG, AGA and AUA to exist in such AT-rich DNA regions (Sauer *et al.*, 1999; Sibold *et al.*, 1988). In the case of the *P. furiosus trpA* and *trpB* genes, which do contain significant numbers of rare AGG/AGA/AUA codons (23 rare codons for *trpA* and 32 rare codons for *trpB*), overexpression of these genes in *E. coli* may be promoted by the presence of the leader ORF rather than by base substitution of the rare codons or by coexpression of rare tRNAs (Ishida *et al.*, 2002). It was further suggested that the addition of leader sequences could provide a general expression vector construction method for the expression of AT-rich genes that would obviate the need for standard optimization of codon usage procedures (Ishida *et al.*, 2002). Given the preponderance of literature showing the effectiveness of codon usage optimization techniques in recombinant protein expression, further investigation will obviously need to be done to address this issue in the case of recombinant extremophile expression.

Like AT-rich genes, the recombinant production of GC-rich genes also tends to be inefficient in *E. coli*. GC-rich genes are fairly common in thermophilic bacteria, such as *Thermus thermophilus* (70% G+C content of genomic DNA), and a number of genes that have been cloned from *T. thermophilus* express poorly in *E. coli* (Ishida *et al.*, 1994). A procedure was devised for highly expressing GC-rich *T. thermophilus* genes in *E. coli*, similar to the method for increased expression of AT-rich genes. It relies on the addition of a short upstream leader ORF that contains a RBS for the target gene and a terminator sequence overlapping the start codon for the cloned gene (Ishida *et al.*, 1997). The efficacy of this approach for the overexpression of GC-rich genes was demonstrated for the expression of the *T. thermophilus leuB* gene (codes for 3-isopropylmalate dehydrogenase) in *E. coli* under the control of the *tac* promoter (Ishida *et al.*, 1996; Ishida *et al.*, 1997). Again, it was suggested that use of the leader sequence can be adopted as a general strategy for the successful production of any GC-rich gene in *E. coli* (Ishida *et al.*, 1997).

Protein Folding Issues

As mentioned previously, when genes from extremophilic organisms were first cloned, it was believed that creating recombinant proteins would be extremely difficult due to the differences between the environment from which the extremophile was isolated and the culture conditions in which the host organism thrives. Surprisingly, the recombinant proteins that were produced had nearly identical properties to the native proteins (Antoine *et al.*, 1999; Bauer *et al.*, 1998; Ghosh *et al.*, 1998).

In order for recombinant proteins to be utilized for biochemical analysis and technological purposes, they must be expressed in active, folded form. As with recombinant mesophilic proteins, misfolding of extremophilic recombinant protein can occur. Formation of inclusion bodies is a common problem associated with the over-expression of recombinant proteins (Carrio *et al.*, 2002; Middelberg, 2002; Rajan *et al.*, 2001). As with mesophilic proteins that form inclusion bodies, extremophilic proteins can often be solublized and re-natured. A subtilisin-like protease from *Thermococcus kodakaraensisi* KOD1 was over-produced in *E. coli* and found to form inclusion bodies (Kannan *et al.*, 2001). The protein was solubilized in 8 M urea and properly folded into active form. Glucose dehydrogenase was cloned from the halophilic archaeon *Haloferax mediterranei* into *E. coli* and overexpressed using the T7 promoter which resulted in the formation of inclusion bodies (Pire *et al.*, 2001). In order to obtain refolded protein that was active, the inclusion bodies needed to be resolubilized in a saline environment. The formation of inclusion bodies is not limited to the host *E. coli*; the α-amylase gene from *Pyrococcus woesei* was cloned into a shuttle vector and expressed in the halophilic bacterium *Halomonas elongata* (Frillingos *et al.*, 2000). The expressed protein was recovered completely in the crude membrane fraction, suggesting formation of inclusion bodies.

There is some evidence to indicate that temperature may affect folding of recombinant proteins in expression organisms. It was shown that misfolded proteins caused a protein secretion bottleneck by not being properly package for secretion by the ER (Smith *et al.*, 2002). It was observed that raising the temperature from 30 °C to 37 °C increased the amount of intracellular soluble protein and secreted protein.

The recent interest in generating crystal structures for all the *P. furiosus* proteins has created the need to quickly produce soluble proteins from proteins which are not readily soluble. For gene expression not leading to properly folded proteins, a novel high through-put approach to generate soluble proteins for crystallographic studies was reported which involved creating chimeric proteins by attaching a green fluorescent protein to the *P. furiosus* protein (Pedelacq *et al.*, 2002). The genes were subjected to random mutagenesis and increasing protein solubility was monitored by an increase of fluorescence.

Conclusion

At present, the recombinant expression of extremophilic proteins relies mostly on the use of commonly used mesophilic expression systems. This has no doubt been successful in enabling the characterization of many extremophilic proteins for technological or scientific ends. Current efforts to develop expression systems in extremophilic hosts foreshadow future endeavors in which extremophilic proteins can be produced in genetically and environmentally compatible organisms. For now, optimization of currently used hosts with respect to codon bias and other limitations will be needed to foster further development of this area of research.

Acknowledgements

RMK acknowledges support in part from the National Science Foundation and the Department of Energy for this work.

References

Abreu, I.A., Saraiva, L.M., Soares, C.M., Teixeira, M., and Cabelli, D.E. 2001. The mechanism of superoxide scavenging by *Archaeoglobus fulgidus* neelaredoxin. J. Biol. Chem. 276: 38995-39001.

Adams, M.W., and Kelly, R.M. 1994. Thermostability and thermoactivity of enzymes from hyperthermophilic Archaea. Bioorg. Med. Chem. 2: 659-667.

Adams, M.W., Perler, F.B., and Kelly, R.M. 1995. Extremozymes: expanding the limits of biocatalysis. Biotechnology (N Y). 13: 662-668.

Adams, M.W.W., and Kelly, R.M. 1998. Finding and using hyperthermophilic enzymes. Trends Biotechnol. 16: 329-332.

Adul Rahman, R.N., Jongsareejit, B., Fujiwara, S., and Imanaka, T. 1997. Characterization of recombinant glutamine synthetase from the hyperthermophilic archaeon *Pyrococcus* sp. strain KOD1. Appl. Environ. Microbiol. 63: 2472-2476.

Ahuja, A., Purcarea, C., Guy, H.I., and Evans, D.R. 2001. A novel carbamoyl-phosphate synthetase from *Aquifex aeolicus*. J. Biol. Chem. 276: 45694-45703.

Akimkina, T., Ivanov, P., Kostrov, S., Sokolova, T., Bonch-Osmolovskaya, E., Firman, K., Dutta, C.F., and McClellan, J.A. 1999. A highly conserved plasmid from the extreme thermophile *Thermotoga maritima* MC24 is a member of a family of plasmids distributed worldwide. Plasmid. 42: 236-240.

Alex, L.A., Reeve, J.N., Orme-Johnson, W.H., and Walsh, C.T. 1990. Cloning, sequence determination, and expression of the genes encoding the subunits of the nickel-containing 8-hydroxy-5-deazaflavin reducing hydrogenase from *Methanobacterium thermoautotrophicum* delta H. Biochemistry. 29: 7237-7244.

Ando, S., Ishida, H., Kosugi, Y., and Ishikawa, K. 2002. Hyperthermostable endoglucanase from *Pyrococcus horikoshii*. Appl. Environ. Microbiol. 68: 430-433.

Ando, S., Ishikawa, K., Ishida, H., Kawarabayasi, Y., Kikuchi, H., and Kosugi, Y. 1999. Thermostable aminopeptidase from *Pyrococcus horikoshii*. FEBS Lett. 447: 25-28.

Antoine, E., Rolland, J.L., Raffin, J.P., and Dietrich, J. 1999. Cloning and over-expression in *Escherichia coli* of the gene encoding NADPH group III alcohol dehydrogenase from *Thermococcus hydrothermalis*. Characterization and comparison of the native and the recombinant enzymes. Eur. J. Biochem. 264: 880-889.

Aparicio, T., Lorenzo, P., and Perera, J. 2000. pT3.2I, the smallest plasmid of *Thiobacillus* T3.2. Plasmid. 44: 1-11.

Aravalli, R.N., and Garrett, R.A. 1997. Shuttle vectors for hyperthermophilic archaea. Extremophiles. 1: 183-191.

Arnold, H.P., She, Q., Phan, H., Stedman, K., Prangishvili, D., Holz, I., Kristjansson, J.K., Garrett, R., and Zillig, W. 1999. The genetic element pSSVx of the extremely thermophilic crenarchaeon *Sulfolobus* is a hybrid between a plasmid and a virus. Mol. Microbiol. 34: 217-226.

Arnott, M.A., Michael, R.A., Thompson, C.R., Hough, D.W., and Danson, M.J. 2000. Thermostability and thermoactivity of citrate synthases from the thermophilic and hyperthermophilic archaea, *Thermoplasma acidophilum* and *Pyrococcus furiosus*. J. Mol. Biol. 304: 657-668.

Ashiuchi, M., Zakaria, M.M., Sakaguchi, Y., and Yagi, T. 1999. Sequence analysis of a cryptic plasmid from *Flavobacterium* sp. KP1, a psychrophilic bacterium. FEMS Microbiol. Lett. 170: 243-249.

Baca, A.M., and Hol, W.G. 2000. Overcoming codon bias: A method for high-level overexpression of *Plasmodium* and other AT-rich parasite genes in *Escherichia coli*. Int. J. Parasitology. 30: 113-118.

Baneyx, F. 1999. Recombinant protein expression in *Escherichia coli*. Curr. Opin. Biotechnol. 10: 411-421.

Bao, Q., Tian, Y., Li, W., Xu, Z., Xuan, Z., Hu, S., Dong, W., Yang, J., Chen, Y., Xue, Y., Xu, Y., Lai, X., Huang, L., Dong, X., Ma, Y., Ling, L., Tan, H., Chen, R., Wang, J., Yu, J., and Yang, H. 2002. A complete sequence of the *T. tengcongensis* genome. Genome Res. 12: 689-700.

Bartosik, D., Wlodarczyk, M., and Thomas, C.M. 1997. Complete nucleotide sequence of the replicator region of *Paracoccus (Thiobacillus) versutus* pTAV1 plasmid and its correlation to several plasmids of *Agrobacterium* and *Rhizobium* species. Plasmid. 38: 53-59.

Bartsch, K., Schneider, R., and Schulz, A. 1996. Stereospecific production of the herbicide phosphinothricin (glufosinate):

purification of aspartate transaminase from *Bacillus stearothermophilus*, cloning of the corresponding gene, *aspC*, and application in a coupled transaminase process. Appl. Environ. Microbiol. 62: 3794-3799.

Bauer, M.W., Driskill, L.E., Callen, W., Snead, M.A., Mathur, E.J., and Kelly, R.M. 1999. An endoglucanase, EglA, from the hyperthermophilic archaeon *Pyrococcus furiosus* hydrolyzes beta-1,4 bonds in mixed-linkage (1—>3),(1—>4)-beta-D-glucans and cellulose. J. Bacteriol. 181: 284-290.

Bauer, M.W., and Kelly, R.M. 1998. The family 1 beta-glucosidases from *Pyrococcus furiosus* and *Agrobacterium faecalis* share a common catalytic mechanism. Biochemistry. 37: 17170-17178.

Bell, G.S., Russell, R.J., Kohlhoff, M., Hensel, R., Danson, M.J., Hough, D.W., and Taylor, G.L. 1998. Preliminary crystallographic studies of triosephosphate isomerase (TIM) from the hyperthermophilic Archaeon *Pyrococcus woesei*. Acta Crystallogr. D Biol. Crystallogr. 54: 1419-1421.

Benbouzid-Rollet, N., Lopez-Garcia, P., Watrin, L., Erauso, G., Prieur, D., and Forterre, P. 1997. Isolation of new plasmids from hyperthermophilic Archaea of the order Thermococcales. Res. Microbiol. 148: 767-775.

Bhuiya, M.W., Sakuraba, H., Kujo, C., Nunoura-Kominato, N., Kawarabayasi, Y., Kikuchi, H., and Ohshima, T. 2000. Glutamate dehydrogenase from the aerobic hyperthermophilic archaeon *Aeropyrum pernix* K1: enzymatic characterization, identification of the encoding gene, and phylogenetic implications. Extremophiles. 4: 333-341.

Birolo, L., Tutino, M.L., Fontanella, B., Gerday, C., Mainolfi, K., Pascarella, S., Sannia, G., Vinci, F., and Marino, G. 2000. Aspartate aminotransferase from the Antarctic bacterium *Pseudoalteromonas haloplanktis* TAC 125. Cloning, expression, properties, and molecular modelling. Eur. J. Biochem. 267: 2790-2802.

Bitan-Banin, G., Ortenberg, R., and Mevarech, M. 2003. Development of a gene knockout system for the halophilic archaeon *Haloferax volcanii* by use of the *pyrE* gene. J. Bacteriol. 185: 772-778.

Blecher, O., Goldman, S., and Mevarech, M. 1993. High expression in *Escherichia coli* of the gene coding for dihydrofolate reductase of the extremely halophilic archaebacterium *Haloferax volcanii*.

Reconstitution of the active enzyme and mutation studies. Eur. J. Biochem. 216: 199-203.

Blondal, T., Thorbjarnardottir, S.H., Kieleczawa, J., Einarsson, J.M., Hjorleifsdottir, S., Kristjansson, J.K., and Eggertsson, G. 1999. Cloning, sequence analysis and overexpression of a rhodothermus marinus gene encoding a thermostable thymidine kinase. FEMS Microbiol. Lett. 179: 311-316.

Britton, K.L., Stillman, T.J., Yip, K.S., Forterre, P., Engel, P.C., and Rice, D.W. 1998. Insights into the molecular basis of salt tolerance from the study of glutamate dehydrogenase from *Halobacterium salinarum*. J. Biol. Chem. 273: 9023-9030.

Brosius, J., Erfle, M., and Storella, J. 1985. Spacing of the -10 and -35 regions in the tac promoter. Effect on its *in vivo* activity. J. Biol. Chem. 260: 3539-3541.

Brown, M.V., and Bowman, J.P. 2001. A molecular phylogenetic survey of sea-ice microbial communities (SIMCO). FEMS Microbiol. Ecol. 35: 267-275.

Brunner, N.A., Brinkmann, H., Siebers, B., and Hensel, R. 1998. NAD$^+$-dependent glyceraldehyde-3-phosphate dehydrogenase from *Thermoproteus tenax*. The first identified archaeal member of the aldehyde dehydrogenase superfamily is a glycolytic enzyme with unusual regulatory properties. J. Biol. Chem. 273: 6149-6156.

Buchanan, C.L., Connaris, H., Danson, M.J., Reeve, C.D., and Hough, D.W. 1999. An extremely thermostable aldolase from *Sulfolobus solfataricus* with specificity for non-phosphorylated substrates. Biochem. J. 343: 563-570.

Buhler, C., Gadelle, D., Forterre, P., Wang, J.C., and Bergerat, A. 1998. Reconstitution of DNA topoisomerase VI of the thermophilic archaeon *Sulfolobus shibatae* from subunits separately overexpressed in *Escherichia coli*. Nucleic Acids Res. 26: 5157-5162.

Bujard, H., Gentz, R., Lanzer, M., Stueber, D., Mueller, M., Ibrahimi, I., Haeuptle, M.T., and Dobberstein, B. 1987. A T5 promoter-based transcription-translation system for the analysis of proteins *in vitro* and *in vivo*. Methods Enzymol. 155: 416-433.

Bult, C.J., White, O., Olsen, G.J., Zhou, L., Fleischmann, R.D., Sutton, G.G., Blake, J.A., FitzGerald, L.M., Clayton, R.A., Gocayne, J.D., Kerlavage, A.R., Dougherty, B.A., Tomb, J.F., Adams, M.D., Reich, C.I., Overbeek, R., Kirkness, E.F., Weinstock, K.G., Merrick, J.M.,

Glodek, A., Scott, J.L., Geoghagen, N.S., and Venter, J.C. 1996. Complete genome sequence of the methanogenic archaeon, *Methanococcus jannaschii*. Science. 273: 1058-1073.

Camacho, M., Rodriguez-Arnedo, A., and Bonete, M.J. 2002. NADP-dependent isocitrate dehydrogenase from the halophilic archaeon *Haloferax volcanii*: cloning, sequence determination and overexpression in *Escherichia coli*. FEMS Microbiol. Lett. 209: 155-160.

Cann, I.K., Ishino, S., Yuasa, M., Daiyasu, H., Toh, H., and Ishino, Y. 2001. Biochemical analysis of replication factor C from the hyperthermophilic archaeon *Pyrococcus furiosus*. J. Bacteriol. 183: 2614-2623.

Cannio, R., Contursi, P., Rossi, M., and Bartolucci, S. 1998. An autonomously replicating transforming vector for *Sulfolobus solfataricus*. J Bacteriol. 180: 3237-3240.

Carrio, M.M., and Villaverde, A. 2002. Construction and deconstruction of bacterial inclusion bodies. J. Biotechnol. 96: 3-12.

Cavicchioli, R. 2000. Extremophiles. In: Encyclopedia of Microbiology, edn. 2. Lederberg, J., Alexander, M., Bloom, B.R., Hopwood, D., Hull, R., Iglewski, B.H., Laskin, A.I., Oliver, S.G., Schaechter, M. and Summers, W.C., ed. Academic Press, Inc., San Diego. p. 317-337

Cavicchioli, R., Siddiqui, K.S., Andrews, D., and Sowers, K.R. 2002. Low-temperature extremophiles and their applications. Curr Opin Biotechnol. 13: 253-261.

Cavicchioli, R., Thomas, T., and Curmi, P.M.G. 2000. Cold stress response in Archaea. Extremophiles. 4: 321-331.

Cendrin, F., Chroboczek, J., Zaccai, G., Eisenberg, H., and Mevarech, M. 1993. Cloning, sequencing, and expression in *Escherichia coli* of the gene coding for malate dehydrogenase of the extremely halophilic archaebacterium *Haloarcula marismortui*. Biochemistry. 32: 4308-4313.

Chakravarty, L., Zupancic, T.J., Baker, B., Kittle, J.D., Fry, I.J., and Tuovinen, O.H. 1995. Characterization of the pTFI91-family replicon of *Thiobacillus ferrooxidans* plasmids. Can. J. Microbiol. 41: 354-365.

Chakravarty, S., and Varadarajan, R. 2000. Elucidation of determinants of protein stability through genome sequence analysis. FEBS Lett. 470: 65-69.

Charlebois, R.L., Lam, W.L., Cline, S.W., and Doolittle, W.F. 1987. Characterization of pHV2 from *Halobacterium volcanii* and its use in demonstrating transformation of an archaebacterium. Proc. Natl. Acad. Sci. USA. 84: 8530-8534.

Chen, L., and Roberts, M.F. 1999. Characterization of a tetrameric inositol monophosphatase from the hyperthermophilic bacterium *Thermotoga maritima*. Appl. Environ. Microbiol. 65: 4559-4567.

Chen, P., Van Vliet, F., Van De Casteele, M., Legrain, C., Cunin, R., and Glansdorff, N. 1998. Aspartate transcarbamylase from the hyperthermophilic eubacterium *Thermotoga maritima*: fused catalytic and regulatory polypeptides form an allosteric enzyme. J. Bacteriol. 180: 6389-6391.

Chessa, J.P., Petrescu, I., Bentahir, M., Van Beeumen, J., and Gerday, C. 2000. Purification, physico-chemical characterization and sequence of a heat labile alkaline metalloprotease isolated from a psychrophilic *Pseudomonas* species. Biochim. Biophys. Acta. 1479: 265-274.

Clay, M.D., Jenney, F.E., Jr., Hagedoorn, P.L., George, G.N., Adams, M.W., and Johnson, M.K. 2002. Spectroscopic studies of *Pyrococcus furiosus* superoxide reductase: implications for active-site structures and the catalytic mechanism. J. Am. Chem. Soc. 124: 788-805.

Cline, S.W., and Doolittle, W.F. 1992. Transformation of members of the genus *Haloarcula* with shuttle vectors based on *Halobacterium halobium* and *Haloferax volcanii* plasmid replicons. J. Bacteriol. 174: 1076-1080.

Contursi, P., Cannio, R., Prato, S., Fiorentino, G., Rossi, M., and Bartolucci, S. 2003. Development of a genetic system for hyperthermophilic Archaea: expression of a moderate thermophilic bacterial alcohol dehydrogenase gene in *Sulfolobus solfataricus*. FEMS Microbiol Lett. 218: 115-120.

Coombs, J.M., and Brenchley, J.E. 1999. Biochemical and phylogenetic analyses of a cold-active beta-galactosidase from the lactic acid bacterium *Carnobacterium piscicola* BA. Appl. Environ. Microbiol. 65: 5443-5450.

Cowan, D.A. 1992. Biotechnology of the Archaea. Trends Biotechnol. 10: 315-23.

Dabrowski, S., and Kur, J. 1998. Recombinant His-tagged DNA polymerase. I. Cloning, purification and partial characterization of

Thermus thermophilus recombinant DNA polymerase. Acta Biochim. Pol. 45: 653-660.

Dabrowski, S., and Kur, J. 1998. Recombinant His-tagged DNA polymerase. II. Cloning and purification of *Thermus aquaticus* recombinant DNA polymerase (Stoffel fragment). Acta Biochim. Pol. 45: 661-667.

Dahlberg, C., Linberg, C., Torsvik, V.L., and Hermansson, M. 1997. Conjugative plasmids isolated from bacteria in marine environments show various degrees of homology to each other and are not closely related to well-characterized plasmids. Appl. Environ. Microbiol. 63: 4692-4697.

Dams, T., Bohm, G., Auerbach, G., Bader, G., Schurig, H., and Jaenicke, R. 1998. Homo-dimeric recombinant dihydrofolate reductase from *Thermotoga maritima* shows extreme intrinsic stability. Biol. Chem. 379: 367-371.

Dams, T., Ostendorp, R., Ott, M., Rutkat, K., and Jaenicke, R. 1996. Tetrameric and octameric lactate dehydrogenase from the hyperthermophilic bacterium *Thermotoga maritima*. Structure and stability of the two active forms. Eur. J. Biochem. 240: 274-279.

Danner, S., and Soppa, J. 1996. Characterization of the distal promoter element of halobacteria *in vivo* using saturation mutagenesis and selection. Mol. Microbiol. 19: 1265-1276.

DasSarma, S. 1995. In: Archaea- A Laboratory Manual. Robb, F.T., ed. Cold Spring Harbor Laboratory Press, Cold Spring Harbor. p. 241-250

de Boer, H.A., and Kastelein, R.A. 1986. Biased codon usage: an exploration of its role in optimization of translation. In: Maximising Gene Expression. Reznikoff, W.S. and Gold, L., ed. Buttersworth, Boston. p. 225-285.

Decanniere, K., Babu, A.M., Sandman, K., Reeve, J.N., and Heinemann, U. 2000. Crystal structures of recombinant histones HMfA and HMfB from the hyperthermophilic archaeon *Methanothermus fervidus*. J. Mol. Biol. 303: 35-47.

Deckert, G., Warren, P.V., Gaasterland, T., Young, W.G., Lenox, A.L., Graham, D.E., Overbeek, R., Snead, M.A., Keller, M., Aujay, M., Huber, R., Feldman, R.A., Short, J.M., Olsen, G.J., and Swanson, R.V. 1998. The complete genome of the hyperthermophilic bacterium *Aquifex aeolicus*. Nature. 392: 353-358.

Deming, J.W. 2002. Psychrophiles and polar regions. Curr Opin Microbiol. 5: 301-309.

Demirjian, D.C., Moris-Varas, F., and Cassidy, C.S. 2001. Enzymes from extremophiles. Curr. Opin. Chem. Biol. 5: 144-151.

Dennis, P.P., and Shimmin, L.C. 1997. Evolutionary divergence and salinity-mediated selection in halophilic archaea. Microbiol. Mol. Biol. Rev. 61: 90-104.

Diruggiero, J., and Robb, F.T. 1995. Expression and *in vitro* assembly of recombinant glutamate dehydrogenase from the hyperthermophilic archaeon *Pyrococcus furiosus*. Appl. Environ. Microbiol. 61: 159-164.

Doemel, W.N., and T.D., B. 1971. The physiological ecology of *Cyanidium caldarium*. J. Gen. Microbiol. 67: 17-32.

Dominy, C.N., Coram, N.J., and Rawlings, D.E. 1998. Sequence analysis of plasmid pTF5, a 19.8-kb geographically widespread member of the *Thiobacillus ferrooxidans* pTFI91-like plasmid family. Plasmid. 40: 50-57.

Dong, G., Vieille, C., and Zeikus, J.G. 1997. Cloning, sequencing, and expression of the gene encoding amylopullulanase from *Pyrococcus furiosus* and biochemical characterization of the recombinant enzyme. Appl Environ. Microbiol. 63: 3577-3584.

Duewel, H.S., Sheflyan, G.Y., and Woodard, R.W. 1999. Functional and biochemical characterization of a recombinant 3-Deoxy-D-manno-octulosonic acid 8-phosphate synthase from the hyperthermophilic bacterium *Aquifex aeolicus*. Biochem. Biophys. Res. Commun. 263: 346-351.

Duffner, F., Bertoldo, C., Andersen, J.T., Wagner, K., and Antranikian, G. 2000. A new thermoactive pullulanase from *Desulfurococcus mucosus*: cloning, sequencing, purification, and characterization of the recombinant enzyme after expression in *Bacillus subtilis*. J. Bacteriol. 182: 6331-6338.

Ebel, C., Faou, P., Franzetti, B., Kernel, B., Madern, D., Pascu, M., Pfister, C., Richard, S., and Zaccai, G. 1999. Molecular interactions in extreme halophiles- the solvation-stabilization hypothesis for halophilic proteins. In: Microbiology and Biogeochemistry of Hypersaline Environments. Oren, A., ed. CRC Press, Boca Raton. p. 227-237

Edwards, K.J., Bond, P.L., Gihring, T.M., and Banfield, J.F. 2000. An archaeal iron-oxidizing extreme acidophile important in acid mine drainage. Science. 287: 1796-1799.

Eidsness, M.K., Richie, K.A., Burden, A.E., Kurtz, D.M., Jr., and Scott, R.A. 1997. Dissecting contributions to the thermostability of *Pyrococcus furiosus* rubredoxin: beta-sheet chimeras. Biochemistry. 36: 10406-10413.

Eisenberg, H., Mevarech, M., and Zaccai, G. 1992. Biochemical, structural, and molecular genetic aspects of halophilism. Adv. Protein Chem. 43: 1-62.

English, R.S., Lorbach, S.C., Huffman, K.M., and Shively, J.M. 1995. Isolation and characterization of the replicon of a *Thiobacillus intermedius* plasmid. Plasmid. 33: 1-6.

Erauso, G., Marsin, S., Benbouzid-Rollet, N., Baucher, M.F., Barbeyron, T., Zivanovic, Y., Prieur, D., and Forterre, P. 1996. Sequence of plasmid pGT5 from the archaeon *Pyrococcus abyssi*: evidence for rolling-circle replication in a hyperthermophile. J. Bacteriol. 178: 3232-3237.

Feller, G., Le Bussy, O., and Gerday, C. 1998. Expression of psychrophilic genes in mesophilic hosts: assessment of the folding state of a recombinant alpha-amylase. Appl. Environ. Microbiol. 64: 1163-1165.

Fernandez-Castillo, R., Vargas, C., Nieto, J.J., Ventosa, A., and Ruiz-Berraquero, F. 1992. Characterization of a plasmid from moderately halophilic eubacteria. J. Gen. Microbiol. 138: 1133-1137.

Fields, P.A., and Somero, G.N. 1998. Hot spots in cold adaptation: localized increases in conformational flexibility in lactate dehydrogenase A4 orthologs of Antarctic notothenioid fishes. Proc. Natl. Acad. Sci. USA. 95: 11476-11481.

Fiorentino, G., Cannio, R., Rossi, M., and Bartolucci, S. 1998. Decreasing the stability and changing the substrate specificity of the *Bacillus stearothermophilus* alcohol dehydrogenase by single amino acid replacements. Protein Eng. 11: 925-930.

Fish, S.A., Duckworth, A.W., and Grant, W.D. 1999. Novel plasmids from alkaliphilic halomonads. Plasmid. 41: 268-273.

Fitz-Gibbon, S.T., Ladner, H., Kim, U.J., Stetter, K.O., Simon, M.I., and Miller, J.H. 2002. Genome sequence of the hyperthermophilic crenarchaeon *Pyrobaculum aerophilum*. Proc. Natl. Acad. Sci. USA. 99: 984-989.

Forterre, P., Bergerat, A., and Lopez-Garcia, P. 1996. The unique DNA topology and DNA topoisomerases of hyperthermophilic archaea. FEMS Microbiol. Rev. 18: 237-248.

Frankenberg, R.J., Hsu, T.S., Yakota, H., Kim, R., and Clark, D.S. 2001. Chemical denaturation and elevated folding temperatures are required for wild-type activity and stability of recombinant *Methanococcus jannaschii* 20S proteasome. Protein Sci. 10: 1887-1896.

Fridjonsson, O., Watzlawick, H., Gehweiler, A., and Mattes, R. 1999. Thermostable alpha-galactosidase from *Bacillus stearothermophilus* NUB3621: cloning, sequencing and characterization. FEMS Microbiol. Lett. 176: 147-153.

Frillingos, S., Linden, A., Niehaus, F., Vargas, C., Nieto, J.J., Ventosa, A., Antranikian, G., and Drainas, C. 2000. Cloning and expression of alpha-amylase from the hyperthermophilic archaeon *Pyrococcus woesei* in the moderately halophilic bacterium *Halomonas elongata*. J. Appl. Microbiol. 88: 495-503.

Fukui, T., Eguchi, T., Atomi, H., and Imanaka, T. 2002. A membrane-bound archaeal Lon protease displays ATP-independent proteolytic activity towards unfolded proteins and ATP-dependent activity for folded proteins. J. Bacteriol. 184: 3689-3698.

Galinski, E.A. 1995. Osmoadaptation in bacteria. Adv Microb Physiol. 37: 272-328.

Gardner, M.N., Deane, S.M., and Rawlings, D.E. 2001. Isolation of a new broad-host-range IncQ-like plasmid, pTC-F14, from the acidophilic bacterium *Acidithiobacillus caldus* and analysis of the plasmid replicon. J. Bacteriol. 183: 3303-3309.

Gerday, C., Aittaleb, M., Bentahir, M., Chessa, J.P., Claverie, P., Collins, T., D'Amico, S., Dumont, J., Garsoux, G., Georlette, D., Hoyoux, A., Lonhienne, T., Meuwis, M.A., and Feller, G. 2000. Cold-adapted enzymes: from fundamentals to biotechnology. Trends Biotechnol. 18: 103-107.

Gerike, U., Danson, M.J., Russell, N.J., and Hough, D.W. 1997. Sequencing and expression of the gene encoding a cold-active citrate synthase from an Antarctic bacterium, strain DS2-3R. Eur. J. Biochem. 248: 49-57.

Ghosh, M., Grunden, A.M., Dunn, D.M., Weiss, R., and Adams, M.W. 1998. Characterization of native and recombinant forms of an unusual cobalt-dependent proline dipeptidase (prolidase) from the hyperthermophilic archaeon *Pyrococcus furiosus*. J. Bacteriol. 180: 4781-4789.

Gianese, G., Argos, P., and Pascarella, S. 2001. Structural adaptation of enzymes to low temperatures. Protein Eng. 14: 141-148.

Goldman, S., Kim, R., Hung, L.W., Jancarik, J., and Kim, S.H. 1998. Purification, crystallization and preliminary X-ray crystallographic analysis of *Pyrococcus furiosus* DNA polymerase. Acta Crystallogr. D Biol. Crystallogr. 54: 986-988.

Gouy, M., and Gautier, C. 1982. Codon usage in bacteria: correlation with gene expressivity. Nucleic Acids Res. 10: 7055-7074.

Grattinger, M., Dankesreiter, A., Schurig, H., and Jaenicke, R. 1998. Recombinant phosphoglycerate kinase from the hyperthermophilic bacterium *Thermotoga maritima*: catalytic, spectral and thermodynamic properties. J. Mol. Biol. 280: 525-533.

Greller, G., Riek, R., and Boos, W. 2001. Purification and characterization of the heterologously expressed trehalose/maltose ABC transporter complex of the hyperthermophilic archaeon *Thermococcus litoralis*. Eur. J. Biochem. 268: 4011-4018.

Grogan, D.W. 1998. Hyperthermophiles and the problem of DNA instability. Mol. Microbiol. 28: 1043-1049.

Gronstad, A., Jaroszewicz, E., Ito, M., Sturr, M.G., Krulwich, T.A., and Kolsto, A.B. 1998. Physical map of alkaliphilic *Bacillus firmus* OF4 and detection of a large endogenous plasmid. Extremophiles. 2: 447-453.

Grunden, A.M., Jenney, F.J., Ma, K., and Adams, M.W.W. 2003. Characterization of Recombinant NAD(P)H:Rubredoxin Oxidoreductase of *Pyrococcus furiosus* and Reconstitution of a NADPH-Dependent, Superoxide Reduction Pathway. Appl. Environ. Microbiol. Submitted.

Guo, D., and Tropp, B.E. 1998. Cloning of the *Bacillus firmus* OF4 cls gene and characterization of its gene product. Biochim. Biophys. Acta. 1389: 34-42.

Haas, B.J., Sandigursky, M., Tainer, J.A., Franklin, W.A., and Cunningham, R.P. 1999. Purification and characterization of *Thermotoga maritima* endonuclease IV, a thermostable apurinic/apyrimidinic endonuclease and 3'-repair diesterase. J. Bacteriol. 181: 2834-2839.

Hack, E.S., Vorobyova, T., Sakash, J.B., West, J.M., Macol, C.P., Herve, G., Williams, M.K., and Kantrowitz, E.R. 2000. Characterization of the aspartate transcarbamoylase from *Methanococcus jannaschii*. J. Biol. Chem. 275: 15820-15827.

Hackett, N.R., and DasSarma, S. 1989. Characterization of the small endogenous plasmid of *Halobacterium* strain SB3 and its use in transformation of *H. halobium*. Can. J. Microbiol. 35: 86-91.

Hakamada, Y., Hatada, Y., Koike, K., Yoshimatsu, T., Kawai, S., Kobayashi, T., and Ito, S. 2000. Deduced amino acid sequence and possible catalytic residues of a thermostable, alkaline cellulase from an Alkaliphilic *Bacillus* strain. Biosci. Biotechnol. Biochem. 64: 2281-2289.

Halio, S.B., Blumentals, II, Short, S.A., Merrill, B.M., and Kelly, R.M. 1996. Sequence, expression in *Escherichia coli*, and analysis of the gene encoding a novel intracellular protease (PfpI) from the hyperthermophilic archaeon *Pyrococcus furiosus*. J. Bacteriol. 178: 2605-2612.

Halldorsdottir, S., Thorolfsdottir, E.T., Spilliaert, R., Johansson, M., Thorbjarnardottir, S.H., Palsdottir, A., Hreggvidsson, G.O., Kristjansson, J.K., Holst, O., and Eggertsson, G. 1998. Cloning, sequencing and overexpression of a *Rhodothermus marinus* gene encoding a thermostable cellulase of glycosyl hydrolase family 12. Appl. Microbiol. Biotechnol. 49: 277-284.

Hansen, T., Reichstein, B., Schmid, R., and Schonheit, P. 2002. The first archaeal ATP-dependent glucokinase, from the hyperthermophilic crenarchaeon *Aeropyrum pernix*, represents a monomeric, extremely thermophilic ROK glucokinase with broad hexose specificity. J. Bacteriol. 184: 5955-5965.

Hansen, T., and Schonheit, P. 2001. Sequence, expression, and characterization of the first archaeal ATP-dependent 6-phosphofructokinase, a non-allosteric enzyme related to the phosphofructokinase-B sugar kinase family, from the hyperthermophilic crenarchaeote *Aeropyrum pernix*. Arch. Microbiol. 177: 62-69.

Harriott, O.T., Huber, R., Stetter, K.O., Betts, P.W., and Noll, K.M. 1994. A cryptic miniplasmid from the hyperthermophilic bacterium *Thermotoga* sp. strain RQ7. J. Bacteriol. 176: 2759-2762.

Haruki, M., Hayashi, K., Kochi, T., Muroya, A., Koga, Y., Morikawa,

M., Imanaka, T., and Kanaya, S. 1998. Gene cloning and characterization of recombinant RNase HII from a hyperthermophilic archaeon. J. Bacteriol. 180: 6207-6214.

Holmes, M., Pfeifer, F., and Dyall-Smith, M. 1994. Improved shuttle vectors for *Haloferax volcanii* including a dual-resistance plasmid. Gene. 146: 117-121.

Holmes, M.L., Pfeifer, F., and Dyall-Smith, M.L. 1995. Analysis of the halobacterial plasmid pHK2 minimal replicon. Gene. 153: 117-121.

Horikoshi, K., and Grant, W.D. 1998. Extremophiles. Microbial Life in Extreme Environments, Wiley-Liss, New York.

Huang, H.S., Ito, K., Yin, C.H., Kabashima, T., and Yoshimoto, T. 1998. Cloning, sequencing, high expression, and crystallization of the thermophile *Thermus aquaticus* glycerol kinase. Biosci. Biotechnol. Biochem. 62: 2375-2381.

Huber, R., Huber, H., and Stetter, K.O. 2000. Towards the ecology of hyperthermophiles: biotopes, new isolation strategies and novel metabolic properties. FEMS Microbiol. Rev. 24: 615-623.

Huber, R., and Stetter, K.O. 2001. Discovery of hyperthermophilic microorganisms. Methods Enzymol. 330: 11-24.

Ichikawa, J.K., and Clarke, S. 1998. A highly active protein repair enzyme from an extreme thermophile: the L-isoaspartyl methyltransferase from *Thermotoga maritima*. Arch. Biochem. Biophys. 358: 222-231.

Iida, T., Iwabuchi, T., Ideno, A., Suzuki, S., and Maruyama, T. 2000. FK506-binding protein-type peptidyl-prolyl cis-trans isomerase from a halophilic archaeum, *Halobacterium cutirubrum*. Gene. 256: 319-326.

Ikawa, K., Araki, H., Tsujino, Y., Hayashi, Y., Igarashi, K., Hatada, Y., Hagihara, H., Ozawa, T., Ozaki, K., Kobayashi, T., and Ito, S. 1998. Hyperexpression of the gene for a *Bacillus* alpha-amylase in *Bacillus subtilis* cells: enzymatic properties and crystallization of the recombinant enzyme. Biosci. Biotechnol. Biochem. 62: 1720-1725.

Ikemura, T. 1985. Codon usage and tRNA content in unicellular and multicellular organisms. Mol. Biol. Evol. 2: 13-34.

Inagaki, K., Tomono, J., Kishimoto, N., Tano, T., and Tanaka, H. 1993. Transformation of the acidophilic heterotroph *Acidiphilium facilis* by electroporation. Biosci. Biotechnol. Biochem. 57: 1770-1771.

Ingledew, W.J. 1982. *Thiobacillus ferrooxidans*. The bioenergetics of an acidophilic chemolithotroph. Biochim. Biophys. Acta. 683: 89-117.

Ishibashi, M., Tokunaga, H., Hiratsuka, K., Yonezawa, Y., Tsurumaru, H., Arakawa, T., and Tokunaga, M. 2001. NaCl-activated nucleoside diphosphate kinase from extremely halophilic archaeon, *Halobacterium salinarum*, maintains native conformation without salt. FEBS Lett. 493: 134-138.

Ishida, M., and Oshima, T. 1994. Overexpression of genes of an extreme thermophile *Thermus thermophilus*, in *Escherichia coli* cells. J Bacteriol. 176: 2767-2770.

Ishida, M., and Oshima, T. 1996. A leader open reading frame is essential for the expression in *Escherichia coli* of GC-rich *leuB* gene of an extreme thermophile, *Thermus thermophilus*. FEMS Microbiol. Lett. 135: 137-142.

Ishida, M., Oshima, T., and Yutani, K. 2002. Overexpression in *Escherichia coli* of the AT-rich trpA and trpB genes from the hyperthermophilic archaeon *Pyrococcus furiosus*. FEMS Microbiol. Lett. 216: 179-183.

Ishida, M., Yoshida, M., and Oshima, T. 1997. Highly efficient production of enzymes of an extreme thermophile, *Thermus thermophilus*: A practical method to overexpress GC-rich genes in *Escherichia coli*. Extremophiles. 1: 157-162.

Ishikawa, K., Ishida, H., Koyama, Y., Kawarabayasi, Y., Kawahara, J., Matsui, E., and Matsui, I. 1998. Acylamino acid-releasing enzyme from the thermophilic archaeon *Pyrococcus horikoshii*. J. Biol. Chem. 273: 17726-17731.

Ishikawa, K., Ishida, H., Matsui, I., Kawarabayasi, Y., and Kikuchi, H. 2001. Novel bifunctional hyperthermostable carboxypeptidase/aminoacylase from *Pyrococcus horikoshii* OT3. Appl. Environ. Microbiol. 67: 673-679.

Ishiwa, H., and Shibahara-Sone, H. 1986. New shuttle vectors for *Escherichia coli* and *Bacillus subtilis*. IV. The nucleotide seqeunce of pHY300PLK and some properties in relation to transformation. Jpn. J. Genet. 61: 515-528.

Isupov, M.N., Fleming, T.M., Dalby, A.R., Crowhurst, G.S., Bourne, P.C., and Littlechild, J.A. 1999. Crystal structure of the glyceraldehyde-3-phosphate dehydrogenase from the hyperthermophilic archaeon *Sulfolobus solfataricus*. J. Mol. Biol. 291: 651-660.

Jaeger, S., Schmuck, R., and Sobek, H. 2000. Molecular cloning, sequency, and expression of the heat-labile uracil-DNA glycosylase from a marine psychrophilic bacterium, strain BMTU3346. Extremophiles. 4: 115-122.

Javor, B. 1989. Hypersaline Environments, Springer, Berlin.

Jeon, B.S., Taguchi, H., Sakai, H., Ohshima, T., Wakagi, T., and Matsuzawa, H. 1997. 4-alpha-glucanotransferase from the hyperthermophilic archaeon *Thermococcus litoralis*—enzyme purification and characterization, and gene cloning, sequencing and expression in *Escherichia coli*. Eur. J. Biochem. 248: 171-178.

Jeon, S.J., Fujiwara, S., Takagi, M., Tanaka, T., and Imanaka, T. 2002. Tk-PTP, protein tyrosine/serine phosphatase from hyperthermophilic archaeon *Thermococcus kodakaraensis* KOD1: enzymatic characteristics and identification of its substrate proteins. Biochem. Biophys. Res. Commun. 295: 508-514.

Jolley, K.A., Rapaport, E., Hough, D.W., Danson, M.J., Woods, W.G., and Dyall-Smith, M.L. 1996. Dihydrolipoamide dehydrogenase from the halophilic archaeon *Haloferax volcanii*: homologous overexpression of the cloned gene. J. Bacteriol. 178: 3044-3048.

Jones, B.E., Grant, W.D., Duckworth, A.W., and Owenson, G.G. 1998. Microbial diversity of soda lakes. Extremophiles. 2: 191-200.

Jongsareejit, B., Rahman, R.N., Fujiwara, S., and Imanaka, T. 1997. Gene cloning, sequencing and enzymatic properties of glutamate synthase from the hyperthermophilic archaeon *Pyrococcus* sp. KOD1. Mol. Gen. Genet. 254: 635-642.

Kaczowka, S.J., and Maupin-Furlow, J.A. 2003. Subunit topology of two 20S proteasomes from *Haloferax volcanii*. J. Bacteriol. 185: 165-174.

Kamekura, M., Seno, Y., and Dyall-Smith, M. 1996. Halolysin R4, a serine proteinase from the halophilic archaeon *Haloferax mediterranei*; gene cloning, expression and structural studies. Biochim. Biophys. Acta. 1294: 159-167.

Kane, J.F. 1995. Effects of rare codon clusters on high-level expression of heterologous proteins in *Escherichia coli*. Curr. Opin. Biotechnol. 6: 494-500.

Kannan, Y., Koga, Y., Inoue, Y., Haruki, M., Takagi, M., Imanaka, T.,

Morikawa, M., and Kanaya, S. 2001. Active subtilisin-like protease from a hyperthermophilic archaeon in a form with a putative prosequence. Appl. Environ. Microbiol. 67: 2445-2452.

Karginov, A.V., Karginova, O.A., Spiridonova, V.A., and Kopylov, A.M. 1995. *In vivo* assembly of plasmid-expressed ribosomal protein S7 of *Thermus thermophilus* into *Escherichia coli* ribosomes and conditions of its overexpression. FEBS Lett. 369: 158-160.

Karlsson, E.N., Bartonek-Roxa, E., and Holst, O. 1998. Evidence for substrate binding of a recombinant thermostable xylanase originating from *Rhodothermus marinus*. FEMS Microbiol. Lett. 168: 1-7.

Karner, M.B., DeLong, E.F., and Karl, D.M. 2001. Archaeal dominance in the mesopelagic zone of the Pacific Ocean. Nature. 409: 507-510.

Kawarabayasi, Y., Hino, Y., Horikawa, H., Jin-no, K., Takahashi, M., Sekine, M., Baba, S., Ankai, A., Kosugi, H., Hosoyama, A., Fukui, S., Nagai, Y., Nishijima, K., Otsuka, R., Nakazawa, H., Takamiya, M., Kato, Y., Yoshizawa, T., Tanaka, T., Kudoh, Y., Yamazaki, J., Kushida, N., Oguchi, A., Aoki, K., Masuda, S., Yanagii, M., Nishimura, M., Yamagishi, A., Oshima, T., and Kikuchi, H. 2001. Complete genome sequence of an aerobic thermoacidophilic crenarchaeon, *Sulfolobus tokodaii* strain7. DNA Res. 8: 123-140.

Kawarabayasi, Y., Hino, Y., Horikawa, H., Yamazaki, S., Haikawa, Y., Jin-no, K., Takahashi, M., Sekine, M., Baba, S., Ankai, A., Kosugi, H., Hosoyama, A., Fukui, S., Nagai, Y., Nishijima, K., Nakazawa, H., Takamiya, M., Masuda, S., Funahashi, T., Tanaka, T., Kudoh, Y., Yamazaki, J., Kushida, N., Oguchi, A., Kikuchi, H., and *et al*. 1999. Complete genome sequence of an aerobic hyper-thermophilic crenarchaeon, *Aeropyrum pernix* K1. DNA Res. 6: 83-101, 145-152.

Kawarabayasi, Y., Sawada, M., Horikawa, H., Haikawa, Y., Hino, Y., Yamamoto, S., Sekine, M., Baba, S., Kosugi, H., Hosoyama, A., Nagai, Y., Sakai, M., Ogura, K., Otsuka, R., Nakazawa, H., Takamiya, M., Ohfuku, Y., Funahashi, T., Tanaka, T., Kudoh, Y., Yamazaki, J., Kushida, N., Oguchi, A., Aoki, K., and Kikuchi, H. 1998. Complete sequence and gene organization of the genome of a hyper-thermophilic archaebacterium, *Pyrococcus horikoshii* OT3. DNA Res. 5: 55-76.

Kawashima, T., Amano, N., Koike, H., Makino, S.-i., Higuchi, S., Kawashima-Ohya, Y., Watanabe, K., Yamazaki, M., Kanehori, K., Kawamoto, T., Nunoshiba, T., Yamamoto, Y., Aramaki, H., Makino,

K., and Suzuki, M. 2000. Archaeal adaptation to higher temperatures revealed by genomic sequence of *Thermoplasma volcanium*. Proc. Natl. Acad. Sci. USA. 97: 14257-14262.

Keeling, P.J., Klenk, H.P., Singh, R.K., Schenk, M.E., Sensen, C.W., Zillig, W., and Doolittle, W.F. 1998. *Sulfolobus islandicus* plasmids pRN1 and pRN2 share distant but common evolutionary ancestry. Extremophiles. 2: 391-393.

Kengen, S.W., Bikker, F.J., Hagen, W.R., de Vos, W.M., and van der Oost, J. 2001. Characterization of a catalase-peroxidase from the hyperthermophilic archaeon *Archaeoglobus fulgidus*. Extremophiles. 5: 323-332.

Kim, B.C., Lee, Y.H., Lee, H.S., Lee, D.W., Choe, E.A., and Pyun, Y.R. 2002. Cloning, expression and characterization of L-arabinose isomerase from *Thermotoga neapolitana*: bioconversion of D-galactose to D-tagatose using the enzyme. FEMS Microbiol. Lett. 212: 121-126.

Kim, J.O., Park, S.R., Lim, W.J., Ryu, S.K., Kim, M.K., An, C.L., Cho, S.J., Park, Y.W., Kim, J.H., and Yun, H.D. 2000. Cloning and characterization of thermostable endoglucanase (Cel8Y) from the hyperthermophilic *Aquifex aeolicus* VF5. Biochem. Biophys. Res. Commun. 279: 420-426.

Kim, K.K., Yokota, H., Santoso, S., Lerner, D., Kim, R., and Kim, S.H. 1998. Purification, crystallization, and preliminary X-ray crystallographic data analysis of small heat shock protein homolog from *Methanococcus jannaschii*, a hyperthermophile. J. Struct. Biol. 121: 76-80.

Klenk, H.P., Clayton, R.A., Tomb, J.F., White, O., Nelson, K.E., Ketchum, K.A., Dodson, R.J., Gwinn, M., Hickey, E.K., Peterson, J.D., Richardson, D.L., Kerlavage, A.R., Graham, D.E., Kyrpides, N.C., Fleischmann, R.D., Quackenbush, J., Lee, N.H., Sutton, G.G., Gill, S., Kirkness, E.F., Dougherty, B.A., McKenney, K., Adams, M.D., Loftus, B., Venter, J.C., and *et al.* 1997. The complete genome sequence of the hyperthermophilic, sulphate-reducing archaeon *Archaeoglobus fulgidus*. Nature. 390: 364-370.

Kletzin, A., Lieke, A., Urich, T., Charlebois, R.L., and Sensen, C.W. 1999. Molecular analysis of pDL10 from *Acidianus ambivalens* reveals a family of related plasmids from extremely thermophilic and acidophilic archaea. Genetics. 152: 1307-1314.

Kobayashi, T., Hatada, Y., Suzumatsu, A., Saeki, K., Hakamada, Y., and Ito, S. 2000. Highly alkaline pectate lyase Pel-4A from alkaliphilic *Bacillus* sp. strain P-4-N: its catalytic properties and deduced amino acid sequence. Extremophiles. 4: 377-383.

Koga, Y., Morikawa, M., Haruki, M., Nakamura, H., Imanaka, T., and Kanaya, S. 1998. Thermostable glycerol kinase from a hyperthermophilic archaeon: gene cloning and characterization of the recombinant enzyme. Protein Eng. 11: 1219-1227.

Komori, K., Fujita, N., Ichiyanagi, K., Shinagawa, H., Morikawa, K., and Ishino, Y. 1999. PI-PfuI and PI-PfuII, intein-coded homing endonucleases from *Pyrococcus furiosus*. I. Purification and identification of the homing-type endonuclease activities. Nucleic Acids Res. 27: 4167-4174.

Krah, M., Misselwitz, R., Politz, O., Thomsen, K.K., Welfle, H., and Borriss, R. 1998. The laminarinase from thermophilic eubacterium *Rhodothermus marinus*—conformation, stability, and identification of active site carboxylic residues by site-directed mutagenesis. Eur. J. Biochem. 257: 101-111.

Krah, R., O'Dea, M.H., and Gellert, M. 1997. Reverse gyrase from *Methanopyrus kandleri*. Reconstitution of an active extremozyme from its two recombinant subunits. J. Bio.l Chem. 272: 13986-13990.

Krebs, M.P., Hauss, T., Heyn, M.P., RajBhandary, U.L., and Khorana, H.G. 1991. Expression of the bacterioopsin gene in *Halobacterium halobium* using a multicopy plasmid. Proc. Natl. Acad. Sci. USA. 88: 859-863.

Krebs, M.P., Mollaaghababa, R., and Khorana, H.G. 1993. Gene replacement in *Halobacterium halobium* and expression of bacteriorhodopsin mutants. Proc. Natl. Acad. Sci. USA. 90: 1987-1991.

Krulwich, T.A., Ito, M., Hicks, D.B., Gilmour, R., and Guffanti, A.A. 1998. pH homeostasis and ATP synthesis: studies of two processes that necessitate inward proton translocation in extremely alkaliphilic *Bacillus* species. Extremophiles. 2: 217-222.

Kulkarni, N., Lakshmikumaran, M., and Rao, M. 1999. Xylanase II from an alkaliphilic thermophilic *Bacillus* with a distinctly different structure from other xylanases: evolutionary relationship to alkaliphilic xylanases. Biochem. Biophys. Res. Commun. 263: 640-645.

Kurland, C., and Gallant, J. 1996. Errors of heterologous protein expression. Curr. Opin. Biotechnol. 7: 489-493.

Kushner, D.J. 1978. Life in high salt and solute concentrations. In: Microbial life in extreme environments. Kushner, D.J., ed. Academic Press, London. p. 317-368

Ladenstein, R., and Antranikian, G. 1998. Proteins from hyperthermophiles: stability and enzymatic catalysis close to the boiling point of water. Adv. Biochem. Eng. Biotechnol. 61: 37-85.

Laderman, K.A., Asada, K., Uemori, T., Mukai, H., Taguchi, Y., Kato, I., and Anfinsen, C.B. 1993. Alpha-amylase from the hyperthermophilic archaebacterium *Pyrococcus furiosus*. Cloning and sequencing of the gene and expression in *Escherichia coli*. J. Biol. Chem. 268: 24402-24407.

Lam, W.L., and Doolittle, W.F. 1989. Shuttle vectors for the archaebacterium *Halobacterium volcanii*. Proc. Natl. Acad. Sci. USA. 86: 5478-5482.

Lam, W.L., and Doolittle, W.F. 1992. Mevinolin-resistant mutations identify a promoter and the gene for a eukaryote-like 3-hydroxy-3-methylglutaryl-coenzyme A reductase in the archaebacterium *Haloferax volcanii*. J. Biol. Chem. 267: 5829-5834.

Lanyi, J.K. 1974. Salt-Dependent Properties of Proteins from Extremely Halophilic Bacteria. Bacteriol. Rev. 38: 272-290.

Lauer, G., Rudd, E.A., McKay, D.L., Ally, A., Ally, D., and Backman, K.C. 1991. Cloning, nucleotide sequence, and engineered expression of *Thermus thermophilus* DNA ligase, a homolog of *Escherichia coli* DNA ligase. J. Bacteriol. 173: 5047-5053.

Lebbink, J.H., Knapp, S., van der Oost, J., Rice, D., Ladenstein, R., and de Vos, W.M. 1999. Engineering activity and stability of *Thermotoga maritima* glutamate dehydrogenase. II: construction of a 16-residue ion-pair network at the subunit interface. J. Mol. Biol. 289: 357-369.

Leclere, M.M., Nishioka, M., Yuasa, T., Fujiwara, S., Takagi, M., and Imanaka, T. 1998. The O6-methylguanine-DNA methyltransferase from the hyperthermophilic archaeon *Pyrococcus* sp. KOD1: a thermostable repair enzyme. Mol. Gen. Genet. 258: 69-77.

Lee, B.I., Chang, C., Cho, S.J., Han, G.W., Yu, Y.G., Eom, S.H., and Suh, S.W. 2000. Lactate dehydrogenase from the hyperthermophilic archaeon *Methanococcus jannaschii*: overexpression, crystallization and preliminary X-ray analysis. Acta Crystallogr. D Biol. Crystallogr. 56: 81-83.

Lee, P.J., and Stock, A.M. 1996. Characterization of the genes and proteins of a two-component system from the hyperthermophilic bacterium *Thermotoga maritima*. J. Bacteriol. 178: 5579-5585.

Lee, S.P., Morikawa, M., Takagi, M., and Imanaka, T. 1994. Cloning of the aapT gene and characterization of its product, alpha-amylase-pullulanase (AapT), from thermophilic and alkaliphilic *Bacillus* sp. strain XAL601. Appl. Environ. Microbiol. 60: 3764-3773.

Legrain, C., Villeret, V., Roovers, M., Gigot, D., Dideberg, O., Pierard, A., and Glansdorff, N. 1997. Biochemical characterisation of ornithine carbamoyltransferase from *Pyrococcus furiosus*. Eur. J. Biochem. 247: 1046-1055.

Leveque, E., Haye, B., and Belarbi, A. 2000. Cloning and expression of an alpha-amylase encoding gene from the hyperthermophilic archaebacterium *Thermococcus hydrothermalis* and biochemical characterisation of the recombinant enzyme. FEMS Microbiol. Lett. 186: 67-71.

Li, J., Wang, J., and Bachas, L.G. 2002. Biosensor for asparagine using a thermostable recombinant asparaginase from *Archaeoglobus fulgidus*. Anal. Chem. 74: 3336-3341.

Liebl, W., Ruile, P., Bronnenmeier, K., Riedel, K., Lottspeich, F., and Greif, I. 1996. Analysis of a *Thermotoga maritima* DNA fragment encoding two similar thermostable cellulases, CelA and CelB, and characterization of the recombinant enzymes. Microbiology. 142: 2533-2542.

Liebl, W., Stemplinger, I., and Ruile, P. 1997. Properties and gene structure of the *Thermotoga maritima* alpha-amylase AmyA, a putative lipoprotein of a hyperthermophilic bacterium. J. Bacteriol. 179: 941-948.

Liebl, W., Wagner, B., and Schellhase, J. 1998. Properties of an alpha-galactosidase, and structure of its gene galA, within an alpha-and beta-galactoside utilization gene cluster of the hyperthermophilic bacterium *Thermotoga maritima*. Syst .Appl. Microbiol. 21: 1-11.

Lipps, G., Ibanez, P., Stroessenreuther, T., Hekimian, K., and Krauss, G. 2001. The protein ORF80 from the acidophilic and thermophilic archaeon *Sulfolobus islandicus* binds highly site-specifically to double-stranded DNA and represents a novel type of basic leucine zipper protein. Nucleic Acids Res. 29: 4973-4982.

Lipps, G., Stegert, M., and Krauss, G. 2001. Thermostable and site-specific DNA binding of the gene product ORF56 from the *Sulfolobus islandicus* plasmid pRN1, a putative archael plasmid copy control protein. Nucleic Acids Res. 29: 904-913.

Logan, C., and Mayhew, S.G. 2000. Cloning, overexpression, and characterization of peroxiredoxin and NADH peroxiredoxin reductase from *Thermus aquaticus*. J. Biol. Chem. 275: 30019-30028.

Lonhienne, T., Gerday, C., and Feller, G. 2000. Psychrophilic enzymes: revisiting the thermodynamic parameters of activation may explain local flexibility. Biochim. Biophys. Acta. 1543: 1-10.

Lonhienne, T., Zoidakis, J., Vorgias, C.E., Feller, G., Gerday, C., and Bouriotis, V. 2001. Modular structure, local flexibility and cold-activity of a novel chitobiase from a psychrophilic Antarctic bacterium. J. Mol. Biol. 310: 291-297.

Lopez-Garcia, P. 1998. DNA topoisomerases, temperature adaptation, and early diversification of life. In: Thermophiles. Wiegel, J. and Adams, M.W.W., ed. Taylor and Francis, London. p. 201-216

Lopez-Garcia, P., Forterre, P., van der Oost, J., and Erauso, G. 2000. Plasmid pGS5 from the hyperthermophilic archaeon *Archaeoglobus profundus* is negatively supercoiled. J. Bacteriol. 182: 4998-5000.

Louis, P., and Galinski, E.A. 1997. Identification of plasmids in the genus *Marinococcus* and complete nucleotide sequence of plasmid pPL1 from *Marinococcus halophilus*. Plasmid. 38: 107-114.

Lucas, S., Toffin, L., Zivanovic, Y., Charlier, D., Moussard, H., Forterre, P., Prieur, D., and Erauso, G. 2002. Construction of a shuttle vector for, and spheroplast transformation of, the hyperthermophilic archaeon *Pyrococcus abyssi*. Appl. Environ. Microbiol. 68: 5528-5536.

Madern, D., Ebel, C., and Zaccai, G. 2000. Halophilic adaptation of enzymes. Extremophiles. 4: 91-98.

Madigan, M.T., and Marrs, B.L. 1997. Extremophiles. Sci Am. 276: 82-87.

Madigan, M.T., and Oren, A. 1999. Thermophilic and halophilic extremophiles. Curr. Opin. Microbiol. 2: 265-269.

Maeda, N., Kitano, K., Fukui, T., Ezaki, S., Atomi, H., Miki, K., and Imanaka, T. 1999. Ribulose bisphosphate carboxylase/oxygenase from the hyperthermophilic archaeon *Pyrococcus kodakaraensis*

KOD1 is composed solely of large subunits and forms a pentagonal structure. J. Mol. Biol. 293: 57-66.

Mai, V., and Wiegel, J. 2000. Advances in development of a genetic system for *Thermoanaerobacterium* spp.: expression of genes encoding hydrolytic enzymes, development of a second shuttle vector, and integration of genes into the chromosome. Appl. Environ. Microbiol. 66: 4817-4821.

Makrides, S.C. 1996. Strategies for achieving high-level expression of genes in *Escherichia coli*. Microbiol. Rev. 60: 512-538.

Manco, G., Giosue, E., D'Auria, S., Herman, P., Carrea, G., and Rossi, M. 2000. Cloning, overexpression, and properties of a new thermophilic and thermostable esterase with sequence similarity to hormone-sensitive lipase subfamily from the archaeon *Archaeoglobus fulgidus*. Arch. Biochem. Biophys. 373: 182-192.

Mandal, A.K., Cheung, W.D., and Arguello, J.M. 2002. Characterization of a thermophilic P-type Ag^+/Cu^+-ATPase from the extremophile *Archaeoglobus fulgidus*. J. Biol. Chem. 277: 7201-7208.

Mankin, A.S., Zyrianova, I.M., Kagramanova, V.K., and Garrett, R.A. 1992. Introducing mutations into the single-copy chromosomal 23S rRNA gene of the archaeon *Halobacterium halobium* by using an rRNA operon-based transformation system. Proc. Natl. Acad. Sci. USA. 89: 6535-6539.

Margesin, R., and Schinner, F. 2001. Potential of halotolerant and halophilic microorganisms for biotechnology. Extremophiles. 5: 73-83.

Marguet, E., and Forterre, P. 2001. Stability and manipulation of DNA at extreme temperatures. Methods Enzymol. 334: 205-215.

Matsuda, T., Morikawa, M., Haruki, M., Higashibata, H., Imanaka, T., and Kanaya, S. 1999. Isolation of TBP-interacting protein (TIP) from a hyperthermophilic archaeon that inhibits the binding of TBP to TATA-DNA. FEBS Lett. 457: 38-42.

Matsui, E., Kawasaki, S., Ishida, H., Ishikawa, K., Kosugi, Y., Kikuchi, H., Kawarabayashi, Y., and Matsui, I. 1999. Thermostable flap endonuclease from the archaeon, *Pyrococcus horikoshii*, cleaves the replication fork-like structure endo/exonucleolytically. J. Biol. Chem. 274: 18297-18309.

Matsui, I., Matsui, E., Sakai, Y., Kikuchi, H., Kawarabayasi, Y., Ura, H., Kawaguchi, S., Kuramitsu, S., and Harata, K. 2000. The molecular structure of hyperthermostable aromatic aminotransferase with novel substrate specificity from *Pyrococcus horikoshii*. J. Biol. Chem. 275: 4871-4879.

McLean, M.A., Maves, S.A., Weiss, K.E., Krepich, S., and Sligar, S.G. 1998. Characterization of a cytochrome P450 from the acidothermophilic archaea *Sulfolobus solfataricus*. Biochem. Biophys. Res. Commun. 252: 166-172.

Mellado, E., Asturias, J.A., Nieto, J.J., Timmis, K.N., and Ventosa, A. 1995. Characterization of the basic replicon of pCM1, a narrow-host-range plasmid from the moderate halophile *Chromohalobacter marismortui*. J. Bacteriol. 177: 3443-3450.

Mellado, E., Nieto, J.J., and Ventosa, A. 1995. Construction of novel shuttle vectors for use between moderately halophilic bacteria and *Escherichia coli*. Plasmid. 34: 157-164.

Meng, L., Ruebush, S., D'Souza V, M., Copik, A.J., Tsunasawa, S., and Holz, R.C. 2002. Overexpression and divalent metal binding properties of the methionyl aminopeptidase from *Pyrococcus furiosus*. Biochemistry. 41: 7199-7208.

Menon, A.L., Hendrix, H., Hutchins, A., Verhagen, M.F., and Adams, M.W. 1998. The delta-subunit of pyruvate ferredoxin oxidoreductase from *Pyrococcus furiosus* is a redox-active, iron-sulfur protein: evidence for an ancestral relationship with 8Fe-type ferredoxins. Biochemistry. 37: 12838-12846.

Merz, A., Knochel, T., Jansonius, J.N., and Kirschner, K. 1999. The hyperthermostable indoleglycerol phosphate synthase from *Thermotoga maritima* is destabilized by mutational disruption of two solvent-exposed salt bridges. J. Mol. Biol. 288: 753-763.

Middelberg, A.R. 2002. Preparative protein refolding. Trends Biotechnol. 20: 437-443.

Mijts, B.N., and Patel, B.K. 2002. Cloning, sequencing and expression of an alpha-amylase gene, amyA, from the thermophilic halophile *Halothermothrix orenii* and purification and biochemical characterization of the recombinant enzyme. Microbiology. 148: 2343-2349.

Min, K., Song, H.K., Chang, C., Lee, J.Y., Eom, S.H., Kim, K.K., Yu, Y.G., and Suh, S.W. 2000. Nucleoside diphosphate kinase from the hyperthermophilic archaeon *Methanococcus jannaschii*: overexpression, crystallization and preliminary X-ray crystallographic analysis. Acta Crystallogr. D Biol. Crystallogr. 56: 1485-1487.

Minakhin, L., Nechaev, S., Campbell, E.A., and Severinov, K. 2001. Recombinant *Thermus aquaticus* RNA polymerase, a new tool for structure-based analysis of transcription. J. Bacteriol. 183: 71-76.

Minuth, T., Henn, M., Rutkat, K., Andra, S., Frey, G., Rachel, R., Stetter, K.O., and Jaenicke, R. 1999. The recombinant thermosome from the hyperthermophilic archaeon *Methanopyrus kandleri*: *in vitro* analysis of its chaperone activity. Biol. Chem. 380: 55-62.

Miura, Y., Kettoku, M., Kato, M., Kobayashi, K., and Kondo, K. 1999. High level production of thermostable alpha-amylase from *Sulfolobus solfataricus* in high-cell density culture of the food yeast *Candida utilis*. J. Mol. Microbiol. Biotechnol. 1: 129-134.

Miyazaki, K. 1996. Isocitrate dehydrogenase from *Thermus aquaticus* YT1: purification of the enzyme and cloning, sequencing, and expression of the gene. Appl. Environ. Microbiol. 62: 4627-4631.

Miyazaki, K., Yaoi, T., and Oshima, T. 1994. Expression, purification, and substrate specificity of isocitrate dehydrogenase from *Thermus thermophilus* HB8. Eur. J. Biochem. 221: 899-903.

Moll, R., Schmidtke, S., Petersen, A., and Schafer, G. 1997. The signal recognition particle receptor alpha subunit of the hyperthermophilic archaeon *Acidianus ambivalens* exhibits an intrinsic GTP-hydrolyzing activity. Biochim. Biophys. Acta. 1335: 218-230.

Moracci, M., Capalbo, L., Ciaramella, M., and Rossi, M. 1996. Identification of two glutamic acid residues essential for catalysis in the beta-glycosidase from the thermoacidophilic archaeon *Sulfolobus solfataricus*. Protein Eng. 9: 1191-1195.

Moracci, M., La Volpe, A., Pulitzer, J.F., Rossi, M., and Ciaramella, M. 1992. Expression of the thermostable beta-galactosidase gene from the archaebacterium *Sulfolobus solfataricus* in *Saccharomyces cerevisiae* and characterization of a new inducible promoter for heterologous expression. J. Bacteriol. 174: 873-882.

Morana, A., Moracci, M., Ottombrino, A., Ciaramella, M., Rossi, M., and De Rosa, M. 1995. Industrial-scale production and rapid purification of an archaeal beta-glycosidase expressed in *Saccharomyces cerevisiae*. Biotechnol. Appl. Biochem. 22: 261-268.

Morsomme, P., Chami, M., Marco, S., Nader, J., Ketchum, K.A., Goffeau, A., and Rigaud, J.L. 2002. Characterization of a hyperthermophilic P-type ATPase from *Methanococcus jannaschii* expressed in yeast. J. Biol. Chem. 277: 29608-29616.

Mukhopadhyay, B., Purwantini, E., Pihl, T.D., Reeve, J.N., and Daniels, L. 1995. Cloning, sequencing, and transcriptional analysis of the coenzyme F420-dependent methylene-5,6,7,8-tetrahydromethanopterin dehydrogenase gene from *Methanobacterium thermoautotrophicum* strain Marburg and functional expression in *Escherichia coli*. J. Biol. Chem. 270: 2827-2832.

Murakawa, T., Yamagata, H., Tsuruta, H., and Aizono, Y. 2002. Cloning of cold-active alkaline phosphatase gene of a psychrophile, *Shewanella* sp., and expression of the recombinant enzyme. Biosci. Biotechnol. Biochem. 66: 754-761.

Musfeldt, M., and Schonheit, P. 2002. Novel type of ADP-forming acetyl coenzyme A synthetase in hyperthermophilic archaea: heterologous expression and characterization of isoenzymes from the sulfate reducer *Archaeoglobus fulgidus* and the methanogen *Methanococcus jannaschii*. J. Bacteriol. 184: 636-644.

Musfeldt, M., Selig, M., and Schonheit, P. 1999. Acetyl coenzyme A synthetase (ADP forming) from the hyperthermophilic Archaeon *Pyrococcus furiosus*: identification, cloning, separate expression of the encoding genes, acdAI and acdBI, in *Escherichia coli*, and *in vitro* reconstitution of the active heterotetrameric enzyme from its recombinant subunits. J. Bacteriol. 181: 5885-5888.

Nelson, K.E., Clayton, R.A., Gill, S.R., Gwinn, M.L., Dodson, R.J., Haft, D.H., Hickey, E.K., Peterson, J.D., Nelson, W.C., Ketchum, K.A., McDonald, L., Utterback, T.R., Malek, J.A., Linher, K.D., Garrett, M.M., Stewart, A.M., Cotton, M.D., Pratt, M.S., Phillips, C.A., Richardson, D., Heidelberg, J., Sutton, G.G., Fleischmann, R.D., Eisen, J.A., Fraser, C.M., and *et al.* 1999. Evidence for lateral gene transfer between Archaea and bacteria from genome sequence of *Thermotoga maritima*. Nature. 399: 323-329.

Ng, W.V., Kennedy, S.P., Mahairas, G.G., Berquist, B., Pan, M., Shukla, H.D., Lasky, S.R., Baliga, N.S., Thorsson, V., Sbrogna, J., Swartzell, S., Weir, D., Hall, J., Dahl, T.A., Welti, R., Goo, Y.A., Leithauser, B., Keller, K., Cruz, R., Danson, M.J., Hough, D.W., Maddocks, D.G., Jablonski, P.E., Krebs, M.P., Angevine, C.M., Dale, H., Isenbarger, T.A., Peck, R.F., Pohlschroder, M., Spudich, J.L., Jung, K.-H., Alam, M., Freitas, T., Hou, S., Daniels, C.J., Dennis, P.P., Omer, A.D.,

Ebhardt, H., Lowe, T.M., Liang, P., Riley, M., Hood, L., and DasSarma, S. 2000. From the Cover: Genome sequence of *Halobacterium* species NRC-1. Proc. Natl. Acad. Sc. USA. 97: 12176-12181.

Noll, K.M., and Vargas, M. 1997. Recent advances in genetic analyses of hyperthermophilic archaea and bacteria. Arch. Microbiol. 168: 73-80.

Numata, K., Muro, M., Akutsu, N., Nosoh, Y., Yamagishi, A., and Oshima, T. 1995. Thermal stability of chimeric isopropylmalate dehydrogenase genes constructed from a thermophile and a mesophile. Protein Eng. 8: 39-43.

Ohshima, T., Nunoura-Kominato, N., Kudome, T., and Sakuraba, H. 2001. A novel hyperthermophilic archaeal glyoxylate reductase from *Thermococcus litoralis*. Characterization, gene cloning, nucleotide sequence and expression in *Escherichia coli*. Eur. J. Biochem. 268: 4740-4747.

Ohto, C., Ishida, C., Koike-Takeshita, A., Yokoyama, K., Muramatsu, M., Nishino, T., and Obata, S. 1999. Gene cloning and overexpression of a geranylgeranyl diphosphate synthase of an extremely thermophilic bacterium, *Thermus thermophilus*. Biosci Biotechnol. Biochem. 63: 261-270.

Okubo, Y., Yokoigawa, K., Esaki, N., Soda, K., and Kawai, H. 1999. Characterization of psychrophilic alanine racemase from *Bacillus psychrosaccharolyticus*. Biochem. Biophys. Res. Commun. 256: 333-340.

Oren, A. 1999. Microbiology and Biogeochemistry of Hypersaline Environments. CRC Press, Boca Raton, FL.

Osipiuk, J., and Joachimiak, A. 1997. Cloning, sequencing, and expression of dnaK-operon proteins from the thermophilic bacterium *Thermus thermophilus*. Biochim. Biophys. Acta. 1353: 253-265.

Ouhammouch, M., and Geiduschek, E.P. 2001. A thermostable platform for transcriptional regulation: the DNA-binding properties of two Lrp homologs from the hyperthermophilic archaeon *Methanococcus jannaschii*. EMBO J. 20: 146-156.

Palmer, J.R., and Daniels, C.J. 1994. A transcriptional reporter for *in vivo* promoter analysis in the archaeon *Haloferax volcanii*. Appl. Environ. Microbiol. 60: 3867-3869.

Pantazaki, A.A., Anagnostopoulos, C.G., Lioliou, E.E., and Kyriakidis,

D.A. 1999. Characterization of ornithine decarboxylase and regulation by its antizyme in *Thermus thermophilus*. Mol. Cell. Biochem. 195: 55-64.

Patenge, N., and Soppa, J. 1999. Extensive proteolysis inhibits high-level production of eukaryal G protein-coupled receptors in the archaeon *Haloferax volcanii*. FEMS Microbiol. Lett. 171: 27-35.

Pedelacq, J., Piltch, E., Liong, E., Berendzen, J., Kim, C., Rho, B., Park, M., Terwilliger, T., and Waldo, G. 2002. Engineering soluble proteins for structural genomics. Nat. Biotechnol. 20: 927-932.

Pereira, S.L., Grayling, R.A., Lurz, R., and Reeve, J.N. 1997. Archaeal nucleosomes. Proc. Natl. Acad. Sci. USA. 94: 12633-12637.

Pereira, S.L., and Reeve, J.N. 1998. Histones and nucleosomes in Archaea and Eukarya: a comparative analysis. Extremophiles. 2: 141-148.

Pick, U. 1999. *Dunaliella acidophila*: a most extreme acidophilic alga. In: Enigmatic Microorganisms and Life in Extreme Environments. Seckbach, J., ed. Kluwer, Dordrecht. p. 467-478

Pire, C., Esclapez, J., Ferrer, J., and Bonete, M.J. 2001. Heterologous overexpression of glucose dehydrogenase from the halophilic archaeon *Haloferax mediterranei*, an enzyme of the medium chain dehydrogenase/reductase family. FEMS Microbiol. Lett. 200: 221-227.

Pisani, F.M., De Felice, M., Carpentieri, F., and Rossi, M. 2000. Biochemical characterization of a clamp-loader complex homologous to eukaryotic replication factor C from the hyperthermophilic archaeon *Sulfolobus solfataricus*. J .Mol. Biol. 301: 61-73.

Plosser, P., and Pfeifer, F. 2002. A bZIP protein from halophilic archaea: structural features and dimer formation of cGvpE from *Halobacterium salinarum*. Mol. Microbiol. 45: 511-520.

Politz, O., Krah, M., Thomsen, K.K., and Borriss, R. 2000. A highly thermostable endo-(1,4)-beta-mannanase from the marine bacterium *Rhodothermus marinus*. Appl. Microbiol. Biotechnol. 53: 715-721.

Porcelli, M., Fusco, S., Inizio, T., Zappia, V., and Cacciapuoti, G. 2000. Expression, purification, and characterization of recombinant S-adenosylhomocysteine hydrolase from the thermophilic archaeon *Sulfolobus solfataricus*. Protein Express. Purif. 18: 27-35.

Prangishvili, D., Albers, S.V., Holz, I., Arnold, H.P., Stedman, K., Klein, T., Singh, H., Hiort, J., Schweier, A., Kristjansson, J.K., and

Zillig, W. 1998. Conjugation in archaea: frequent occurrence of conjugative plasmids in *Sulfolobus*. Plasmid. 40: 190-202.

Purcarea, C., Herve, G., Cunin, R., and Evans, D.R. 2001. Cloning, expression, and structure analysis of carbamate kinase-like carbamoyl phosphate synthetase from *Pyrococcus abyssi*. Extremophiles. 5: 229-239.

Raffaelli, N., Pisani, F.M., Lorenzi, T., Emanuelli, M., Amici, A., Ruggieri, S., and Magni, G. 1997. Characterization of nicotinamide mononucleotide adenylyltransferase from thermophilic archaea. J. Bacteriol. 179: 7718-7723.

Rajan, R.S., Illing, M.E., Bence, N.F., and Kopito, R.R. 2001. Specificity in intracellular protein aggregation and inclusion body formation. Proc. Natl. Acad. Sci.USA. 98: 13060-13065.

Ravenschlag, K., Sahm, K., and Amann, R. 2001. Quantitative molecular analysis of the microbial community in marine arctic sediments (Svalbard). Appl. Environ. Microbiol. 67: 387-395.

Rawlings, D.E., and Kusano, T. 1994. Molecular genetics of *Thiobacillus ferrooxidans*. Microbiol. Rev. 58: 39-55.

Reed, D.W., and Hartzell, P.L. 1999. The *Archaeoglobus fulgidus* D-lactate dehydrogenase is a $Zn(2^+)$ flavoprotein. J. Bacteriol. 181: 7580-7587.

Rentier-Delrue, F., Mande, S.C., Moyens, S., Terpstra, P., Mainfroid, V., Goraj, K., Lion, M., Hol, W.G., and Martial, J.A. 1993. Cloning and overexpression of the triosephosphate isomerase genes from psychrophilic and thermophilic bacteria. Structural comparison of the predicted protein sequences. J. Mol. Biol. 229: 85-93.

Richard, S.B., Madern, D., Garcin, E., and Zaccai, G. 2000. Halophilic adaptation: Novel solvent protein interactions observed in the 2.9 and 2.6 a resolution structures of the wild type and a mutant of malate dehydrogenase from *Haloarcula marismortui*. Biochemistry. 39: 992-1000.

Rina, M., Pozidis, C., Mavromatis, K., Tzanodaskalaki, M., Kokkinidis, M., and Bouriotis, V. 2000. Alkaline phosphatase from the Antarctic strain TAB5. Properties and psychrophilic adaptations. Eur. J. Biochem. 267: 1230-1238.

Robb, F.T., Maeder, D.L., Brown, J.R., DiRuggiero, J., Stump, M.D., Yeh, R.K., Weiss, R.B., and Dunn, D.M. 2001. Genomic sequence of hyperthermophile, *Pyrococcus furiosus*: implications for physiology and enzymology. Methods Enzymol. 330: 134-157.

Robinson, H., Gao, Y.G., McCrary, B.S., Edmondson, S.P., Shriver, J.W., and Wang, A.H. 1998. The hyperthermophile chromosomal protein Sac7d sharply kinks DNA. Nature. 392: 202-205.

Rolfsmeier, M., Haseltine, C., Bini, E., Clark, A., and Blum, P. 1998. Molecular characterization of the alpha-glucosidase gene (malA) from the hyperthermophilic archaeon *Sulfolobus solfataricus*. J. Bacteriol. 180: 1287-1295.

Rothschild, L.J., and Mancinelli, R.L. 2001. Life in extreme environments. Nature. 409: 1092-1101.

Ruepp, A., Graml, W., Santos-Martinez, M.L., Koretke, K.K., Volker, C., Mewes, H.W., Frishman, D., Stocker, S., Lupas, A.N., and Baumeister, W. 2000. The genome sequence of the thermoacidophilic scavenger *Thermoplasma acidophilum*. Nature. 407: 508-513.

Russell, N.J., and Hamamoto, T. 1998. Psychrophiles. In: Extremophiles: Microbial Life in Extreme Environments. Horikoshi, K. and Grant, W.D., ed. Wiley-Liss, New York. p. 25-45

Sakamoto, S., Terada, I., Iijima, M., Ohta, T., and Matsuzawa, H. 1995. Expression of aqualysin I (a thermophilic protease) in soluble form in *Escherichia coli* under a bacteriophage T7 promoter. Biosci. Biotechnol. Biochem. 59: 1438-1443.

Sakuraba, H., Yoshioka, I., Koga, S., Takahashi, M., Kitahama, Y., Satomura, T., Kawakami, R., and Ohshima, T. 2002. ADP-dependent glucokinase/phosphofructokinase, a novel bifunctional enzyme from the hyperthermophilic archaeon *Methanococcus jannaschii*. J. Biol. Chem. 277: 12495-12498.

Sandigursky, M., Faje, A., and Franklin, W.A. 2001. Characterization of the full length uracil-DNA glycosylase in the extreme thermophile *Thermotoga maritima*. Mutat. Res. 485: 187-195.

Sandman, K., Bailey, K.A., Pereira, S.L., Soares, D., Li, W.T., and Reeve, J.N. 2001. Archaeal histones and nucleosomes. Methods Enzymol. 334: 116-129.

Satoh, T., Samejima, T., Watanabe, M., Nogi, S., Takahashi, Y., Kaji, H., Teplyakov, A., Obmolova, G., Kuranova, I., and Ishii, K. 1998. Molecular cloning, expression, and site-directed mutagenesis of inorganic pyrophosphatase from *Thermus thermophilus* HB8. J. Biochem. (Tokyo). 124: 79-88.

Satoh, T., Shinoda, H., Ishii, K., Koyama, M., Sakurai, N., Kaji, H.,

Hachimori, A., Irie, M., and Samejima, T. 1999. Primary structure, expression, and site-directed mutagenesis of inorganic pyrophosphatase from *Bacillus stearothermophilus*. J. Biochem. (Tokyo). 125: 48-57.

Sauer, J., and Nygaard, P. 1999. Expression of the *Methanobacterium thermoautotrophicum* hpt gene, encoding hypoxanthine (Guanine) phosphoribosyltransferase, in *Escherichia coli*. J. Bacteriol. 181: 1958-1962.

Schleper, C., Holz, I., Janekovic, D., Murphy, J., and Zillig, W. 1995. A multicopy plasmid of the extremely thermophilic archaeon *Sulfolobus* effects its transfer to recipients by mating. J. Bacteriol. 177: 4417-4426.

Schleper, C., Puehler, G., Holz, I., Gambacorta, A., Janekovic, D., Santarius, U., Klenk, H.P., and Zillig, W. 1995. *Picrophilus* gen. nov., fam. nov.: a novel aerobic, heterotrophic, thermoacidophilic genus and family comprising archaea capable of growth around pH 0. J. Bacteriol. 177: 7050-7059.

Schleper, C., Puhler, G., Kuhlmorgen, B., and Zillig, W. 1995. Life at extremely low pH. Nature. 375: 741-742.

Schleper, C., Swanson, R.V., Mathur, E.J., and DeLong, E.F. 1997. Characterization of a DNA polymerase from the uncultivated psychrophilic archaeon *Cenarchaeum symbiosum*. J. Bacteriol. 179: 7803-7811.

Schon, U., and Schumann, W. 1995. Overproduction, purification and characterization of GroES and GroEL from thermophilic *Bacillus stearothermophilus*. FEMS Microbiol. Lett. 134: 183-188.

Schramm, A., Siebers, B., Tjaden, B., Brinkmann, H., and Hensel, R. 2000. Pyruvate kinase of the hyperthermophilic crenarchaeote *Thermoproteus tenax*: physiological role and phylogenetic aspects. J. Bacteriol. 182: 2001-2009.

Schreier, H.J., Robinson-Bidle, K.A., Romashko, A.M., and Patel, G.V. 1999. Heterologous expression in the Archaea: transcription from *Pyrococcus furiosus* gdh and mlrA promoters in *Haloferax volcanii*. Extremophiles. 3: 11-19.

Schwermann, B., Pfau, K., Liliensiek, B., Schleyer, M., Fischer, T., and Bakker, E.P. 1994. Purification, properties and structural aspects of a thermoacidophilic alpha-amylase from *Alicyclobacillus acidocaldarius* atcc 27009. Insight into acidostability of proteins. Eur. J. Biochem. 226: 981-991.

Seckbach, J. 2000. Journey to Diverse Microbial Worlds: Adaption to Exotic Environments. Kluwer, Dordrecht.

She, Q., Phan, H., Garrett, R.A., Albers, S.V., Stedman, K.M., and Zillig, W. 1998. Genetic profile of pNOB8 from *Sulfolobus*: the first conjugative plasmid from an archaeon. Extremophiles. 2: 417-425.

She, Q., Singh, R.K., Confalonieri, F., Zivanovic, Y., Allard, G., Awayez, M.J., Chan-Weiher, C.C.-Y., Clausen, I.G., Curtis, B.A., De Moors, A., Erauso, G., Fletcher, C., Gordon, P.M.K., Heikamp-de Jong, I., Jeffries, A.C., Kozera, C.J., Medina, N., Peng, X., Thi-Ngoc, H.P., Redder, P., Schenk, M.E., Theriault, C., Tolstrup, N., Charlebois, R.L., Doolittle, W.F., Duguet, M., Gaasterland, T., Garrett, R.A., Ragan, M.A., Sensen, C.W., and Van der Oost, J. 2001. The complete genome of the crenarchaeon *Sulfolobus solfataricus* P2. Proc. Natl. Acad. Sci. USA. 98: 7835-7840.

Shima, S., Weiss, D.S., and Thauer, R.K. 1995. Formylmethanofuran:tetrahydromethanopterin formyltransferase (Ftr) from the hyperthermophilic *Methanopyrus kandleri*. Cloning, sequencing and functional expression of the ftr gene and one-step purification of the enzyme overproduced in *Escherichia coli*. Eur. J. Biochem. 230: 906-913.

Sibold, L., and Henriquet, M. 1988. Cloning of the trp genes from the archaebacterium *Methanococcus voltae*: nucleotide sequence of the trpBA genes. Mol. Gen. Genet. 214: 439-450.

Siddiqui, M.A., Fujiwara, S., Takagi, M., and Imanaka, T. 1998. *In vitro* heat effect on heterooligomeric subunit assembly of thermostable indolepyruvate ferredoxin oxidoreductase. FEBS Lett. 434: 372-376.

Siebers, B., Brinkmann, H., Dorr, C., Tjaden, B., Lilie, H., van der Oost, J., and Verhees, C.H. 2001. Archaeal fructose-1,6-bisphosphate aldolases constitute a new family of archaeal type class I aldolase. J. Biol. Chem. 276: 28710-28718.

Singleton, M.R., Taylor, S.J., Parrat, J.S., and Littlechild, J.A. 2000. Cloning, expression, and characterization of pyrrolidone carboxyl peptidase from the archaeon *Thermococcus litoralis*. Extremophiles. 4: 297-303.

Slesarev, A.I., Mezhevaya, K.V., Makarova, K.S., Polushin, N.N., Shcherbinina, O.V., Shakhova, V.V., Belova, G.I., Aravind, L., Natale, D.A., Rogozin, I.B., Tatusov, R.L., Wolf, Y.I., Stetter, K.O., Malykh, A.G., Koonin, E.V., and Kozyavkin, S.A. 2002. The complete genome of hyperthermophile *Methanopyrus kandleri* AV19 and monophyly of archaeal methanogens. Proc. Natl. Acad. Sci. USA. 99: 4644-4649.

Smith, A.S., and Rawlings, D.E. 1997. The poison-antidote stability system of the broad-host-range *Thiobacillus ferrooxidans* plasmid pTF-FC2. Mol. Microbiol. 26: 961-970.

Smith, J.D., and Robinson, A.S. 2002. Overexpression of an archaeal protein in yeast: Secretion bottleneck at the ER. Biotechnol. Bioeng. 79: 713-723.

Smith, K.S., and Ferry, J.G. 1999. A plant-type (beta-class) carbonic anhydrase in the thermophilic methanoarchaeon *Methanobacterium thermoautotrophicum*. J. Bacteriol. 181: 6247-6253.

Soares, D., Dahlke, I., Li, W.T., Sandman, K., Hethke, C., Thomm, M., and Reeve, J.N. 1998. Archaeal histone stability, DNA binding, and transcription inhibition above 90 degrees C. Extremophiles. 2: 75-81.

Sobecky, P.A., Mincer, T.J., Chang, M.C., and Helinski, D.R. 1997. Plasmids isolated from marine sediment microbial communities contain replication and incompatibility regions unrelated to those of known plasmid groups. Appl. Environ. Microbiol. 63: 888-895.

Sorensen, H.P., Sperling-Petersen, H.U., and Mortensen, K.K. 2003. Production of recombinant thermostable proteins expressed in *Escherichia coli*: completion of protein synthesis is the bottleneck. J. Chromatogr. B. In Press:

Sowers, K.R., and Schreier, H.J. 1999. Gene transfer systems for the Archaea. Trends Microbiol. 7: 212-219.

Sperling, D., Kappler, U., Wynen, A., Dahl, C., and Truper, H.G. 1998. Dissimilatory ATP sulfurylase from the hyperthermophilic sulfate reducer *Archaeoglobus fulgidus* belongs to the group of homo-oligomeric ATP sulfurylases. FEMS Microbiol. Lett. 162: 257-264.

Sriskanda, V., Kelman, Z., Hurwitz, J., and Shuman, S. 2000. Characterization of an ATP-dependent DNA ligase from the thermophilic archaeon *Methanobacterium thermoautotrophicum*. Nucleic Acids Res. 28: 2221-2228.

Stedman, K.M., Schleper, C., Rumpf, E., and Zillig, W. 1999. Genetic requirements for the function of the archaeal virus SSV1 in *Sulfolobus solfataricus*: construction and testing of viral shuttle vectors. Genetics. 152: 1397-1405.

Stedman, K.M., She, Q., Phan, H., Holz, I., Singh, H., Prangishvili,

D., Garrett, R., and Zillig, W. 2000. pING family of conjugative plasmids from the extremely thermophilic archaeon *Sulfolobus islandicus*: insights into recombination and conjugation in Crenarchaeota. J. Bacteriol. 182: 7014-7020.

Sterner, R., Kleemann, G.R., Szadkowski, H., Lustig, A., Hennig, M., and Kirschner, K. 1996. Phosphoribosyl anthranilate isomerase from *Thermotoga maritima* is an extremely stable and active homodimer. Protein Sci. 5: 2000-2008.

Story, S.V., Grunden, A.M., and Adams, M.W. 2001. Characterization of an aminoacylase from the hyperthermophilic archaeon *Pyrococcus furiosus*. J. Bacteriol. 183: 4259-4268.

Studier, F.W., and Moffatt, B.A. 1986. Use of bacteriophage T7 RNA polymerase to direct selective high-level expression of cloned genes. J. Mol. Biol. 189: 113-130.

Tabor, S., and Richardson, C.C. 1985. A bacteriophage T7 RNA polymerase/promoter system for controlled exclusive expression of specific genes. Proc. Natl. Acad. Sci.USA. 82: 1074-1078.

Tachibana, A., Yano, Y., Otani, S., Nomura, N., Sako, Y., and Taniguchi, M. 2000. Novel prenyltransferase gene encoding farnesylgeranyl diphosphate synthase from a hyperthermophilic archaeon, *Aeropyrum pernix*. Molecularevolution with alteration in product specificity. Eur. J. Biochem. 267: 321-328.

Takami, H., and Horikoshi, K. 2000. Analysis of the genome of an alkaliphilic *Bacillus* strain from an industrial point of view. Extremophiles. 4: 99-108.

Takami, H., Kobayashi, T., Aono, R., and Horikoshi, K. 1992. Molecular cloning, nucleotide sequence and expression of the structural gene for a thermostable alkaline protease from *Bacillus* sp. no. AH-101. Appl. Microbiol. Biotechnol. 38: 101-108.

Takami, H., Takaki, Y., and Uchiyama, I. 2002. Genome sequence of *Oceanobacillus iheyensis* isolated from the Iheya Ridge and its unexpected adaptive capabilities to extreme environments. Nucleic Acids Res. 30: 3927-3935.

Tamakoshi, M., Uchida, M., Tanabe, K., Fukuyama, S., Yamagishi, A., and Oshima, T. 1997. A new *Thermus-Escherichia coli* shuttle integration vector system. J. Bacteriol. 179: 4811-4814.

Tanaka, T., Fujiwara, S., Nishikori, S., Fukui, T., Takagi, M., and Imanaka, T. 1999. A unique chitinase with dual active sites and triple substrate binding sites from the hyperthermophilic archaeon *Pyrococcus kodakaraensis* KOD1. Appl. Environ. Microbiol. 65: 5338-5344.

Tang, X.F., Ezaki, S., Atomi, H., and Imanaka, T. 2001. Anthranilate synthase without an LLES motif from a hyperthermophilic archaeon is inhibited by tryptophan. Biochem. Biophys. Res. Commun. 281: 858-865.

Taupin, C.M., Hartlein, M., and Leberman, R. 1997. Seryl-tRNA synthetase from the extreme halophile *Haloarcula marismortui*— isolation, characterization and sequencing of the gene and its expression in *Escherichia coli*. Eur. J. Biochem. 243: 141-150.

Taylor, A., Brown, D.P., Kadam, S., Maus, M., Kohlbrenner, W.E., Weigl, D., Turon, M.C., and Katz, L. 1992. High-level expression and purification of mature HIV-1 protease in *Escherichia coli* under control of the *araBAD* promoter. Appl. Microbiol. Biotechnol. 37: 205-210.

Te'o, V.S., Cziferszky, A.E., Bergquist, P.L., and Nevalainen, K.M. 2000. Codon optimization of xylanase gene xynB from the thermophilic bacterium *Dictyoglomus thermophilum* for expression in the filamentous fungus *Trichoderma reesei*. FEMS Microbiol. Lett. 190: 13-19.

Tetsch, L., and Kunte, H.J. 2002. The substrate-binding protein TeaA of the osmoregulated ectoine transporter TeaABC from *Halomonas elongata*: purification and characterization of recombinant TeaA. FEMS Microbiol. Lett. 211: 213-218.

Thompson, M.J., and Eisenberg, D. 1999. Transproteomic evidence of a loop-deletion mechanism for enhancing protein thermostability. J. Mol. Biol. 290: 595-604.

Tomschy, A., Glockshuber, R., and Jaenicke, R. 1993. Functional expression of D-glyceraldehyde-3-phosphate dehydrogenase from the hyperthermophilic eubacterium *Thermotoga maritima* in *Escherichia coli*. Authenticity and kinetic properties of the recombinant enzyme. Eur. J. Biochem. 214: 43-50.

Tsiboli, P., Triantafillidou, D., Franceschi, F., and Choli-Papadopoulou, T. 1998. Studies on the Zn-containing S14 ribosomal protein from *Thermus thermophilus*. Eur. J. Biochem. 256: 136-141.

Tsunasawa, S., Nakura, S., Tanigawa, T., and Kato, I. 1998. Pyrrolidone carboxyl peptidase from the hyperthermophilic Archaeon *Pyrococcus furiosus*: cloning and overexpression in *Escherichia coli* of the gene, and its application to protein sequence analysis. J. Biochem. (Tokyo). 124: 778-783.

Tutino, M.L., Duilio, A., Moretti, M.A., Sannia, G., and Marino, G. 2000. A rolling-circle plasmid from *Psychrobacter* sp. TA144: evidence for a novel rep subfamily. Biochem. Biophys. Res. Commun. 274: 488-495.

Tutino, M.L., Duilio, A., Parrilli, R., Remaut, E., Sannia, G., and Marino, G. 2001. A novel replication element from an Antarctic plasmid as a tool for the expression of proteins at low temperature. Extremophiles. 5: 257-264.

Tutino, M.L., Parrilli, E., Giaquinto, L., Duilio, A., Sannia, G., Feller, G., and Marino, G. 2002. Secretion of alpha-amylase from *Pseudoalteromonas haloplanktis* TAB23: two different pathways in different hosts. J. Bacteriol. 184: 5814-5817.

Tutino, M.L., Tosco, A., Marino, G., and Sannia, G. 1997. Expression of *Sulfolobus solfataricus* trpE and trpG genes in *E. coli*. Biochem. Biophys. Res. Commun. 230: 306-310.

Uriarte, M., Marina, A., Ramon-Maiques, S., Fita, I., and Rubio, V. 1999. The carbamoyl-phosphate synthetase of *Pyrococcus furiosus* is enzymologically and structurally a carbamate kinase. J. Biol. Chem. 274: 16295-16303.

van de Vossenberg, J.L., Driessen, A.J., and Konings, W.N. 1998. The essence of being extremophilic: the role of the unique archaeal membrane lipids. Extremophiles. 2: 163-170.

Vargas, C., Fernandez-Castillo, R., Canovas, D., Ventosa, A., and Nieto, J.J. 1995. Isolation of cryptic plasmids from moderately halophilic eubacteria of the genus *Halomonas*. Characterization of a small plasmid from *H. elongata* and its use for shuttle vector construction. Mol. Gen. Genet. 246: 411-418.

Verhagen, M.F., O'Rourke, T.W., Menon, A.L., and Adams, M.W. 2001. Heterologous expression and properties of the gamma-subunit of the Fe-only hydrogenase from *Thermotoga maritima*. Biochim. Biophys. Acta. 1505: 209-219.

Vieille, C., and Zeikus, G.J. 2001. Hyperthermophilic enzymes: sources, uses, and molecular mechanisms for thermostability. Microbiol. Mol. Biol. Rev. 65: 1-43.

Walsh, D.J., Gibbs, M.D., and Bergquist, P.L. 1998. Expression and secretion of a xylanase from the extreme thermophile, *Thermotoga* strain FjSS3B.1, in *Kluyveromyces lactis*. Extremophiles. 2: 9-14.

Ward, D.E., Kengen, S.W., van Der Oost, J., and de Vos, W.M. 2000. Purification and characterization of the alanine aminotransferase from the hyperthermophilic Archaeon *Pyrococcus furiosus* and its role in alanine production. J. Bacteriol. 182: 2559-2566.

Ward, D.E., Revet, I.M., Nandakumar, R., Tuttle, J.H., de Vos, W.M., van der Oost, J., and DiRuggiero, J. 2002. Characterization of plasmid pRT1 from *Pyrococcus* sp. strain JT1. J. Bacteriol. 184: 2561-2556.

Wassenberg, D., Schurig, H., Liebl, W., and Jaenicke, R. 1997. Xylanase XynA from the hyperthermophilic bacterium *Thermotoga maritima*: structure and stability of the recombinant enzyme and its isolated cellulose-binding domain. Protein Sci. 6: 1718-1726.

Wayne, J., and Xu, S.Y. 1997. Identification of a thermophilic plasmid origin and its cloning within a new *Thermus-E-coli* shuttle vector. Gene. 195: 321-328.

Welch, T.J., and Bartlett, D.H. 1997. Cloning, sequencing and overexpression of the gene encoding malate dehydrogenase from the deep-sea bacterium *Photobacterium* species strain SS9. Biochim. Biophys. Acta. 1350: 41-46.

Welker, C., Bohm, G., Schurig, H., and Jaenicke, R. 1999. Cloning, overexpression, purification, and physicochemical characterization of a cold shock protein homolog from the hyperthermophilic bacterium *Thermotoga maritima*. Protein Sci. 8: 394-403.

Wright, D.B., Banks, D.D., Lohman, J.R., Hilsenbeck, J.L., and Gloss, L.M. 2002. The effect of salts on the activity and stability of *Escherichia coli* and *Haloferax volcanii* dihydrofolate reductases. J. Mol. Biol. 323: 327-344.

Xu, Y., Kobayashi, T., and Kudo, T. 1996. Molecular cloning and nucleotide sequence of the groEL gene from the alkaliphilic *Bacillus* sp. strain C-125 and reactivation of thermally inactivated alpha-glucosidase by recombinant GroEL. Biosci. Biotechnol. Biochem. 60: 1633-1636.

Yamamoto, N., Kato, R., and Kuramitsu, S. 1996. Cloning, sequencing and expression of the uvrA gene from an extremely thermophilic bacterium, *Thermus thermophilus* HB8. Gene. 171: 103-106.

Yang, H., Chiang, J.H., Fitz-Gibbon, S., Lebel, M., Sartori, A.A., Jiricny, J., Slupska, M.M., and Miller, J.H. 2002. Direct interaction between uracil-DNA glycosylase and a proliferating cell nuclear antigen homolog in the crenarchaeon *Pyrobaculum aerophilum*. J. Biol. Chem. 277: 22271-22278.

Yang, H., Phan, I.T., Fitz-Gibbon, S., Shivji, M.K., Wood, R.D., Clendenin, W.M., Hyman, E.C., and Miller, J.H. 2001. A thermostable endonuclease III homolog from the archaeon *Pyrobaculum aerophilum*. Nucleic Acids Res. 29: 604-613.

Yernool, D.A., McCarthy, J.K., Eveleigh, D.E., and Bok, J.D. 2000. Cloning and characterization of the glucooligosaccharide catabolic pathway beta-glucan glucohydrolase and cellobiose phosphorylase in the marine hyperthermophile *Thermotoga neapolitana*. J. Bacteriol. 182: 5172-5179.

Yu, J.S., and Noll, K.M. 1997. Plasmid pRQ7 from the hyperthermophilic bacterium *Thermotoga* species strain RQ7 replicates by the rolling-circle mechanism. J. Bacteriol. 179: 7161-4.

Zavodszky, P., Kardos, J., Svingor, and Petsko, G.A. 1998. Adjustment of conformational flexibility is a key event in the thermal adaptation of proteins. Proc. Natl. Acad. Sci. USA. 95: 7406-7411.

Zecchinon, L., Claverie, P., Collins, T., D'Amico, S., Delille, D., Feller, G., Georlette, D., Gratia, E., Hoyoux, A., Meuwis, M.A., Sonan, G., and Gerday, C. 2001. Did psychrophilic enzymes really win the challenge? Extremophiles. 5: 313-321.

Zillig, W., Arnold, H.P., Holz, I., Prangishvili, D., Schweier, A., Stedman, K., She, Q., Phan, H., Garrett, R., and Kristjansson, J.K. 1998. Genetic elements in the extremely thermophilic archaeon *Sulfolobus*. Extremophiles. 2: 131-140.

Zillig, W., Prangishvilli, D., Schleper, C., Elferink, M., Holz, I., Albers, S., Janekovic, D., and Gotz, D. 1996. Viruses, plasmids and other genetic elements of thermophilic and hyperthermophilic Archaea. FEMS Microbiol. Rev. 18: 225-236.

2

Expression, Folding, and Degradation in *Escherichia coli*

Mirna Mujacic and François Baneyx

Abstract

The Gram-negative bacterium *Escherichia coli* has been and remains extensively used for the production of heterologous proteins at both laboratory and industrial scales. Over the past decade, sophisticated genetic tools and a growing understanding of the function of the *E. coli* proteome have been exploited to improve the synthesis of complex proteins of therapeutic or commercial interest in a soluble and biologically active form. This chapter reviews the chief aspects of heterologous protein production in *E. coli* with a focus on the mechanisms of folding, degradation and export, and the practical implications of recent discoveries in these areas.

Introduction

The enteric bacterium *Escherichia coli* is one of the most studied prokaryotic organism and is widely used as a shuttle host for genetic manipulations as well as for the industrial production of proteins of therapeutic or commercial interest. *E. coli* is a facultative, Gram-negative, rod-shaped bacterium approximately 2 μm in length by 0.5 to 1 μm in diameter. It is relatively tolerant of environmental variations and remains viable between 10 and 50°C and pH 5 to 9, although its growth rate decreases significantly at extremes of temperature and pH (Ingraham and Marr, 1996). Under optimal conditions (rich medium, 37°C and pH 7), the doubling time of *E. coli* is approximately 20 min which allows for rapid biomass accumulation and reduces the likelihood of contamination. Additional advantages include the facts that *E. coli* can be grown to high density on a variety of inexpensive carbon sources, that fermentation scale-up is straightforward, and that product yield (as a percentage of biomass) is high. Because many therapeutic proteins including insulin, human growth hormone and a number of cytokines are already produced in *E. coli* (Swartz, 1996), approval of *E. coli*-based processes by the US Food and Drug Administration (FDA) or other regulatory agencies is relatively routine. Finally, the availability of genetic tools and a wide number of mutant strains, cloning and expression vectors allows one to custom-tailor the host in order to optimize the production or folding of a desired polypeptide.

On the negative side, *E. coli* is the subject of an extensive patent portfolio, and although many are approaching their expiration date, the prospect of negotiating multiple licensing agreements may deter companies that do not possess patented in-house technologies from selecting this bacterium as a production host. The other major disadvantage of *E. coli* is that it is unable to glycosylate proteins and is relatively inefficient at carrying out a number of other post-translational modifications including the formation and isomerization of disulfide bonds (although much progress has been made in this respect as will be discussed later). Thus, if glycosylation of a recombinant gene product is required to maximize *in vivo* efficacy, an alternative expression host such as CHO or insect cells should be selected. A final drawback is that, like all Gram-negative bacteria, *E. coli* synthesizes endotoxic lipopolysaccharides (LPS) that may contaminate recombinant protein preparations and cause pyrogenic and shock reactions in humans if not removed. The development of

LPS mutant strains exhibiting strongly reduced endotoxicity (Somerville *et al.*, 1996; Cognet *et al.*, 2002) together with recent progress in the design of ion exchange matrices for one-step adsorption of LPS (McNeff *et al.*, 1999) should help circumvent this problem.

This chapter reviews the chief aspects of heterologous protein production in *E. coli* with a particular emphasis on protein folding and degradation issues.

Plasmids

Although chromosomal expression systems have been described (Peredelchuk and Bennett, 1997; Olson *et al.*, 1998), plasmid-based expression remains the preferred means of producing recombinant proteins in *E. coli*. A large number of vectors have been constructed and optimized over the years. The most commonly used plasmids are derivatives of pBR322 (Bolivar *et al.*, 1977) which make use of the ColE1/pMB1 origin of replication. Under balanced growth conditions, ColE1 derivatives are present at 15-30 copies per cell. However, plasmid copy number – and hence gene dosage – can be increased to several hundreds by interfering with the regulation of replication. This is examplified by the pUC series (Yanisch-Perron *et al.*, 1985) which are present at 500-700 copies per cell due to the deletion of the *rom/rop* gene (which encodes a short protein stabilizing the interaction between the priming RNAII and the antisense RNAI) and a mutation that shortens RNAI and weakens its interaction with RNAII. Conversely, the copy number of ColE1 derivatives can be lowered to a few copies per cell if the host strain carries a mutation in the *pcnB* gene. This is due to the fact that in strains missing poly(A) polymerase (the *pcnB* gene product), RNAI is not polyadenylated and therefore not turned over as efficiently by host RNAses. Accumulation of high levels of non-adenylated RNAI, which associates with the target RNAII more efficiently than polyadenylated RNAI, is ultimately responsible for repression of replication and copy number decrease (Xu *et al.*, 1993; Xu *et al.*, 2002).

Plasmids carrying the p15A origin of replication (*e.g.* the pACYC series, Chang and Cohen, 1978) are present at 10-15 copies per cell and belong to a different incompatibility group than ColE1 derivatives. Because they replicate via independent mechanisms, both types of plasmids can be stably maintained in a single cell. This feature is often used to co-overproduce accessory proteins such as foldases or molecular chaperones (see below) on a p15A derivative along with a target protein encoded on a ColE1 derivative. If a third replicon is needed, pSC101-based cloning vectors that are present at about 5 copies per cell and can co-exist with both ColE1 and p15A-derived plasmids are also available (Lerner and Inouye, 1990).

Stable maintenance of plasmids is usually achieved by supplementing the growth medium with antibiotics that are inactivated by plasmid-encoded resistance genes. The most common markers are ampicillin (Amp^r), chloramphenicol (Cat^r), kanamycin/neomycin (Kan^r/Neo^r) and Spectinomycin (Spc^r) resistances. Ampicillin, along with its more stable derivative carbenicillin, binds to several enzymes involved in cell wall synthesis. Ampicillin resistance is conferred by the periplasmic enzyme β-lactamase (a 29-kDa monomeric protein) which cleaves β-lactam ring antibiotics. One of the disadvantages of using Amp^r is that β-lactamase may be released into the growth medium in high density cultures or if the integrity of the outer membrane is compromised, thereby conferring protection to plasmid-free cells that may take over the culture due to their faster growth rate. Chloramphenicol binds to the large ribosomal subunit and inhibits protein synthesis. Resistance to this antibiotic is provided by chloramphenicol acetyl transferase, a homotetrameric enzyme of 23-kDa subunits encoded by the *cat* gene. In the presence of AcetylCoA, chloramphenicol acetyl transferase catalyzes the formation of inactive hydroxyl acetoxyl derivatives of chloramphenicol. Of interest to medium formulation is the fact that chloramphenicol acetyl transferase expression increases 5-10 fold when bacteria are grown on a carbon source other than glucose due to release of catabolite represion (Malke and Ferretti, 1985). Kanamycin and neomycin are deoxystreptamine aminoglycosides that bind to ribosomal components and inhibit translation. Both antibiotics are inactivated by a periplasmic aminoglycoside phosphotransferase of molecular mass 25-kDa.

Phosphorylation of kanamycin and neomycin appears to interfere with their active transport into the cell cytoplasm. Kan^r plasmids are commonly used for industrial production processes. Spectinomycin and streptomycin are aminoglycosides that irreversibly inhibit protein synthesis. Resistance to these antibiotics is conferred by an adenyltransferase (the *aadA* gene product). In general, spectinomycin supplementation should be used for maintaining plasmids containing a Spc^r/Str^r marker since many expression strains contain the *rpsL* mutation in 30S ribosomal protein S12 that makes them resistant to streptomycin. It should finally be noted that alternatives to the use of antibiotic resistance markers are available. These include programmed cell death of plasmid-free cells using proteic killer systems (Jensen and Gerdes, 1995) and the use of complementation strategies in which expression plasmids bear a copy of an essential gene that is missing in the chromosome of the host strain.

Promoters

Promoters are DNA sequences from which RNA polymerase initiates transcription. *E. coli* promoters consist of two hexanucleotide sequences located approximately 35 and 10 bases away from the transcriptional start site (the -35 and -10 boxes, respectively). The two hexamers are separated by a short spacer region, the length and composition of which greatly affect promoter strength. RNA polymerase recognizes promoter regions with the help of transcriptional factors, known as sigma factors. *E. coli* synthesizes at least six such sigma factors. The most extensively studied is the vegetative σ^{70} (aka σ^D) factor that directs RNA polymerase to the vast majority of *E. coli* promoters. Two alternative sigma factors, σ^{32} (aka σ^H) and σ^{24} (aka σ^E), are responsible for the transcription of heat shock genes which are upregulated when unfolded or misfolded proteins accumulate in the cytoplasm or the periplasm, respectively (Gross, 1996; Missiakas and Raina, 1997). The stationary phase sigma factor σ^{38} (aka σ^S) is involved in the transcription of 50-100 genes implicated in bacterial resistance to global stress, including starvation (Hengge-Aronis, 1996). Finally, two specialized sigma factors are responsible for the transcription of nitrogen assimilation genes (σ^{54} or σ^N), and flagellum and chemotaxis genes (σ^{28} or σ^F).

Table 1. Examples of commonly used promoters and their properties.

Promoter[a]	Regulation[b]	Induction	Expression	Source or Reference
E. coli-derived promoters				
lac	*lacI* or *lacI^q*	IPTG	Low	Clontech
tac	*lacI* or *lacI^q*	IPTG	Moderate	Clontech, Invitrogen, Life Technologies, Promega, New England Biolabs
trc	*lacI* or *lacI^q*	IPTG	Moderate	Clontech, Invitrogen, Life Technologies, Promega, New England Biolabs
trp	Attenuation, *trpR*	Tryptophan starvation	High	Stratagene
phoA	*phoB* (positive), *phoR* (negative)	Phosphate starvation	Variable	(Miyake et al., 1985)
araBAD	*araC*	L-arabinose	Moderate	Invitrogen
lpp-lac	*lacI* or *lacI^q*	IPTG	Moderate	MoBiTec
tetA	*tetR*	Anhydrotetracycline	Variable	Biometra
$P_{LtetO-1}$	*tetR*	Anhydrotetracycline	Variable	Clontech
cadA	*cadR*	pH	Variable	(Chou et al., 1995)
cspA	5' untranslated region (UTR)	Shift to less than 20°C	Low to moderate	(Baneyx and Mujacic, 2003)
Phage-derived promoters				
λp_L	λ cI857(ts)	Shift to 42°C	Moderate	Invitrogen
λp_L-9G-50	IHF protein	Shift to less than 20°C	Moderate	(Giladi et al., 1995)
T7	Cascade system	IPTG	Very high	Novagen
T7-*lacO*	Cascade system, *lacI^q*	IPTG	Very high	Novagen
T3-*lacO*	Cascade system, *lacI^q*	IPTG	Moderate	(Yamada et al., 1991)
T5-*lacO*	Cascade system, *lacI* or *lacI^q*	IPTG	Very high	Qiagen

[a] The *tac* and *trc* promoters consist of the −35 box of the *trp* promoter and the −10 box of the *lac* promoter and only differ by one nucleotide in their spacing sequence. They are of equivalent strength.

[b] The *lacI^q* allele contains a single mutation in the −35 box of the *lacI* promoter leading to a 5-fold increase in the number of LacI repressor molecules relative to authentic *lacI*.

Promoter strength and transcriptional efficiency can be significantly affected by the sequences flanking the promoter region and by the amount of negative DNA supercoiling. For example, the presence of an AT-rich UP element upstream of certain core promoters (*e.g. rrnB* P1) significantly enhances transcription by facilitating interactions with the α subunit of RNA polymerase (Aiyar *et al.*, 1998). Certain promoters also contain sites to which repressor proteins (*e.g.* the LacI repressor protein) or the catabolic activator protein (CAP) complexed to cAMP bind, thus decreasing or increasing promoter strength, respectively (Record *et al.*, 1996).

The selection of a promoter is a key step in recombinant protein production. If high-level synthesis is desired, the promoter should be sufficiently strong to allow the heterologous protein to accumulate at a level corresponding to at least 30% of the total cellular protein. The bacteriophage T7 and T5 promoters (Table 1) are often used when such high concentrations are required. However, strong promoters have their pitfalls. For example, high level overexpression can burden bacterial metabolism, hinder cell growth and ultimately lead to ribosome destruction and cell death. In addition, titration of the cellular supply of molecular chaperones and foldases by a large number of nascent polypeptide chains could (and often does) result in misfolding and the formation of inclusion bodies. Although this is not a problem if the target protein can be easily refolded *in vitro*, the use of bacteriophage promoters should be evaluated with care if a soluble protein is desired.

A second important property that a promoter should possess is tight regulation (*e.g.* low basal transcriptional activity in the absence of inducer). This feature is very important when the target gene product is toxic to cells. The arabinose (*araBAD* or P_{BAD}) and tetracycline (*tetA*) promoters are often used if tight repression of promoter activity is required.

Third, promoter induction should be cost-effective and simple. One of the most popular inducers is the lactose analog isopropyl-β-D-thiogalactopyranoside (IPTG). IPTG binds to the LacI repressor protein to release it and prevent its subsequent association with *lacO* operator sequences. This leads to the induction of *lac* and *lac*-derived promoters (*e.g. tac* and *trc*) via loss of negative regulation. IPTG is also used to indirectly induce bacteriophage promoters (since the T7, T5 or T3 RNA polymerases genes are usually placed under control of a *lac*-derived promoter within the host chromosome), and to simultaneously release negative regulation from T7, T5 or T3 promoters that have been fused to a *lacO* sequence in order to reduce basal transcription (Table 1). Nevertheless, the use of IPTG in large-scale production may be problematic due to its toxicity and high cost. Nutritionally regulated (*e.g. trp* and *phoA*) and thermally inducible promoters (*e.g.* λ p_L) are the most widely used systems for industrial protein production. Induction upon tryptophan or phosphate starvation has the advantage of separating biomass accumulation and protein production phases but the accumulation levels of the desired gene product may be low. The use of a temperature upshift for induction has the potential disadvantage of enhancing target protein misfolding and of promoting its degradation since high temperatures lead to the upregulation of heat shock proteases (see below). If this is an issue, the *cspA* promoter, which is induced by temperature downshift from 37°C to below 20°C (Vasina *et al.*, 1998), may be useful since low temperatures improve protein folding and reduce proteolysis (Mujacic *et al.*, 1999; Baneyx and Mujacic, 2003). A more extensive discussion of *E. coli* promoters can be found elsewhere (Goldstein and Doi, 1995; Makrides, 1996).

Protein Folding and Degradation in the Cytoplasm

Cytoplasmic Molecular Chaperones and Foldases

Molecular chaperones represent a class of proteins that help other polypeptides reach a proper conformation or cellular location by binding to solvent-exposed hydrophobic patches in partially folded protein substrates, thereby preventing their self-association and subsequent aggregation. These folding helpers play multiple roles in the conformational quality control of the proteome including in *de novo* protein folding, refolding of damaged proteins, disaggregation of thermally aggregated proteins, export from the cytoplasm and degradation of irreversibly damaged polypeptides (Goldberg *et al.*, 1981; Gross, 1996; Murphy and Beckwith, 1996; Wickner *et al.*, 1999; Gottesman and Hendrickson, 2000; Thirumalai and Lorimer, 2001; Hartl and Hayer-Hartl, 2002). Because heat shock and other forms of stress (including recombinant protein overexpression) cause protein structural damage, most – but not all – of the cytoplasmic molecular chaperones are heat shock proteins (Hsps) that are transcribed at high level by RNA polymerase complexed to the alternative sigma factor σ^{32}. Functionally, molecular chaperones can be divided into three subclasses: those that actively promote the net unfolding/refolding of bound substrates (*e.g.* the DnaK-DnaJ-GrpE and GroEL-GroES systems) are classified as folding chaperones; those that maintain partially folded client proteins on their surface to await the availability of folding chaperones following stress abatement (*e.g.* IbpAB) are referred to as holdases; those that disentangle thermal aggregates (*e.g.* ClpB) are known as disaggregases. Foldases, on the other hand, are true enzymes that accelerate rate-limiting steps along protein folding pathways. However, they may combine both activities as is the case with trigger factor (TF), the principal foldase of the *E. coli* cytoplasm. A brief review of the main characteristics of the cytosolic folding helpers as well as their potential in improving the solubility of heterologous proteins is provided in the following paragraphs.

The DnaK-DnaJ-GrpE System

DnaK is a 69-kDa, monomeric, two-domain protein that plays a central role in *de novo* protein folding, host protein refolding, translocation and in the general management of the deleterious effects of stress. The DnaK N-terminus domain is responsible for its ATPase activity while the C-terminus domain is involved in substrate binding. The substrate specificity of DnaK has been inferred from the crystal structure of its C-terminal domain (Zhu *et al.*, 1996) and by assessing the ability of the chaperone to bind to immobilized random peptide libraries (Rüdiger *et al.*, 1997). DnaK was found to exhibit a preference for heptameric stretches of amino acids composed of a 4-5 residues-long hydrophobic core enriched in leucines and flanked by basic residues. Statistical analyses indicate that these motifs are quite common, occurring every 36 residues on the average protein (Rüdiger, et al., 1997). It is therefore not surprising that DnaK binds to a large number of structurally and functionally unrelated proteins provided that they are in a nonnative (partially unfolded) state. However, some sites are preferred as binding constants can vary between 5 nM and 5 μM (Bukau and Horwich, 1998).

To properly function in vivo, DnaK requires the assistance of two additional cofactors, DnaJ and GrpE. DnaJ is a 41-kDa protein that triggers ATP-hydrolysis dependent substrate association of partially folded proteins to the substrate binding cavity of DnaK, and contains a conserved J domain that is required for its association with DnaK (Karzai and McMacken, 1996). DnaJ can independently bind unfolded proteins with low affinity and is believed to scan partially folded substrates in order to direct DnaK to high affinity binding sites (Rüdiger et al., 2001). GrpE, a homodimer of 20-kDa subunits, triggers ADP release from DnaK and subsequent substrate release as ATP rebinds DnaK. It is generally accepted that substrate proteins ejected from DnaK either fold into a proper conformation, are recaptured by DnaK-DnaJ for additional cycles of binding and release or are transferred in a partially folded form to GroEL-GroES for subsequent folding.

The GroEL-GroES System

The chaperonin GroEL is an ≈ 800-kDa hollow cylinder consisting of two seven-subunit rings stacked back to back (Braig et al., 1994). GroEL uses a set of hydrophobic residues located at either end of the cylinder to capture non-native substrate and to interact with its co-chaperone GroES, a 70-kDa, dome-shaped homoheptamer (Fenton et al., 1994; Hunt et al., 1996). In vivo, binding of client proteins occurs on the ring of GroEL that is not complexed to GroES (known as the trans ring) and involves multiple contacts with the substrate that lead to nanomolar dissociation constants. GroEL can fold proteins up to 60-kDa in size and appears to exhibit a preference for compact intermediates consisting of two or more domains with α/β-folds that are enriched in hydrophobic and basic residues (Coyle et al., 1997; Houry et al., 1999).

GroEL-mediated protein folding is believed to involve the following sequence of events: (1) substrate binding to the nucleotide-free *trans* ring; (2) binding of 7 ATP molecules and encapsulation by GroES which results in substrate release in an enlarged and now hydrophilic cavity as well as ejection of bound ADP and GroES from the opposite ring; (3) substrate folding timed by the hydrolysis of ATP ($\approx$ 10s); and (4) release of GroES, ADP and either properly folded protein or folding intermediate caused by the binding of 7 ATPs and fresh substrate to the opposite ring (Hartl and Hayer-Hartl, 2002). If the ejected substrate still exhibits significant surface hydrophobicity, it will be re-captured by GroEL and the above cycle repeated until a correct conformation is reached.

The ClpB ATPase

ClpB is an hexameric ATPase that collaborates with the DnaK-DnaJ-GrpE system to reactivate large insoluble aggregates *in vitro* (Goloubinoff *et al.*, 1999; Zolkiewski, 1999) and thermally aggregated proteins *in vivo* (Mogk *et al.*, 1999). This operation is thought to involve the ATP-dependent shearing of protein aggregates and the transfer of partially folded proteins to the DnaK-DnaJ-GrpE system for subsequent folding (Ben-Zvi and Goloubinoff, 2001). In addition, ClpB facilitates *de novo* protein folding (Thomas and Baneyx, 2000), possibly by using its remodeling activity to disentangle early aggregates consisting of a few associated proteins which remain partially folded but have packed DnaK recognition sequences away from the solvent. While inactivation of *clpB* only affects *E. coli* growth above 44-46°C, *clpB* null mutants experience a significant loss in their ability to survive transient exposure to 50°C (the maximal growth temperature of *E. coli*), presumably due to their inability to process and clear thermosensitive protein aggregates (Squires *et al.*, 1991; Thomas and Baneyx, 1998). Interestingly, the *clpB* gene contains an internal translation start site and is synthesized as two products of approximate molecular masses 95 and 80-kDa. Both proteins are able to bind model substrates *in vitro* (Tek and Zolkiewski, 2002) and can substitute for each other *in vivo* although ClpB80 is less efficient than ClpB95 (I.-T. Chow and F. Baneyx, unpublished data). At present, the precise function of the truncated form of ClpB remains unclear although it has been proposed to act as a regulator of ClpB95 by affecting its ability to carry out ATP hydrolysis (Park *et al.*, 1993).

The IbpA-IbpB Small Heat Shock Proteins

IbpA and IbpB, two homologous 16-kDa proteins belonging to the α-crystallin (aka small heat shock) protein family (Naberhaus, 2002), were first identified as contaminants present in recombinant protein inclusion bodies (Allen *et al.*, 1992). IbpB exhibits chaperone function *in vitro* and forms large amorphous aggregates that dissociate into ≈ 600-kDa oligomers following incubation at high temperatures (Shearstone and Baneyx, 1999). IbpA/B are thought to function as holdases that maintain partially folded polypeptides on their surface until stress has abated and the DnaK-DnaJ-GrpE system becomes available for processing of bound substrates (Veinger *et al.*, 1998). Massive overproduction of IbpA/B can alleviate host protein aggregation (Kuczynska-Wisnik *et al.*, 2002) and protect cell and enzymes from heat and superoxide stress (Kitagawa *et al.*, 2000; Kitagawa *et al.*, 2002). However, there is little increase in cellular protein misfolding when *ibp* null mutants are incubated at high temperatures (Thomas and Baneyx, 1998; Mogk, *et al.*, 1999), and an increase in the intracellular levels of IbpA/B does not improve the solubility of several aggregation-prone heterologous proteins (Thomas and Baneyx, 2000, unpublished data). Very recent evidence suggests that the physiological role of small Hsps is to intercalate themselves within large protein aggregates that form upon exposure to stress, thereby facilitating their disaggregation and refolding by DnaK-DnaJ-GrpE alone or in collaboration with ClpB (Mogk *et al.*, 2003).

HtpG

E. coli HtpG is a 71-kDa homodimer that exhibits chaperone activity *in vitro* (Spence and Georgopoulos, 1989; Nemoto *et al.*, 2001). Substrate binding appears to involve elements from both the N- and C-terminal domains of the protein (Nemoto, *et al.*, 2001; Yamada *et al.*, 2003). Although HtpG is dispensable in *E. coli* (Bardwell and Craig, 1988), its eukaryotic homolog is essential (Borkovich *et al.*, 1989). In eukaryotes, Hsp90 family members associate with other chaperones and foldases to form super-chaperone complexes involved in controlling the activation and preventing the misfolding of steroid receptors and kinases (Buchner, 1999). The physiological function of HtpG is much more obscure although it appears to participate in *de novo* folding events by enhancing the ability of the DnaK-DnaJ-GrpE system to interact with partially folded proteins (Thomas and Baneyx, 2000) and has been implicated in secretion (Ueguchi and Ito, 1992; Shirai *et al.*, 1996).

Hsp33

Hsp33 is a redox-regulated molecular chaperones that was identified based on its induction by heat shock (Chuang *et al.*, 1993; Jakob *et al.*, 1999). The primary function of Hsp33 is to alleviate cellular damage following oxidative stress. To carry out this role, Hsp33 relies on an elaborate regulatory mechanism. Under balanced growth conditions, the protein accumulates in a monomeric and reduced form in which four conserved C-terminal cysteines coordinate a zinc atom. Following exposure to oxidizing conditions and heat, two intramolecular disulfide bonds are formed, zinc is ejected and the protein dimerizes to become active as a molecular chaperone (Jakob, *et al.*, 1999; Barbirz *et al.*, 2000; Graumann *et al.*, 2001; Kim *et al.*, 2001a; Vijayalakshmi *et al.*, 2001). Truncation of C-terminal residues completely abolishes redox regulation and results in a dimeric protein that exhibits constitutive chaperone activity (Kim *et al.*, 2001).

Hsp31

Like Hsp33, Hsp31 was first identified as a heat shock gene (Richmond *et al.*, 1999). Hsp31 is organized as a homodimer of 31-kDa subunits and exhibits molecular chaperone activity *in vitro* (Sastry *et al.*, 2002; Malki *et al.*, 2003). Similar to small heat shock proteins, Hsp31 relies on temperature-driven exposure of structured hydrophobic domains to capture nonnative protein substrates. However, Hsp31 rapidly releases a fraction of bound proteins in an active form following temperature downshift suggesting that it transiently stabilizes early unfolding intermediates to prevent overloading of the DnaK-DnaJ-GrpE system in times of stress (Sastry, *et al.*, 2002). There is indeed genetic evidence that these chaperones cooperate (M. Mujacic and F. Baneyx, unpublished data). Although it is not an ATPase, Hsp31 is subject to regulation by ATP, the binding of which reduces the size or availability of the client protein binding site (Sastry, *et al.*, 2002). The crystal structure of Hsp31 has revealed the presence of a poorly accessible catalytic triad in each monomer (Quigley *et al.*, 2003). Although no proteolytic activity has been detected to date, it is conceivable that Hsp31 combines chaperone and hydrolase functions. *In vivo*, Hsp31 is dispensable but its absence impairs the ability of cells to recover from severe heat shock (M. Mujacic and F. Baneyx, unpublished data).

Trigger Factor

Trigger factor (TF) is an abundant 48-kDa protein that associates with 1:1 stoichiometry with about half of the cell ribosomes and docks next to the L23 and L29 protein of the 50S ribosomal subunit in the vicinity of the peptide exit site (Kramer *et al.*, 2002; Blaha *et al.*, 2003). TF is a modular protein consisting of three independent folding units (Zarnt *et al.*, 1997). The N-terminus domain is responsible for ribosome binding. The central domain shares weak homology with the FKBP family and contains a peptidyl *cis/trans* isomerase (PPIase) active site that catalyzes the conversion of X-Pro peptide bonds from *trans* to *cis* conformation, a rate-limiting step in the folding of certain proteins. Scarce information is available on the C-terminal domain which may play a role in mediating the association of ribosome-bound TF with nascent polypeptides.

In addition to functioning as a PPIase, TF displays chaperone activity *in vitro* (Scholz *et al.*, 1997). Because of its physical localization, TF has been postulated to play an important role in *de novo* protein folding and to cooperate with DnaK in this process. Indeed, mutations in the trigger factor gene (*tig*) and the *dnaKJ* operon are synthetically lethal (Deuerling *et al.*, 1999; Teter *et al.*, 1999), and TF and DnaK have overlapping substrate pools *in vivo* (Deuerling *et al.*, 2003). That the two chaperones bind similar substrates is well explained by the fact that TF recognizes an eight amino acid motif enriched in aromatic and basic residues (Patzelt *et al.*, 2001; Deuerling, *et al.*, 2003) which is similar to the DnaK recognition motif (Zhu, *et al.*, 1996; Rüdiger, *et al.*, 1997). Nevertheless, there appears to be a hierarchy with a competitive advantage of TF over DnaK for the binding of nascent polypeptide chains (Deuerling *et al.*, 2003).

SlyD and SlpA

The *slyD* gene encodes an abundant 196 amino-acid protein belonging to the FKBP family of PPIases (Hottenrott *et al.*, 1997; Roof *et al.*, 1997). SlyD is involved in the stabilization of the E lysis protein of bacteriophage φX174 (Bernhardt *et al.*, 2002), but little else is known about its function in protein folding. SlpA (FkpB) is a 16-kDa protein homologous to SlyD that exhibits PPIase activity *in vitro* (Hottenrott, *et al.*, 1997). The physiological function of SlpA is unknown. However, the fact that an increase in its intracellular concentration suppresses phenotypes associated with the lack of protease Lon (Trempy *et al.*, 1994) suggests that this PPIase plays a role in protein folding.

Practical Applications of Molecular Chaperones and Foldases

There is abundant evidence that co-expression of molecular chaperones and foldases can improve the solubility of heterologous proteins overexpressed in the cytoplasm of *E. coli*, and many ColE1-compatible plasmids encoding molecular chaperones and foldases have been constructed for this purpose (Thomas *et al.*, 1997; Baneyx and Palumbo, 2003). To date, most success stories involve an increase in the intracellular concentration of DnaK-DnaJ (with or without GrpE), TF or GroEL-GroES. As previously discussed, both DnaK and TF can engage nascent polypeptides as they exit from the ribosome. Under conditions of high level heterologous protein expression, the cellular supply of these folding helpers may become limiting, leading to the aggregation of unchaperoned newly synthesized polypeptides. It therefore stands to reason that this situation would be corrected upon co-expression of TF or DnaK-DnaJ. Significantly, those substrates whose folding is enhanced upon TF overexpression are also rescued from misfolding upon DnaK-DnaJ overproduction (Nishihara *et al.*, 2000), bringing support to the idea that DnaK and TF can functionally substitute for one another (Deuerling *et al.*, 2003).

For those foreign proteins that transit rapidly through the DnaK and/or TF system but require the assistance of chaperonins to reach a proper conformation, co-expression of the *groESL* operon can greatly enhance solubility whereas that of the *dnaKJ* operon has little positive effect. In principle, this would be expected of multidomain proteins that are smaller than 60-kDa (so that they can fit within the GroEL chamber) and that rapidly acquire secondary and tertiary structural elements but stall in their progress towards the native state. However, some reports indicate that the folding of proteins larger than 60-kDa is also enhanced in cells overexpressing GroEL-GroES (Roman *et al.*, 1995). This presumably results from the GroES-independent capture and stabilization of hydrophobic subdomains by GroEL which facilitates the isomerization of solvent-exposed portion of the polypeptide chain (Ayling and Baneyx, 1996). Finally, in some cases, the highest levels of heterologous protein solubility are observed upon coordinated expression of DnaK-DnaJ (or TF) and GroEL-GroES (Nishihara *et al.*, 1998; Nishihara, *et al.*, 2000). For these substrates, interactions with both chaperone systems may be an absolute requirement, or DnaK may time the transfer of partially folded isoforms to the chaperonin for efficient folding.

Comparatively little attention has been paid to the influence of other chaperones or foldases on heterologous protein folding. One, for example, could imagine that the disaggregase activity of ClpB may be harnessed to solubilize aggregated proteins. However, the ultrastructure of thermosensitive protein aggregates which likely consist of improperly oligomerized chains may be very different from that of heterologous proteins inclusion bodies. Indeed, an increase in the intracellular levels of ClpB does not rescue the misfolding or enhance the solubility of a number of aggregation-prone proteins (Thomas and Baneyx, 2000; M. Mujacic and F. Baneyx, unpublished data). A challenge will be to engineer ClpB to perform this function, possibly by directing it to an earlier step in the heterologous protein misfolding process. Other alternatives such as the co-expression of signal sequenceless versions of periplasmic chaperones or foldases to improve the folding of cytoplasmic proteins (Levy *et al.*, 2001) or the use of a C-terminally truncated version of Hsp33 will likely prove useful to address the problem of inclusion body formation.

What is clear from the available data, however, is that it remains impossible to predict in advance which chaperone will enhance the solubility of a particular target protein since interactions with host molecular chaperones likely depend on both the folding pathway and the kinetic partitioning of partially folded species. Thus, the most effective approach remains the screening of various chaperone-overproducing strains for enhanced solubility of a desired polypeptide.

Disulfide Bond Formation in the Cytoplasm

Stable disulfide bonds do not form in the cytoplasm of wild type *E. coli* owing to its reducing environment. The thioredoxin and glutaredoxin pathways are primarily responsible for the reduction of disulfide bridges in this cellular compartment and both derive reducing power from NADPH (Figure 1). The thioredoxin pathway involves two thioredoxins (TrxA and TrxC) and a thioredoxin reductase (TrxB), while three glutaredoxins (GrxA, GrxB and GrxC), glutathione reductase (Gor) and tripeptide glutathione make up the glutaredoxin system. Enzymes from both pathways share the thioredoxin fold, composed minimally of three α-helices and a four-stranded β-sheet, and contain an active site with a CXXC motif, where X is any amino acid. The physiological function of both systems is to recycle enzymes such as ribonucleotide reductase, phosphoadenosine-phosphosulfate reductase, methionine sulfoxide reductase, and arsenate reductase which transiently form disulfide bridges during catalysis (Ritz and Beckwith, 2001). This is achieved through the formation of mixed disulfide species between oxidoreductases and their substrates.

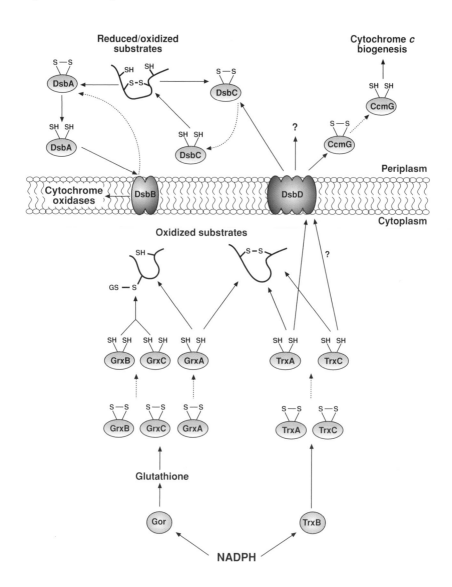

Figure 1. Cytoplasmic and periplasmic disulfide bond pathways in *E. coli*. Two reducing pathways exist in the cytoplasm. Thioredoxins (TrxA and TrxB) and glutaredoxins (GrxA, GrxB and GrxC) reduce oxidized substrates. They are kept in a reduced form via the action of the NADPH-dependent enzymes thioredoxin reductase (TrxB) and glutaredoxins reductase (Gor). In the oxidizing environment of the periplasm, DsbA catalyzes the formation of disulfide bridges, while DsbC acts primarily as an isomerase. DsbA is re-oxidized by DsbB, while DsbD is responsible for keeping DsbC in a reduced form. DsbD also reduces CcmG, which has been implicated in cytochrome c biogenesis. Solid arrows represent electron flow, while dotted arrows indicate products obtained after oxidation or reduction. GS-S represents a glutathione mixed disulfide bond.

The Thioredoxin Pathway

The major thioredoxin of *E. coli* is thioredoxin 1 (TrxA). With a redox potential of -270 mV and a CGPC active site motif, TrxA is an efficient thiol-disulfide reductase both *in vitro* and *in vivo* (Ritz and Beckwith, 2001). In addition to its role in the reduction of disulfide bonds, TrxA is also required for maturation of filamentous phage f1 and acts as a processivity factor for the replication of bacteriophage T7 DNA (Russel and Model, 1986; Tabor *et al.*, 1987). It is one of the most highly expressed genes of *E. coli*, and its synthesis increases following entry into stationary phase but not upon oxidative stress. The second *E. coli* thioredoxin, thioredoxin 2 or TrxC, shares 38% sequence identity with TrxA, and it is predicted to have a similar structure. TrxC expression is regulated by the transcriptional factor OxyR, whose activation during oxidative stress induces expression of antioxidant genes (Ritz and Beckwith, 2001). Strains harboring deletions in *trxA* alone or in both *trxA* and *trxC* are viable (Russel and Model, 1986; Ritz and Beckwith, 2001), implying that cells can circumvent the absence of these enzymes, most likely via the glutaredoxin pathway. The final component of the thioredoxin reducing machinery is the dimeric flavoenzyme TrxB which reduces TrxA and TrxC in a NADPH-dependent manner after they become oxidized following reduction of their target proteins.

The Glutaredoxin Pathway

In the glutaredoxin pathway (Figure 1), disulfide bonds are reduced by glutaredoxins GrxA, GrxB and GrxC, which are maintained in their reduced state by glutathione. Tripeptide glutathione, synthesized from either the *gshA* or *gshB* genes, is present in the cytoplasm in a mostly reduced form at a concentration of approximately 5 mM and is not essential for *E. coli* survival. Glutathione is kept in a reduced state by glutathione reductase (Gor) that shares significant homolog to thioredoxin reductases from higher eukaryotes. However, *gor* null mutants do not exhibit altered ratios of reduced to oxidized glutathione (Tuggle and Fuchs, 1985), suggesting that other glutathione reductases may be encoded in the *E. coli* genome. While not as efficient as thioredoxins in disulfide bond reduction (Holmgren, 1985), glutaredoxins, which contain a CPYC active site, can efficiently reduce glutathione mixed disulfides (Ritz and Beckwith, 2001). GrxA was isolated based on its ability to reduce ribonucleotide reductase in a mutant strain lacking TrxA (Holmgren, 1976). Triple mutants lacking TrxA, TrxC and GrxA cannot survive without plasmid complementation with one of the three hydrogen donors (Stewart *et al.*, 1998). GrxB, which is produced at twice the levels of GrxC and 25 times the levels of GrxA, is the major *E. coli* glutaredoxin (Ritz and Beckwith, 2001). While GrxB is incapable of reducing ribonucleotide reductase *in vitro* and *in vivo* (Åslund *et al.*, 1994), it is the most efficient hydrogen donor for the arsenate reductase ArsC (Shi *et al.*, 1999). GrxC has a redox potential that is 35 mV higher than that of GrxA, and is only able to reduce ribonucleotide reductase *in vitro* (Ritz and Beckwith, 2001). Finally, a glutaredoxin homolog (NrdH) has also been identified in *E. coli*. When NrdH is overexpressed, it is capable of acting as a hydrogen donor to ribonucleotide reductase *in vitro* and *in vivo* (Jordan *et al.*, 1997; Stewart *et al.*, 1998). However, even though the sequence of NrdH resembles that of glutaredoxins, it has a thioredoxin-like activity profile and is reduced by TrxB rather than Gor (Jordan, *et al.*, 1997).

Promoting Disulfide Bond Formation in the Cytoplasm

Due to the reducing nature of *E. coli* cytoplasm, the production of correctly folded recombinant proteins can be hindered if multiple disulfide bridges are required for biological activity. However, formation of stable disulfide bonds is possible in *trxB* mutants (Derman *et al.*, 1993), owing to the accumulation of oxidized TrxA and TrxC and their reversal of function from reductases to oxidases (Stewart *et al.*, 1998). Cells harboring deletions in *trxB* and *gor* or *trxB* and *gshA* allow even more efficient disulfide bond formation (Prinz *et al.*, 1997). Since the double mutants grow slowly under aerobic conditions, *trxB gor* cells were subjected to a round of selection for suppressor mutations that restore normal growth. One such *trxB gor supp* isolate, FA113 (commercialized under the trade name Origami by Novagen), grows normally and has been used to increase the production of active vtPA (a truncated form of tissue plasminogen activator) and single-chain Fv (ScFv) antibodies (Bessette *et al.*, 1999; Jurado *et al.*, 2002). Interestingly, and for unclear reasons, transfer of wild type or *trxB* cells to low temperatures makes their cytoplasm more oxidizing and this strategy can be used as an alternative route to promote disulfide bond formation in heterologous proteins (Derman and Beckwith, 1995; Schneider *et al.*, 1997).

For proteins requiring a complex pattern of disulfide bridges to exhibit biological activity, the breaking and reshuffling of incorrect disulfide bridges is often necessarry. *In vivo*, this function is carried out by protein disulfide isomerases (PDIs). Co-overproduction of active site variants of TrxA exhibiting PDI activity or of a signal sequence-less version of the periplasmic PDI DsbC (see below) has proven useful in increasing the yields of active vtPA and ScFv in *trxB gor supp* cells (Bessette *et al.*, 1999a; Jurado *et al.*, 2002). As the field progresses, cytoplasmic expression may prove a viable alternative to periplasmic expression for the production of eukaryotic proteins containing multiple disulfide bonds.

Cytoplasmic Proteases and Peptidases

In the cytoplasm of *E. coli*, early degradation steps of misfolded and prematurely terminated proteins are carried out principally by five ATP-dependent heat shock proteases Lon, ClpYQ (HslUV), ClpAP, ClpXP, and FtsH (HflB) (Miller, 1996). Completion of the protein degradation process is believed to involve peptidases which attack 2-5 residue long sequences (Miller, 1996). The main characteristics of the proteases and peptidases present in the *E. coli* cytoplasm are summarized in the following paragraphs.

Lon

Lon (La) is a homotetramer of 87-kDa subunits classified as a serine protease but exhibiting chymotrypsin-like activity. Each protomer is composed of three functional domains: a N-terminal region of unknown function, a central ATPase domain containing a typical Walker sequence and a C-terminal catalytic domain containing the active site serine. Lon exhibits regulatory function by degrading a class of proteins designed to be unstable (*e.g.* SulA, λN, RcsA...) and performs a general housekeeping function by contributing to the hydrolysis of misfolded proteins (Goldberg *et al.*, 1994; Gottesman, 1996). Lon also appears to play a central role in the degradation of thermosensitive proteins that accumulate following heat shock as evidenced by a significant increase in the amount of aggregated proteins in *lon* mutants (Tomoyasu *et al.*, 2001; Rosen *et al.*, 2002).

ClpAP and ClpXP

ClpA is an 84-kDa protein belonging to the Hsp100 family of heat shock proteins. It assembles as a hexamer in the presence of Mg-ATP. Like ClpB, ClpA contains two nucleotide-binding modules that are connected head to tail in a conformation consistent with a role in promoting the vectorial translation of protein substrates through the central pore of the hexamer (Guo *et al.*, 2002a). When serving as a protease, ClpA associates on both side of ClpP, a serine proteolytic component of 14 identical subunits arranged as two stacked heptameric rings (Ishikawa *et al.*, 2001). ClpX, a smaller (46-kDa) member of the Hsp100 family that contains a single ATP binding domain also binds ClpP. The interaction of both ClpA and ClpX with ClpP is mediated by "ClpP" loops containing a [LIV]-G-[FL] tripeptide (Kim *et al.*, 2001a; Singh *et al.*, 2001). This feature is missing in ClpB, which explains why it does not interact with ClpP. The ClpAP and ClpXP proteases have been shown to degrade proteins that have been modified at their C-terminus by the addition of a non-polar tail through the action of the SsrA system (Gottesman *et al.*, 1998; Herman *et al.*, 1998). However, ClpAP and ClpXP exhibit distinct substrate specificity (Gottesman *et al.*, 1997; Porankiewicz *et al.*, 1999; Wickner, *et al.*, 1999; Gottesman and Hendrickson, 2000). Finally, a small adaptor protein termed ClpS binds to the N-terminal domain of ClpA thereby redirecting ClpAP protease activity from soluble proteins to aggregated species (Dougan *et al.*, 2002; Guo *et al.*, 2002b; Zeth *et al.*, 2002).

ClpYQ (HslUV)

ClpY (HslU) is a 49.5-kDa Hsp100 family member that contains a single nucleotide binding domain and forms homohexamers. ClpY rings bind to either or both ends of ClpQ (HslV), a homododecamer organized as two stacked hexameric rings of 19-kDa subunits, to form a structure that resembles the 26S proteasome (Bochtler *et al.*, 2000). Together with Lon, the ClpYQ protease appears to be responsible for bulk degradation of misfolded proteins (Missiakas *et al.*, 1996).

FtsH

Among the *E. coli* ATP-dependent proteases, FtsH has the distinctive characteristic of being associated with the cytoplasmic membrane through two N-terminal transmembrane segments. The large cytoplasmic domain of the protease includes a 200 amino acid-long module common to the AAA+ superfamily of ATPases followed by a HEAGH sequence at the C-terminus that is a signature of zinc metalloproteases (Tomoyasu *et al.*, 1995). FtsH is believed to form ring-like hexameric structures (Shotland *et al.*, 1997; Karata *et al.*, 2001) and degrades a number of soluble cytoplasmic proteins including the heat shock sigma factor σ^{32} (Herman *et al.*, 1995; Tomoyasu, *et al.*, 1995) and SsrA-tagged polypeptides (Herman, *et al.*, 1998). FtsH also attacks integral membrane proteins (see for example (Akiyama *et al.*, 1996; Kihara *et al.*, 1999)) and retains an ability to do so even under denaturing conditions (Cooper and Baneyx, 2001).

Other Possible Endoproteases

Four additional endoproteolytic activities have been identified biochemically in the *E. coli* cytoplasm (Miller, 1996). Protease II is an 82-kDa serine protease specific for classic trypsin substrates (Pacaud, 1976; Kanatani *et al.*, 1991). Protease In is a 66-kDa monomeric protease that degrades chromogenic peptide substrates for trypsin-like enzymes (Kato *et al.*, 1992). Protease Fa is a possible 110-kDa metallo-protease (Goldberg, *et al.*, 1981). Finally, protease

So is a dimeric serine protease composed of 77-kDa subunits that is capable of degrading oxidatively damaged glutamine synthetase (Lee *et al.*, 1988) and the Ada protein (Lee *et al.*, 1990) *in vitro*.

Peptidases

Peptidases of *E. coli* are all able to hydrolyze small peptides although they may have other functions in the cell. Peptidases have been reviewed extensively by Miller (Miller, 1996) and only a brief description is presented here. Dipeptidases which can only hydrolyze dipeptides include the broad specificity Peptidase D, Peptidase Q which recognizes X-Pro dipeptides, and Peptidase E which is specific for Asp-X. Peptidase T is an aminotripeptidase that exhibits specificity for tripeptides but is rather sequence nonspecific.

In addition to Peptidase M, which is responsible for the removal of N-terminal methionines if the adjacent residue is small (Sherman *et al.*, 1985) and is the only peptidase essential for cell viability (Chang *et al.*, 1989), there are three known aminopeptidase activities in the cytoplasm of *E. coli* (Miller, 1996). Peptidase A is a leucine aminopeptidase that removes many N-terminal amino acids from peptides if the neighboring residue is not proline. Peptidase B is a poorly characterized broad-specificity aminopeptidase. Peptidase N is another broad specificity metallo-aminopeptidase belonging to the alanyl aminopeptidase family. Finally, Peptidase P specifically removes N-terminal amino acids adjacent to a proline. By contrast, only one carboxypeptidase, Dipeptylcarboxypeptidase, has been found in *E. coli*. This enzyme removes dipeptides from the C-terminus of its substrates and is believed to be involved in the degradation of intracellular proteins.

Of final interest is Oligopeptidase A (OpdA), a 680 residues long zinc metalloprotease. This peptidase is heat-inducible, contains a σ^{32}-dependent promoter and is transcribed as an operon with *yhiQ*, a gene of unknown function (Conlin and Miller, 2000). OpdA degrades the proliprotein signal peptide and may be involved in the degradation of damaged proteins.

Practical Implications

The degradation of recombinant proteins in the cell cytoplasm usually results from their inability to reach a native-like conformation and/or because they lack essential stabilizing interactions such as disulfide bridges. Thus, an obvious strategy to alleviate degradation is to improve folding as discussed in the above sections. Many strains deficient in cytoplasmic proteases have also been constructed. Although these may be employed for laboratory-scale production (particularly if the target protein is known to be predominantly degraded by a single protease), their large-scale use is not without problems. For instance, *ftsH* is an essential gene and although thermosensitive mutants have been constructed (Kihara *et al.*, 1995), they grow rather poorly and lack feedback regulation of σ^{32}-dependent genes which leads to an increase in the intracellular concentration of other heat-shock proteases. Similarly, K-12 *lon* mutants are filamentous and not well suited for high-density fermentations. *E. coli* B strains such as BL21 are naturally *lon*-deficient. However, it is likely that the lack of Lon is compensated by higher intracellular levels of other ATP-dependent proteases such as ClpYQ, ClpAP, ClpXP. On the other hand, *clpP* cells may be useful for the production of recombinant proteins containing nonpolar C-termini that are naturally recognized as ClpAP or ClpXP substrates.

Periplasmic Expression

If the production of a recombinant protein proves difficult in the cytoplasm, it may be useful to target it to the periplasm. Periplasmic expression has a number of advantages over cytoplasmic synthesis. First, if the protein is exported by the Sec pathway via fusion of a N-terminal signal sequence, the correct N-terminus can be obtained. Second, the periplasm is an oxidizing environment and therefore naturally conducive to the formation of disulfide bridges. Third, this cellular compartment contains a lower number of proteases than the cytoplasm and many of these deal with specific substrates. Finally, because only a fraction of host proteins are exported to the cell envelope, purification of the target protein is usually simplified. The following sections review the main issues associated with periplasmic expression.

Protein Translocation

Localization of proteins to different cellular compartments plays an important role in cell survival. In Gram-negative bacteria, a number of proteins are exported from the cytoplasm to the cytoplasmic membrane, periplasmic space, outer membrane and/or into the growth medium. Aside from specialized pathways used by certain pathogenic strains to export virulence factors, insertion or transport of proteins across the *E. coli* inner membrane involves four distinct mechanisms: the Sec-dependent, twin arginine (Tat), SRP-dependent and spontaneous translocation pathways (Figure 2).

Sec-dependent Export

Twenty percent of the approximately 4300 proteins in the *E. coli* proteome are exported from the cytoplasm in a Sec-dependent manner (Berks *et al.*, 2000). Proteins naturally secreted via this pathway are synthesized with an amino-terminal extension, the signal sequence, that is 20-to-30 residues in length and consists of a hydrophobic core followed by a proteolytic cleavage site (Manting and Driessen, 2000). Foreign cDNAs can be fused downstream of heterologous signal sequences to direct them to the periplasm and vectors suitable for this purpose are available from a variety of commercial sources (*e.g.* Novagen, New England Biolabs and Invitrogen). The most commonly used signal sequences have been borrowed from *E. coli* (OmpA, OmpT, MalE), *Erwinia* (PelB) and phage (gIII) proteins.

The Sec translocon of *E. coli* (Figure 2) consists of three core membrane proteins, SecY, SecE and SecG, and requires substoichiometric amounts of SecD, SecF and yajC for efficient export (Settles and Martienssen, 1998). The major feature of Sec-dependent translocation is export of proteins in an extended conformation. Premature folding of preproteins (polypeptides which are still tethered to a signal sequence) is usually prevented via interactions with molecular chaperones that can be either generic (*e.g.* GroEL and DnaK) or specialized (*e.g.*, SecB). This process is best understood in the case of SecB, a homotetrameric secretory chaperone that is required for

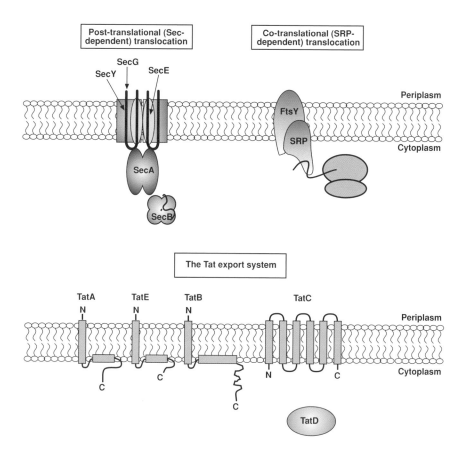

Figure 2. Major bacterial translocation pathways. Top panel: in Sec-dependent, post translational export, SecB interacts with hydrophobic domains of a partially folded protein and delivers it to SecA. SecA and hetero-trimeric SecYEG complexes act in an ATP-dependent manner to translocate the protein across the inner membrane. The signal sequence is cleaved off by leader peptidases during export. In SRP-dependent co-translational export, the signal recognition particle (SRP) interacts with a protein that is still associated with the ribosome. Translocation occurs with the help of membrane-embedded FtsY and SecYEG. Bottom panel: sequence-derived topology predictions of components of the twin-arginine (Tat) translocation pathway through which folded redox proteins are exported (Redrawn from Berks, *et al.*, 2000).

efficient export of a subset of periplasmic and outer membrane proteins. SecB retards preprotein folding by binding to hydrophobic stretches of amino acids that would normally be buried within the core of the properly folded polypeptide. As a result, SecB-bound preproteins neither fold nor aggregate and remain in a translocation-competent state (Dalbey and Robinson, 1999). Binary complexes are targeted to the translocon through the high affinity binding of SecB to the C-terminus of SecYEG-bound peripheral membrane protein SecA. The signal sequence of the preprotein next binds to SecA, which triggers its release from SecB and simultaneous transfer to SecA (Driessen *et al.*, 2001). Upon ATP binding, SecA inserts itself into the membrane along with the bound preprotein. Cycles of substrate binding and ATP hydrolysis drive the protein through the translocon and are accompanied by topological changes in SecG (Settles and Martienssen, 1998). During the export process, the signal sequence is removed by leader peptidases (see below). The driving forces of the Sec translocation machinery are ATP and the proton motive force, with azide as their common inhibitor.

Components of the *E. coli* Sec-dependent translocation pathway were identified by mutagenizing the signal sequences of known Sec-dependent substrates. The resulting genes were termed *prl* (protein localization) by the Silhavy group and *sec* by the Beckwith group. These synonymous denominations are still in use. For example, the *prlA* gene encodes SecY, while *prlG* gene encodes SecE. Interestingly, some of the genes identified contain mutations that enable the correct localization of proteins with defective signal sequences. Of particular practical interest are *prlA4* strains which contain mutations in two transmembrane segments (TMS) of SecY and allow more efficient export of heterologous proteins that have a tendency to misfold and accumulate as precursor inclusion bodies in the cell cytoplasm (van der Wolk *et al.*, 1998; Manting and Driessen, 2000).

Tat-dependent Export

The twin-arginine translocation (Tat) pathway has recently been recognized as a major export mechanism. The Tat pathway was first identified in chloroplasts following the observations that a subset of proteins are transferred across the thylakoid membrane in Sec-independent fashion and in a complete dependence on a pH difference across the inner membrane (Dalbey and Robinson, 1999). Proteins exported via the Tat pathway are synthesized with signal sequences that are longer, less hydrophobic and contain more basic residues in the C-terminal region than the leader sequences of Sec-dependent proteins (Cristobal *et al.*, 1999). Furthermore, both plant and prokaryotic signal sequences contain a conserved (S/T)RRXFLK signature before their hydrophobic core although the presence of the arginine dipeptide is not absolutely required for Tat-dependent export (DeLisa *et al.*, 2002). The majority of natural Tat-dependent substrates belong to a group of periplasmic proteins that bind redox cofactors and are required for anaerobic growth of cells. The most remarkable characteristic of the Tat pathway is that, unlike the Sec export system, its function is to transport *folded* or at least partially folded proteins of variable dimensions across the cytoplasmic membrane, without rendering the membrane freely permeable to protons and other ions (Berks, *et al.*, 2000). The integral membrane proteins TatA, TatB, TatC and TatE and the soluble cytoplasmic protein TatD have been identified as components of Tat system (Figure 2). Among these, TatA and TatE appear to be functionally interchangeable (Sargent *et al.*, 1998), while TatC and TatB seem to be core components of the system (Sargent *et al.*, 1999; Ize *et al.*, 2002). TatD does not seem to be obligatory since its inactivation has no detectable defect on Tat-dependent secretion (Berks *et al.*, 2000). Although the precise mode of action of the Tat machinery is still unclear, the following steps might be involved: interaction of the preprotein with DnaK or another chaperone; cofactor insertion and interaction with enzyme-specific accessory protein; binding of the signal peptide to a leader-binding protein; transfer of the protein to the membrane bound Tat translocon; and translocation and cleavage of the signal peptide by the leader peptidases. Since TatC orthologs share considerable identity and contain conserved polar residues, it has been proposed that TatC acts as the receptor of the Tat translocon by binding to the signal peptide of preproteins (Berks *et al.*, 2000). Based on the observation that folded GFP can be partially exported to the periplasm when fused to a Tat-dependent signal sequence (Santini *et al.*, 2001; Thomas *et al.*, 2001), the Tat translocation system will likely prove useful for the export of other recombinant proteins. Indeed, it has recently been shown that a fraction of ScFv and heterodimeric F_{ab} fragments can be exported to the periplasm via the Tat pathway, provided that they first form disulfide bonds in the cytoplasm and thus adopt a native conformation (DeLisa *et al.*, 2003).

SRP-dependent Export

Signal recognition particle (SRP)-dependent transport was first identified as playing an important role in protein export to the endoplasmic reticulum of eukaryotes. Bacterial homologs of eukaryotic SRP components are the 48-kDa GTPase Ffh and a 4.5S RNA (Figure 2), which primarily target a subset of hydrophobic, inner-membrane proteins for co-translational translocation (Settles and Martienssen, 1998). SRP binds to either the signal sequence or a transmembrane segment of a protein emerging from the ribosome. The SRP-bound ribosome nascent chain complex (RNC) is then targeted to the membrane receptor FtsY. Binding of GTP to FtsY releases the RNC from SRP and leads to its transfer to SecYEG translocon (Driessen *et al.*, 2001). Elongation of the polypeptide chain continues until translation is completed. Even though SecA is not required for the targeting of SRP-dependent substrates to the SecYEG translocon, it is necessary for a successful translocation event. It is tempting to hypothesize that overproduction of members of the SRP system will prove useful in improving the yields and alleviating the toxicity of heterologous membrane proteins synthesized in *E. coli*.

Spontaneous Insertion

Certain proteins, such as the M13-virus coat protein spontaneously insert within the *E. coli* membrane. The viral M13 pro-coat protein has a signal sequence that mediates electrostatic interactions between positively charged residues at the termini of the protein and acidic phospholipids in the membrane bilayer (Settles and Martienssen, 1998; Dalbey and Robinson, 1999). The protein inserts itself into the membrane by a mechanism that involves the hydrophobic leader peptide and a membrane-anchor domain. Translocation of acidic residues in the central loop seems to be driven by the proton motive force (Dalbey and Robinson, 1999). This mechanism appears to be tightly linked to structural features and is unlikely to find widespread applications for protein production.

Cell Envelope Chaperones and Foldases

Although it lacks an ATP pool, the periplasmic space contains a subset of proteins that function as folding helpers. Some of these are specialized chaperones, while others have a more general role in assisting outer membrane biogenesis and the folding of periplasmic proteins. The chaperone activity of the Dsb proteins and that of the protease DegP are discussed in later sections of this chapter.

Specialized Periplasmic Chaperones

The *E. coli* periplasm contains several substrate-specific molecular chaperones. One of these is LolA (aka p20), a protein that specifically recognizes the sorting sequences of outer membrane lipoproteins and promotes their incorporation into the outer membrane (Yokota *et al.*, 1999). Another class of specialized periplasmic chaperones is required for the assembly of pili (fimbriae), which are hairlike appendages mediating the attachment of enteric bacteria to target cells in eukaryotic hosts. Pilus-specific chaperones prevent the aggregation and degradation of their substrates by capping interactive surfaces (Hultgren *et al.*, 1991). Each pilus subunit exhibits an incomplete immunoglobulin-like (Ig) fold lacking its seventh C-terminal β strand. This creates a hydrophobic groove on the surface of a subunit. Pilus-specific chaperones occupy this groove to complete the Ig fold (Sauer *et al.*, 1999; Barnhart *et al.*, 2000). In the presence of the chaperone, the groove is held in an open and activated conformation. Following disassociation of the chaperone, the subunit undergoes a topological transition to attain its closed, ground state conformation as a part of the pilus. The transition, induced by the priming action of the chaperone, drives the assembly of pilus fibers (Sauer *et al.*, 2002). The PapD and FimC chaperones are specific for type P and type 1 pili, respectively. They also interact with outer membrane usher proteins, which form channels through which pilus subunits are translocated (Oliver, 1996).

Skp

Also known as OmpH, Skp is a non-essential 17-kDa periplasmic protein regulated by the alternative sigma factor σ^{24} (Roy and Coleman, 1994; Dartigalongue *et al.*, 2001). Skp has been suggested to play a role in the transport and assembly of outer membrane proteins and/or that of lipopolysaccharides (Chen and Henning, 1996; Bothmann and Plückthun, 2000). It also appears to serve as a generic chaperone based on the observations that it facilitates phage display and the folding of ScFv expressed in the periplasm (Bothmann and Plückthun, 1998; Hayhurst and Harris, 1999; Bothmann and Plückthun, 2000). The fact that co-overexpression of a signal sequence-less version of Skp improves the recovery yields of antibody fragments produced in the *E. coli* cytoplasm (Levy *et al.*, 2001) supports the idea that Skp is a promiscuous chaperone.

PPIases

To date, four PPIases (SurA, FkpA, PpiA/RotA and PpiD) have been identified in the periplasm of *E. coli*. Like Skp, FkpA and SurA are members of the σ^{24} regulon (Dartigalongue *et al.*, 2001), while *ppiD* transcription relies on the cytoplasmic heat-shock transcriptional factor σ^{32} (Dartigalongue and Raina, 1998). SurA and PpiD catalyze the isomerization of peptidyl-prolyl bonds in outer membrane β-barrel proteins (Lazar and Kolter, 1996; Dartigalongue and Raina, 1998; Rizzitello *et al.*, 2001) and improve the folding of unstable or aggregation-prone proteins in the periplasm (Missiakas *et al.*, 1996). Mutations in *surA* and *ppiD* are synthetically lethal (Dartigalongue and Raina, 1998). On the other hand, PpiA is dispensible, does not play a crucial role in folding events, and its absence or overproduction has little influence on the recovery yields of recombinant proteins expressed in the periplasm (Knappik *et al.*, 1993; Battistoni *et al.*, 1995; Kleerebezem *et al.*, 1995; Bothmann and Plückthun, 2000). FkpA, a homodimer that combines PPIase and chaperone activities (Ramm and Plückthun, 2000; Arie *et al.*, 2001), appears to have the broadest folding helper role among all periplasmic PPIase (Missiakas *et al.*, 1996; Bothmann and Plückthun, 2000).

Practical Implications

Skp and FkpA have recently emerged as the major folding modulators in the periplasm of *E. coli*. Although past performance is never a guarantee of future success, it is likely that increasing the concentration of these folding helpers either individually or in concert will be useful to enhance the solubility of other recombinant proteins targeted to the periplasmic space. It should however be kept in mind that folding and degradation are intimately linked. Thus, careful attention should be paid to the nature of the events responsible for low recovery yields. For instance, if the desired polypeptide must form multiple disulfide bridges to reach a native conformation, facilitating oxidation and disulfide bridge reshuffling using the strategies outlined in the following section may be more effective than foldase/PPIase co-expression. Similarly, if a heterologous protein exhibits slow folding kinetics, low yields of full-length product could be explained by the hydrolysis of folding intermediates that are typically more vulnerable to proteases than the native polypeptide. Under these circumstances, it may be useful to combine folding helper co-expression with the use of protease-deficient strains.

Disulfide Bond Formation in the Periplasm

E. coli proteins containing stable disulfide bonds are only found in the oxidizing environment of the cell envelope. This is due to the fact that the periplasm contains a set of enzymes (the Dsb proteins, Missiakas and Raina, 1997; Figure 1) which work together to catalyze the formation and isomerization of disulfide bridges (recently reviewed by Fabianek *et al.*, 2000). As a result, eukaryotic proteins with multiple disulfide bonds, such as htPA, have been successfully produced in the *E. coli* periplasm in an active form (Vlahos *et al.*, 1991).

DsbA and DsbB

DsbA is a soluble, monomeric 21-kDa protein containing a thioredoxin fold and the CXXC motif commonly found in cytoplasmic disulfide bond reductases. DsbA has a redox potential of –125 mV and is present in a completely oxidized state *in vivo* suggesting that its principal function is to act as a thiol oxidant. Deletion of *dsbA* has pleiotropic effects including defects in enterotoxin I secretion, inefficient folding of periplasmic copper and zinc superoxide dismutases, sensitivity to reduced DTT and benzylpencillin, and depressed levels of alkaline phosphatase, β-lactamase, OmpA and holocytochrome *c*. Additionally, cell motility is compromised and functional P-type, type I and type IV fimbriae are missing. In the case of P fimbriae assembly, DsbA not only acts as an oxidizing agent for disulfide bond formation, but also serves as a chaperone in the folding of the protein (Jacob-Dubuisson *et al.*, 1994). Other *in vivo* and *in vitro* studies have confirmed that DsbA exhibits chaperone-like activity (Zheng *et al.*, 1997; Sauvonnet and Pugsley, 1998).

DsbB is a cytoplasmic membrane protein that contains two periplasmic loops, each with a pair of cysteine residues. Its function is to maintain DsbA in an oxidized state. Although it was proposed that intramolecular transfer of a disulfide bond between the two cysteine pairs leads to DsbA oxidation (Kishigami *et al.*, 1995), a recent study raises the possibility that oxidation of DsbA by DsbB occurs directly via quinone reduction (Regeimbal and Bardwell, 2002). After it oxidizes DsbA, DsbB becomes reduced and must be reoxidized to regain functionality. This is accomplished by using quinones and nitrate or nitrite and fumarate as electron acceptors under aerobic and anaerobic conditions, respectively (Kobayashi *et al.*, 1997; Bader *et al.*, 1999). The phenotypes of *dsbB* mutants are very similar to those of *dsbA* cells and most likely stem from the fact that DsbA accumulates in a reduced form in *dsbB* mutants and is therefore unable to promote disulfide bond formation.

DsbC and DsbD

DsbC is a homodimeric protein composed of two 23.5-kDa subunits, each assuming a thioredoxin-like fold with a CXXC active site motif. It is in charge of the isomerization of incorrect disulfide bonds which can form in proteins containing more than two cysteine residues. In addition to functioning as a PDI, purified DsbC can promote the refolding of model proteins *in vitro*. The C-terminal domain appears to be important in DsbC chaperone-like activity (Chen *et al.*, 1999). *E. coli dsbC* mutants are impaired in disulfide bond formation, but the defects are not as pronounced as in *dsbA* and *dsbB* mutants, probably due to the fact that only few periplasmic proteins contain more than one disulfide bond per monomer and require a PDI for proper folding (Joly and Swartz, 1997). Overexpression of DsbC in *dsbA* and/or *dsbB* genetic backgrounds results in an almost complete restoration of the wild-type phenotype, indicating that DsbC is capable of acting as a thiol oxidant. Conversely, DsbA overproduction complements the lack of DsbC implying that DsbA can also act as a PDI, as has been observed *in vitro* (Fabianek, *et al.*, 2000).

For DsbC to be functional as a PDI, it must be maintained in a reduced form. DsbD, a cytoplasmic membrane protein with two large periplasmic domains, is in charge of DsbC reduction. The reductive function of DsbD depends on six conserved cysteine residues that are involved in a cascade of disulfide bond reduction steps in which electrons are passed from cytoplasmic thioredoxin to the active site cysteines of DsbC (Katzen and Beckwith, 2000). Deletion of the *dsbD* gene results in a partial phenotypic compensation of *dsbA* and *dsbB* mutants that possess an intact copy of the *dsbC* gene (Missiakas *et al.*, 1995; Rietsch *et al.*, 1996). This is most likely due to the fact that in *dsbD* cells, DsbC accumulates in an oxidized form and is able to catalyze disulfide bond formation.

Other Periplasmic Proteins Implicated in Disulfide Bond Formation

Three additional thiol-redox proteins, CcmG (aka DsbE), CcmH and DsbG have been identified in the periplasm of *E. coli* (Ritz and

Beckwith, 2001). The CcmGH proteins are involved in cytochrome *c* biogenesis, during which they are believed to be in charge of maintaining cysteine residues to which the cytochrome *c* heme lyase attaches the heme group in a reduced state (Grove *et al.*, 1996). CcmG shares homology with TrxA and is kept in a reduced state by DsbD. It appears to exhibit high substrate specificity (Fabianek *et al.*, 2000).

DsbG is a homodimeric PDI that shares 29% homology with DsbC, contains a reactive disulfide bond in its CXXC active site motif (Bessette *et al.*, 1999b), and is reduced by DsbD (van Straaten *et al.*, 1998). At present, the physiological function of DsbG remains unclear. DsbG has been shown to function as a chaperone in the oxidative refolding of ribonuclease, and, like DsbC, exhibits a general chaperone activity (Shao *et al.*, 2000). Lastly, a study by Raina and colleagues has raised a possibility of the existence of low-molecular-weight periplasmic redox agents which may be involved in disulfide bond formation in the periplasm (Dartigalongue *et al.*, 2000).

Practical Implications

Manipulation of Dsb proteins levels in the periplasm has proven a succesful route to improve the folding of heterologous proteins that must form a complex pattern of disulfide bonds to exhibit biologically activity (Qiu *et al.*, 1998; Kurokawa *et al.*, 2000; Kurokawa *et al.*, 2001). Coordinated expression of DsbC and its cognate reducing agent DsbD is desirable since it further increases the PDI activity of DsbC (Kurokawa *et al.*, 2001) whereas that of all DsbABCD enzymes can synergistically improve the transport and folding of proteins targeted to the periplasm (Kurokawa *et al.*, 2000). It is likely that further improvements will be achieved through co-expression of chaperones and foldases along with the desired protein.

Cell Envelope Proteases and Peptidases

The *E. coli* cell envelope contains a number of proteases and peptidases that process and degrade signal sequences from exported proteins and catabolize misfolded proteins that accumulate in the periplasm. In the absence of high-energy phosphates, periplasmic proteases have evolved to function in an ATP-independent manner. The following paragraphs summarize the major properties of known cell envelope proteases. More information can be found in excellent reviews by Miller and Gottesman (Gottesman, 1996; Miller, 1996).

Lep and Lsp

Lep and Lsp are signal (leader) peptidases whose role is to cleave signal peptides from precursor proteins. Even though these two proteases do not seem to be structurally related, both of them are integral membrane proteins with active sites located in the periplasm. Lsp is an 18-kDa protein that cleaves peptides from glyceride-modified prolipoproteins, while the 37-kDa Lep peptidase processes all other signal peptide-containing preproteins (Miller, 1996).

SppA and SohB

SppA (aka protease IV) is a homotetrameric serine protease of 67-kDa subunits that digests the cleaved signal peptide of the major lipoprotein (Ichihara *et al.*, 1984) and that of other preproteins (Palmer and St. John, 1987). *In vitro*, SppA hydrolyzes synthetic chromogenic endoprotease substrates with a preference for those containing hydrophobic amino acid residues (Pacaud, 1982). Cells lacking or overexpressing SppA do not exhibit a deleterious phenotype, and there is no excess signal peptides in *sppA* null mutants (Suzuki *et al.*, 1987). SppA shares substantial homology with SohB, a 37-kDa periplasmic protein capable of suppressing the temperature-sensitive phenotype of *degP* mutants (Baird *et al.*, 1991). As in the case of SppA, *sohB* null mutants do not exhibit a discernible phenotype (Suzuki *et al.*, 1987).

DegP

The serine protease DegP (aka HtrA or Do) is necessary for cell survival above 42°C, presumably because it degrades misfolded and unfolded periplasmic proteins. Natural DegP substrates include MalS, the colicin A lysis protein and the PapA and PapG pilus subunits (Cavard *et al.*, 1989; Spiess *et al.*, 1999; Jones *et al.*, 2002), but DegP can also processes recombinant proteins (Betton *et al.*, 1988; Baneyx and Georgiou, 1990). DegP appears to cleave its substrates at discrete locations, and exhibits a possible preference for paired hydrophobic residues (Jones *et al.*, 2002). Transcription of the *degP* gene is under the dual control of the extracytoplasmic σ^{24} sigma factor and the Cpx two-component pathway. DegP and the related proteases DegQ and DegS (see below) have similar structural features including an N-terminal segment believed to have regulatory function, a conserved trypsin-like protease domain, and one (DegS) or two (DegP and DegQ) PDZ domains (Clausen *et al.*, 2002). The recently solved crystal structure of the open and closed conformations of DegP has shed insights on its mechanism of action (Krojer *et al.*, 2002). The DegP hexamer is formed by the staggered association of trimeric rings and its proteolytic sites are located in a central cavity bounded by mobile side-walls formed by twelve PDZ domains. Motion of the PDZ domains controls the opening and closing of the particle and probably intial steps in substrate binding (Krojer *et al.*, 2002).

An interesting feature of DegP is its ability to switch functions between chaperone and protease in a temperature-dependent manner. Using MalS as a substrate, Spiess *et al* demonstrated that DegP improves the folding of this natural substrate at low temperature (28°C), while at elevated temperature (42°C), it degrades the same protein (Spiess *et al.*, 1999).

DegQ and DegS

DegQ (aka HhoA) and DegS (aka HhoB) are σ^{70}-regulated serine proteases that share sequence homology with DegP (Waller and Sauer, 1996) and were isolated as suppressors of the high temperature salt sensitivity phenotype exhibited by mutants lacking protease Prc (Bass *et al.*, 1996). Overexpression of DegQ, the closest relative of DegP, suppresses the temperature-sensitive phenotype of *degP* mutants and both proteases cleave denatured substrates at discrete V-X or I-X locations, where X is any amino acid (Kolmar *et al.*, 1996; Waller and Sauer, 1996). Whereas deletion of *degQ* has no obvious physiological consequences (Bass, *et al.*, 1996; Waller and Sauer, 1996), *degS* is an essential gene implicated in the regulation of σ^{24} activity (Alba *et al.*, 2001). DegS is an inner-membrane bound homotrimer that binds Y-X-F and Y-Q-F motifs exposed at the C-terminus of immature outer membrane porins via its PDZ domains, thereby acting as a sensor of periplasmic misfolding (Walsh *et al.*, 2003). Movement of PDZ domains is believed to unmask the protease active site leading to cleavage of RseA, a transmembrane protein that sequesters σ^{24} and prevents it from interacting with RNA polymerase. This ultimatelay leads to σ^{24} release and transcription of σ^{24}-dependent promoters including that of DegP (Walsh *et al.*, 2003).

Ptr

Ptr, also known as protease III, protease Pi and pitrilysin, is a zinc metalloprotease with high sequence specificity that is active against small peptides (10-30 amino acids in length). Although the majority of Ptr substrates seem to be protein fragments, this protease has been implicated in the *in vivo* breakdown of larger substrates (Betton *et al.*, 1988; Baneyx and Georgiou, 1991). *ptr* null mutants do not exhibit any detectable phenotypic alterations except for a slight reduction in growth rate (Baneyx and Georgiou, 1991).

Prc

Prc (<u>p</u>rocessing <u>C</u>-terminal) protease, also known as Tsp (<u>t</u>ail-<u>s</u>pecific) or Protease Re, is a serine protease that selectively degrades proteins with nonpolar C-terminal regions, as well as membrane proteins such as TonB (Gottesman, 1996; Oliver, 1996). Part of this 660 amino-acid-long protease, which contains a signal peptide for export into the periplasm, seems to be associated with the cytoplasmic membrane, but its active site faces the periplasmic space. Prc was first identified as a protease responsible for cleavage of the carboxy-terminal 11 amino acid residues of Pencillin Binding Protein 3 *in vivo* (Hara *et al.*, 1991). Prc is also capable of degrading casein, oxidized glutamine synthetase, fragments of λ cI repressor, and the B chain of human relaxin *in vitro* (Bass, *et al.*, 1996; Gottesman, 1996; Miller, 1996). It endoproteolytically cleaves its substrates at a limited number of peptide bonds with rather broad sequence specificity (Silber *et al.*, 1992; Keiler *et al.*, 1995). Strains lacking the *prc* gene are temperature-sensitive for growth in low-salt medium and leak periplasmic proteins, leading to a suggestion that Trc plays a role in the degradation of proteins damaged by osmotic shock or thermal stress (Hara *et al.*, 1991). Revertants that are no longer thermosensitive carry an extragenic suppressor in the *spr* gene (Hara *et al.*, 1996).

OmpT and OmpP

OmpT (aka protease VII) is a 317-residues long serine endoprotease that localizes in the outer membrane and is organized as a 10-stranded antiparallel β-barrel that protrudes into the extracellular space. The active site is located at the top of the vase-like structure and faces the growth milieu (Vandeputte-Rutten *et al.*, 2001). The levels of OmpT are temperature-regulated, with low expression at 30°C and high expression at 42°C and its activity is dependent on the presence of LPS (Kramer *et al.*, 2000). OmpT was originally suspected of processing a variety of proteins *in vivo*. However, it was later discovered that degradation was the result of the exposure of susceptible proteins to OmpT during cell disruption since this protease remains highly active under most lysis conditions and in the presence of denaturants (White *et al.*, 1995). OmpT exhibits specificity for paired basic residues and is inhibited by $ZnCl_2$ (Sugimura and Higashi, 1988; Sugimura and Nishihara, 1988). *E. coli* K-12 also contains a related protease, OmpP (aka OmpX), which shares 72% sequence identity with OmpT, and whose synthesis is also thermoregulated and controlled by cAMP (Kaufmann *et al.*, 1994). Members of the omptin family of proteases are involved in bacterial pathogenicity, ranging from bacterial defense and plasmin-mediated tissue infiltration to motility inside infected cells (Stathopoulos, 1998).

Other Periplasmic Proteases

Three poorly characterized periplasmic or membrane-associated proteases termed Mi, V and VI have been described. Protease Mi hydrolizes casein and globin but not small N-terminal fragments of β-galactosidase (Goldberg *et al.*, 1981). Protease V preferentially hydrolyzes N-blocked p-nitrophenyl esters of glycine, leucine, and valine and is also capable of degrading radiolabelled *E. coli* cytoplasmic proteins as well as casein. Finally, protease VI is capable of hydrolyzing membrane proteins but unlike protease V, it is unable to hydrolyze chromogenic ester substrates and is sensitive to an inhibitor of trypsine-like enzymes (Miller, 1996).

Practical Implications

The above paragraphs suggest that a few simple rules should be followed to minimize proteolysis of heterologous proteins synthesized in the periplasmic space of *E. coli*. First, *ompT* deficient K-12 strains (or B strains such as BL21 which do not encode *ompT* and *ompP*) should be used as production hosts in order to minimize degradation of the target protein during purification. Second, the desired gene product should have a relatively polar C-terminus to circumvent Prc-mediated hydrolysis. A YXF sequence at the C-terminus is particularly undesirable since this motif is recognized by DegS. Finally, since DegP appears to be the main housekeeping protease of the cell envelope, *degP*-deficient strains should be used for the production of proteolytically sensitive proteins in the periplasm. If problems persist, strains containing various combinations of deletions in the *ompT*, *degP*, *ptr* and *prc* gene are available (Baneyx and Georgiou, 1990; Baneyx and Georgiou, 1991; Meerman and Georgiou, 1994) and should be evaluated as expression hosts. However, since the growth rate decreases with the number of cell envelope protease mutations, the number of deletions should be kept to a minimum.

Concluding Remarks

With a long and proven history of industrial use and an unsurpassed degree of genetic and biochemical characterization, *E. coli* remains as attractive as ever for the expression of proteins of therapeutic or commercial interest. Over the past decade, research aimed at uncovering fundamental aspects of protein folding, export and degradation has been harnessed for practical applications. Unexpected findings (*e.g.* the possibility of producing disulfide-bonded proteins in the cell cytoplasm) have been exploited to improve the production of complex eukaryotic proteins. Future progress in understanding the function of hypothetical *E. coli* proteins and the interplay between protein folding pathways and folding/degradation decisions will likely suggest additional improvements in the area of heterologous protein expression. We anticipate that the availability of simple techniques to mutate any *E. coli* gene (Datsenko and Wanner, 2000) along with the possibility of evolving molecular chaperones dedicated to the folding of a desired substrate (Wang *et al.*, 2002) will play an increasingly important role in future *E. coli* expression technologies.

Acknowledgements

This work was supported by grants from the National Science Foundation (BES-0097430) and the American Cancer Society (MBC-99-335-01).

References

Aiyar, S.E., Gourse, R.L., and Ross, W. 1998. Upstream A-tracts increase bacterial promoter activity through interactions with the RNA polymerase a subunit. Proc. Natl. Acad. Sci. USA. 95: 14652-14657.

Akiyama, Y., Kihara, A., and Ito, K. 1996. Subunit a of proton ATPase Fo sector is a substrate of the FtsH protease in *Escherichia coli*. FEBS Lett. 399: 26-28.

Alba, B.M., Zhong, H.J., Pelayo, J.C., and Gross, C.A. 2001. *degS (hhoB)* is an essential *Escherichia coli* gene whose indispensable function is to provide sigma (E) activity. Mol. Microbiol. 40: 1323-33.

Allen, S.P., Polazzi, J.O., Gierse, J.K., and Easton, A.M. 1992. Two novel heat shock genes encoding proteins produced in response to heterologous protein expression in *Escherichia coli*. J. Bacteriol. 174: 6938-6947.

Arie, J.P., Sassoon, N., and Betton, J.M. 2001. Chaperone function of FkpA, a heat shock prolyl isomerase, in the periplasm of *Escherichia coli*. Mol. Microbiol. 39: 199-210.

Åslund, F., Ehn, B., Miranda-Vizuete, A., Pueyo, C., and Holmgren, A. 1994. Two additional glutaredoxins exist in *Escherichia coli*: glutaredoxin 3 is a hydrogen donor for ribonucleotide reductase in a thioredoxin/glutaredoxin 1 double mutant. Proc. Natl. Acad. Sci. USA. 91: 9813-9817.

Ayling, A., and Baneyx, F. 1996. Influence of the GroE molecular chaperone machine on the *in vitro* folding of *Escherichia coli* β-galactosidase. Protein Sci. 5: 478-487.

Bader, M., Muse, W., Ballou, D.P., Gassner, C., and Bardwell, J.C. 1999. Oxidative protein folding is driven by the electron transport system. Cell. 98: 217-227.

Baird, L., Lipinska, B., Raina, S., and Georgopoulos, C. 1991. Identification of the *Escherichia coli sohB* gene, a multicopy suppressor of the HtrA (DegP) null phenotype. J. Bacteriol. 173: 5763-5770.

Baneyx, F., and Georgiou, G. 1990. *In vivo* degradation of secreted fusion proteins by the *Escherichia coli* outer membrane protease OmpT. J. Bacteriol. 172: 491-494.

Baneyx, F., and Georgiou, G. 1991. Construction and characterization of *Escherichia coli* strains deficient in multiple secreted proteases: protease III degrades high-molecular-weight substrates *in vivo*. J. Bacteriol. 173: 2696-2703.

Baneyx, F., and Mujacic, M. 2003. Cold-inducible promoters for heterologous protein expression. Methods Mol. Biol. 205: 1-18.

Baneyx, F., and Palumbo, J.L. 2003. Improving heterologous protein folding via molecular chaperone and foldase co-expression. Methods Mol. Biol. 205: 171-197.

Barbirz, S., Jakob, U., and Glocker, M.O. 2000. Mass spectrometry unravels disulfide bond formation as the mechanism that activates a molecular chaperone. J. Biol. Chem. 275: 18759-18766.

Bardwell, J.C., and Craig, E.A. 1988. Ancient heat shock gene is dispensable. J. Bacteriol. 170: 2977-2983.

Barnhart, M.M., Pinkner, J.S., Soto, G.E., Sauer, F.G., Langermann, S., Waksman, G., Frieden, C., and Hultgren, S.J. 2000. PapD-like chaperones provide the missing information for folding of pilin proteins. Proc. Natl. Acad. Sci. USA. 97: 7709-7714.

Bass, S., Gu, Q., and Christen, A. 1996. Multicopy suppressors of *prc* mutant *Escherichia coli* include two HtrA (DegP) protease homologs (HhoAB), DksA, and a truncated R1pA. J. Bacteriol. 178: 1154-1161.

Battistoni, A., Mazzetti, A.P., Petruzzelli, R., Muramatsu, M., Federici, G., Ricci, G., and Lo Bello, M. 1995. Cytoplasmic and periplasmic production of human placental glutathione transferase in *Escherichia coli*. Protein Express. Purif. 6: 579-587.

Ben-Zvi, A.P., and Goloubinoff, P. 2001. Review: mechanisms of disaggregation and refolding of stable protein aggregates by molecular chaperones. J. Struct. Biol. 135: 84-93.

Berks, B.C., Sargent, F., and Palmer, T. 2000. The Tat protein export pathway. Mol. Microbiol. 35: 260-274.

Bernhardt, T.G., Roof, W.D., and Young, R. 2002. The *Escherichia coli* FKBP-type PPIase SlyD is required for the stabilization of the E lysis protein of bacteriophage phi X174. Mol. Microbiol. 45: 99-108.

Bessette, P.H., Åslund, F., Beckwith, J., and Georgiou, G. 1999a. Efficient folding of proteins with multiple disulfide bonds in the *Escherichia coli* cytoplasm. Proc. Natl. Acad. Sci. USA.. 96: 13703-13708.

Bessette, P.H., Cotto, J.J., Gilbert, H.F., and Georgiou, G. 1999b. *In vivo* and *in vitro* function of the *Escherichia coli* periplasmic cysteine oxidoreductase DsbG. J. Biol. Chem. 274: 7784-7792.

Betton, J.M., Sassoon, N., Hofnung, M., and Laurent, M. 1988. Degradation versus aggregation of misfolded maltose-binding protein in the periplasm of *Escherichia coli*. J. Biol. Chem. 273: 8897-8902.

Blaha, G., Wilson, D.N., Stoller, G., Fischer, G., Willumeit, R., and Nierhaus, K.H. 2003. Localization of the trigger factor binding site on the ribosomal 50S subunit. J. Mol. Biol. 326: 887-897.

Bochtler, M., Hartmann, C., Song, H.K., Bourenkov, G.P., Bartunik, H.D., and Huber, R. 2000. The structure of HslU and the ATP-dependent protease HslU-HslV. Nature. 403: 800-805.

Bolivar, F., Rodriguez, R.L., Greene, P.J., Betlach, M.C., Heyneker, H.L., and Boyer, H.W. 1977. Construction and characterization of new cloning vehicles. II. A multipurpose cloning system. Gene. 2: 95-113.

Borkovich, K.A., Farrelly, F.W., Finkelstein, D.B., Taulien, J., and Lindquist, S. 1989. Hsp82 is an essential protein that is required in higher concentrations for growth of cells at higher temperatures. Mol. Cell. Biol. 9: 3919-3930.

Bothmann, H., and Plückthun, A. 1998. Selection for a periplasmic factor improving phage display and functional periplasmic expression. Nat. Biotech. 16: 376-380.

Bothmann, H., and Plückthun, A. 2000. The periplasmic *Escherichia coli* peptidylprolyl *cis,trans*-isomerase FkpA. I. Increased functional expression of antibody fragments with and without *cis*-prolines. J. Biol. Chem. 275: 17100-17105.

Braig, K., Otwinowski, Z., Hedge, R., Boisvert, D.C., Joachimiak, A., Horwich, A.L., and Sigler, P.B. 1994. The crystal structure of the bacterial chaperonin GroEL at 2.8 Å. Nature. 371: 578-586.

Buchner, J. 1999. Hsp90 and Co. - a holding for folding. Trends Biochem. Sci. 24: 136-141.

Bukau, B., and Horwich, A.L. 1998. The Hsp70 and Hsp60 chaperone machines. Cell. 92: 351-366.

Cavard, D., Lazdunski, C., and Howard, S.P. 1989. The acylated precursor form of the colicin A lysis protein is a natural substrate of the DegP protease. J. Bacteriol. 171: 6316-6322.

Chang, A.C.Y., and Cohen, S.N. 1978. Construction and characterization of amplifiable multicopy DNA cloning vehicles derived from the P15A cryptic miniplasmid. J. Bacteriol. 134: 1141-1156.

Chang, S.Y., McGary, E.C., and Chang, S. 1989. Methionine aminopeptidase gene of *Escherichia coli* is essential for cell growth. J. Bacteriol. 171: 4071-4072.

Chen, J., Song, J.L., Zhang, S., Wang, Y., Cui, D.F., and Wang, C.C. 1999. Chaperone activity of DsbC. J. Biol. Chem. 274: 19601-19605.

Chen, R., and Henning, U. 1996. A periplasmic protein (Skp) of *Escherichia coli* selectively binds a class of outer membrane proteins. Mol. Microbiol. 19: 1287-1294.

Chou, C.H., Aristidou, A.A., Meng, S.-Y., Bennett, G.N., and San, K.-Y. 1995. Characterization of a pH-inducible promoter system for high-level expression of recombinant proteins in *Escherichia coli*. Biotechnol. Bioeng. 47: 186-192.

Chuang, S.-E., Burland, V., Plunkett III, G., Daniels, D.L., and Blattner, F.R. 1993. Sequence analysis of four new heat-shock genes constituting the *hslTS/ibpAB* and *hslVU* operons in *Escherichia coli*. Gene. 134: 1-6.

Clausen, T., Southan, C., and Ehrmann, M. 2002. The HtrA family of proteases: implications for protein composition and cell fate. Mol. Cell. 10: 443-455.

Cognet, I., Benoit de Coignac, A., Magistrelli, G., Jeannin, P., Aubry, J.-P., Maisnier-Patin, K., Caron, G., Chevalier, S., Humbert, F., Nguyen, T., Beck, A., Velin, D., Delneste, Y., Malissard, M., and Gauchat, J.-F. 2002. Expression of recombinant proteins in a lipid A mutant of *Escherichia coli* BL21 with a strongly reduced capacity to induce dendritic cell activation and maturation. J. Immunol. Methods. 272: 199-210.

Conlin, C.A., and Miller, C.G. 2000. *opdA*, a *Salmonella enterica* serovar typhimurium gene encoding a protease, is part of an operon regulated by heat shock. J. Bacteriol. 182: 518-521.

Cooper, K.W., and Baneyx, F. 2001. *Escherichia coli* FtsH (HflB) degrades a membrane-associated TolAI-II-β-lactamase fusion protein under highly denaturing conditions. Protein Express. Purif. 21: 323-332.

Coyle, J.E., Jaeger, J., Gross, M., Robinson, C.V., and Radford, S.E. 1997. Structural and mechanistic consequences of polypeptide binding by GroEL. Fold. Des. 2: 93-104.

Cristobal, S., de Gier, J.W., Nielsen, H., and von Heijne, G. 1999. Competition between Sec- and TAT-dependent protein translocation in *Escherichia coli*. EMBO J. 18: 2982-2990.

Dalbey, R.E., and Robinson, C. 1999. Protein translocation into and across the bacterial plasma membrane and the plant thylakoid membrane. Trends Biochem. Sci. 24: 17-22.

Dartigalongue, C., Missiakas, D., and Raina, S. 2001. Characterization of the *Escherichia coli* sigma E regulon. J. Biol. Chem. 276: 20866-20875.

Dartigalongue, C., Nikaido, H., and Raina, S. 2000. Protein folding in the periplasm in the absence of primary oxidant DsbA: modulation of redox potential in periplasmic space via OmpL porin. EMBO J. 19: 5980-5988.

Dartigalongue, C., and Raina, S. 1998. A new heat-shock gene, *ppiD*, encodes a peptidyl-prolyl isomerase required for folding of outer membrane proteins in *Escherichia coli*. EMBO J. 17: 3968-3980.

Datsenko, K.A., and Wanner, B.L. 2000. One-step inactivation of chromosomal gene in *Escherichia coli* K-12 using PCR products. Proc. Natl. Acad. Sci. USA. 97: 6640-6645.

DeLisa, M.P., Samuelson, P., Palmer, T., and Georgiou, G. 2002. Genetic analysis of the twin arginine translocator secretion pathway in bacteria. J. Biol. Chem. 277: 29825-29831.

DeLisa, M.P., Tullman, D., and Georgiou, G. 2003. Folding quality control in the export of proteins by the bacterial twin-arginine translocation pathway. Proc. Natl. Acad. Sci. USA 100: 6115-6120.

Derman, A.I., and Beckwith, J. 1995. *Escherichia coli* alkaline phosphatase localized to the cytoplasm acquires enzymatic activity in cells whose growth has been suspended: a caution for gene fusion studies. J. Bacteriol. 177: 3764-3770.

Derman, A.I., Prinz, W.A., Belin, D., and Beckwith, J. 1993. Mutations that allow disulfide bond formation in the cytoplasm of *Escherichia coli*. Science. 262: 1744-1747.

Deuerling, E., Patzelt, H., Vorderwulbecke, S., Rauch, T., Kramer, G., Schaffitzel, E., Mogk, A., Schulze-Specking, A., Langen, A., and Bukau, B. 2003. Trigger factor and DnaK possess overlapping substrate pools and binding specificities. Mol. Microbiol. 47: 1317-1328.

Deuerling, E., Schulze-Specking, A., Tomoyasu, T., Mogk, A., and Bukau, B. 1999. Trigger factor and DnaK cooperate in folding of newly synthesized proteins. Nature. 400: 693-696.

Dougan, D.A., Reid, B.G., Horwich, A.L., and Bukau, B. 2002. ClpS, a substrate modulator of the ClpAP machine. Mol. Cell 9: 673-683.

Driessen, A.J.M., Manting, E.H., and van der Does, C. 2001. The structural basis of protein targeting and translocation in bacteria. Nat. Struct. Biol. 8: 492-498.

Fabianek, R.A., Hennecke, H., and Thony-Meyer, L. 2000. Periplasmic protein thiol:disulfide oxidoreductases of *Escherichia coli*. FEMS Microbiol. Rev. 24: 303-316.

Fenton, A., Kashi, Y., Furtak, K., and Horwich, A.L. 1994. Residues in chaperonin GroEL required for polypeptide binding and release. Nature. 371: 614-619.

Giladi, H., Goldenberg, D., Koby, S., and Oppenheim, A.B. 1995. Enhanced activity of the bacteriophage lambda P_L promoter at low temperature. Proc. Natl. Acad. Sci. USA. 92: 2184-2188.

Goldberg, A.L., Moerschell, R.P., Chung, C.H., and Maurizi, M.R. 1994. ATP-dependent protease La (Lon) from *Escherichia coli*. Methods Enzymol. 244: 350-375.

Goldberg, A.L., Swamy, K.H., Chung, C.H., and Larimore, F.S. 1981. Proteases in *Escherichia coli*. Methods Enzymol. 80: 680-702.

Goldstein, M.A., and Doi, R.H. 1995. Prokaryotic promoters in biotechnology. Biotechnol. Annu. Rev. 1: 105-128.

Goloubinoff, P., Mogk, A., Ben Zvi, A.P., Tomoyasu, T., and Bukau, B. 1999. Sequential mechanism of solubilization and refolding of stable protein aggregates by a bichaperone network. Proc. Natl. Acad. Sci. USA. 96: 13732-13737.

Gottesman, M.E., and Hendrickson, W.A. 2000. Protein folding and unfolding by *Escherichia coli* chaperones and chaperonins. Curr. Opin. Microbiol. 3: 197-202.

Gottesman, S. 1996. Proteases and their targets in *Escherichia coli*. Annu. Rev. Genet. 30: 465-506.

Gottesman, S., Maurizi, M.R., and Wickner, S. 1997. Regulatory subunits of energy-dependent proteases. Cell. 91: 435-438.

Gottesman, S., Roche, E., Zhou, Y.N., and Sauer, R.T. 1998. The ClpXP and ClpAP proteases degrade proteins with carboxyl-terminal peptide tails added by the SsrA-tagging system. Genes Dev. 12: 1338-1347.

Graumann, J., Lilie, H., Tang, X., Tucker, K.A., Hoffmann, J.H., Vijayalakshmi, J., Saper, M., Bardwell, J.C., and Jakob, U. 2001. Activation of the redox-regulated molecular chaperone Hsp33—a two-step mechanism. Structure. 9: 377-87.

Gross, C.A. 1996. Function and regulation of the heat shock proteins. In: *Escherichia coli* and *Salmonella* Cellular and Molecular Biology. F.C. Neidhardt, ed. ASM Press, Washington, D. C. p. 1382-1399.

Grove, J., Tanapongpipat, S., Thomas, G., Griffiths, L., Crooke, H., and Cole, J. 1996. *Escherichia coli* K-12 genes essential for the synthesis of c-type cytochromes and a third nitrate reductase located in the periplasm. Mol. Microbiol. 19: 467-481.

Guo, F., Maurizi, M.R., Esser, M., and Xia, D. 2002a. Crystal structure of ClpA, an Hsp100 chaperone and regulator of ClpAP protease. J. Biol. Chem. 277: 46743-46752.

Guo, F., Esser, L., Singh, S.K., Maurizi, M.R., and Xia, D. 2002b. Crystal structure of the heterodimeric complex of the adaptor, ClpS, with the N-domain of the AAA+ chaperone, ClpA. J. Biol. Chem. 277: 46753-46762.

Hara, H., Yamamoto, Y., Higashitani, A., Suzuki, H., and Nishimura, Y. 1991. Cloning, mapping, and characterization of the *Escherichia coli prc* gene, which is involved in C-terminal processing of penicillin-binding protein 3. J. Bacteriol. 173: 4799-4813.

Hara, H., Abe, N., Nakakouji, M., Nishimura, Y., and Hiriuchi, K. 1996. Overproduction of penicillin-binding protein 7 suppresses thermosensitive growth defect at low osmolarity due to an *spr* mutation of *Escherichia coli*. Microbiol. Drug Res. 2: 63-72.

Hartl, F.U., and Hayer-Hartl, M. 2002. Molecular chaperones in the cytosol: from nascent chain to folded protein. Science. 295: 1852-1858.

Hayhurst, A., and Harris, W.J. 1999. *Escherichia coli* Skp chaperone coexpression improves solubility and phage display of single-chain antibody fragments. Protein Express. Purif. 15: 336-343.

Hengge-Aronis, R. 1996. Regulation of gene expression during entry into stationary phase. In: *Escherichia coli* and *Salmonella* Cellular and Molecular Biology. F.C. Neidhardt, ed. ASM Press, Washington, D. C. p. 1497-1512.

Herman, C., Thévenet, D., Bouloc, P., Walker, G.C., and D'Ari, R. 1998. Degradation of carboxy-terminal-tagged cytoplasmic proteins by the *Escherichia coli* protease HflB (FtsH). Genes Dev. 12: 1348-1355.

Herman, C., Thévenet, D., D'Ari, R., and Bouloc, P. 1995. Degradation of σ^{32}, the heat shock regulator in *Escherichia coli*, is governed by HflB. Proc. Natl. Acad. Sci. USA. 92: 3516-3520.

Holmgren, A. 1976. Hydrogen donor system for *Escherichia coli* ribonucleoside-diphosphate reductase dependent upon glutathione. Proc. Natl. Acad. Sci. USA. 73: 2275-2279.

Holmgren, A. 1985. Thioredoxin. Annu. Rev. Biochem. 54: 237-271.

Hottenrott, S., Schumann, T., Pluckthun, A., Fischer, G., and Rahfeld, J.U. 1997. The *Escherichia coli* SlyD is a metal ion-regulated peptidyl-prolyl cis/trans-isomerase. J. Biol. Chem. 272: 15697-15701.

Houry, W.A., Frishman, D., Eckerskorn, C., Lottspeich, F., and Hartl, F.U. 1999. Identification of *in vivo* substrates of the chaperonin GroEL. Nature. 402: 147-154.

Hultgren, S.J., Normark, S., and Abraham, S.N. 1991. Chaperone-assisted assembly and molecular architecture of adhesive pili. Annu. Rev. Microbiol. 45: 383-415.

Hunt, J.F., Weaver, A.J., Landry, S.J., Gierasch, L., and Deisenhofer, J. 1996. The crystal structure of the GroES co-chaperonin at 2.8 Å resolution. Nature. 379: 37-45.

Ichihara, S., Beppu, N., and Mizushima, S. 1984. Protease IV, a cytoplasmic membrane protein of *Escherichia coli*, has signal peptide peptidase activity. J. Biol. Chem. 259: 9853-9857.

Ingraham, J.L., and Marr, A.G. 1996. Effect of temperature, pressure, pH and osmotic stress on growth. In: *Escherichia coli* and *Salmonella* Cellular and Molecular Biology. F.C. Neidhardt, ed. ASM Press, Washington, D. C. p. 1570-1578.

Ishikawa, T., Beuron, F., Kessel, M., Wickner, S., Maurizi, M.R., and Steven, A.C. 2001 Translocation pathway of protein substrates in ClpAP protease. Proc. Natl. Acad. Sci. USA 98: 4328-4333.

Ize, B., Gerard, F., Zhang, M., Chanal, A., Voulhoux, R., Palmer, T., Filloux, A., and Wu, L.F. 2002. *In vivo* dissection of the Tat translocation pathway in *Escherichia coli*. J. Mol. Biol. 317: 327-335.

Jacob-Dubuisson, F., Pinkner, J., Xu, Z., Striker, R., Padmanhaban, A., and Hultgren, S.J. 1994. PapD chaperone function in pilus biogenesis depends on oxidant and chaperone-like activities of DsbA. Proc. Natl. Acad. Sci. USA. 91: 11552-11556.

Jakob, U., Muse, W., Eser, M., and Bardwell, J.C.A. 1999. Chaperone activity with a redox switch. Cell. 96: 341-352.

Jensen, R.B., and Gerdes, K. 1995. Programmed cell death in bacteria: proteic plasmid stabilization systems. Mol. Microbiol. 17: 205-210.

Joly, J.C., and Swartz, J.R. 1997. *In vitro* and *in vivo* redox states of the *Escherichia coli* periplasmic oxidoreductases DsbA and DsbC. Biochemistry. 36: 10067-10072.

Jones, C.H., Dexter, P., Evans, A.K., Liu, C., Hultgren, S.J., and Hruby, D.E. 2002. *Escherichia coli* DegP protease cleaves between paired hydrophobic residues in a natural substrate: the PapA pilin. J. Bacteriol. 184: 5762-5771.

Jordan, A., Åslund, F., Pontis, E., Reichard, P., and Holmgren, A. 1997. Characterization of *Escherichia coli* NrdH. A glutaredoxin-like protein with a thioredoxin-like activity profile. J. Biol. Chem. 272: 18044-18050.

Jurado, P., Ritz, D., Beckwith, J., de Lorenzo, V., and Fernandez, L.A. 2002. Production of functional single-chain Fv antibodies in the cytoplasm of *Escherichia coli*. J. Mol. Biol 320: 1-10.

Kanatani, A., Masuda, T., Shimoda, T., Misoka, F., Lin, X.S., Yoshimoto, T., and Tsuru, D. 1991. Protease II from *Escherichia coli*: sequencing and expression of the enzyme gene and characterization of the expressed enzyme. J. Biochem. (Tokyo). 110: 315-320.

Karata, K., Verma, C.S., Wilkinson, A.J., and Ogura, T. 2001. Probing the mechanism of ATP hydrolysis and substrate translocation in the AAA protease FtsH by modelling and mutagenesis. Mol. Microbiol. 39: 890-903.

Karzai, A.W., and McMacken, R. 1996. A bipartite signaling mechanism involved in DnaJ-mediated activation of the *Escherichia coli* DnaK protein. J. Biol. Chem. 271: 11236-11246.

Kato, M., Irisawa, T., Ohtani, M., and Muramatu, M. 1992. Purification and characterization of protease In, a trypsin-like proteinase, in *Escherichia coli*. Eur J. Biochem. 210: 1007-1014.

Katzen, F., and Beckwith, J. 2000. Transmembrane electron transfer by the membrane protein DsbD occurs via a disulfide bond cascade. Cell. 103: 769-779.

Kaufmann, A., Stierhof, Y.D., and Henning, U. 1994. New outer membrane-associated protease of *Escherichia coli* K-12. J. Bacteriol. 176: 359-367.

Keiler, K.C., Silber, K.R., Downard, K.M., Papayannopoulos, I.A., Biemann, K., and Sauer, R.T. 1995. C-terminal specific protein degradation: activity and substrate specificity of the Tsp protease. Protein Sci. 4: 1507-1515.

Kihara, A., Akiyama, Y., and Ito, K. 1995. FtsH is required for proteolytic elimination of uncomplexed forms of SecY, an essential protein translocase subunit. Proc. Natl. Acad. Sci. USA. 92: 4532-4536.

Kihara, A., Akiyama, Y., and Ito, K. 1999. Dislocation of membrane proteins in FtsH-mediated proteolysis. EMBO J. 18: 2970-2981.

Kim, S.J., Jeong, D.G., Chi, S.W., Lee, J.S., and Ryu, S.E. 2001a. Crystal structure of proteolytic fragments of the redox-sensitive Hsp33 with constitutive chaperone activity. Nat. Struct. Biol. 8: 459-466.

Kim, Y.-I., Levchenko, I., Fraczkowska, K., Woodruff, R.V., Sauer, R.T., and Baker, T.A. 2001b. Molecular determinants of complex formation between Clp/Hsp100 ATPases and the ClpP peptidase. Nat. Struct. Biol. 8: 230-233.

Kishigami, S., Kanaya, E., Kikuchi, M., and Ito, K. 1995. DsbA-DsbB interaction through their active site cysteines. Evidence from an odd cysteine mutant of DsbA. J. Biol. Chem. 270: 17072-17074.

Kitagawa, M., Matsumara, Y., and Tsuchido, T. 2000. Small heat shock proteins, IbpA and IbpB, are involved in resistances to heat and superoxide stresses in *Escherichia coli*. FEMS Microbiol. Lett. 184: 165-171.

Kitagawa, M., Miyakawa, M., Matsumara, Y., and Tsuchido, T. 2002. *Escherichia coli* small heat shock proteins, IbpA and IbpB, protect enzymes from inactivation by heat and oxidants. Eur. J. Biochem. 269: 2907-2917.

Kleerebezem, M., Heutink, M., and Tommassen, J. 1995. Characterization of an *Escherichia coli rotA* mutant, affected in periplasmic peptidyl-prolyl *cis/trans* isomerase. Mol. Microbiol. 18: 313-320.

Knappik, A., Krebber, C., and Plückthun, A. 1993. The effect of folding catalysts on the *in vivo* folding process of different antibody fragments expressed in *Escherichia coli*. Biotechnology (N.Y.). 11: 77-83.

Kobayashi, T., Kishigami, S., Sone, M., Inokuchi, H., Mogi, T., and Ito, K. 1997. Respiratory chain is required to maintain oxidized states of the DsbA-DsbB disulfide bond formation system in aerobically growing *Escherichia coli cells*. Proc. Natl. Acad. Sci. USA. 94: 11857-11862.

Kolmar, H., Waller, P.R., and Sauer, R.T. 1996. The DegP and DegQ periplasmic endoproteases of *Escherichia coli*: specificity for cleavage sites and substrate conformation. J. Bacteriol. 178: 5925-5929.

Kramer, G., Rauch, T., Rist, W., Vordewulbecke, S., Patzelt, H., Schulze-Specking, A., Ban, N., Deuerling, E., and Bukau, B. 2002. L23 functions as a chaperone docking site on the ribosome. Nature. 419: 171-174.

Kramer, R.A., Zandwijken, D., Egmond, M.R., and Dekker, N. 2000. *In vitro* folding, purification and characterization of *Escherichia coli* outer membrane protease OmpT. Eur. J. Biochem. 267: 885-893.

Krojer, T., Garrido-Franco, M., Huber, R., Ehrmann, M., and Clausen, T. 2002. Crystal structure of DegP (HtrA) reveals a new protease-chaperone machine. Nature. 416: 455-459.

Kuczynska-Wisnik, D., Kedzierska, S., Matuszewska, E., Lund, P., Taylor, A., Lipinska, B., and Laskowska, E. 2002. The *Escherichia coli* small heat-shock proteins IbpA and IbpB prevent the aggregation of endogenous proteins denatured *in vivo* during extreme heat shock. Microbiology. 148: 1757-1765.

Kurokawa, Y., Yanagi, H., and Yura, T. 2000. Overexpression of protein disulfide isomerase DsbC stabilizes multiple-disulfide-bonded recombinant protein produced and transported to the periplasm in *Escherichia coli*. Appl. Environ. Microbiol. 66: 3960-3965.

Kurokawa, Y., Yanagi, H., and Yura, T. 2001. Overproduction of bacterial protein disulfide isomerase (DsbC) and its modulator (DsbD) markedly enhances periplasmic production of human nerve growth factor in *Escherichia coli*. J. Biol. Chem. 276: 14393-14399.

Lazar, S.W., and Kolter, R. 1996. SurA assists the folding of *Escherichia coli* outer membrane proteins. J. Bacteriol. 178: 1770-1773.

Lee, C.S., Hahm, J.K., Hwang, B.J., Park, K.C., Ha, D.B., and Chung, C.H. 1990. Processing of Ada protein by two serine endoproteinases Do and So. FEBS Lett. 262: 310-312.

Lee, Y.S., Park, S.C., Goldberg, A.L., and Chung, C.H. 1988. Protease So from *Escherichia coli* preferentially degrades oxidatively damaged glutamine synthetase. J. Biol. Chem. 263: 6643-6646.

Lerner, C.G., and Inouye, M. 1990. Low copy number plasmids for regulated low-level expression of cloned genes in *Escherichia coli* with blue/white insert screening capability. Nucleic Acids Res. 18: 4631.

Levy, R., Weiss, R., Chen, G., Iverson, B.L., and Georgiou, G. 2001. Production of correctly folded Fab antibody fragments in the cytoplasm of *Escherichia coli trxB gor* mutants via the coexpression of molecular chaperones. Protein Express. Purif. 23: 338-347.

Makrides, S.C. 1996. Strategies for achieving high-level expression of genes in *Escherichia coli*. Mircobiol. Rev. 60: 512-538.

Malke, H., and Ferretti, J.J. 1985. Expression in *Escherichia coli* of streptococcal plasmid-determined erythromycin resistance directed by the *cat* gene promoter of pACYC184. J. Basic. Microbiol. 25: 393-400.

Malki, A., Kern, R., Abdallah, J., and Richarme, G. 2003. Characterization of the *Escherichia coli* YedU protein as a molecular chaperone. Biochem. Biophys. Res. Commun. 301: 430-436.

Manting, E.H., and Driessen, A.J.M. 2000. *Escherichia coli* translocase: the unraveling of a molecular machine. Mol. Microbiol. 37: 226-238.

McNeff, C., Zhao, Q., Almlof, E., Flickinger, M., and Carr, P.W. 1999. The efficient removal of endotoxins from insulin using quaternized polyethyleneimine-coated porous zirconia. Anal. Biochem. 274: 181-187.

Meerman, H.J., and Georgiou, G. 1994. Construction and characterization of a set of *E. coli* strains deficient in all known loci affecting the proteolytic stability of secreted recombinant proteins. Biotechnology (N.Y.). 12: 1107-1110.

Miller, C.G. 1996. Protein degradation and proteolytic modification. In: *Escherichia coli* and *Salmonella* Cellular and Molecular Biology. F.C. Neidhardt, ed. ASM Press, Washington, D. C. p. 938-954.

Missiakas, D., Betton, J.M., and Raina, S. 1996. New components of protein folding in extracytoplasmic compartments of *Escherichia coli* SurA, FkpA and Skp/OmpH. Mol. Microbiol. 21: 871-884.

Missiakas, D., and Raina, S. 1997. Protein folding in the bacterial periplasm. J. Bacteriol. 179: 2465-2471.

Missiakas, D., Schwager, F., Betton, J.M., Georgopoulos, C., and Raina, S. 1996. Identification and characterization of HslV HslU (ClpQ ClpY) proteins involved in overall proteolysis of misfolded proteins in *Escherichia coli*. EMBO J. 15: 6899-6909.

Missiakas, D., Schwager, F., and Raina, S.f. 1995. Identification and characterization of a new disulfide isomerase-like protein (DsbD) in *Escherichia coli*. EMBO J. 14: 3415-3424.

Miyake, T., Oka, T., Nishizawa, T., Misoka, F., Fuwa, T., Yoda, K., Yamasaki, M., and Tamura, G. 1985. Secretion of human interferon-alpha induced by using secretion vectors containing a promoter and signal sequence of alkaline phosphatase gene of *Escherichia coli*. J Biochem (Tokyo). 97: 1429-1436.

Mogk, A., Tomoyasu, T., Goloubinoff, P., Rüdiger, S., Röder, D., Langen, A., and Bukau, B. 1999. Identification of thermolabile *Escherichia coli* proteins: prevention of aggregation by DnaK and ClpB. EMBO J. 18: 6934-6949.

Mogk, A., Schlieker, C., Friedrich, K. L., Schönfeld, H.-J., Vierling, E., and Bukau, B. 2003. Refolding of substrates bound to small Hsps relies on a disaggregation reaction mediated most efficiently by ClpB/DnaK. J. Biol. Chem. 278: 31033-31042.

Mujacic, M., Cooper, K.W., and Baneyx, F. 1999. Cold-inducible cloning vectors for low-temperature protein expression in *Escherichia coli*: application to the production of a toxic and proteolytically sensitive fusion protein. Gene. 238: 325-332.

Murphy, C.K., and Beckwith, J. 1996. Export of proteins to the cell envelope in *Escherichia coli*. In: *Escherichia coli* and *Salmonella* Cellular and Molecular Biology. F.C. Neidhardt, ed. ASM Press, Washington, D. C. p. 967-978.

Naberhaus, F. 2002. Alpha-crystallin-type heat shock proteins: socializing minichaperones in the context of a multichaperone network. Microbiol. Mol. Biol. Rev. 66: 64-93.

Nemoto, T.K., Ono, T., and Tanaka, K. 2001. Substrate-binding characteristics of proteins in the 90 kDa heat shock protein family. Biochem. J. 354: 663-670.

Nishihara, K., Kanemori, M., Kitagawa, M., Yanaga, H., and Yura, T. 1998. Chaperone coexpression plasmids: differential and synergistic roles of DnaK-DnaJ-GrpE and GroEL-GroES in assisting folding of an allergen of Japanese cedar pollen, Cryj2, in *Escherichia coli*. Appl. Environ. Microbiol. 64: 1694-1699.

Nishihara, K., Kanemori, M., Yanagi, H., and Yura, T. 2000. Overexpression of trigger factor prevents aggregation of recombinant proteins in *Escherichia coli*. Appl. Environ. Microbiol. 66: 884-889.

Oliver, D.B. 1996. Periplasm. In: *Escherichia coli* and *Salmonella* Cellular and Molecular Biology. F.C. Neidhardt, ed. ASM Press, Washington, D. C. p. 88-103.

Olson, P., Zhang, Y., Olsen, D., Owens, A., Cohen, P., Nguyen, K., Ye, J.-J., Bass, S., and Mascarenhas, D. 1998. High-level expression of eukaryotic polypeptides from bacterial chromosomes. Protein Express. Purif. 14: 160-166.

Pacaud, M. 1976. Purification of protease II from *Escherichia coli* by affinity chromatography and separation of two enzyme species from cells harvested at late log phase. Eur. J. Biochem. 64: 199-204.

Pacaud, M. 1982. Purification and characterization of two novel proteolytic enzymes in membranes of *Escherichia coli*. Protease IV and protease V. J. Biol. Chem. 257: 4333-9.

Palmer, S.M., and St. John, A.C. 1987. Characterization of a membrane-associated serine protease in *Escherichia coli*. J. Bacteriol. 169: 1474-1479.

Park, S.K., Kim, K.I., Woo, K.M., Seol, J.H., Tanaka, K., Ichihara, A., Ha, D.B., and Chung, C.H. 1993. Site-directed mutagenesis of the dual translational initiation sites of the *clpB* gene of *Escherichia coli* and characterization of its gene products. J. Biol. Chem. 268: 20170-20174.

Patzelt, H., Rudiger, S., Brehmer, D., Kramer, G., Vorderwulbecke, S., Schaffitzel, E., Waitz, A., Hesterkamp, T., Dong, L., Schneider-Mergener, J., Bukau, B., and Deuerling, E. 2001. Binding specificity of *Escherichia coli* trigger factor. Proc. Natl. Acad. Sci. USA. 98: 14244-14249.

Peredelchuk, M.Y., and Bennett, G.N. 1997. A method for construction of *E. coli* strains with multiple DNA insertions in the chromosome. Gene. 187: 231-238.

Porankiewicz, J., Wang, J., and Clarke, A.K. 1999. New insights into the ATP-dependent Clp protease: *Escherichia coli* and beyond. Mol. Microbiol. 32: 449-458.

Prinz, W.A., Åslund, F., Holmgren, A., and Beckwith, J. 1997. The role of the thioredoxin and glutaredoxin pathways in reducing protein disulfide bonds in the *Escherichia coli* cytoplasm. J. Biol. Chem. 272: 15661-15667.

Qiu, J., Swartz, J.R., and Georgiou, G. 1998. Expression of active human tissue-type plasminogen activator in *Escherichia coli*. Appl. Environ. Microbiol. 64: 4891-4896.

Quigley, P.M., Korotkov, K., Baneyx, F., and Hol, W.G.J. 2003. The 1.6-Å crystal structure of the class of chaperones represented by

Escherichia coli Hsp31 reveals a putative catalytic triad. Proc. Natl. Acad. Sci. USA. 100: 3137-3142.

Ramm, K., and Plückthun, A. 2000. The periplasmic *Escherichia coli* peptidylprolyl cis,trans-isomerase FkpA. II. Isomerase-independent chaperone activity *in vitro*. J. Biol. Chem. 275: 17106-17113.

Record, M.T.J., Reznikoff, W.S., Craig, M.L., McQuade, K.L., and Schlax, P.J. 1996. *Escherichia coli* RNA Polymerase (Eσ^{70}), Promoters, and the Kinetics of the Steps of Transcription Initiation. In: *Escherichia coli* and *Salmonella* Cellular and Molecular Biology. F.C. Neidhardt, ed. ASM Press, Washington, D. C. p. 792-820.

Regeimbal, J., and Bardwell, J.C. 2002. DsbB catalyzes disulfide bond formation *de novo*. J. Biol. Chem. 277: 32706-32713.

Richmond, C.S., Glasner, J.D., Mau, R., Jin, H., and Blattner, F.R. 1999. Genome-wide expression profiling in *Escherichia coli* K-12. Nucleic Acids Res. 27: 3821-3835.

Rietsch, A., Belin, D., Martin, N., and Beckwith, J. 1996. An *in vivo* pathway for disulfide bond isomerization in *Escherichia coli*. Proc. Natl. Acad. Sci. USA. 93: 13048-13053.

Ritz, D., and Beckwith, J. 2001. Roles of thiol-redox pathways in bacteria. Annu. Rev. Microbiol. 55: 21-48.

Rizzitello, A.E., Harper, J.R., and Silhavy, T.J. 2001. Genetic evidence for parallel pathways of chaperone activity in the periplasm of *Escherichia coli*. J. Bacteriol. 183: 6794-6800.

Roman, L.J., Sheta, E.A., Martasek, P., Gross, S.S., Liu, Q., and Masters, B.S.S. 1995. High-level expression of functional rat neuronal nitric oxide synthase in *Escherichia coli*. Proc. Natl. Acad. Sci. USA. 92: 8428-8432.

Roof, W.D., Fang, H.Q., Young, K.D., Sun, J., and Young, R. 1997. Mutational analysis of *slyD*, an *Escherichia coli* gene encoding a protein of the FKBP immunophilin family. Mol. Microbiol. 25: 1031-1046.

Rosen, R., Biran, D., Gur, E., Becher, D., Hecker, M., and Ron, E.Z. 2002. Protein aggregation in *Escherichia coli*: role of proteases. FEMS Microbiol. Lett. 207: 9-12.

Roy, A.M., and Coleman, J. 1994. Mutations in *firA*, encoding the second acyltransferase in lipopolysaccharide biosynthesis, affect multiple steps in lipopolysaccharide biosynthesis. J. Bacteriol. 176: 1639-1646.

Rüdiger, S., Germeroth, L., Schneider-Mergener, J., and Bukau, B. 1997. Substrate specificity of the DnaK chaperone determined by screening cellulose-bound peptide libraries. EMBO J. 16: 1501-1507.

Rüdiger, S., Schneider-Mergener, J., and Bukau, B. 2001. Its substrate specificity characterizes the DnaJ co-chaperone as a scanning factor for the DnaK chaperone. EMBO J. 20: 1042-1050.

Russel, M., and Model, P. 1986. The role of thioredoxin in filamentous phage assembly. Construction, isolation, and characterization of mutant thioredoxins. J. Biol. Chem. 261: 14997-5005.

Santini, C.L., Bernadac, A., Zhang, M., Chanal, A., Ize, B., Blanco, C., and Wu, L.F. 2001. Translocation of jellyfish green fluorescent protein via the Tat system of *Escherichia coli* and change of its periplasmic localization in response to osmotic up-shock. J. Biol. Chem. 276: 8159-8164.

Sargent, F., Bogsch, E.G., Stanley, N.R., Wexler, M., Robinson, C., Berks, B.C., and Palmer, T. 1998. Overlapping functions of components of a bacterial Sec-independent protein export pathway. EMBO J. 17: 3640-3650.

Sargent, F., Stanley, N.R., Berks, B.C., and Palmer, T. 1999. Sec-independent protein translocation in *Escherichia coli*. A distinct and pivotal role for the TatB protein. J. Biol. Chem. 274: 36073-36082.

Sastry, M.S.R., Korotkov, K., Brodsky, Y., and Baneyx, F. 2002. Hsp31, the *Escherichia coli yedU* gene product, is a molecular chaperone whose activity is inhibited by ATP at high temperatures. J. Biol. Chem. 277: 46026-46034.

Sauer, F.G., Futterer, K., Pinkner, J.S., Dodson, K.W., Hultgren, S.J., and Waksman, G. 1999. Structural basis of chaperone function and pilus biogenesis. Science. 285: 1058-1061.

Sauer, F.G., Pinkner, J.S., Waksman, G., and Hultgren, S.J. 2002. Chaperone priming of pilus subunits facilitates a topological transition that drives fiber formation. Cell. 111: 543-551.

Sauvonnet, N., and Pugsley, A.P. 1998. The requirement for DsbA in pullulanase secretion is independent of disulphide bond formation in the enzyme. Mol. Microbiol. 27: 661-667.

Schneider, E.L., Thomas, J.G., Bassuk, J.A., Sage, E.H., and Baneyx, F. 1997. Manipulating the aggregation and oxidation of human SPARC in the cytoplasm of *Escherichia coli*. Nat. Biotechnol. 15: 581-585.

Scholz, C., Stoller, G., Zarnt, T., Fischer, G., and Schmid, F.X. 1997. Cooperation of enzymatic and chaperone functions of trigger factor in the catalysis of protein folding. EMBO J. 16: 54-58.

Settles, A.M., and Martienssen, R. 1998. Old and new pathways of protein export in chloroplasts and bacteria. Trends Cell Biol. 8: 494-501.

Shao, F., Bader, M.W., Jakob, U., and Bardwell, J.C. 2000. DsbG, a protein disulfide isomerase with chaperone activity. J. Biol. Chem. 275: 13349-13352.

Shearstone, J.R., and Baneyx, F. 1999. Biochemical characterization of the small heat shock protein IbpB from *Escherichia coli*. J. Biol. Chem. 274: 9937-9945.

Sherman, F., Stewart, J.W., and Tsunasawa, S. 1985. Methionine or not methionine at the begining of a protein. Bioessays. 3: 27-31.

Shi, J., Vlamis-Gardikas, A., Åslund, F., Holmgren, A., and Rosen, B.P. 1999. Reactivity of glutaredoxins 1, 2, and 3 from *Escherichia coli* shows that glutaredoxin 2 is the primary hydrogen donor to ArsC-catalyzed arsenate reduction. J. Biol. Chem. 274: 36039-36042.

Shirai, Y., Akiyama, Y., and Ito, K. 1996. Suppression of *ftsH* mutant phenotypes by overproduction of molecular chaperones. J. Bacteriol. 178: 1141-1145.

Shotland, Y., Koby, S., Teff, D., Mansur, N., Oren, D.A., Tatematsu, K., Tomoyasu, T., Kessel, M., Bukau, B., Ogura, T., and Oppenheim, A.B. 1997. Proteolysis of the phage l CII regulatory protein by FtsH (HflB) of *Escherichia coli*. Mol. Microbiol. 24: 1303-1310.

Silber, K.R., Keiler, K.C., and Sauer, R.T. 1992. Tsp: a tail-specific protease that selectively degrades proteins with nonpolar C-termini. Proc. Natl. Acad. Sci. USA 89: 295-299.

Singh, S.K., Rozycki, J., Ortega, J., Ishikawa, T., Lo, J., Steven, A.C., and Maurizi, M.R. 2001 Functional domains of the ClpA and ClpX molecular chaperones identified by limited proteolysis and deletion analysis. J. Biol. Chem. 276: 29420-29429.

Somerville, J.E.J., Cassiano, L., Bainbridge, B., Cunningham, M.D., and Darveau, R.P. 1996. A novel *Escherichia coli* lipid A mutant that produces an antiinflammatory lipopolysaccharide. J. Clin. Invest. 97: 359-365.

Spence, J., and Georgopoulos, C. 1989. Purification and properties of the *Escherichia coli* heat shock protein, HtpG. J. Biol. Chem. 264: 4398-4403.

Spiess, C., Beil, A., and Ehrmann, M. 1999. A temperature-dependent switch from chaperone to protease in a widely conserved heat shock protein. Cell. 97: 339-347.

Squires, C.L., Pedersen, S., Ross, B.M., and Squires, C. 1991. ClpB is the *Escherichia coli* heat shock protein F84.1. J. Bacteriol. 173: 4254-4262.

Stathopoulos, C. 1998. Structural features, physiological roles and biotechnological applications of the membrane proteases of the OmpT bacterial endopeptidase family: a micro-review. Membr. Cell Biol. 12: 1-8.

Stewart, E.J., Åslund, F., and Beckwith, J. 1998. Disulfide bond formation in the *Escherichia coli* cytoplasm: an *in vivo* role reversal for the thioredoxins. EMBO J. 17: 5543-5550.

Sugimura, K., and Higashi, N. 1988. A novel outer-membrane-associated protease in *Escherichia coli*. J. Bacteriol. 170: 3650-3654.

Sugimura, K., and Nishihara, T. 1988. Purification, characterization and primary structure of *Escherichia coli* protease VII with specificity for paired basic residues: identity of protease VII and OmpT. J. Bacteriol. 170: 5625-5632.

Suzuki, T., Itoh, A., Ichihara, S., and Mizushima, S. 1987. Characterization of the *sppA* gene coding for protease IV, a signal peptide peptidase of *Escherichia coli*. J. Bacteriol. 169: 2523-2528.

Swartz, J.R. 1996. *Escherichia coli* recombinant DNA technology. In: *Escherichia coli* and *Salmonella* Cellular and Molecular Biology. F.C. Neidhardt, ed. ASM Press, Washington, D. C. p. 1693-1711.

Tabor, S., Huber, H.E., and Richardson, C.C. 1987. *Escherichia coli* thioredoxin confers processivity on the DNA polymerase activity of the gene 5 protein of bacteriophage T7. J. Biol. Chem. 262: 16212-16223.

Tek, V., and Zolkiewski, M. 2002. Stability and interactions of the amino-terminal domain of ClpB from *Escherichia coli*. Protein Sci. 11: 1192-1198.

Teter, S.A., Houry, W.A., Ang, D., Tradier, T., Rockabrand, D., Fischer, G., Blum, P., Georgopoulos, C., and Hartl, F.U. 1999. Polypeptide flux through bacterial Hsp70: DnaK cooperates with trigger factor in chaperoning nascent chains. Cell. 97: 755-765.

Thirumalai, D., and Lorimer, G.H. 2001. Chaperonin-mediated protein folding. Annu. Rev. Biophys. Biomol. Struct. 30: 245-269.

Thomas, J.D., Daniel, R.A., Errington, J., and Robinson, C. 2001. Export of active green fluorescent protein to the periplasm by the twin-arginine translocase (Tat) pathway in *Escherichia coli*. Mol. Microbiol. 39: 47-53.

Thomas, J.G., Ayling, A., and Baneyx, F. 1997. Molecular chaperones, folding catalysts and the recovery of biologically active recombinant proteins from *E. coli*: to fold or to refold. Appl. Biochem. Biotechnol. 66: 197-238.

Thomas, J.G., and Baneyx, F. 1998. Roles of the *Escherichia coli* small heat shock proteins IbpA and IbpB in thermal stress management: comparison with ClpA, ClpB, and HtpG *in vivo*. J. Bacteriol. 180: 5165-5172.

Thomas, J.G., and Baneyx, F. 2000. ClpB and HtpG facilitate *de novo* protein folding in stressed *Escherichia coli* cells. Mol. Microbiol. 36: 1360-1370.

Tomoyasu, T., Gamer, J., Bukau, B., Kanemori, M., Mori, H., Rutman, A.J., Oppenheim, A.B., Yura, T., Yamanaka, K., Niki, H., Hiraga, S., and Ogura, T. 1995. *Escherichia coli* FtsH is a membrane-bound, ATP-dependent protease which degrades the heat-shock transcription factor s32. EMBO J. 14: 2551-2560.

Tomoyasu, T., Mogk, A., Langen, H., Goloubinoff, P., and Bukau, B. 2001. Genetic dissection of the roles of chaperones and proteases in protein folding and degradation in the *Escherichia coli* cytosol. Mol. Microbiol. 40: 397-413.

Trempy, J.E., Kirby, J.E., and Gottesman, S. 1994. Alp suppression of Lon: dependence on the *slpA* gene. J. Bacteriol. 176: 2061-2067.

Tuggle, C.K., and Fuchs, J.A. 1985. Glutathione reductase is not required for maintenance of reduced glutathione in *Escherichia coli* K-12. J. Bacteriol. 162: 448-450.

Ueguchi, C., and Ito, K. 1992. Multicopy suppression: an approach to understanding intracellular functioning of the protein export system. J. Bacteriol. 174: 1454-1461.

van der Wolk, J.P.W., Fekkes, P., Boorsma, A., Huie, J.L., Silhavy, T.J., and Driessen, A.J.M. 1998. *PrlA4* prevents the rejection of signal sequence defective preproteins by stabilizing the SecA-SecY interaction during the initiation of translocation. EMBO J. 17: 3631-3639.

van Straaten, M., Missiakas, D., Raina, S., and Darby, N.J. 1998. The functional properties of DsbG, a thiol-disulfide oxidoreductase from the periplasm of *Escherichia coli*. FEBS Lett. 428: 255-258.

Vandeputte-Rutten, L., Kramer, R.A., Kroon, J., Dekker, N., Egmond, M.R., and Gros, P. 2001. Crystal structure of the outer membrane protease OmpT from *Escherichia coli* suggests a novel catalytic site. EMBO J. 20: 5033-5039.

Vasina, J.A., Peterson, M.S., and Baneyx, F. 1998. Scale-up and optimization of the low-temperature inducible *cspA* promoter system. Biotechnol. Prog. 14: 714-721.

Veinger, L., Diamant, S., Buchner, J., and Goloubinoff, P. 1998. The small heat-shock protein IbpB from *Escherichia coli* stabilizes stress-denatured proteins for subsequent refolding by a multichaperone network. J. Biol. Chem. 273: 11032-11037.

Vijayalakshmi, J., Mukhergee, M.K., Graumann, J., Jakob, U., and Saper, M.A. 2001. The 2.2 A crystal structure of Hsp33: a heat shock protein with redox- regulated chaperone activity. Structure. 9: 367-375.

Vlahos, C.J., Wilhelm, O.G., Hassell, T., Jaskunas, S.R., and Bang, N.U. 1991. Disulfide pairing of the recombinant kringle-2 domain of tissue plasminogen activator produced in *Escherichia coli*. J. Biol. Chem. 266: 10070-10072.

Waller, P.R., and Sauer, R.T. 1996. Characterization of *degQ* and *degS*, *Escherichia coli* genes encoding homologs of the DegP protease. J. Bacteriol. 178: 1146-1153.

Walsh, N.P., Alba, B.M., Bose, B., Gross, C.A., and Sauer, R.T. 2003. OMP Peptide Signals Initiate the Envelope-Stress Response by Activating DegS Protease via Relief of Inhibition Mediated by Its PDZ Domain. Cell. 113: 61-71.

Wang, J.D., Herman, C., Tipton, K.A., Gross, C.A., and Weissman, J.S. 2002. Directed evolution of substrate-optimized GroEL/S chaperonins. Cell. 111: 1027-1039.

White, C.B., Chen, Q., Kenyon, G.L., and P.C., B. 1995. A novel activity of OmpT. Proteolysis under extreme denaturing conditions. J. Biol. Chem. 270: 12990-12994.

Wickner, A., Maurizi, M.R., and Gottesman, S. 1999. Posttranslational quality control: folding, refolding, and degrading proteins. Science. 286: 1888-1893.

Xu, F., Lin-Chao, S., and Cohen, S.N. 1993. The *Escherichia coli pcnB* gene promotes adenylation of antisense RNAI of ColE1-type plasmids *in vivo* and degradation of RNAI decay intermediates. Proc. Natl. Acad. Sci. USA. 90: 6756-6760.

Xu, F.F., Gaggero, C., and Cohen, S.N. 2002. Polyadenylation can regulate ColE1 type plasmid copy number independently of any effect on RNAI decay by decreasing the interaction of antisense RNAI with its RNAII target. Plasmid. 48: 49-58.

Yamada, M., Kubo, M., Miyake, T., Sakaguchi, R., Higo, Y., and Imanaka, T. 1991. Promoter sequence analysis in *Bacillus* and *Escherichia*: construction of strong promoters in *E. coli*. Gene. 99: 109-114.

Yamada, S., Ono, T., Mizuno, A., and Nemoto, T.K. 2003. A hydrophobic segment within the C-terminal domain is essential for both client-binding and dimer formation of the HSP90-family molecular chaperone. Eur. J. Biochem. 270: 146-154.

Yanisch-Perron, C., Vieira, J., and Messing, J. 1985. Improved M13 phage cloning vectors and host strains: nucleotide sequences of the M13mp18 and pUC19 vectors. Gene. 33: 103-119.

Yokota, N., Kuroda, T., Matsuyama, S., and Tokuda, H. 1999. Characterization of the LolA-LolB system as the general lipoprotein localization mechanism of *Escherichia coli*. J. Biol. Chem. 274: 30995-30999.

Zarnt, T., Tradler, T., Stoller, G., Scholz, C., Schmid, F.X., and Fischer, G. 1997. Modular structure of the trigger factor is required for high activity in protein folding. J. Mol. Biol. 271: 827-837.

Zeth, K., Ravelli, R.B., Paal, K., Cusack, S., Bukau, B., and Dougan, D.A. 2002. Structural analysis of the adaptor protein ClpS in complex with the N-terminal domain of ClpA. Nat. Struct. Biol. 9: 906-911.

Zheng, W.D., Quan, H., Song, J.L., Yang, S.L., and Wang, C.C. 1997. Does DsbA have chaperone-like activity? Arch. Biochem. Biophys. 337: 326-331.

Zhu, X., Zhao, X., Burkholder, W.F., Gragerov, A., Ogata, C.M., Gottesman, M.E., and Hendrickson, W.A. 1996. Structural analysis of substrate binding by the molecular chaperone DnaK. Nature. 272: 1606-1614.

Zolkiewski, M. 1999. ClpB cooperates with DnaK, DnaJ and GrpE in suppressing protein aggregation. J. Biol. Chem. 274: 28083-28086.

3

Tools for Metabolic Engineering in *Escherichia coli*

Christina D. Smolke, Vincent J.J. Martin, and Jay D. Keasling

Abstract

Metabolic engineering is the redirection of cellular metabolism for the production of valuable or novel chemicals or to remediate toxic chemicals in the environment. To genetically modify cellular metabolism, particular gene expression tools are required for the precise and balanced expression of multiple genes, so that flux through a heterologous pathway is not limited by any one enzyme and cellular resources are not wasted. These tools target many aspects of gene expression control and host development including stable maintenance of heterologous DNA in a host cell (chromosome engineering, low-copy number plasmids); stringent, homogeneous, and timed control over gene expression (promoter design); coordinated and differential production of multiple enzymes in a pathway (directed RNA processing and decay, translational control); techniques to generate genetic diversity (directed evolution strategies); and tools that enable the quantification of metabolism and biochemical pathways (genome-wide analytical tools, metabolic modeling). This chapter reviews the current tools available for metabolic engineering applications in *E. coli* and points to potential areas where further developments are needed.

Introduction

Organic chemistry has been the work horse in the traditional manufacturing of many of the industrial bulk and fine chemicals and pharmaceuticals. While much effort has been devoted to the development of chemical synthesis methods, some organic compounds, particularly those with multiple stereochemical centers, remain difficult to synthesize on an industrial scale. The significant progress made in understanding and manipulating enzymatic reactions and biochemical pathways over the past twenty years has made biological synthesis a viable option to chemical synthesis for some compounds. The potential advantages of using enzymes or enzyme cascades for synthesis of chemicals include higher yields, production of enantiomerically pure molecules, and less toxic waste byproducts. In performing these reactions inside the cell, one can take advantage of the cell's ability to replenish both enzymes and cofactors and to provide valuable precursors from inexpensive starting materials. This redirection of cellular metabolism for the production of useful or novel chemicals or to remediate toxic chemicals in the environment is known as metabolic engineering.

Metabolic engineering often requires the introduction and elimination of one or multiple enzymes to reroute the carbon to a biochemical pathway of interest. One goal of metabolic engineering is to optimize flux through a particular pathway. This requires very controlled and balanced expression of the individual enzymes of the pathway so that no single step limits product formation. In addition, tight control of gene expression minimizes over-expression that often results in wasted cellular energy and product precursors. The sub-optimal expression of genes in a pathway can lead to metabolic burden, accumulation of metabolic intermediates, and inefficient use of cellular resources. Consequently, the gene expression systems required should be designed for accurate and reproducible control (usually at a low level), including the coordination and differential regulation of multiple genes, a feature usually not required for high-level heterologous protein production.

E. coli is a well-characterized and genetically tractable bacterium and therefore an important industrial microorganism. As a result, sophisticated metabolic engineering tools have been developed for its use. This chapter describes the gene expression tools that have been developed for metabolic engineering applications in *E. coli* and addresses areas where there is still a need for development.

Chromosome Engineering

Advantages of Inserting Heterologous DNA into the Chromosome

The use of plasmids in metabolic engineering is a convenient method to express genes or pathways of interest. However, the use of high- or low-copy number plasmids can be problematic. For example, plasmids can be unstable (caused by rearrangements and segregational instability) and usually require a resistance marker to maintain a positive selective pressure. For the engineering of organisms for use in food products or as biocatalysts for chemical synthesis, the requirement for plasmid maintenance is undesirable. In addition, limitations in the choice of resistance markers and compatible origins of replication generally restrict the use of more than three plasmids in a single host. Finally, multi-copy plasmids and over-expression of proteins or pathways from these plasmids may produce undesirable phenotypes such as reduced growth or imbalances in metabolic pathways that will result in reduced yields. Most of the detrimental effects associated with gene expression from plasmids can be avoided by single-copy expression from the bacterial chromosome or from single-copy plasmids.

Numerous methods have been developed to insert into and express genes from the chromosome of *E. coli*. These methods can be categorized into site-directed or homology-dependent and randomized

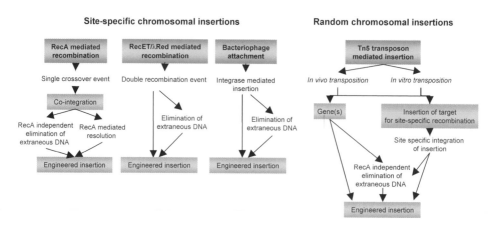

Figure 1. Summary of genetic systems developed for site-specific and random insertions in the chromosome of *E. coli*.

or homology-independent methods (Figure 1). Using these methods, foreign DNA can be inserted into the chromosome of *E. coli* with the help of a host or transiently-expressed, plasmid-encoded recombinogenic function. In the most recent generation of integration vectors, exogenous DNA sequences, such as delivery plasmids or selection markers, can be excised by site-specific recombination events. In a few examples, the chromosomal insertion can be excised and rescued if desired. It should be noted that although this section describes strategies for insertion of genes into and expression from the chromosome, similar approaches can be used to create mutations and deletions.

RecA-Mediated Homologous Recombination

The use of RecA function in homologous recombination has been exploited extensively in *E. coli* genetics. In order to insert a DNA sequence to be expressed from the chromosome a homologous sequence, used as a guide, must be present on the delivery vector (Figure 2). The extra step required to insert the guide sequence into the delivery vector may be time-consuming, but it allows for the precise, site-specific insertion of large DNA sequences at a high frequency. The frequency of recombination between the guide and

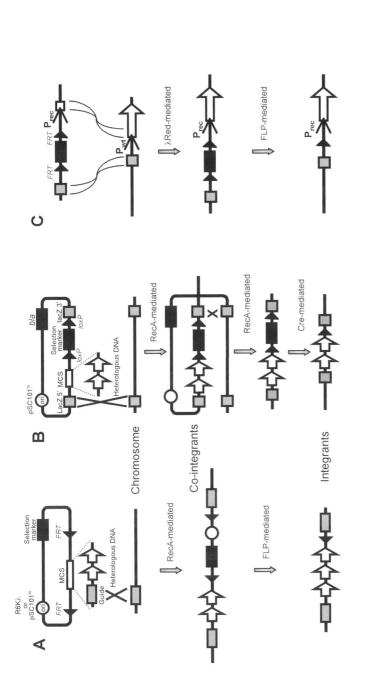

Figure 2. Mechanism for chromosomal integration by RecA-dependent and RecA-independent homologous recombination. (A): Single region of homology combined with FLP-*FRT* recombinase system for the precise elimination of vector and selection marker DNA subsequent to chromosomal integration. The integration vectors contain conditional replicons (R6Kλ or pSC101) and a multi-cloning site (MCS) flanked by tandem *FRT* sites (Martinez-Morales et al., 1999). (B): Integration of heterologous DNA using the pBRINT-Ts plasmid family (Le Borgne *et al.*, 2001). Two crossover events in the *lacZ* gene results in β-galactosidase mutants that are sensitive to carbenicillin (*bla*). (C): Chromosomal integration of a linear PCR product using λRed and elimination of the selection marker using the site-specific FLP-*FRT* recombinase. This example illustrates the exchange of a wild-type promoter (P_{wt}) for a recombinant promoter (P_{rec}).

the target DNA are directly dependent on the length of the homologous sequence. For example, in *E. coli*, very low ($<10^{-8}$) frequencies were observed when using a 25-bp region of homology but rose dramatically to $>10^{-5}$ with increasing length and appeared to approach a plateau for homologies of >400-bp (Lovett *et al.*, 2002).

Two general types of RecA-mediated homologous recombination systems have been developed for chromosomal insertions of heterologous DNA. One method requires a single region of homology (Figure 2A) (Martinez-Morales *et al.*, 1999), whereas the other requires homologous sequences flanking both sides of the heterologous DNA to be inserted into the chromosome (Figure 2B) (Palmeros *et al.*, 2000; Merlin *et al.*, 2002). Using a single region of homology, the integration of DNA encoding a desired function can be targeted to a specific site from a single crossover event where the entire plasmid is incorporated into the chromosome. Plasmids with conditional origins of replication (*ori*$_{R6K\lambda}$ and *ori*$_{101}$) were designed to promote integration into the host chromosome. Integrants are subsequently transformed with a helper plasmid expressing the yeast FLP recombinase gene to excise from the chromosome the DNA bracketed by the *FRT* sites. Once the helper plasmid is cured from the host, this system leaves a single *FRT* site, the heterologous DNA, and a second copy of the guide DNA on the chromosome. The utility of this chromosome integration system was demonstrated with the expression of the pyruvate decarboxylase (*pdc*) and alcohol dehydrogenase II (*adh*B) genes from *Zymomonas mobilis* for the metabolic engineering of ethanol production in *E. coli* (Martinez-Morales *et al.*, 1999).

Similar chromosomal integration systems requiring two crossover events have also been developed (Le Borgne *et al.*, 2001; Merlin *et al.*, 2002). The pBRINT-Ts plasmids, for example (Figure 2B), were designed with a temperature sensitive origin of replication and several antibiotic cassettes flanked by *lox*P sites. Double crossover integrants at the *lac*Z gene, which are sensitive to carbenicillin and *lac*Z⁻, are selected at elevated temperature. Unlike the single-site recombination system (Figure 2A), integration does not result in the duplication of the guide sequence; however, the location for chromosomal insertion is limited to *lac*Z. The resistance marker flanked by two *lox*P sites is subsequently removed with the help of transiently-expressed Cre recombinase (Palmeros *et al.*, 2000).

Recombination Using λRed or RecET

An alternative to the RecA-dependent homologous recombination strategy was developed whereby integration of linear DNA into the chromosome of *E. coli* is mediated by three phage-derived protein functions. This recombination strategy relies on the transient expression of these proteins to induce a RecA-independent, hyper-recombinogenic state in *E. coli*. The expression of a protein pair, either Redλ/Redβ from lambda or RecE/RecT from the Rac phage, promotes single-stranded annealing and assimilation into the chromosome (reviewed in Court *et al.*, 2002). The co-expression of the *gam* protein, which binds to the host's RecBCD enzyme and inhibits nuclease activity, is used to enhance recombination efficiency. In *E. coli*, recombination between transformed linear DNA and the chromosome can occur with homologous sequences as short as 30-bp (Datsenko and Wanner, 2000). This technology offers several advantages over traditional recombination methods in that (i) integration is achieved in a single step since resolution of the co-integrant is not required (Figure 2C); (ii) prior cloning or sub-cloning of the gene(s) to be integrated is optional since PCR products can be integrated; and (iii) only small regions of homology are required for recombination to take place. While λRed recombination is used principally in deletions or mutations, a few examples of incorporation of heterologous DNA encoding promoter and reporter fusions have been reported (Khlebnikov *et al.*, 2001; Ellermeier *et al.*, 2002). Large DNA insertions can be achieved using this technique as demonstrated by the integration of 5.7-kb gene cassettes used for *lac* fusions (Ellermeier *et al.*, 2002).

Chromosomal Integration Using Phage Attachment Sites

Site-specific integration and excision of any DNA fragment into bacteriophage attachment sites (*att*B) located on the *E. coli* chromosome can be achieved using specially designed vectors. These systems are based upon delivery vectors containing a bacteriophage attachment site (*att*P), a host containing the corresponding *att*B site, and a helper plasmid expressing the integrase (Int) to promote site-specific integration. Several systems based on site-specific integration of bacteriophage have been developed (Boyd *et al.*, 2000; Haldimann and Wanner, 2001; Rossignol *et al.*, 2002). One recent study reported on the construction of conditional-replication, integration, excision and modular (CRIM) plasmids (Haldimann and Wanner, 2001). These plasmids allow for the precise excision and retrieval of the inserted heterologous sequence, an attribute that may be useful in screening single-copy library clones. Since phage integrase and excisionase (Xis) are supplied in *trans*, the resulting integrants are stably maintained on the host's chromosome. An alternative to this approach is to integrate the entire bacteriophage into the chromosome and subsequently delete the phage genome, leaving behind the heterologous sequence to be expressed, as described by Boyd *et al.* (2000).

Random Integration Using Tn*5* Transposon Systems

Transposons are specific DNA sequences that catalyze their own movement to alternative sites within a chromosome. Original systems designed to use natural transposons as tools for inserting heterologous DNA into a chromosome were plagued by several problems: difficulty in genetic manipulation due to the large size of the transposon sequences, frequent DNA rearrangements due to the repetitive nature of these sequences, and inability for multiple rounds of insertion into a single strain. The development of mini-transposon cloning vectors, based on the Tn*10*, Tn*5*, and Tn*7* systems, circumvented many of the problems associated with the natural transposon systems. The Tn*5*-based vectors have found the widest use as a genetic cloning tool, largely due to their small size and broad range transposition mechanisms (de Lorenzo *et al.*, 1990). This section will focus on Tn*5*-based cloning vectors and their applications for random chromosomal insertions of heterologous DNA.

The mobility of the Tn*5* transposon is determined by two insertion sequences, and the mobile DNA is flanked by two 19-bp I and O termini (reviewed in de Lorenzo *et al.*, 1998). An important property of this transposon is that the Tn5 transposase will mobilize a sequence flanked by I and O termini even when it is located outside of the mobile element. Recombinant transposons have been constructed in which the transposase gene is outside of, but adjacent to, the mobile DNA element. In these mini-transposon systems, the transposase gene is lost after insertion. Therefore, insertions resulting from this system are stably inherited, are not subject to further genetic rearrangements, and allow for multiple insertions into the same strain. To increase the utility of the Tn5 transposon derivatives as cloning tools, vectors have been developed with a variety of cloning sites, selection markers, and recombinase recognition sites that allow for removal of markers after insertion and selection of the heterologous DNA into the chromosome. For instance, vectors that utilize site-specific recombination based upon the multimer resolution system (*mrs*) have been constructed. Precise removal of DNA flanked by 140-bp resolution (*res*) sites can be accomplished by the ParA resolvase. Incorporation of the ParA/*res* recombination system into Tn5 vectors allowed the construction of "quasi-natural" recombinant strains, in which segments of heterologous DNA harboring markers or reporters can be precisely removed after selection (Kristensen *et al.*, 1995). This Tn5-*mrs* system was utilized to engineer a *Pseudomonas putida* strain for growth on toluene as the sole carbon source (Panke *et al.*, 1998). The upper TOL pathway of *P. putida* was inserted into the chromosome of the recipient strain with a reporter gene (*xyl*E) and a selection marker (*npt*) flanked between two *res* sites. This design allowed for the subsequent removal of both markers and resulted in the conversion of toluene to benzoate, which could then be further metabolized through the native *ortho*-ring cleavage pathway of the catechol intermediate.

An *in vitro* transposon system was also designed to introduce a strong transcription terminator, the *ara*C gene, and the P_{BAD} promoter into the chromosome of *E. coli* (Grant *et al.*, 2001). This arabinose regulator-promoter pair insertion system can be used to investigate the phenotypic effects of altered levels of gene expression resulting from random insertions. The TnΩP_{BAD} system was used to study the effects of expression levels of BipA, which affects growth of *E. coli* on solid media at low temperatures and resulted in a strain of *E. coli* whose growth at low temperatures on solid media was dependent on threshold levels of arabinose. One advantage of this tool over traditional cloning procedures is that it does not require prior knowledge of the location of the gene of interest. Furthermore, it allows for optimization of gene transcription by generating a range of derivatives, in which the promoter is located at varying distances from the gene of interest.

Tn5-based transposon systems have also been used to insert into the chromosome of *E. coli* target sequences used for site-specific recombination. A Tn5-based delivery of *FRT* sites recognized by the FLP recombinase of yeast has been developed to reversibly introduce heterologous DNA at random *FRT* sites located on the chromosome (Huang *et al.*, 1997). The heterologous DNA segment is flanked with FRT sites (minimum 34-bp in length), and integration occurs when FLP recombinase is present in the strain.

Plasmids

Advantages of Using Plasmids for Faithful DNA Replication

The major benefit of integrating genes of interest into the chromosome of *E. coli* is stable maintenance through replication and cell division cycles. This benefit is balanced by the sometimes unpredictable behavior associated with heterologous gene expression from the chromosome, and the difficulty in removing, retrieving, or modifying these insertions for subsequent rounds of strain engineering. In addition, random gene insertion into a chromosome will usually produce variable and unpredictable gene expression levels.

Plasmids represent a common alternative to permanently inserting DNA into the chromosome. They are extra-chromosomal circular pieces of DNA that replicate in the cell autonomously of the chromosome. Important features to consider when selecting plasmids include incompatibility type (determined by the replication system in the plasmid); structural stability (faithful replication inside the cell without alterations in the DNA sequence); segregational stability (equivalent segregation of plasmids into daughter cells at division); and copy number (number of copies of plasmid per chromosome equivalent). For applications requiring multiple plasmids to be stably maintained in a single cell, it is important to select plasmids with different incompatibility types, as plasmids with similar replication systems will compete for replication functions.

Desirable Features of Plasmids for Metabolic Engineering

In metabolic engineering, plasmid selection represents a convenient method of changing gene dosage in the cell. High- and medium-copy number plasmids are popular choices for applications where the goal is to maximize protein production in the cell (Makrides, 1996). Therefore, much work was initially done in the development of high- and medium-copy number plasmids in *E. coli*. These plasmids are small, easily manipulated, yield large quantities of DNA when isolated from cells, and generally yield high gene expression levels. Because of these advantages, many different types of high- and medium-copy number plasmids have been created for expression studies by altering promoters, selection markers, origins of replication, incompatibility types, multi-cloning sites (MCS), and transcription terminators (Balbas and Bolivar, 1990). Unfortunately, these types of plasmids are typically characterized by having very poor expression control and structural or segregational stability in cell populations. Furthermore, the benefits of high gene copy number on protein production are often offset by the metabolic burden placed on the cells for the energy required to replicate the plasmids and to translate plasmid-encoded genes (Bentley *et al.*, 1990; Birnbaum and Bailey, 1991; Vind *et al.*, 1993).

To overcome the limitations of high-copy gene expression, low- to single-copy plasmid derivatives that behave similarly to artificial chromosomes were developed for use in *E. coli*. These plasmids are ideal for metabolic engineering applications in which lower gene copy number and thus lower levels of enzymes produced from the plasmid-encoded genes minimize the metabolic burden placed on the cell. In addition, the plasmid's capacity to faithfully replicate large pieces of DNA will allow one to clone and express genes encoding an entire metabolic pathway. This section describes the recent advances made in the development of these specialized single-copy gene expression vectors.

Low-Copy Number Plasmids

Several research groups have developed low-copy plasmids (defined here as 1-5 copies per cell). A single-copy number plasmid based on the F plasmid from *E. coli* was devised, which behaves as an artificial chromosome in the cell. This mini-F plasmid was constructed from a 9-kb fragment of the F-plasmid in *E. coli*, which contains several genes required for faithful replication and segregation (Figure 3) (Jones and Keasling, 1998). Replication from the *ori*V and *ori*S origins results in the synchronization of the plasmid's replication with the cell cycle. Inclusion of the partition elements (*par/sop*) results in the reproducible

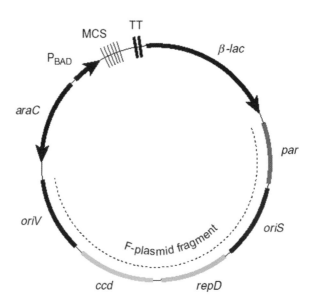

Figure 3. Low-copy number plasmid developed from the F plasmid in *E. coli*. The fragment from the F plasmid is indicated by the dotted line. Abbreviations: *araC*, arabinose repressor; β-*lac*, β-lactamase gene; *ccd*, cell-killing locus; MCS, multi-cloning site; *ori*S, F-plasmid origin of replication; P$_{BAD}$, arabinose-inducible *ara*BAD promoter; *par*, F-plasmid partition element; *repD*, resolution-of-multimers locus; TT, transcription terminators (reproduced from Keasling, 1999).

segregation of the plasmid into daughter cells at division. The combination of these origins and partition elements results in a plasmid that is stably maintained at 1-2 copies per chromosome equivalent. The coupled cell division locus (*ccd*) provides a back-up "kill" mechanism to the active partitioning mechanism, which inhibits growth in plasmid-free segregants and ensures stable maintenance of the plasmid. To help in cloning, the ampicillin resistance selection marker encoded by the β-lactamase gene was included in this plasmid. For ease of cloning and heterologous gene expression, a MCS, transcription terminators, as well as the *ara*C-P$_{BAD}$ regulator-promoter pair were designed into this plasmid.

Expression of the *lac*Z gene from the mini-F plasmid has shown the stable maintenance of this low-copy plasmid in induced cell populations for up to 150 generations even in the absence of ampicillin selection (Jones and Keasling, 1998). In contrast, cells grown under identical conditions lost multi-copy plasmids harboring the identical promoter, transcription terminator, and selection marker after 53 generations in the absence of selective pressure. Cell doubling time was not affected by gene expression from this low-copy plasmid. In comparison, increases in doubling time were observed from gene expression in cells harboring multi-copy plasmids, which was indicative of metabolic burden. In fact, productivity (defined as enzyme activity divided by cell doubling time) was greater under certain inducer concentrations from cells harboring the low-copy plasmid versus cells harboring multi-copy plasmids (Carrier *et al.*, 1998; Jones and Keasling, 1998). In addition, these types of plasmids can harbor very large fragments of cloned DNA, a property useful for multiple gene expression applications, such as pathway engineering.

Low-Copy Number Plasmids in Metabolic Engineering

The benefits of low-copy plasmid systems have been demonstrated by their successful application in metabolic engineering. The performance of the low-copy mini-F-based system, described above, was compared to that of a high-copy pMB1-based plasmid for the production in *E. coli* of polyphosphate (polyP) and lycopene, which is derived from the precursor isopentenyl diphosphate (Jones *et al.*, 2000). The genes encoding enzymes involved in the early steps in these metabolic pathways, polyP kinase (*ppk*) and 1-deoxyxylulose-5-phosphate synthase (*dxs*) were cloned into the mini-F- and pMB1-based plasmids and expressed in *E. coli* from the P_{tac} or P_{BAD} promoters. Rather that monitoring changes in reporter protein levels from low- and high-copy systems, these studies examined the effects of altering recombinant enzyme levels in *E. coli* by altering plasmid copy number. The results demonstrated that the enhanced synthesis of polyP and lycopene were comparable when using low- or high-copy plasmids. These examples point out and emphasize that gratuitous over-production of recombinant enzymes can alter the energy state of the cell, compromising overall productivity. In addition to the burden incurred on the cell to replicate plasmid DNA and translate plasmid-encoded genes, there may be physiological effects associated with metabolites consumed or accumulated in the cell caused by the excess enzyme expression. Low-copy plasmids are useful gene expression tools in metabolic engineering applications where the introduced heterologous pathways produce novel products, while using substrates involved in the primary metabolism of the host organism. Low-copy plasmid systems have also been used for expressing the lactate dehydrogenase gene (*ldh*) from *Streptococcus bovis* for L-lactate production in *E. coli* (Dien *et al.*, 2001).

To increase the utility of these plasmids in metabolic engineering it will be imperative to expand the existing low-copy plasmids to include promoter-regulator systems more suitable for metabolic engineering applications (see section below). Since large DNA fragments are difficult to clone in these rather large low-copy plasmids, the incorporation of a recombinase-derived cloning system in lieu of the MCS would greatly improve their efficacy.

Promoters

Desirable Features of Promoters for Metabolic Engineering

The regulation of natural biochemical pathways often relies on fine-tuned gene expression through the use of regulated promoter systems. Therefore, desirable features of promoter systems for use in metabolic engineering differ from those of recombinant protein expression systems. Stringent control over the level of gene expression is a critical feature of any metabolic engineering strategy. Control of gene expression can be achieved by choosing constitutive or inducible promoters of varying strength and by varying the concentration of the inducing compound. The best promoter for metabolic engineering applications would show no expression in the absence of the inducer and a linear gene expression level proportional to the amount of inducer present over a large range of inducer concentrations.

The timing of gene expression may also have a substantial effect on the success of a pathway engineering strategy. For example, in certain circumstances it may be optimal for the expression of a gene or a pathway to occur in a particular growth phase (*i.e.* exponential or stationary). The precise timing of gene expression can be regulated by the addition of an inducer to the culture at a predetermined period. However, more elegant systems exist whereby the cell self-regulates gene expression based upon a shift in its metabolic state. These promoter systems are more tractable and cost effective since the addition of a sometimes expensive inducer, which must be distributed quickly and evenly throughout the culture, is not essential.

Finely-Tunable Engineered Promoters

Early endeavors in gene expression technology were aimed at the production of recombinant proteins in model hosts such as *E. coli*. For this task, strong inducible promoters such as the T7, *lac, trc,* and P_L were recruited, and these promoters produced high levels of the desired protein. Unfortunately, these expression systems lack the finesse required for metabolic engineering. Overexpression of recombinant proteins is often associated with metabolic burden manifested by a severe reduction in growth. In metabolic engineering, proteins are expressed for their activity and not as the end product. Therefore, methods to control the expression of these strong promoters were sought to overcome these problems. Regulated promoters systems were developed and are now commonly used in research and industrial applications. These promoters rely on the co-expression of an activator or repressor to minimize gene expression in the absence of the inducer. A drawback of these promoters is that the inducing molecule often induces its own transport, *e.g. lac*Z expression under the control of the P_{lacUV5} promoter (Figure 4). This autocatalytic behavior, coined the "all-or-none" phenomenon (Novick and Weiner, 1957), leads to sharp induction curves, where small additions of inducer produces large increases in protein production, an undesirable feature for fine-tuning gene expression. In addition, at non-saturating levels of the

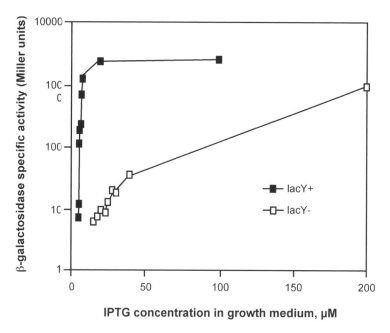

Figure 4. Steady state β-galactosidase expression from the P_{lacUV5} promoter vs. IPTG concentration in a *lac*Y⁺ and lactose transport mutant *lac*Y⁻ strain of *E. coli* (reproduced from Jensen *et al.*, 1993).

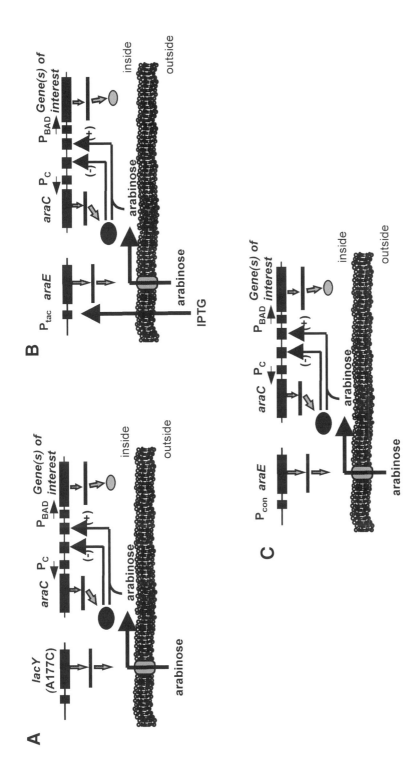

Figure 5. Elimination of the "all-or-none" gene expression by the *ara*C-P$_{BAD}$ regulator-promoter by uncoupling the expression of the arabinose permease (AraE) from the induction of the gene(s) of interest. In these engineered systems, arabinose can be transported by a lactose permease with relaxed substrate specificity (lacY A177C) A an IPTG inducible promoter (P$_{tac}$) B or constitutive promoters of varying strength (P$_{con}$) C.

inducer, this "all-or-none" autocatalytic gene expression behavior results in a fraction of the cells in the culture that is fully induced while the remaining subpopulation is not induced (Siegele and Hu, 1997). Hosts with mutations in the transport system can extend the linear range of induction, but this requires an inducer that can diffuse freely across the cell membrane (*e.g.* IPTG).

An alternative method to circumvent the "all-or-none" effect is to uncouple the expression of the transporter from the inducer concentration. Three such gene expression systems were developed using the arabinose-inducible *ara*C-P$_{BAD}$ system (Figure 5). In the first two examples, the arabinose transporter encoded by the *ara*E gene was expressed from an IPTG inducible promoter (Khlebnikov *et al.*, 2000) (Figure 5B) or constitutive promoter (Khlebnikov *et al.*, 2001) (Figure 5C). In this way, it is possible to control the expression of *ara*E by varying the concentration of IPTG or the strength of the constitutive promoter thereby controlling arabinose transport and activity from P$_{BAD}$. In this case, cultures of cells induced for expression of a gene under control of P$_{BAD}$ had a single population at all arabinose concentrations. Alternatively, the AraE arabinose transporter can be replaced by a lactose transporter with relaxed specificity (Figure 5A) (Morgan-Kiss *et al.*, 2002). This *E. coli* strain, which also harbored deletions in the arabinose degradation (*ara*BAD) and transport (*ara*FGH) gene clusters showed a tight distribution of gene expression in the population over a 100-fold range of arabinose concentration.

Engineered Promoters for Metabolic Optimization

Significant control over the expression of metabolic pathways can be accomplished by adjusting gene expression from promoters of varying strength. The consensus promoter in *E. coli* consists of two hexanucleotide sequence regions (-35 and -10) separated by a short spacer sequence (Figure 6). The structural features of promoters as they relate to strength have been studied extensively (Record *et al.*, 1996). Deviation from the -10 TTGACA and -35 TATAAT consensus sequences (Kobayashi *et al.*, 1990) or spacer length (Mulligan *et al.*, 1985) leads to altered contact or open complex formation by the RNA

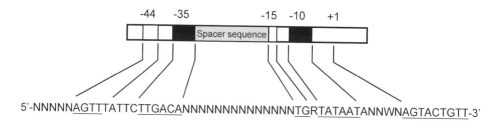

Figure 6. Schematic representation of the consensus promoter and oligonucleotide sequence used for generating a library of synthetic constitutive promoters for expression in *L. lactis* and *E. coli*. R = 50% each A and G; W = 50% each A and T.

polymerase to the promoter and significant decrease in promoter activity. In contrast, substitutions in the spacer region sequence result in moderate changes in promoter activity and therefore, allows the fine-tuning of gene expression (Auble *et al.*, 1986). Based on this observation, Jensen and Hammer (1998b; 1998a) developed a library of randomized synthetic promoters derived from the consensus promoter sequence of *Lacotococcus lactis* (Figure 6). This promoter library, which included variations in the length (16- or 17-bp) and sequence of the spacer region as well as the -10 and -35 regions, produced a wide range of activities in small step increments in *E. coli* and *L. lactis*.

Metabolically-Regulated Promoters and Sensors

In spite of the large collection of natural promoters known to control the metabolic activity and physiological state of bacteria, their successful application in metabolic engineering is limited to a few examples. This may be the result of the often obscure, yet intimate association between the engineered pathway and the metabolic and physiological status of the cell. In this section, we describe examples of promoter systems that were designed to address this issue. These metabolically-regulated promoters respond to an innate inducer, whose production is triggered by the cell's metabolic state.

Since the expression of a recombinant pathway can result in reduced cell growth, one possible strategy to solve this problem is to uncouple pathway expression from biomass production. *E. coli* has evolved sophisticated mechanisms to cope with the challenges of rapid growth, cell starvation, and viability in stationary phase. Many promoters of proteins produced by the cell in response to these challenges have been identified and exploited to produce recombinant proteins in *E. coli*. Examples include the promoters of the *fic* gene, expressed in stationary phase (Utsumi *et al.*, 1993) and the growth rate dependent Ribosomal Modulation Factor gene, *rmf* (Chao *et al.*, 2001).

Promoters induced by nutrient starvation have also been recruited for uncoupling growth and protein production. For example, the expression of the alkaline phosphatase (*pho*A) gene, which is regulated by PhoB, reaches high level when cells are starved for phosphate (Wanner, 1993). A protein production system based on gene expression from the carbon starvation inducible *cst-1* locus was also reported (Tunner *et al.*, 1992). Although examples of the use of these classes of promoters for the metabolic control of an engineered *E. coli* pathway do not yet exist, in view of the potential benefits it becomes evident that these gene expression systems deserve further studies.

The metabolically regulated gene expression systems described thus far were designed to respond to a metabolic trigger sensed by the cell, but do not truly auto-regulate. Once the gene or pathway is induced, there is no mechanism to control the level and duration of expression. A more attractive design would sense the metabolic state of the cell and auto-regulate the expression of the recombinant gene or pathway in a method inherent to natural metabolic networks. Farmer and Liao (2000) engineered a dynamic gene expression loop which they coined metabolic control engineering (Figure 7). This promoter-sensor feedback system is designed to fine-tune the expression of a recombinant pathway in response to excess glycolytic flux, during which acetyl phosphate accumulates within the cell. In a NRII (the nitrogen sensor of the Ntr regulon) mutant *E. coli* strain, excess acetyl phosphate phosphorylates NRI, the response regulator of this regulon (Figure 7). Phosphorylated NRI (NRI-P) in turn, positively regulates the expression of the recruited *glnAp2* promoter in the engineered

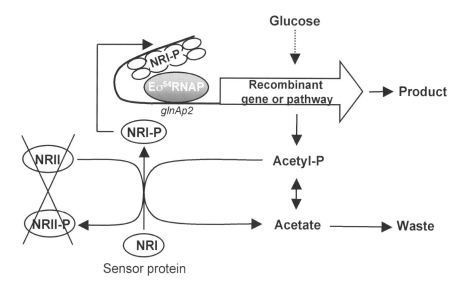

Figure 7. Features of a promoter-sensor feedback dynamic controller designed to respond to excess glycolytic flux (adapted from Farmer and Liao, 2000).

cell. Although the use and design of metabolically regulated gene expression systems are still in their infancy, the knowledge gained from the current integrated systems biology studies will undoubtedly lead to more complex and refined metabolic engineering control designs.

Directed RNA Processing and Decay

Utility of RNA Design to Pathway Engineering

Cells have evolved mRNA as a relatively unstable intermediate in gene expression allowing for a relatively quick response to changes in their environments. In gene expression, cells use different types of post-transcriptional regulation mechanisms as efficient means of controlling protein levels in the cell. Varying mRNA stabilities, or half-lives, is one form of a post-transcriptional mechanism used by bacteria to regulate protein levels. The majority of transcripts in *E. coli* have half-lives ranging from 1 to 30 minutes. These differences in half-lives are a result of the different rates at which ribonucleases (RNases) act to inactivate and degrade nascent transcripts. *E. coli* is an ideal organism for the development of gene expression tools based on mRNA design. A great deal of progress has been made in elucidating the mechanisms of mRNA decay, the players involved in the general decay pathways, as well as in identifying *cis*-acting elements responsible for the range of mRNA half-lives observed in this organism. This section will focus on RNA design strategies for altering mRNA processing and stability and their application in the development of alternative pathway engineering tools for *E. coli*.

A limiting factor of the metabolic or pathway engineering technologies described thus far is the added complexity and increased metabolic burden placed on the cell when used to control the differential expression of multiple proteins. Applications requiring precise expression levels of multiple enzymes involved in a heterologous pathway have traditionally relied on multiple plasmids or inducible promoter systems (Figure 8A). A more efficient and less taxing method for the simultaneous control of multiple genes can be gleaned from natural systems. For instance, bacteria have evolved a type of directed RNA processing and segmental stability operon design to differentially control levels of related proteins, such as enzymes in a metabolic pathway. Through evolution, the genes encoding the enzymes of a particular pathway are often found under the control of a single promoter. In this way, the genes can be transcribed on a single multi-cistronic transcript. The cell then directs the processing of this long primary transcript and the secondary transcripts containing coding regions for one or more of the enzymes in this pathway. This strategy can result in vastly different protein levels produced from a single transcript (Figure 8B).

Coordinating the Expression of Multiple Proteins by RNA Design

Recent studies have examined the effects of single- and multi-gene transcript design on the resulting protein levels in *E. coli*. This work focused on the design of various RNA stability control elements in

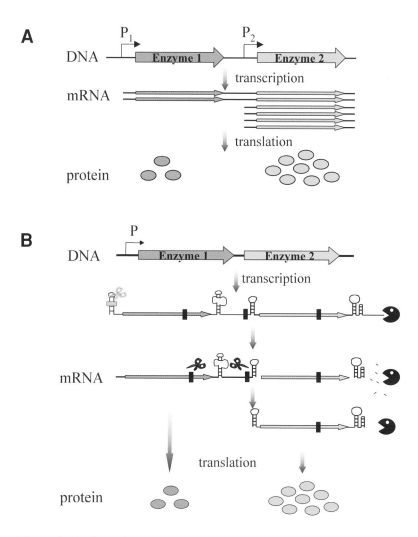

Figure 8. Design schemes for multi-gene expression systems. (A): Design of a multiple promoter system where P_1 and P_2 are either different promoters or similarly-controlled promoters of different strengths. Expression from both promoters simultaneously results in higher production of the downstream gene. (B): Design features for direct mRNA processing and decay. Both genes are transcribed on a single dual-gene transcript from a single promoter (P). Boxes denote endoribonuclease sites engineered into the transcript. This transcript is initially processed by a double-strand-specific endoribonuclease, RNase III (scissor), at an engineered hairpin containing the cleavage site for this enzyme and degraded by a 3' to 5' exonuclease. After this initial cleavage the upstream gene is inactivated by a single-strand-specific endonuclease, RNase E, (scissors), at cleavage sites for this enzyme. The resulting secondary transcript is protected at both ends from further immediate inactivation by RNases. Both of these designs result in higher production of the protein encoded by the second gene (adapted from Martin *et al.*, 2002).

the form of RNase cleavage sites (differing in sequence), secondary structures (differing in free energy of folding, shape, and sequence), and relative gene location (Smolke *et al.*, 2000; Smolke and Keasling, 2002b). The effects of these control elements on RNA processing, segmental stability, and protein levels were tested in a two-gene transcript system expressing *lacZ* and *gfp$_{uv}$* from the arabinose-inducible *ara*BAD promoter (Smolke *et al.*, 2001a). By varying the combination and location of the RNA stability control elements, relative steady-state transcript and protein levels could be varied 500- and 1,000-fold, respectively (Smolke and Keasling, 2002b; Smolke and Keasling, 2002a). These initial advances in multi-cistronic mRNA engineering hint at the potential impact this technology may have on pathway engineering applications. Larger libraries of mRNA stability control elements constructed using directed molecular evolution strategies have been generated that increase the stability element design combination and consequently the gene expression range of this technology (J. D. Keasling, unpublished).

RNA Engineering can Alter Flux Through Metabolic Pathways

Directed mRNA processing and stability was applied to the design of a heterologous operon encoding a carotenoid biosynthetic pathway in *E. coli* (Smolke *et al.*, 2001b). The lycopene synthase (*crt*I) and β-carotene cyclase (*crt*Y) genes from *Erwinia herbicola* were cloned into a synthetic two-gene operon. The mRNA control elements characterized in the previous studies were engineered into the transcript design of this operon to direct differential segmental stability of the transcripts and thus produce varying levels of these enzymes. Results from this study indicated that this technology could be successfully applied to alter the flux through this metabolic pathway as well as intermediates accumulated.

Guidelines for Operon Design

Using the results of these mRNA stability studies, guidelines for operon design strategies have been proposed (Smolke and Keasling, 2002b): (i) secondary structures at the 5' end of genes can have a dramatic effect on the stability of endonuclease-susceptible transcripts and protein production from those transcripts; (ii) secondary structures at the 3' end of genes increase transcript stability and consequently, protein production; (iii) directing mRNA cleavage between coding regions is an effective means of uncoupling gene expression to allow for independent control of transcript stability and protein production; (iv) gene location can strongly influence transcript stability and protein production, particularly for those transcripts susceptible to endonuclease inactivation. Although the effectiveness of this mRNA design strategy is obvious, additional experimental data are required to construct better models for predictive design.

Translational Control

Engineering Translation

In addition to engineering transcript stability, one can also manipulate heterologous protein expression by altering translation initiation and elongation rates. Translation initiation and elongation, and the factors influencing them, are well characterized in *E. coli*. Modulation of translation can be achieved by altering the ribosome binding site (RBS), the spacer sequence between the RBS and the start codon, or by modifying the codon usage of the gene. While these strategies have been successfully applied to alter the production of a single heterologous protein, gene expression design strategies using this type of control for fine-tuning the relative expression of multiple genes are less common. Approaches targeting translational efficiency present a potentially powerful way to alter relative expression levels of multiple genes.

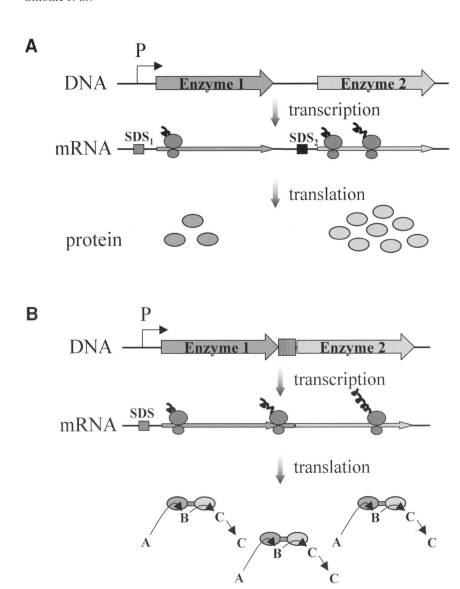

Figure 9. Methods for coordinating multiple protein production using translation-based control schemes. (A): Systems designed to regulate multiple gene expression by engineering relative RBS strength. In this scheme, relative protein levels in the cell are regulated by engineering the relative translation initiation rates of the two genes. (B): Systems designed to regulate the flux through a metabolic pathway by synthesizing artificial bi-functional enzymes using protein fusion technologies. In this scheme, relative activities of sequential enzymes are controlled by engineering substrate channeling into the system.

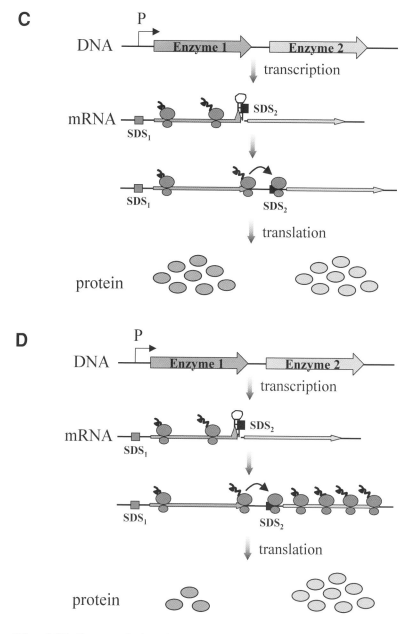

(C) and (D): Systems designed to control the relative levels of two protein products based upon translational coupling. In this scheme, the translatability of the distal gene is under the control of the translatability of the proximal gene. This regulatory mechanism can be engineered as shown in C to produce equimolar amounts of the two proteins by designing "reinitiating" translational coupling systems. Alternatively, it can be engineered as shown in D to produce higher levels of the distal gene by designing "facilitated binding" translational coupling systems.

The effects of mutating or altering the RBS as well as varying the distance between the RBS and the start codon on heterologous protein production have been examined in detail. By replacing the native RBS upstream of the coding region with a consensus RBS or by varying the distance between the RBS and the translation start site, one can vary the translational efficiency several fold (Chapon, 1982; de Boer *et al.*, 1983). Other alterations known to affect translation initiation in *E. coli* include changes in the sequence at the immediate N-terminus of the protein or changes in the operator sequences (Looman *et al.*, 1987). Taken together, these various strategies can be used to alter protein production significantly (Figure 9A).

Engineering the "translatability" or the translational efficiency of heterologous genes by optimizing amino acid codon usage for the host organism may be used to a varying degree. Cells, including *E. coli*, use a specific subset of the available 61 codons. Major codons are those for which there is an abundance of tRNA species whereas minor codons are those for which the tRNA species are present in low levels in the organism (Makrides, 1996). Minor codons tend to be found in genes expressed at low levels in the cell, whereas major codons are more prevalent in highly expressed genes (Kane, 1995). This codon bias may present a problem in that recombinant genes from distant organisms can have codon usage profiles much different from those found in *E. coli*. These problems manifest themselves in poor translational efficiency, resulting in a reduction in the quantity of the protein (Kane, 1995). While the detrimental effects of codon biases on heterologous protein synthesis are widely recognized, optimizing the codon usage of heterologous genes is not common practice. Directed evolution and *de novo* gene synthesis are simple and sometimes efficient methods to optimize the translational efficiency of an enzyme in a heterologous host (Martin *et al.*, 2003). For example, it was demonstrated that the microbial production of plant sesquiterpenes suffers from poor expression of the plant genes in *E. coli* (Martin *et al.*, 2001). To increase the expression levels of terpene synthases and produce high-levels of the artemisinin precursor amorphadiene, a synthetic codon-optimized amorphadiene synthase (ADS) gene was assembled *in vitro* from oligonucleotides and expressed in *E. coli* (Martin *et al.*, 2003). When coupled to the recombinant isoprenoid pathway from *Saccharomyces cerevisiae*, expression of this codon-optimized ADS gene resulted in a 10- to 300-fold improvement in terpene synthesis. By optimizing the expression of the terpene synthase, the limitation of terpene synthesis in this *E. coli* strain was shifted from expression of the synthase gene to supply of the precursors by the isoprenoid pathway.

Protein Fusion as a Means of Controlling Metabolic Flux

To optimize cellular efficiency or channel metabolic intermediates, enzymes catalyzing sequential reactions in a metabolic pathway are sometimes co-localized (*i.e.* in organelles, membranes, or enzyme complexes) or fused. The process by which two enzymes in a pathway interact to transfer their substrate directly between their active sites without allowing free diffusion of the intermediate is called substrate channeling (Meek *et al.*, 1985; Hyde *et al.*, 1988). Another form of substrate channeling called "the proximity effect" occurs when free diffusion of the intermediate is allowed but the close proximity of intermediate-producing and intermediate-consuming enzymes in a cell offers catalytic advantages, such as enhanced steady-state levels. These substrate channeling processes are believed to play an important role in the metabolic regulation and control of enzymatic activities.

Protein fusion strategies are commonly used to direct protein folding (Georgiou and Valax, 1996), solubility (Makrides, 1996), and protein quantification (Cha *et al.*, 2000) and have more recently been applied to achieve enzyme co-localization (Figure 9B) (Bulow, 1987; Bulow and Mosbach, 1991; Brodelius *et al.*, 2002). An important design consideration of protein fusions is the presence and size of a linker between the two fused coding regions. An artificial bi-functional protein was recently constructed by fusing two enzymes, farnesyl diphosphate synthase (FPPS) from *Artemisia annua* and *epi*-aristolochene synthase (*e*AS) from tobacco using a small linker (Brodelius *et al.*, 2002). In this way, a bi-functional enzyme that catalyzes the conversion of the C_{15}-substrate farnesyl diphosphate (FPP) to the sesquiterpene *epi*-aristolochene was created. Production levels of FPP and *epi*-aristolochene *in vitro* from the bi-functional enzyme fusion were compared to those from the combination of the two single enzymes. While both the fusion and single enzyme systems generated equivalent amounts of FPP, the production of sesquiterpene was much higher from the bi-functional enzyme. The authors attributed this result to proximity effect, which limited the FPP diffusion seen when the two enzymes were expressed independently.

Protein fusions can potentially be used to control relative protein expression levels in a cell. By fusing two genes that encode enzymes in a pathway, the two domains (now one protein) will be produced in a 1:1 ratio. However, it is not clear that the two enzyme activities will be unchanged and remain at the levels found in the mono-functional proteins. Because of the many unknowns, protein production control through protein fusions has not yet been used in the design of heterologous pathways.

Translational Regulation Strategies for Pathway Engineering

Translational coupling of two genes may be a potentially powerful but yet untapped gene expression control strategy. In this process, the translation of one gene in an operon (typically the downstream gene) is coupled to the translation of another gene in the operon (typically the upstream gene). In its most common form, the RBS of a downstream gene is sequestered in a mRNA secondary structure, which inhibits ribosome access and thus represses expression of the gene (Figure 9C and D) (Kiser and Schmidt, 1999). However, as ribosomes translate the upstream gene, the mRNA secondary structure is unfolded, thereby allowing ribosomes access to the RBS and the translation of the downstream coding region. By this process, translation of the downstream gene is coupled to translation of the upstream gene. In some instances, the ribosomes terminating at the stop codon of the upstream gene reinitiate at the adjacent start codon of the second gene rather than dissociating from the mRNA (Figure 9C) (Aksoy *et al.*, 1984). Designs based upon these types of reinitiation control strategies should result in equimolar amounts of the two proteins, assuming that all of the ribosomes reinitiate. In other instances, such as the geneX/

*sec*A genes in *E. coli*, facilitated binding of ribosomes at the distal gene allows it to be expressed at higher levels than the proximal gene (Figure 9D) (Petersen, 1989; Schmidt and Oliver, 1989; Rex *et al.*, 1994).

Directed Molecular Evolution Strategies

Molecular Evolution as a Tool for Generating Genetic Diversity

The method of directed molecular evolution is a very useful metabolic engineering tool. Directed molecular evolution utilizes a combined approach of DNA shuffling, random PCR mutagenesis, and predictive modeling to generate novel genes, pathways, and genomes. DNA shuffling (Stemmer, 1994) is an *in vitro* gene recombination method that mimics natural recombination, but at a much higher frequency. In this method, overlapping DNA fragments of genes are generated through a variety of techniques such as DNase I treatment, or overlapping synthetic oligonucleotides. These fragments of DNA are reassembled randomly to create a large pool of recombinant products, from which a very small number may exhibit an improved or novel function. Often, DNA shuffling is used to accelerate recombination events between families of related genes from different species, where homologies are significant enough to support recombination events. However, alternative methods for performing shuffling of DNA exhibiting less homology have also been developed (Shao *et al.*, 1998; Zhao *et al.*, 1998).

PCR mutagenesis takes advantage of the natural mutation rate of DNA polymerases and utilizes the cycling process of PCR to accumulate point mutations in a gene. Point mutations can affect either the specific activity or expression of an enzyme by changing its catalytic properties or the transcriptional and translational efficiency. Rational mutagenesis requires a prior knowledge of protein structure/function relationship to design specific mutations and predict their outcome on enzyme activity and expression. In contrast, random mutagenesis benefits from searching a wider sequence space of potential designs relying less on our current understanding of protein structure/function. However, both DNA shuffling and random PCR mutagenesis require the use of selection or high-throughput screens to identify, from a large number of clones, rare mutation events with interesting or desirable properties.

The desire to increase the efficiency of DNA shuffling and random mutagenesis techniques have led some laboratories to build models to predict areas in the DNA segments where recombination or mutagenesis techniques should be targeted (Voigt *et al.*, 2001a; Voigt *et al.*, 2001b). One such model breaks proteins into building blocks, or schemas, that retain structure and thus function when swapped (Voigt *et al.*, 2002). These models allow a combined rational and random approach strategy, which when applied should optimize the mutant libraries generated by these techniques (Voigt *et al.*, 2001a).

Directed Evolution and Metabolic Engineering of Carotenoid Biosynthesis

Directed molecular evolution is most commonly used in generating proteins with novel functions or improved activities. *In vitro* evolution of genes of heterologous metabolic pathways can be used to optimize production of a compound of interest or to produce new compounds in *E. coli*. For example, a geranylgeranyl diphosphate (GGPP) synthase from *Archaeoglobus fulgidus* was evolved to enhance precursor production in this pathway (Wang *et al.*, 2000). This led to the isolation of an enzyme with several mutations, which exhibited enhanced activity and expression levels. The latter were due to mutations in the 5' untranslated region, such as the RBS, which affected the efficiency of translation initiation. Another research group evolved two genes, a phytoene desaturase and a lycopene cyclase (*crt*I and *crt*Y from *E. uredovora*) located at branch points of the carotenoid biosynthetic pathways (Schmidt-Dannert *et al.*, 2000). Novel carotenoid compounds were produced in *E. coli* by expressing the mutated *crt*I and *crt*Y gene products in a strain engineered to synthesize phytoene. The same laboratory also evolved the C_{30} carotene synthase gene (*crt*M) from *Staphylococcus aureus* and the C_{40} carotene synthase gene (*crt*B) from *E. uredovora* (Umeno *et al.*, 2002). The evolved *crt*M gene functioned as a C_{40} carotene (phytoene) synthase when expressed heterologously in *E. coli*, whereas CrtB could not be evolved to a C_{30} carotene (dehydrosqualene) synthase. The mutant *crt*M genes displayed altered substrate specificity, hinting at the potential of creating new carotenoid backbone structures by evolving the specificity of these enzymes.

Evolution of Regulatory Circuits and Gene Expression

Simple genetic circuits can be engineered by directed evolution. These circuits may serve as useful model systems to study genetic regulatory networks in cells. Directed evolution was applied to study the connectivity of a simple model network and to characterize the logic functions that emerged (Guet *et al.*, 2002). In this example, the evolution of genetic circuits resulted in a library of networks with varying connectivity. The fine-tuning and performance optimization of rationally constructed networks may also be achieved by directed evolution. For example, a library of genetic devices that could be used to construct more complex circuits was developed in *E. coli* (Yokobayashi *et al.*, 2002; Yokobayashi *et al.*, 2003). Given the success in manipulating these simple control networks, one can foresee a time when directed evolution may be used to understand and engineer more complex genetic regulatory networks, such as cellular stress response pathways.

Evolving Metabolic Pathways and Genomes

Larger target DNA sequences such as entire operons and genomes have also been tackled by directed evolution. For example, the efficiency of an arsenate detoxification pathway was evolved in *E. coli*, resulting in an operon encoding a 40-fold greater resistance to arsenate (Crameri *et al.*, 1997). Examination of the mutations accumulated in this evolved operon revealed point mutations localized in the genes as well as in the promoter sequence. These results illustrated that evolving entire pathways can be an effective alternative to rational designs strategies for operons. On a larger scale, genome shuffling based on recursive protoplast fusion (Hopwood *et al.*, 1977; Baltz, 1978) and recombination between bacterial genomes was used to increase the acid tolerance of a strain of *Lactobacillus* (Patnaik *et al.*, 2002) and to improve tylosin production from *Streptomyces fradiae* (Zhang *et al.*, 2002).

Genome-Wide Analytical Tools

The quantification and regulation of metabolism is essential for the progress of metabolic engineering. This knowledge is required to effectively engineer strains and improve the yield of biosynthetic pathways. The increasing availability of complete genome sequences provides the opportunity to investigate in detail the functioning of these organisms. The past ten years has brought the development of technologies that allow the rapid analysis of large sets of transcripts (transcriptome), proteins (proteome), and low molecular weight metabolites (metabolome) present in the cell at any given time. Each data set provides slightly different information - measurements of mRNA provide information on the transcriptional status of the cell; measurements of protein provide information on how mRNA ends up as functional machines within the cell; measurements of metabolites provide information on how proteins act to produce energy and process metabolites. Analysis of this type of global information will lead to a detailed understanding of the mechanisms resulting in any given phenotype. Each "ome" is a progression along the gene expression pathway to phenotype, and comparison of these profiles provides clues to the regulatory mechanisms in the cell.

Technology development in these areas is still rapidly progressing as tools designed to measure the proteome and metabolome of the cell are much less mature than those developed to measure the transcriptome.

Global Technologies Useful for Determining the Metabolic State of a Cell

DNA or oligonucleotide microarrays are used to track the global cellular transcript levels of an organism. Both techniques spot DNA onto glass slides or membranes and then detect fluorescently or radioactively labeled RNA or cDNA (by reverse transcription) using hybridization. DNA microarrays spot onto the slide or membrane denatured double stranded PCR products encoding open reading frames regions, whereas oligonucleotide arrays spot fragments of coding regions in the form of 50 to 100-bp oligonucleotides.

Technologies for tracking global protein and metabolite levels inside a cell are less developed than those for tracking transcript levels. Two-dimensional (2D) gel electrophoresis separates cellular proteins based on their pI and size. Once separated on the gel, the proteins are excised and identified by mass spectrometry methods such as matrix-assisted laser desorption ionization time-of-flight (MALDI-TOF) MS. Significant efforts are currently aimed at developing alternatives to 2D gels to achieve higher throughput and reproducibility as well as to increase proteome coverage. Most of these efforts are directed at tandem LC-MS methods as well as protein arrays, which are still in their infancy. Metabolites can be identified using a combination of techniques such as nuclear magnetic resonance spectroscopy, mass spectrometry, and chromatographic analysis.

Applications Toward Understanding Cellular Metabolic States

Microarrays have been used to track the transcript profiles of *E. coli* cells growing under assorted nutritional conditions, metabolic burdens, and environments. Transcript profiles from a subset of the transcriptome were quantified from *E. coli* growing on glucose, acetate, or glycerol as a carbon source (Oh and Liao, 2000b; Oh *et al.*, 2002). The microarray data did not probe the entire transcript set, but focused on the transcript profiles of genes involved in central metabolism, key biosynthetic pathways, and some regulatory functions. The data interpretation centered on correlating the transcript profiles and established flux ratios to the current understanding of the central metabolic pathways of *E. coli*. While transcript levels from this study did not correlate with fluxes in a quantitative manner, the study pointed out that changes in transcript levels could indicate important regulatory pathways on which to focus. This group also characterized a subset of transcript profiles for different *E. coli* strains growing under the metabolic burden of excess heterologous protein production (Oh and Liao, 2000a). These studies revealed that gratuitous overproduction of a heterologous protein in *E. coli* results in a complex cellular response, which cannot be explained by the existing knowledge of global regulatory mechanisms.

It is becoming increasingly apparent that merely understanding the transcript profile of a cell does not translate to a complete or accurate understanding of the protein or metabolite profile of that cell (Hatzimanikatis *et al.*, 1999; Hatzimanikatis and Lee, 1999). Parallel studies of transcript and protein profiling to obtain a more thorough understanding of the relationship between metabolism and phenotypic changes in the cell should yield valuable information. Combined transcriptome and proteome analyses of *E. coli* and *Bacillus subtilis* were recently reported. The *B. subtilis* study analyzed the global response of cells grown under various amino acid limiting conditions, as well as under conditions invoking the stringent response (Eymann *et al.*, 2002; Mader *et al.*, 2002). These studies identified sets of genes up- or down-regulated in response to these conditions. Only a subset of changes observed in transcript levels were observed at the protein level, indicating a possible difference in the sensitivity of the two approaches. The *E. coli* study looked at the global response of cells during high cell density cultivation (Yoon *et al.*, 2003). This study identified sets of genes involved in energy metabolism and stress response (*i.e.* chaperones) that were up-regulated or down-regulated over a range of growth conditions, including exponential and stationary phase.

Recently, advances have been made in the development of analyses to quantify the small molecular weight metabolites of *E. coli* as a means of understanding cell metabolism and regulation. Researchers recently applied a combination of enzyme assays, HPLC, and HPLC-MS methods to analyze over 30 different metabolites including nucleotides and cofactors in *E. coli* (Buchholz *et al.*, 2002). While metabolomics has yet to be developed to the point of global analysis of metabolites in *E. coli*, the potential applications and utility of this field to metabolic engineering are significant (Phelps *et al.*, 2002) and warrant further development.

Models to Guide Metabolic Engineering

Quantitative analysis of microorganisms and their metabolic pathways can provide useful information to guide metabolic engineering. Flux-based metabolic models quantify the metabolic fluxes through all reactions included in a model of an organism's metabolic network, and as such are important analytical tools for metabolic engineering (Pramanik and Keasling, 1997; Stephanopoulos *et al.*, 1998; Pedersen *et al.*, 1999). By modeling the metabolism of a given organism, one can quantify the effect of genetic manipulations or changes to growth conditions on the cell's entire metabolic network. The existence of pathways can be tested; theoretical yields can be determined; the rigidity of network branch points can be examined; pathways can be compared for their yield of product; and the effects of changes to genes or growth conditions on the entire metabolic network can be identified. Such information has been useful for the identification of candidate targets for genetic manipulation. Many references describe the theory and implementation of this technique (Zupke and Stephanopoulos, 1994; Schmidt *et al.*, 1997; Stephanopoulos *et al.*, 1998).

Metabolic flux analysis can be implemented as an over-determined, exactly-determined, or under-determined system (Stephanopoulos *et al.*, 1998). When an under-determined model is developed, an objective function must be postulated; *e.g.* maximization of growth, minimization of energy utilization, or maximization of substrate consumption. However, an organism's true objective may be difficult or impossible to know. Finally, under-determined models result in many solutions and the optimal solution may never by known.

Inputs to flux-based models are the set of potentially active metabolic reactions and measurements of the steady-state production rates of metabolites such as DNA, RNA, chitin, carbohydrates, fatty acids, and protein (Pramanik and Keasling, 1997; Pramanik and Keasling, 1998). Improved estimates of the cell's fluxes can be obtained by feeding a ^{13}C-labeled carbon source and measuring the isotopic label state of resulting metabolites, such as amino acids (Stephanopoulos *et al.*, 1998). Such an approach is termed isotopomer analysis and often leads to an over-determined system of equations.

The network of reactions is represented by mole balances around each metabolite, which are contained in a stoichiometric matrix. Inverting this matrix and multiplying it by a vector of intracellular (zero) and extracellular (nonzero) net metabolite production rates gives the solution to the model: a vector containing flux values for each reaction (Stephanopoulos *et al.*, 1998). When isotopomer data are included, nonlinearities are introduced into the model, and iterative solution methods must be used (Stephanopoulos *et al.*, 1998; Forbes *et al.*, 2000; Forbes *et al.*, 2001).

Recently, large, under-determined models of metabolism for *E. coli* and a host of other microorganisms have been constructed from annotated genomic sequences (Schilling *et al.*, 2000; Covert *et al.*, 2001; Covert and Palsson, 2002; Schilling *et al.*, 2002). These 'complete' models of metabolism have been used to determine the minimal number of reactions necessary for an organism to function under a given set of growth conditions (Burgard and Maranas, 2001; Burgard *et al.*, 2001) and are currently being used to direct minimization of the *E. coli* genome.

The changes in an organism's genetic programming are difficult, if not impossible to account for in these models and make accurate flux predictions even more unlikely. Nevertheless, the models can be extremely powerful in calculating intracellular fluxes with little input information. As these models become more sophisticated and are integrated with the metabolome and fluxome of an organism, they should become extremely useful tools for engineering metabolism.

Conclusions: Design Principles for Metabolic Engineering

The following design principles can be deduced from the research that has been summarized above:

1. Low-copy is generally better that high-copy. Most of the metabolic genes in a cell exist at low-copy number; single- or low-copy number of the heterologous genes should be sufficient for producing catalytic amounts of metabolic enzymes. The benefits of low-copy number include robust and accurate control of gene expression, minimal metabolic burden, and high structural and segregational stability.

2. Consistent control of gene expression in all cells of a culture is essential to ensure product consistency. In addition, consistent control of gene expression in the culture will ensure stability of the culture so that non-producing organisms do not outgrow producing organisms.

3. Promoters that can sense the state of the cell will allow the metabolic engineer to time production of the desired product when the cell is in the best physiological state for production.

4. Synthetic operons are an efficient way to control expression of genes that encode all of the enzymes in the heterologous metabolic pathway. Synthetic operons enable simultaneous and stoichiometric control of expression of multiple genes using a minimum number of promoters. To make synthetic operons, one needs to be able to control the stability of the various coding regions in the multi-cistronic operon as well as translational efficiency.

5. Laboratory evolution can be used to evolve metabolic pathways to produce variants of the natural molecules or to fine-tune a pathway once it has been constructed.

6. Metabolic models allow one to analyze the resulting flux through a heterologous metabolic pathway.

References

Aksoy, S., Squires, C.L., and Squires, C. 1984. Translational coupling of the *trpB* and *trpA* genes in the *Escherichia coli* tryptophan operon. J. Bacteriol. 157: 363-367.

Auble, D.T., Allen, T.L., and deHaseth, P.L. 1986. Promoter recognition by *Escherichia coli* RNA polymerase. Effects of substitutions in the spacer DNA separating the -10 and -35 regions. J. Biol. Chem. 261: 11202-6.

Balbas, P., and Bolivar, F. 1990. Design and construction of expression plasmid vectors in *Escherichia coli*. Methods. Enzymol. 185: 14-37.

Baltz, R.H. 1978. Genetic recombination in *Streptomyces fradiae* by protoplast fusion and cell regeneration. J. Gen. Microbiol. 107: 93-102.

Bentley, W.E., Mirjalili, N., Andersen, D., Davis, R., and Kompala, D. 1990. Plasmid-encoded protein: The principal factor in the metabolic burden associated with recombinant bacteria. Biotechnol. Bioeng. 35: 668-681.

Birnbaum, S., and Bailey, J.E. 1991. Plasmid presence changes the relative levels of many host cell proteins and ribosome components in recombinant *Escherichia coli*. Biotechnol. Bioeng. 37: 736-745.

Boyd, D., Weiss, D.S., Chen, J.C., and Beckwith, J. 2000. Towards single-copy gene expression systems making gene cloning physiologically relevant: lambda InCh, a simple *Escherichia coli* plasmid-chromosome shuttle system. J. Bacteriol. 182: 842-847.

Brodelius, M., Lundgren, A., Mercke, P., and Brodelius, P.E. 2002. Fusion of farnesyldiphosphate synthase and *epi*-aristolochene synthase, a sesquiterpene cyclase involved in capsidiol biosynthesis in *Nicotiana tabacum*. Eur. J. Biochem. 269: 3570-3577.

Buchholz, A., Hurlebaus, J., Wandrey, C., and Takors, R. 2002. Metabolomics: quantification of intracellular metabolite dynamics. Biomol. Eng. 19: 5-15.

Bulow, L. 1987. Characterization of an artificial bifunctional enzyme, beta- galactosidase/galactokinase, prepared by gene fusion. Eur. J. Biochem. 163: 443-448.

Bulow, L., and Mosbach, K. 1991. Multienzyme systems obtained by gene fusion. Trends Biotechnol. 9: 226-231.

Burgard, A.P., and Maranas, C.D. 2001. Probing the performance limits of the *Escherichia coli* metabolic network subject to gene additions or deletions. Biotechnol. Bioeng. 74: 364-375.

Burgard, A.P., Vaidyaraman, S., and Maranas, C.D. 2001. Minimal reaction sets for *Escherichia coli* metabolism under different growth requirements and uptake environments. Biotechnol. Prog. 17: 791-797.

Carrier, T., Jones, K.L., and Keasling, J.D. 1998. mRNA stability and plasmid copy number effects on gene expression from an inducible promoter system. Biotechnol. Bioeng. 59: 666-672.

Cha, H.J., Wu, C.-F., Valdes, J.J., Rao, G., and Bentley, W.E. 2000. Observations of green fluorescent protein as a fusion partner in genetically engineered *Escherichia coli*: Monitoring protein expression and solubility. Biotechnol. Bioeng. 67: 565-574.

Chao, Y.-P., Wen, C.-S., Chiang, C.-J., and Wang, J.-J. 2001. Construction of the expression vector based on the growth phase- and growth rate *rmf* promoter: use of cell growth rate to control the expression of cloned genes in *Escherichia coli*. Biotechnol. lett. 23: 5-11.

Chapon, C. 1982. Expression of *mal*T, the regulator gene of the maltose region in *Escherichia coli*, is limited both at transcription and translation. EMBO J. 1: 369-74.

Court, D.L., Sawitzke, J.A., and Thomason, L.C. 2002. Genetic engineering using homologous recombination. Annu. Rev. Genet. 36: 361-388.

Covert, M.W., and Palsson, B.O. 2002. Transcriptional regulation in constraints-based metabolic models of *Escherichia coli*. J. Biol. Chem. 277: 28058-28064.

Covert, M.W., Schilling, C.H., Famili, I., Edwards, J.S., Goryanin, II, Selkov, E., and Palsson, B.O. 2001. Metabolic modeling of microbial strains *in silico*. Trends Biochem. Sci. 26: 179-86.

Crameri, A., Dawes, G., Rodriguez, E., Silver, S., and Stemmer, W.P.C. 1997. Molecular evolution of an arsenate detoxification pathway by DNA shuffling. Nat. Biotechnol. 15: 436-438.

Datsenko, K.A., and Wanner, B.L. 2000. One-step inactivation of chromosomal genes in *Escherichia coli* K-12 using PCR products. Proc. Natl. Acad. Sci. USA. 97: 6640-6645.

de Boer, H.A., Comstock, L.J., Hui, A., Wong, E., and Vasser, M. 1983. A hybrid promoter and portable Shine-Dalgarno regions of *Escherichia coli*. Biochem. Soc. Symp. 48: 233-244.

de Boer, H.A., Hui, A., Comstock, L.J., Wong, E., and Vasser, M. 1983. Portable Shine-Dalgarno regions: a system for a systematic study of defined alterations of nucleotide sequences within *E. coli* ribosome binding sites. DNA. 2: 231-235.

de Lorenzo, V., Herrero, M., Jakubzik, U., and Timmis, K.N. 1990. Mini-Tn5 transposon derivatives for insertion mutagenesis, promoter probing, and chromosomal insertion of cloned DNA in gram-negative eubacteria. J. Bacteriol. 172: 6568-6572.

de Lorenzo, V., Herrero, M., Sanchez, J.M., and Timmis, K.N. 1998. Mini-transposons in microbial ecology and environmental biotechnology. FEMS Microbiol. Ecol. 27: 211-224.

Dien, B.S., Nichols, N.N., and Bothast, R.J. 2001. Recombinant *Escherichia coli* engineered for production of L-lactic acid from hexose and pentose sugars. J. Indus. Microbiol. Biotechnol. 27: 259-264.

Ellermeier, C.D., Janakiraman, A., and Slauch, J.M. 2002. Construction of targeted single copy *lac* fusions using lambda Red and FLP-mediated site-specific recombination in bacteria. Gene. 290: 153-161.

Eymann, C., Homuth, G., Scharf, C., and Hecker, M. 2002. *Bacillus subtilis* functional genomics: global characterization of the stringent response by proteome and transcriptome analysis. J. Bacteriol. 184: 2500-2520.

Farmer, W.R., and Liao, J.C. 2000. Improving lycopene production in *Escherichia coli* by engineering metabolic control. Nat. Biotechnol. 18: 533-537.

Forbes, N.S., Clark, D.S., and Blanch, H.W. 2000. Analysis of metabolic fluxes in mammalian cells. In: Schugerl, K. and Bellgart, K., ed. Bioreaction Engineering: Modeling and Control. Berlin Heidelberg: Springer-Verlag.

Forbes, N.S., Clark, D.S., and Blanch, H.W. 2001. Using isotopomer path tracing to quantify metabolic fluxes in pathway models containing reversible reactions. Biotechnol. Bioeng. 74: 196-211.

Georgiou, G., and Valax, P. 1996. Expression of correctly folded proteins in *Escherichia coli*. Curr. Opin. Biotechnol. 7: 190-197.

Grant, A.J., Haigh, R., Williams, P., and O'Connor, C.D. 2001. An *in vitro* transposon system for highly regulated gene expression: construction of *Escherichia coli* strains with arabinose-dependent growth at low temperatures. Gene. 280: 145-151.

Guet, C.C., Elowitz, M.B., Hsing, W., and Leibler, S. 2002. Combinatorial synthesis of genetic networks. Science. 296: 1466-1470.

Haldimann, A., and Wanner, B.L. 2001. Conditional-replication, integration, excision, and retrieval plasmid-host systems for gene structure-function studies of bacteria. J. Bacteriol. 183: 6384-6393.

Hatzimanikatis, V., Choe, L.H., and Lee, K.H. 1999. Proteomics: theoretical and experimental considerations. Biotechnol. Prog. 15: 312-318.

Hatzimanikatis, V., and Lee, K.H. 1999. Dynamical analysis of gene networks requires both mRNA and protein expression information. Metab. Eng. 1: 275-281.

Hopwood, D.A., Wright, H.M., Bibb, M.J., and Cohen, S.N. 1977. Genetic recombination through protoplast fusion in *Streptomyces*. Nature. 268: 171-174.

Huang, L.C., Wood, E.A., and Cox, M.M. 1997. Convenient and reversible site-specific targeting of exogenous DNA into a bacterial chromosome by use of the FLP recombinase: the FLIRT system. J. Bacteriol. 179: 6076-6083.

Hyde, C.C., Ahmed, S.A., Padlan, E.A., Miles, E.W., and Davies, D.R. 1988. Three-dimensional structure of the tryptophan synthase alpha 2 beta 2 multienzyme complex from *Salmonella typhimurium*. J. Biol. Chem. 263: 17857-17871.

Jensen, P.R., and Hammer, K. 1998a. Artificial promoters for metabolic optimization. Biotechnol. Bioeng. 58: 191-195.

Jensen, P.R., and Hammer, K. 1998b. The sequence of spacers between the consensus sequences modulates the strength of prokaryotic promoters. Appl. Environ. Microbiol. 64: 82-87.

Jensen, P.R., Westerhoff, H.V., and Michelsen, O. 1993. The use of *lac*-type promoters in control analysis. Eur. J. Biochem. 211: 181-191.

Jones, K.L., and Keasling, J.D. 1998. Construction and characterization of F plasmid-based expression vectors. Biotechnol. Bioeng. 59: 659-665.

Jones, K.L., Kim, S.W., and Keasling, J.D. 2000. Low-copy plasmids can perform as well as or better than high-copy plasmids for metabolic engineering of bacteria. Metab. Eng. 2: 328-338.

Kane, J.F. 1995. Effects of rare codon clusters on high-level expression of heterologous proteins in *Escherichia coli*. Curr. Opin. Biotechnol. 6: 494-500.

Keasling, J.D. 1999. Gene-expression tools for the metabolic engineering of bacteria. Trends Biotechnol. 17: 452-460.

Khlebnikov, A., Datsenko, K.A., Skaug, T., Wanner, B.L., and Keasling, J.D. 2001. Homogeneous expression of the P(BAD) promoter in *Escherichia coli* by constitutive expression of the low-affinity high-capacity AraE transporter. Microbiol. 147: 3241-3247.

Khlebnikov, A., Risa, O., Skaug, T., Carrier, T.A., and Keasling, J.D. 2000. Regulatable arabinose-inducible gene expression system with consistent control in all cells of a culture. J. Bacteriol. 182: 7029-7034.

Kiser, K.B., and Schmidt, M.G. 1999. Regulation of the *Escherichia coli secA* gene is mediated by two distinct RNA structural conformations. Curr. Microbiol. 38: 113-121.

Kobayashi, M., Nagata, K., and Ishihama, A. 1990. Promoter selectivity of *Escherichia coli* RNA polymerase: effect of base substitutions in the promoter -35 region on promoter strength. Nucleic Acids Res. 18: 7367-7372.

Kristensen, C.S., Eberl, L., Sanchez-Romero, J.M., Givskov, M., Molin, S., and De Lorenzo, V. 1995. Site-specific deletions of chromosomally located DNA segments with the multimer resolution system of broad-host-range plasmid RP4. J. Bacteriol. 177: 52-58.

Le Borgne, S., Palmeros, B., Bolivar, F., and Gosset, G. 2001. Improvement of the pBRINT-Ts plasmid family to obtain marker-free chromosomal insertion of cloned DNA in *E. coli*. Biotechniques. 30: 252-254, 256.

Looman, A.C., Bodlaender, J., Comstock, L.J., Eaton, D., Jhurani, P., de Boer, H.A., and van Knippenberg, P.H. 1987. Influence of the codon following the AUG initiation codon on the expression of a modified *lacZ* gene in *Escherichia coli*. EMBO J. 6: 2489-2492.

Lovett, S.T., Hurley, R.L., Sutera, V.A., Jr., Aubuchon, R.H., and Lebedeva, M.A. 2002. Crossing over between regions of limited homology in *Escherichia coli*. RecA-dependent and RecA-independent pathways. Genetics. 160: 851-859.

Mader, U., Homuth, G., Scharf, C., Buttner, K., Bode, R., and Hecker, M. 2002. Transcriptome and proteome analysis of *Bacillus subtilis* gene expression modulated by amino acid availability. J. Bacteriol. 184: 4288-4295.

Makrides, S.C. 1996. Strategies for achieving high-level expression of genes in *Escherichia coli*. Microbiol. Rev. 60: 512-538.

Martin, V.J., Yoshikuni, Y., and Keasling, J.D. 2001. The *in vivo* synthesis of plant sesquiterpenes by *Escherichia coli*. Biotechnol. Bioeng. 75: 497-503.

Martin, V.J.J., Pitera, D., Withers, S., Newman, J., and Keasling, J.D. 2003. Production of amorpha-4,11-diene via an engineered mevalonate pathway in *Escherichia coli*. Nat. Biotechnol. 21: 796-802.

Martin, V.J.J., Smolke, C.D., and Keasling, J.D. 2002. Redesigning Cells for Production of Complex Organic Molecules. ASM News. 68: 336-343.

Martinez-Morales, F., Borges, A.C., Martinez, A., Shanmugam, K.T., and Ingram, L.O. 1999. Chromosomal integration of heterologous DNA in *Escherichia coli* with precise removal of markers and replicons used during construction. J. Bacteriol. 181: 7143-7148.

Meek, T.D., Garvey, E.P., and Santi, D.V. 1985. Purification and characterization of the bifunctional thymidylate synthetase-dihydrofolate reductase from methotrexate-resistant *Leishmania tropica*. Biochem. 24: 678-686.

Merlin, C., McAteer, S., and Masters, M. 2002. Tools for characterization of *Escherichia coli* genes of unknown function. J. Bacteriol. 184: 4573-4581.

Morgan-Kiss, R.M., Wadler, C., and Cronan, J.E., Jr. 2002. Long-term and homogeneous regulation of the *Escherichia coli ara*BAD promoter by use of a lactose transporter of relaxed specificity. Proc. Natl. Acad. Sci. USA. 99: 7373-7377.

Mulligan, M.E., Brosius, J., and McClure, W.R. 1985. Characterization *in vitro* of the effect of spacer length on the activity of *Escherichia coli* RNA polymerase at the TAC promoter. J. Biol. Chem. 260: 3529-3538.

Novick, A., and Weiner, M. 1957. Enzyme induction as an all-or-none phenomenon. Proc. Natl. Acad. Sci. USA. 43: 553-556.

Oh, M.K., and Liao, J.C. 2000a. DNA microarray detection of metabolic responses to protein overproduction in *Escherichia coli*. Metab. Eng. 2: 201-209.

Oh, M.K., and Liao, J.C. 2000b. Gene expression profiling by DNA microarrays and metabolic fluxes in *Escherichia coli*. Biotechnol. Prog. 16: 278-286.

Oh, M.K., Rohlin, L., Kao, K.C., and Liao, J.C. 2002. Global expression profiling of acetate-grown *Escherichia coli*. J. Biol. Chem. 277: 13175-13183.

Palmeros, B., Wild, J., Szybalski, W., Le Borgne, S., Hernandez-Chavez, G., Gosset, G., Valle, F., and Bolivar, F. 2000. A family of removable cassettes designed to obtain antibiotic-resistance-free genomic modifications of *Escherichia coli* and other bacteria. Gene. 247: 255-264.

Panke, S., Sanchez-Romero, J.M., and de Lorenzo, V. 1998. Engineering of quasi-natural *Pseudomonas putida* strains for toluene metabolism through an ortho-cleavage degradation pathway. Appl. Environ. Microbiol. 64: 748-751.

Patnaik, R., Louie, S., Gavrilovic, V., Perry, K., Stemmer, W.P., Ryan, C.M., and del Cardayre, S. 2002. Genome shuffling of *Lactobacillus* for improved acid tolerance. Nat. Biotechnol. 20: 707-712.

Pedersen, H., Carlsen, M., and Nielsen, J. 1999. Identification of enzymes and quantification of metabolic fluxes in the wild type and in a recombinant *Aspergillus oryzae* strain. Appl. Environ. Microbiol. 65: 11-19.

Petersen, C. 1989. Long-range translational coupling in the *rpl*JL-*rpo*BC operon of *Escherichia coli*. J. Mol. Biol. 206: 323-332.

Phelps, T.J., Palumbo, A.V., and Beliaev, A.S. 2002. Metabolomics and microarrays for improved understanding of phenotypic characteristics controlled by both genomics and environmental constraints. Curr. Opin. Biotechnol. 13: 20-24.

Pramanik, J., and Keasling, J.D. 1997. Stoichiometric model of *Escherichia coli* metabolism: Incorporation of growth-rate dependent biomass composition and mechanistic energy requirements. Biotechnol. Bioeng. 56: 398-421.

Pramanik, J., and Keasling, J.D. 1998. Effect of *Escherichia coli* biomass composition on central metabolic fluxes predicted by a stoichiometric model. Biotechnol. Bioeng. 60: 230-238.

Record M.T. Jr., Reznikoff, W.S., Craig, M.L., McQuade, K.L., and Schlax, P.A. 1996. *Escherichia coli* RNA polymerase (Eσ70), promoters, and the kinetics of the steps of transcription initiation. In: Neidhardt, F.C., ed. *Escherichia coli* and *Salmonella* : cellular and molecular biology. Washington, D.C.: ASM Press.

Rex, G., Surin, B., Besse, G., Schneppe, B., and McCarthy, J.E. 1994. The mechanism of translational coupling in *Escherichia coli*. Higher order structure in the *atp*HA mRNA acts as a conformational switch regulating the access of de novo initiating ribosomes. J. Biol. Chem. 269: 18118-18127.

Rossignol, M., Moulin, L., and Boccard, F. 2002. Phage HK022-based integrative vectors for the insertion of genes in the chromosome of multiply marked *Escherichia coli* strains. FEMS Microbiol. Lett. 213: 45-49.

Schilling, C.H., Covert, M.W., Famili, I., Church, G.M., Edwards, J.S., and Palsson, B.O. 2002. Genome-scale metabolic model of *Helicobacter pylori* 26695. J. Bacteriol. 184: 4582-4593.

Schilling, C.H., Edwards, J.S., Letscher, D., and Palsson, B.O. 2000. Combining pathway analysis with flux balance analysis for the comprehensive study of metabolic systems. Biotechnol. Bioeng. 71: 286-306.

Schmidt, K., Carlsen, M., Nielsen, J., and Villadsen, J. 1997. Modeling isotopomer distributions in biochemical networks using isotopomer mapping matrices. Biotechnol. Bioeng. 55: 831-840.

Schmidt, M.G., and Oliver, D.B. 1989. SecA protein autogenously represses its own translation during normal protein secretion in *Escherichia coli*. J. Bacteriol. 171: 643-649.

Schmidt-Dannert, C., Umeno, D., and Arnold, F.H. 2000. Molecular breeding of carotenoid biosynthetic pathways. Nat. Biotechnol. 18: 750-753.

Shao, Z., Zhao, H., Giver, L., and Arnold, F.H. 1998. Random-priming *in vitro* recombination: an effective tool for directed evolution. Nucleic Acids Res. 26: 681-683.

Siegele, D.A., and Hu, J.C. 1997. Gene expression from plasmids containing the *ara*BAD promoter at subsaturating inducer concentrations represents mixed populations. Proc. Natl. Acad. Sci. USA. 94: 8168-8172.

Smolke, C.D., Carrier, T.A., and Keasling, J.D. 2000. Coordinated, differential expression of two genes through directed mRNA cleavage and stabilization by secondary structures. Appl. Environ. Microbiol. 66: 5399-5405.

Smolke, C.D., and Keasling, J.D. 2002a. Effect of copy number and mRNA processing and stabilization on transcript and protein levels from an engineered dual-gene operon. Biotechnol. Bioeng. 78: 412-424.

Smolke, C.D., and Keasling, J.D. 2002b. Effect of gene location, mRNA secondary structures, and RNase sites on expression of two genes in an engineered operon. Biotechnol. Bioeng. 80: 762-76.

Smolke, C.D., Khlebnikov, A., and Keasling, J.D. 2001a. Effects of transcription induction homogeneity and transcript stability on expression of two genes in a constructed operon. Appl. Microbiol. Biotechnol. 57: 689-696.

Smolke, C.D., Martin, V.J.J., and Keasling, J.D. 2001b. Controlling the metabolic flux through the carotenoid pathway using directed mRNA processing and stabilization. Metab. Eng. 3: 313-321.

Stemmer, W.P. 1994. Rapid evolution of a protein *in vitro* by DNA shuffling. Nature. 370: 389-391.

Stephanopoulos, G., Aristidou, A., and Nielsen, J. 1998. Metabolic Engineering: Principles and Methodologies. San Diego: Academic Press

Tunner, J.R., Robertson, C.R., Schippa, S., and Matin, A. 1992. Use of glucose starvation to limit growth and induce protein production in *Escherichia coli*. Biotechnol. Bioeng. 40: 271-279.

Umeno, D., Tobias, A.V., and Arnold, F.H. 2002. Evolution of the C30 carotenoid synthase CrtM for function in a C40 pathway. J. Bacteriol. 184: 6690-6699.

Utsumi, R., Kusafuka, S., Nakayama, T., Tanaka, K., Takayanagi, Y., Takahashi, H., Noda, M., and Kawamukai, M. 1993. Stationary phase-specific expression of the fic gene in *Escherichia coli* K-12 is controlled by the *rpo*S gene product (sigma 38). FEMS Microbiol. Lett. 113: 273-278.

Vind, J., Sorensen, M.A., Rasmussen, M.D., and Pedersen, S. 1993. Synthesis of proteins in *Escherichia coli* is limited by the concentration of free ribosomes. Expression from reporter genes does not always reflect functional mRNA levels. J. Mol. Biol. 231: 678-688.

Voigt, C.A., Martinez, C., Wang, Z.G., Mayo, S.L., and Arnold, F.H. 2002. Protein building blocks preserved by recombination. Nat. Struct. Biol. 9: 553-558.

Voigt, C.A., Mayo, S.L., Arnold, F.H., and Wang, Z.G. 2001a. Computational method to reduce the search space for directed protein evolution. Proc. Natl. Acad. Sci. USA. 98: 3778-3783.

Voigt, C.A., Mayo, S.L., Arnold, F.H., and Wang, Z.G. 2001b. Computationally focusing the directed evolution of proteins. J. Cell. Biochem. Suppl. Suppl: 58-63.

Wang, C., Oh, M.K., and Liao, J.C. 2000. Directed evolution of metabolically engineered *Escherichia coli* for carotenoid production. Biotechnol. Prog. 16: 922-926.

Wanner, B.L. 1993. Gene regulation by phosphate in enteric bacteria. J. Cell. Biochem. 51: 47-54.

Yokobayashi, Y., Collins, C.H., Leadbetter, J.R., Weiss, R., and Arnold, F.H. 2003. Evolutionary design of genetic circuits and cell-cell communications. Adv. Comp. Sys. In press.

Yokobayashi, Y., Weiss, R., and Arnold, F.H. 2002. Directed evolution of a genetic circuit. Proc. Natl. Acad. Sci. USA. 99: 16587-16591.

Yoon, S.H., Han, M.J., Lee, S.Y., Jeong, K.J., and Yoo, J.S. 2003. Combined transcriptome and proteome analysis of *Escherichia coli* during high cell density culture. Biotechnol. Bioeng. 81: 753-767.

Zhang, Y.X., Perry, K., Vinci, V.A., Powell, K., Stemmer, W.P., and del Cardayre, S.B. 2002. Genome shuffling leads to rapid phenotypic improvement in bacteria. Nature. 415: 644-646.

Zhao, H., Giver, L., Shao, Z., Affholter, J.A., and Arnold, F.H. 1998. Molecular evolution by staggered extension process (StEP) *in vitro* recombination. Nat. Biotechnol. 16: 258-261.

Zupke, C., and Stephanopoulos, G. 1994. Modeling of isotope distributions and intracellular fluxes in metabolic networks using atom mapping matrices. Biotechnol. Prog. 10: 489-498.

4

Expression Systems in *Bacillus*

Rob Meima, Jan Maarten van Dijl, Siger Holsappel, and Sierd Bron

Abstract

Bacillus subtilis is the best-studied Gram-positive bacterium known today. Over the past five decades this species has become the paradigm of Gram-positive genetics and physiology. In addition, *B. subtilis* and several other bacilli have a long history of safe use in traditional food production and industrial scale applications, making them of considerable commercial importance. In this chapter, we describe some salient features of gene expression in this intriguing family of microbes. A general overview of the genus, some of its members and their characteristics is presented first. The next section outlines general characteristics of gene expression, and highlights a number of examples of the expression and regulation repertoire available to the *Bacillus* cell. A section on protein secretion describes the different, well-characterized protein secretion pathways of bacilli, which are capable of secreting a large number of different proteins, often at considerable levels. Next, an overview of commonly used vectors and (inducible) promoters is given, followed by a section devoted to *Bacillus* genomics as an important source of new insights and tools. The chapter concludes with a summary of current topics in the field of *Bacillus* research and perspectives for future developments.

The *Bacillus* Genus

Bacteria of the Gram-positive genus *Bacillus* (type strain *Bacillus subtilis* Marburg ATCC6051) are among the most widely distributed microorganisms in nature, with representatives commonly isolated from soil and water environments (Priest, 1989). Bacilli are rod-shaped bacteria capable of endospore formation. The genus comprises a large number of species representing an extraordinary metabolic diversity, including thermophiles, psychrophiles, alkalophiles and acidophiles. Well-known species include aerobic species like *B. subtilis* (which can also be grown anaerobically using NO_3^- as an alternative electron acceptor; see Nakano and Zuber, 2002) and *B. amyloliquefaciens*, as well as (facultative) anaerobes like *B. licheniformis*. In addition, the genus includes mesophilic and thermophilic species, such as *B. thermoproteolyticus* and *B. stearothermophilus*, and several alkalophilic and acidophilic species. A number of acidophilic thermophiles, such as *B. acidocaldaricus* have recently been re-assigned to a separate genus, *Alicyclobacillus* (Wisotzkey *et al.*, 1992). The genus includes a variety of commercially important species, responsible for the production of a range of products including enzymes, fine biochemicals, antibiotics and insecticides (Harwood, 1992; Priest and Harwood, 1994; see below). Most species are harmless to man and animals, and only a few pathogens are known. The latter include *B. anthracis*, the causative agent of anthrax, *B. cereus*, which causes food poisoning, and several insect pathogens. Bacilli have also been used in several traditional food fermentations including the production in South-East Asia of Natto from soybean by *B. subtilis* var. *natto*. The low level of reported incidence of pathogenicity and the widespread use of its products and those of its close relatives (*B. amyloliquefaciens*, *B. licheniformis*) in the food, beverage and detergent industries, has resulted in the granting of GRAS (generally regarded as safe) status to several enzymes produced in *B. subtilis* by the U.S. Food and Drug Administration (FDA).

Industrial Uses

Besides the use of *B. subtilis* for the production of the soybean fermentation product Natto, bacilli have a history of safe use in industrial production of enzymes for food, feed and detergent applications (Table 1), as well as for the synthesis of fine chemicals, insecticides and antibiotics. The industrial production of enzymes constitutes an annual market of approximately US$ 1 billion. Currently, half of this volume is produced in bacilli, and the majority of enzymes are made in *B. subtilis*, *B. licheniformis* and *B. amyloliquefaciens*. In addition, *B. thuringiensis* is used for the production of crystalline proteins with very potent insecticidal properties. Furthermore, *Bacillus*

Table 1. An overview of some of the most important industrial enzymes produced in *Bacillus* spp. Adapted and updated from Bron *et al.* (1999).

Enzyme	Host strain(s)
α-amylase	*B. licheniformis*, *B. amyloliquefaciens*, *B. circulans*, *B. subtilis*, *B. stearothermophilus*
β-amylase	*B. polymyxa*, *B. cereus*, *B. megaterium*
Alkaline phosphatase	*B. licheniformis*
Cyclodextran glucanotransferase	*B. macerans*, *B. megaterium*, *Bacillus* sp.
β-galactosidase	*B. stearothermophilus*
β-Glucanase	*B. subtilis*, *B. circulans*, *B. amyloliquefaciens*
β-Glucosidase	*Bacillus* sp.
Glucose iosmerase	*B. coagulans*
Glucosyl transferase	*B. megaterium*
Glutaminase	*B. subtilis*
Galactomannase	*B. subtilis*
β-lactamase	*B. licheniformis*
Lipases	*Bacillus* sp.
Neutral (metallo-) protease	*B. lentus*, *B. polymyxa*, *B. subtilis*, *B. thermoproteolyticus*, *B. amyloliquefaciens*
Penicillin acylase	*Bacillus* sp.
Pullulanase	*Bacillus* sp., *B. acidopullulans*
Alkaline (serine-) protease	*B. amyloliquefaciens*, *B. amylosaccharicus*, *B. licheniformis*, *B. subtilis*
Urease	*Bacillus* sp.
Uricase	*Bacillus* sp.
Xylanases	*Bacillus* sp.

spp. are used for the synthesis of peptides with antimicrobial properties, which may either be produced non-ribosomally (*e.g.* gramicidin-S) or via normal ribosomal synthesis (*e.g.* subtilin).

High-level gene expression is of particular importance in industrial applications where commercially relevant production titers are required. Furthermore, detailed studies of gene expression and the mechanisms by which the cell is able to selectively induce or repress (sets of) genes, has significantly contributed to our fundamental understanding of the bacterial cell and provided important tools for researchers. The basic gene expression elements and some of the mechanisms governing gene expression in *Bacillus* sp. are discussed in the following paragraphs.

Regulation of Gene Expression: Common Themes and Examples

The conversion of the genetic code into a biologically active polypeptide involves controlled transcription of the primary DNA sequence into a short-lived messenger molecule, or mRNA, and its subsequent translation into the corresponding amino acid sequence. The genetic elements and enzymatic functions involved in these basic processes of life have all been identified in *B. subtilis*, and some of the well-characterized expression modules and transcriptional regulators are described below.

Transcription and Translation Elements in Bacilli

Temporal and compartmentalized expression of genes in *B. subtilis* is governed by a variety of sigma (σ)-factors which, together with the RNA polymerase core enzyme, constitute the RNA polymerase holoenzyme. The holoenzyme stimulates transcription of genes in a process that is initiated by sequence specific interactions between the -35 and -10 promoter sequences, and specific regions within the sigma factors. For instance, many housekeeping genes expressed during

Table 2. Sigma factors of *B. subtilis*.

Sigma factor	Coding gene	Class	Consensus recognition sequence [a]
σ^A (σ^{43})	*sigA* (*rpoD*)	Vegetative	TTGaca-N$_{14}$-tgnTAtaat
σ^B (σ^{37})	*sigB* (*rpoF*)	General stress	rGGwTTrA-N$_{12-15}$-GGgtAt
σ^D (σ^{28})	*sigD* (*flaB*)	Flagella synthesis, motility, chemotaxis	TAAA-N$_{14-16}$-gCCGATAT
σ^E (σ^{29})	*sigE* (*spoIIGB*)	Early sporulation stages, mothercell	ATa-N$_{16-18}$-cATAcanT
σ^F	*sigF* (*spoIIAC*)	Early sporulation, forespore	GywTA-N$_{15}$-GgnrAnAnTw
σ^G	*sigG* (*spoIIIG*)	Late sporulation, forespore	gnATr-N$_{15}$-cATnmTA
σ^H (σ^{30})	*sigH* (*spoOH*)	Vegetative, early stationary, alkaline stress; minor	RnAGGAwWW-N$_{11-12}$-RnnGAAT
σ^K (σ^{27})	*spoIIIC-spoIVCB* [b]	Late sporulation, mothercell	AC-N$_{16-18}$-CATAnmnT
σ^L (σ^{54}, σ^N)	SigL	Utilization of arginine and ornithine, transport of fructose; minor	TGGcA-N$_5$-CTTGCAT
σ^M	*sigM* (*yhdM*)	ECF [c], salt/osmotic stress	TGCAAC-N$_{16-17}$-CGTGta
σ^V	SigV	Probable ECF [c], minor	Unknown
σ^W	*sigW* (*ybbL*)	ECF [c], alkaline stress	TGAAAC-N$_{16-17}$-CGTa
σ^X	*sigX* (*ypuM*)	ECF [c], peptidoglycan synthesis and turnover; heat shock	tGtAAC-N$_{16-17}$-CGwC
σ^Y	*sigY* (*yxlB*)	Probable ECF [c], minor	Unknown
σ^Z	*sigZ*	ECF [c], minor	Unknown
σ^{ykoZ}	*ykoZ*	Unknown	Unknown
σ^{ylaC}	*ylaC*	ECF [c], hypothetical	Unknown

[a] adapted from Helmann and Moran, Jr. (2002), where N represents any nucleotide; R, A or G; W, A or T; and Y, C or T
[b] the sigma K encoding gene is interrupted by the so-called *skin* element, the excision of which depends on the SpoIVCA recombinase
[c] ECF, extracytoplasmic function-type sigma factor

vegetative growth of bacilli, contain a typical σ^A-dependent promoter, which is characterized by a –35 TTGACA consensus sequence and the –10 TATAAT hexanucleotide (Price *et al.*, 1983; Moran *et al.*, 1982; Table 2). These two critical elements are usually separated by a 17-nucleotide spacer sequence, and transcription is initiated around 5 nucleotides downstream of the –10 box (referred to as transcription start site or +1 nucleotide). Altogether, some 4,000 genes are part of the σ^A regulon of *B. subtilis*, although their relative expression may vary significantly depending primarily, but not exclusively, on the actual sequence of the –35 and –10 elements. Besides this major σ^A-dependent class of promoters, several other classes can be discriminated, based on their sequences and sigma factor-dependence. These include promoters induced under (i) general environmental and energetic stress (σ^B-dependent; Haldenwang and Losick, 1980), (ii) alkaline shock (σ^W; Huang *et al.*, 1998), (iii) salt stress (σ^M; Horsburgh and Moir, 1999), (iv) promoters of gene clusters involved in flagella synthesis and chemotaxis (σ^D; Helmann, 1991), and (v) a large number of promoters, the induction of which requires mother cell-specific (σ^E, Haldenwang *et al.*, 1981; σ^K, Stragier *et al.*, 1989) and forespore-specific (σ^F, Nakayama *et al.*, 1981; σ^G, Sun *et al.*, 1991) sigma factors during the different stages of the sporulation process. Other, minor sigma factors include σ^H, σ^L, σ^V, σ^X, and σ^Y (reviewed in Helmann and Moran, 2002; see Table 1). In addition, the activation of certain sigma factors, especially those active in different compartments during sporulation, depends on the expression and stability of so-called anti-sigma factors (*e.g.* SpoIIAB or anti-σ^F, and RsbW or anti-σ^B) and anti-anti-sigma factors (*e.g.* SpoIIAA or anti-anti-σ^F, and RsbV or anti-anti-σ^B; for a review, see Helmann, 1999).

The regulation of many genes and operons depends on the concerted activities of sigma factors and transcription activators or repressors. For instance, whereas heat-shock induction of the *dnaK* and *groESL* genes of *E. coli* depends on the alternative sigma factor σ^{32}, the *B. subtilis* homologues are transcribed from a σ^A-dependent promoter and their expression is regulated by the HrcA repressor and its cognate CIRCE element (controlling inverted repeat of chaperonin expression; Zuber and Schumann, 1994; Schulz *et al.*, 1995; Yuan and Wong, 1995). Other well-studied examples of transcription regulators include:

(i) CcpA, involved in catabolite repression (Henkin *et al.*, 1991; for review, see Stülke and Hillen, 2000); (ii) ComK, the competence transcription factor (van Sinderen *et al.*, 1994); (iii) SpoOA which, in its phosphorylated form, acts both as an activator and a repressor (depending on the affected gene) during the early stages of sporulation ("phosphorelay"; reviewed in Hoch, 1993); (iv) AbrB, the transition-state regulator which inhibits premature expression of genes involved in competence and sporulation during exponential growth (Perego *et al.*, 1988); and (v) as many as 35 different putative two-component regulatory systems each dedicated to unique signals and (sets of) genes (*e.g.* DegS-DegU [see below], PhoR-PhoP, *etc.*). It should be noted that this list is by no means complete, but merely serves as an illustration of the variety of promoters and regulatory elements which may be considered of potential use in industrial applications for high level expression of target genes.

Translation of mRNA species into polypeptides involves the concerted activities of ribosomes, consisting of rRNA and ribosmal proteins, tRNA[aa] molecules, initiation, elongation, and release factors. Since these are rather common features in various well-studied microorganisms, they will not be further discussed here. As is the case in other eubacteria, translation in *B. subtilis* is initiated by interaction of the 16S RNA, which is part of the ribosome, with the Shine-Dalgarno box or ribosome-binding site (RBS). Computational analysis of a large number of *B. subtilis* ribosome-binding sites has identified the consensus RBS sequence AAAGGAGG, which is separated from the start codon by a 7-nucleotide spacer sequence (Rocha *et al.*, 1999). The most frequently used start codon is ATG (78%), but TTG (13%), GTG (9%), and CTG (<1%) are also used as translation starts.

Two-Component Regulatory Systems

These are an important part of the cell's repertoire enabling it to adequately respond to environmental stimuli. These systems consist of a sensor kinase, which is able to process environmental signals into a cellular response by transferring a phosphate to its cognate response regulator. Once phosphorylated, the latter acts as a transcriptional regulator modulating the expression of its target genes. Based on experimental evidence and the genome sequence, a total of 35 putative two-component systems and one additional orphan sensor kinase have been identified in *B. subtilis* (for review, see Perego and Hoch, 2002). One of the earliest documented two-component systems of *B. subtilis* is DegSU (Kunst *et al.*, 1988). First identified in the mid 70's, the *degU* locus (then called *sacU*) was implicated in extracellular (degradative) enzyme production (Steinmetz *et al.*, 1976). The *degU32*[hy] mutation, resulting in hyper-phosphorylation of DegU, was shown to significantly increase protein secretion levels, while inhibiting competence (Dahl *et al.*, 1992). More recent evidence indicates that non-phosphorylated DegU is involved in the nutritional regulation of genetic competence by acting as a positive regulator of the *comK* gene (see below). The divergent activities of DegU and DegU~P in the stimulation of competence and protein secretion, respectively, effectively ensures a separation in time of these different adaptational strategies as the cell enters the stationary phase.

Another two-component system that was recently identified as being involved in protein secretion and quality control is the CssRS system. Accumulation of misfolded proteins at the extracytoplasmic side of the membrane is signaled by the membrane-associated CssS protein, which results in phosphorylation of CssR. This induces expression of the *htrA* and *htrB* genes, as well as the *cssRS* operon. The former encode membrane-associated and extracellular serine proteases involved in degradation of misfolded proteins (Noone *et al.*, 2000; Darmon *et al.*, 2002).

Interestingly, one of the two-component systems detected in *B. subtilis*, YccGF, was shown to be essential under laboratory growth conditions. This system is particularly well conserved in other Gram-positive species, for which it is also essential (S. Dubrac and T. Msadek; personal communication). The reasons why this is the case remain unknown.

Transcriptional Regulators

In addition to the response regulators described in the above section, many other regulators, acting either as inducers or repressors for the expression of specific (sets of) genes, have been identified. For reasons of simplicity and space, only a limited number of these transcriptional regulators will be discussed here.

A well-documented example of a transcriptional activator is the CTF or competence transcription factor, encoded by the *comK* gene. Genetic competence (*i.e.* the ability to internalize exogenous DNA) involves a cascade of gene activations and repressions, as well as numerous regulatory circuits, altogether involving the products of over 25 different genes (for a comprehensive review, see Dubnau and Lovett, 2002). The development of competence is under control of quorum sensing and nutritional inputs (Figure 1). The competence pheromones, ComX and CSF (competence stimulating factor, PhrC), are involved in the quorum-sensing control of competence by modulating the phosphorylation state of an early competence protein, ComA. Accumulation of the ComX protein is sensed by ComP, which subsequently catalyses phosphorylation of ComA. The mature PhrC pentapeptide (ERGMT) is internalized via the SpoOK (Opp) transporter system and acts as an inhibitor of the RapC protein which negatively controls the binding affinity of ComA~P for the *srfA* promoter (Perego, 2003). Phosphorylation of the ComA protein results in transcription of the *srfA* operon, encoding elements required for the non-ribosomal synthesis of the cyclic peptide antibiotic surfactin. Within the *srfA* operon resides a very small ORF encoding the ComS protein (D'Souza *et al.*, 1994; Hamoen *et al.*, 1995). This protein interacts with the ComK-MecA-ClpCP complex, thus destabilizing it (Turgay *et al.*, 1997). As a result, ComK is released and acts as a transcriptional activator of its own promoter and that of late competence genes, as well as a number of genes required for homologous recombination of internalized DNA. The ComK protein binds to its cognate target sites (AAAAN$_5$TTTT) as a tetramer, each of the two binding sites, which are separated by 2 or 4 helical turns, interacting with a ComK dimer (Hamoen *et al.*, 1998). Other factors that control the expression of competence are SinR, a transition state regulator that acts as an activator of ComS and ComK synthesis, non-phosphorylated DegU (see above), CodY, which represses the promoters of both *srfA* and *comK* in the presence of excess amino acids (Serror and Sonenshein, 1996), and the AbrB transition state regulator. When the concentration of SpoOA~P increases, the level of SinI, which acts as an antagonist of SinR, is stimulated. The resulting inhibition of SinR leads to decreased expression of *comS* and *comK*, and derepression of several sporulation genes. Hence, competence is downregulated and sporulation is induced. As is the case with SinR, the level of AbrB also depends on the phosphorylation state of SpoOA. The role of AbrB in the regulation of competence and sporulation is, however, somewhat more complicated. At high concentrations, AbrB acts as a negative regulator of *comK* expression. Phosphorylation of SpoOA represses transcription of *abrB*, allowing competence to develop. However, since AbrB also seems to act as an activator of *srfA*, *comK* activation is abolished when AbrB levels drop below a certain threshold. Since AbrB also controls transcription of *spoOA* from its σ^H-dependent promoter by acting as a repressor of the σ^H gene (*spoOH*), SpoOA synthesis is simultaneously induced, thus stimulating sporulation.

A common and very well characterized regulatory network in bacteria is carbon catabolite activation (CCA) or repression (CCR) of genes involved in carbohydrate metabolism. CCR in bacteria was first described in *B. subtilis*, and involves binding of the catabolite control protein, CcpA, to its cognate operator sequence, *cre* (catabolite response element; TGWNANCGNTNWCA). The interaction of CcpA alone with the *cre* operator is usually rather weak, and full repression depends on the concerted activity of CcpA and the Ser-46 phosphorylated HPr kinase/phosphatase. Several dozens of functional *cre* operators were found in a genome-wide survey, and it is estimated that around 10% of the genes in *B. subtilis* are regulated by CCR, underlining the importance of catabolite control in this organism (Deutscher *et al.*, 2002).

Although the above overview is by no means complete, it briefly illustrates how *Bacillus* cells are capable of processing developmental signals into, sometimes complex, cellular responses.

Genetic Amenibility

The isolation of naturally transformable bacilli strains in the late 1950's (Spizizen, 1958) represented a landmark in the genetic and physiological exploration of this intriguing organism. Transformation and transduction experiments have been instrumental, not only in early genetic and physiological studies, but also in building a physical map of the genome that has served as a blueprint for the genome-sequencing project (see below).

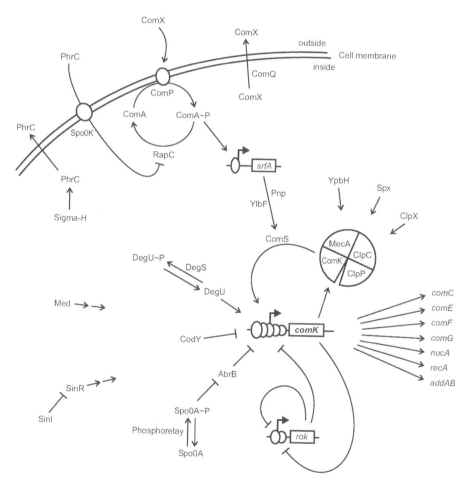

Figure 1. Schematic representation of the nutritional and developmental stimuli resulting in the establishment of genetic competence in *Bacillus subtilis*. See text for details.

Since the development of molecular genetic techniques, a variety of tools for modification and analysis of bacilli have been developed, including autonomous and integrative cloning systems, gene inactivation vectors, inducible expression cassettes, protein localization tools, *etc.* A number of these systems and their uses will be discussed in the Host-Vector Systems section. To date, with the availability of the complete genome sequence, exciting new possibilities have opened up for the holistic characterization of cellular process using genomics, transcriptomics, proteomics, and metabolomics, some of which are discussed in more detail in the Genomics section.

Networks: Multinational Consortia for *Bacillus* Research

Research on bacilli and a number of other Gram-positives has greatly benefited from various international collaborations. Arguably, the most important program was the joint European-Japanese effort to sequence the complete genome of *B. subtilis* 168 which, at the time of publication, represented the first Gram-positive whole genome sequence (Kunst *et al.*, 1997). A major spin-off from the genome sequencing project was the *Bacillus* Systematic gene Function Analysis program (BSFA), which is dedicated to the elucidation of the possible function of each of the genes in the genome that is either unknown or shares similarity with hypothetical genes (Kobayashi *et al.*, 2003). This and several other programs are discussed in the Genomics section.

Protein Secretion Pathways in Bacilli

B. subtilis and closely related bacilli are known for their excellent capacity to secrete proteins (at gram per liter concentrations) into the environment. This property has led to their exploitation as "cell factories" for a variety of secreted proteins (Table 1). Secretory proteins are synthesized in the cytoplasm, translocated across the cytoplasmic membrane, and channeled through the cell wall. In contrast to homologous proteins, heterologous proteins are frequently secreted with low efficiency. This problem, which applies especially to eukaryotic proteins, can be attributed to a variety of secretion pathway bottlenecks, such as poor membrane targeting, inefficient membrane translocation or cell wall passage, slow or incorrect protein folding, and degradation (Bolhuis *et al.*, 1999a). With the aim to eliminate or bypass such bottlenecks, much research addressing protein secretion in *Bacillus* was focused, in particular during the past decade, on the elucidation of mechanisms for protein secretion. Four distinct pathways for protein export from the cytoplasm have thus far been described. The largest number of exported proteins seems to follow the Sec pathway for protein secretion. In fact, the Sec pathway does not only export the majority of secretory proteins, but also most membrane- and cell wall-associated proteins. In contrast, the twin-arginine translocation (Tat) pathway, a pseudopilin export pathway for competence development, and certain ATP-binding cassette (ABC) transporters represent "special-purpose" pathways for the export of a relatively small number of proteins. This chapter will only describe the Sec and Tat pathways as they are most relevant or promising for recombinant protein production.

Signal Peptides

All known *Bacillus* proteins exported *via* the Sec or Tat pathways are synthesized as pre-proteins with amino-terminal signal peptides that direct them to their respective transport system. Based on the presence of a (putative) signal peptide and absence of a known retention signal, it has been estimated that about 4% of all *B. subtilis* proteins have the potential to be secreted into the environment. However, since retention signals can be removed by proteolysis, certain membrane- or cell wall-retained proteins are eventually released into the growth medium as well (Antelmann *et al.*, 2001).

Sec-Type Signal Peptides

Two major classes of Sec-type signal peptides are known to exist in *B. subtilis*. The first class is present in secretory pre-proteins that are processed by type I signal peptidases (SPases; Tjalsma *et al.*, 1998). The second class is present in lipoprotein precursors that are processed by the lipoprotein-specific (type II) SPase (Lsp; Tjalsma *et al.*, 1999a). Typical signal peptides of the first class direct the secretion of degradative enzymes. Signal peptide predictions have resulted in the identification of about 180 potential substrates for type I SPases in *B. subtilis* (Tjalsma *et al.*, 2000a). One of these was shown to be Tat-specific, while 9 other signal peptides are potentially Tat-specific (see below; Jongbloed *et al.*, 2002). This suggests that about 170 predicted signal peptides direct proteins into the Sec pathway. Indeed, 50 of

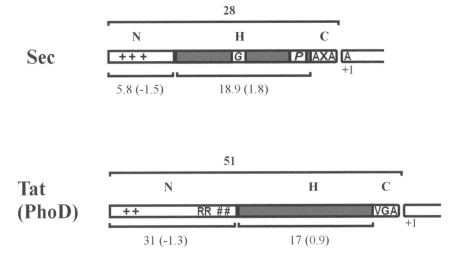

Figure 2. Properties of cleavable amino-terminal signal peptides of *B. subtilis*. Signal peptides consist of three distinct domains: a positively charged N-domain (N; "+"), a hydrophobic H-domain (H; gray box), and a C-domain (C) that includes the SPase recognition and cleavage site. Helix-breaking Gly or Pro residues in the middle of the H-domain allow the formation of hairpin-like structures and helix-breaking residues at the end of the H-domain facilitate SPase cleavage. Signal peptides of potentially secreted Sec-type pre-proteins of *B. subtilis* (top) were identified with the SignalP algorithm as described previously (Nielsen *et al.*, 1997; Tjalsma *et al.*, 2000a). The twin-arginine motif in the N-region of the signal peptide of PhoD (bottom), helix-breaking residues in the H-domains, consensus SPase recognition sites, and the most frequently occurring residues at the "+1" position of mature proteins are indicated. For each type of signal peptide, the average length of complete signal peptides, N-domains, and H-domains is indicated, together with the average hydrophobicity of N- and H-domains (between brackets). Adapted from van Dijl *et al.* (2002).

these proteins were identified in the growth medium of *B. subtilis* (Antelmann *et al.*, 2001). The properties of their signal peptides are summarized in Figure 2. The most frequently encountered (putative) SPase I recognition sequences are Ala-Ser-Ala, Ala-Lys-Ala, and Ala-Glu-Ala.

Twin-Arginine (RR-) Signal Peptides

It was recently predicted that 69 potential signal peptides, with and without SPase cleavage sites, can direct proteins into the Tat pathway of *B. subtilis* (Jongbloed *et al.*, 2002). This prediction was based on the presence of the motif K/R-R-X-#-# (where # is a hydrophobic residue) in the N-terminus of the signal peptides (Berks *et al.*, 2000). However, results of a proteomic verification indicate that the Tat pathway of *B. subtilis* is highly selective (Jongbloed *et al.*, 2002). Only one protein, the phosphodiesterase PhoD, was shown to be secreted in a strictly Tat-dependent fashion. This suggests that highly stringent criteria, such as a low overall hydrophobicity of the hydrophobic H-region (< 2.0) and the presence of hydrophobic residues at *both* +2 and +3 positions (Cristóbal *et al.*, 1999), must be applied for the prediction of RR-signal peptides. Thus, the total number of RR-signal peptides that may direct Tat-dependent protein secretion into the growth medium of *B. subtilis* was reduced to nine (*i.e.* the signal peptides of AlbB, PhoD, YkuE, YuiC, YwbN, PelB, PhrG, SpoIIQ and YwcI). The properties of the PhoD signal peptide are compared to those of Sec-type signal peptides in Figure 2.

The Sec-Dependent Protein Export Machinery

The Sec-dependent protein export machinery of *B. subtilis* is composed of six elements: (i) cytoplasmic chaperones or targeting factors; (ii) the translocation motor SecA; (iii) the translocation channel composed of SecYEG and, most likely, of SecDF-YrbF and SpoIIIJ/YqjG; (iv) SPases; (v) signal peptide peptidases (SPPases); and (vi) folding catalysts.

Chaperones and Targeting Factors

Passage through the translocation channel of the Sec machinery is only possible for pre-proteins in a loosely folded state. The best-characterized targeting factor of *B. subtilis* is the GTPase Ffh (fifty-four homologue), a homologue of the 54-kDa subunit of the eukaryotic signal recognition particle (SRP) (Honda *et al.*, 1993). The essential Ffh protein forms a ribonucleoprotein complex with the small cytoplasmic RNA (scRNA). In addition, the histone-like protein HBsu of *B. subtilis* is bound to the scRNA (Nakamura *et al.*, 1999). Most likely, Ffh binds to signal peptides that emerge from the ribosome. The SRP-nascent chain-ribosome complex is then targeted to the membrane with the aid of the GTPase FtsY. The precise docking site for *B. subtilis* FtsY is not known, but it seems likely that this essential protein will bind to anionic phospholipids, like FtsY of *E. coli* (de Leeuw *et al.*, 2000). As shown for *E. coli*, the release of the nascent chain and ribosome from the SRP-FtsY complex is preceded or accompanied by GTP binding to the Ffh protein in SRP and FtsY. Hydrolysis of GTP bound to Ffh and FtsY results in the dissociation and recycling of these components. Unlike the eukaryotic SRP, bacterial SRP does not seem to cause a translation arrest (Pugsley, 1993). For *E. coli*, it has been demonstrated that the more hydrophobic signal peptides target proteins primarily into the SRP pathway (Valent *et al.*, 1997). Since signal peptides of *B. subtilis* are, on the average, longer and more hydrophobic than those of *E. coli* (Tjalsma *et al.*, 2000), it is conceivable that many *B. subtilis* proteins are secreted in a SRP-dependent-fashion. In fact, the H-domains of 54 predicted signal peptides of *B. subtilis* have a mean hydrophobicity of 2.1 or higher. In *E. coli*, this hydrophobicity seems to be the threshold for SRP-dependent export of proteins with relatively short signal peptides (de Gier *et al.*, 1998; Valent *et al.*, 1997). It was shown by proteomic techniques that the secretion of many extracellular proteins of *B. subtilis* depends on SRP (Hirose *et al.*, 1999). The secretion of these proteins also depends on SecA (Hirose *et al.*, 1999), suggesting that SRP-dependent protein secretion by *B. subtilis* requires the translocation motor SecA. Consistently, Ffh of *B. subtilis* was shown to bind SecA *in vitro*, which may enhance the binding of preproteins to SecA (Bunai *et al.*, 1999). Notably, the *in vivo* effects of Ffh and SecA depletion may be indirect since the assembly of the Sec translocation channel is probably SRP- and/or SecA-dependent as well.

In *E. coli* it has been observed that general chaperones of the GroEL/ES and DnaK/DnaJ/GrpE machinery can maintain certain pre-proteins in a translocation-competent state by preventing their folding or aggregation (Kusukawa *et al.*, 1989; Wild *et al.*, 1996). More importantly, secretion-dedicated chaperones, such as SecB, facilitate protein targeting to the translocase by binding to mature regions of certain pre-proteins and the carboxyl-terminus of SecA (Fekkes and Driessen, 1999). Interestingly, SecB is absent from *B. subtilis* and no evidence has been obtained that general chaperones are involved in protein secretion by *B. subtilis* (T. Wiegert and W. Schumann, personal communication). However, the CsaA protein of *B. subtilis*, which has affinity for SecA, seems to act as a secretion-dedicated chaperone for a subset of secretory proteins (Müller *et al.*, 2000).

The Sec Machinery

The key elements of the Sec machinery of *B. subtilis* are the SecA motor protein and the SecYEG subcomplex that forms the preprotein conducting channel. According to models for preprotein translocation by the *E. coli* Sec machinery, several successive translocation steps can be distinguished (Fekkes and Driessen, 1999). Briefly, SecA dimers bind to acidic phospholipids and SecY, which results in their activation for preprotein recognition. SecA binding results in the insertion of a short pre-protein fragment (2-2.5-kDa) into the SecYEG translocation channel, which is followed by ATP binding. This causes a major conformational change in SecA, resulting in the insertion of the C-terminus of SecA into the membrane. This membrane insertion causes the translocation of a 2-2.5 kDa pre-protein segment through the SecYEG channel. Next, ATP is hydrolyzed by SecA, which leads to pre-protein release and SecA de-insertion from the membrane. Subsequent pre-protein translocation is driven both by repeated cycling of SecA through ATP-binding and hydrolysis, and the proton motive force (PMF). Notably, different secreted proteins of *B. subtilis* show different levels of SecA-dependence. This suggests that SecA may bind different precursors with different affinities (Leloup *et al.*, 1999). Finally, it has been proposed that cytoplasmic SecA of *B. subtilis* may replace a SecB ortholog by acting as an export-specific chaperone (Herbort *et al.*, 1999).

In addition to the indispensable heterotrimeric SecYEG sub-complex, the translocation channel of the *E. coli* Sec machinery contains the heterotrimeric SecDF-YajC sub-complex, which is dispensable for life of the cell and protein translocation (Fekkes and Driessen, 1999). The latter complex appears to be conserved in *B. subtilis*. However, in contrast to the situation in *E. coli* and most known bacteria, which contain separate *secD* and *secF* genes, *B. subtilis* contains a natural *secDF* gene fusion (Bolhuis *et al.*, 1998). Since the SecD and SecF proteins show sequence similarity, the fused SecDF protein has been termed a "molecular siamese twin". Unlike its equivalents in *E. coli*, SecDF of *B. subtilis* is not required for wild-type levels of protein secretion, but only when a secretory protein is overproduced at gram/liter levels (Bolhuis *et al.*, 1998). It has been proposed that SecDF of *B. subtilis*, together with the YajC orthologue YrbF, plays a role in: (i) Sec translocation channel assembly; (ii) PMF-driven protein translocation; or (iii) clearing of the translocation channel from signal peptides or misfolded proteins.

In addition to SecDF-YajC, the YidC protein of *E. coli* is part of the Sec machinery (Scotti *et al.*, 2000). The latter protein, which seems to play a dedicated and essential role in the membrane integration of newly synthesized membrane proteins, is present in a heterotetrameric complex with SecDF-YajC (Nouwen and Driessen, 2002). Two YidC orthologues, denoted SpoIIIJ and YqjG, have been identified in *B. subtilis* (Tjalsma *et al.*, 2003). The presence of either SpoIIIJ or YqjG is required for cell viability. Interestingly, the topology and stability of a variety of membrane proteins is not, or only mildly, affected by SpoIIIJ limitation in the absence of YqjG. By contrast, SpoIIIJ- and YqjG-limiting conditions result in a strong post-translocational defect in the stability of secretory proteins. While SpoIIIJ is required for sporulation, YqjG is dispensable for this process. The presently available data indicate that SpoIIIJ and YqjG have different, but overlapping functions in *B. subtilis*. Most likely, the post-translocational defect in the folding of secretory proteins upon SpoIIIJ/YqjG limitation is due to the malfunction of one or more membrane proteins.

Type I Signal Peptidases (SPases)

Type I SPases catalyse the release of secretory proteins from the membrane by signal peptide removal from the corresponding precursors (Dalbey *et al.*, 1997). Strikingly, *B. subtilis* 168 contains five paralogous type I SPases (SipS, SipT, SipU, SipV and SipW). In addition, plasmid-encoded SPases (SipP) have been identified in certain Natto-producing *B. subtilis* strains (Meijer *et al.*, 1995). Like in other organisms, SPase I activity in *B. subtilis* is required for cell viability and the presence of at least one of the "*major*" SPases SipS, SipT or SipP are essential (Tjalsma *et al.*, 1998, 1999b). In contrast, the "*minor*" SPases SipU, SipV or SipW, are insufficient for preprotein processing in general and for cell viability. The difference between *major* and *minor* SPases appears to be caused by differences in substrate specificity rather than expression levels despite the fact that the various *B. subtilis* SPases have largely overlapping substrate specificities (van Roosmalen *et al.*, 2001). A distinct substrate specificity has only been demonstrated for SipW, which specifically processes the TasA and YqxM pre-proteins (Stöver and Driks, 1999a,b). Interestingly, SipW belongs to the ER-type SPases, which are typically present in the endoplasmic reticulum of eukaryotes and the plasma membrane of archaea. These enzymes seem to employ a Ser-His catalytic dyad, or a Ser-His-Asp catalytic triad (Tjalsma *et al.*, 2000b). In contrast, SipS, SipT, SipU and SipV are prokaryotic-type SPases that make use of a Ser-Lys catalytic dyad (van Dijl *et al.*, 1995). Finally, it should be noted that SipS, SipT, SipV and SipW, but not SipU, are conserved in *B. amyloliquefaciens*. The SPases of this organism seem to have more distinct substrate specificities than their equivalents in *B. subtilis* (Chu *et al.*, 2002).

Signal Peptide Peptidases (SPPases)

Upon preprotein processing by SPase, the cleaved signal peptides are degraded. Two proteases of *E. coli* seem to have SPPase activity: the membrane-bound SppA (also called protease IV) and the cytoplasmic oligopeptidase A (OpdA; Novak and Dev, 1988). SppA appears to cleave membrane-inserted signal peptides and the resulting fragments are subsequently degraded in the cytoplasm by OpdA. While OpdA is not conserved in *B. subtilis*, an orthologue of SppA (YteI) is present in *B. subtilis*. SppA is required for efficient preprotein processing in *B. subtilis*, suggesting that SPase I activity is reduced by signal peptide accumulation in an *sppA* mutant (Bolhuis *et al.*, 1999b). In addition to SppA, the cytoplasmic translocation-enhancing protein TepA (YmfB) of *B. subtilis* has been implicated in signal peptide degradation on the basis of its similarity to SppA. However, TepA, which is not only required for preprotein processing, but also for efficient preprotein translocation, could also have other roles in secretion.

Folding Catalysts

Upon exit from the SecYEG translocation channel and entry in the cell wall, exported proteins must fold into their native conformation. Efficient and precise folding is critical for the activity and stability of exported proteins because the membrane-cell wall interface is a highly proteolytic environment (Bolhuis *et al.*, 1999a; Meens *et al.*, 1997; Stephenson and Harwood, 1998). Extracytoplasmic protein folding is promoted by various folding catalysts, which can be classified into five types in *B. subtilis*.

Peptidyl-prolyl cis-trans isomerases (PPIases)
PPIases catalyze a *cis-trans* isomerization of peptidyl-prolyl bonds. The only known (potential) extracytoplasmic PPIases of *B. subtilis* are PrsA (Kontinen and Sarvas, 1993) and YacD. PrsA is an essential lipoprotein that has been shown to set a limit to the secretion of degradative enzymes. PrsA limitation results in increased proteolysis of translocated proteins. The function of YacD is presently not known.

Thiol-disulfide oxidoreductases

Disulfide bonds are important for the activity and stability of many exported proteins, in particular those of eukaryotes. The *in vivo* formation of disulfide bonds is catalyzed by thiol-disulfide oxidoreductases. In most cases, disulfide bonds in exported proteins are formed upon membrane translocation since the cytoplasm is generally too reducing for this process. Even though the formation of (multiple) disulfide bonds in exported heterologous proteins is frequently inefficient in *B. subtilis*, at least four thiol-disulfide oxidoreductases have been identified in this organism (Bolhuis *et al.*, 1999c; Meima *et al.*, 2002). These are BdbA (YolI), BdbB (YolK), BdbC (YvgU) and BdbD (YvgV). While it is presently not clear whether BdbA and BdbD are secreted or retained in the membrane or cell wall, BdbB and BdbC are typical integral membrane proteins that expose their catalytic Cys residues on the extracytoplasmic membrane side. BdbC and BdbD are very important for the folding of heterologously produced and secreted alkaline phosphatase (PhoA) of *E. coli*, which contains two disulfide bonds. While BdbA is dispensable for this process, BdbB has a minor role in PhoA folding (Bolhuis *et al.*, 1999c). Interestingly, BdbC and BdbD are critical for the development of genetic competence, probably because they catalyze the formation of an intramolecular disulfide bond in the pseudopilin ComGC, which is involved in DNA binding and uptake (Meima *et al.*, 2002). Importantly, none of the four Bdb proteins is required for the secretion of proteins that lack disulfide bonds.

HtrA-like proteins

HtrA of *E. coli* is a periplasmic heat shock-inducible protease that removes "hopelessly" misfolded proteins (Spiess *et al.*, 1999). In addition, HtrA acts as a folding catalyst, in particular at low temperatures. *B. subtilis* contains three orthologs of *E. coli* HtrA: HtrA (YkdA), HtrB (YvtA) and YyxA (Noone *et al.*, 2000). These three proteins have a predicted amino-terminal membrane anchor, their protease active sites being located on the extracytoplasmic side of the membrane. Notably, transcription of the *htrA* and *htrB* genes is controlled by the CssRS two-component system (Darmon *et al.*, 2002), indicating that HtrA and HtrB have important roles in combating protein secretion stress. It is presently not known to what extent this involves their, presumably conserved, protein folding or degrading activities.

Non-proteinaceous folding catalysts
Some secretory proteins require cations for folding. For example, Fe^{3+} acts as a folding catalyst for levansucrase (Chambert *et al.*, 1990), and Ca^{2+} is required for the folding of levansucrase, neutral protease and α-amylase (Stephenson *et al.*, 1998; Veltman *et al.*, 1997). In addition, certain wall components impact on protein folding. Recent studies with *dlt* mutants of *B. subtilis* indicate that negatively charged teichoic acids can increase the rate of folding and/or stability of exported proteins (Hyyryläinen *et al.*, 2000). In these mutants, the negative charge of teichoic acids is not neutralized by ester-linked D-Ala residues. Notably, the increased negative charge of the wall of *dlt* mutant strains may result in the trapping of increased amounts of cations, which may thus become more available for post-translocational protein folding.

Intramolecular folding catalysts
Several secreted proteins of *B. subtilis* contain a dedicated intramolecular folding catalyst in the form of a propeptide. These propeptides are located between the signal peptide and the mature protein. Only the propeptides of secreted proteases have been shown to act in protein folding. Once the protease is active, the propeptide is removed through self-cleavage. Certain propeptides can promote protease activation *in trans* (Wandersman, 1989).

The Tat Machinery

Next to the Sec pathway, *B. subtilis* contains a functional Tat pathway, which is required for secretion of at least one protein, the phosphodiesterase PhoD (Jongbloed *et al.*, 2000). In general, Tat pathways are distinguishable from Sec pathways by two features: (i) they are accessible only to signal peptides with a RR-motif (see above), and (ii) they can transport tightly folded proteins. In fact, the prime function of the Tat pathway is probably the transport of proteins that either fold too rapidly or too tightly to allow their membrane passage through the Sec pathway (Berks *et al.*, 2000). Two major types of Tat machinery components have been identified in *E. coli* and other organisms. These components are termed TatC and TatA/B/E. The TatA/B/E proteins are paralogs (Berks *et al.*, 2000). In contrast to *E. coli*, which contains only one *tatC* gene, *B. subtilis* contains two genes specifying TatC orthologs, known as *tatCd* (*ycbT*) and *tatCy* (*ydiJ*). These *tatC* genes are preceded by *tatA* genes, denoted *tatAd* (*yczB*) and *tatAy* (*ydiI*), respectively. The third *tatA* gene of *B. subtilis*, known as *tatAc* (*ynzA*), is not genetically linked to a *tatC* gene. The presence of two *tatAC* gene clusters suggests that *B. subtilis* may have one Tat pathway with various paralogous components or, alternatively, two parallel Tat routes. Consistent with the latter view, the *tatAdCd* genes, which are located downstream of *phoD*, are only transcribed under conditions of phosphate starvation (Jongbloed *et al.*, 2000). Moreover, TatCd expression is critical for PhoD secretion, while TatCy is not required. So far, it is not clear whether the Tat pathway of *B. subtilis* can be exploited for the export of heterologous or cytoplasmic proteins, as has been demonstrated for the Tat pathway of *E. coli* (Thomas *et al.*, 2001).

Regulation of Secretion Machinery Components

The expression of protein secretion machinery components in *B. subtilis* is subject to a complex regulation that depends on nutrients, growth phase and cell density (van Dijl *et al.*, 2002). This is well exemplified with the *tatAdCd* genes, which are mainly expressed under conditions of phosphate starvation (Jongbloed *et al.*, 2000). Likewise, the expression of different components of the Sec-dependent protein transport pathway is regulated. Surprisingly, however, their expression patterns show major differences, which suggests that the demand for components of the Sec-dependent pathway vary under different conditions. In terms of protein secretion bottlenecks, it is interesting to note that the type I SPases SipS and SipT, which can be limiting for the secretion of certain hybrid precursors, are expressed in concert with genes for secreted degradative enzymes, their transcription being controlled by the DegS-DegU two-component regulatory system (van Dijl *et al.*, 1992, Tjalsma *et al.*, 1998).

Product Degradation

At all stages in the secretion process, heterologous proteins can be subject to degradation by proteases (Tjalsma *et al.*, 2000). Recent studies suggest that, in addition to at least eight secreted proteases, proteases residing at the membrane-cell wall interface can be particularly problematic (Wu *et al.*, 2002). In particular, the wall-bound serine protease CWBP52, specified by the *wprA* gene, is active at the site of preprotein translocation. Importantly, CWBP52 limitation results in increased yields of secreted α-amylase (Stephenson and Harwood, 1998). Thus, proteases of the latter type are, most likely, involved in the degradation of secreted proteins. It is anticipated that, ultimately, the elimination of wall associated proteases and the engineering of extracytoplasmic folding catalysts, such as PPIases and thiol-disulfide oxidoreductases, will lead to the successful secretion of complex heterologous proteins from *Bacillus* (Wu *et al.*, 1998, 2002).

Host-Vector Systems

Two different systems are basically used for gene cloning and gene expression in *B. subtilis*. The first makes use of plasmids that replicate autonomously in *B. subtilis*, and the second employs integrations in the host chromosome. These systems are briefly described in the following sections.

Plasmid-Based Systems

Non-Native Plasmids

Since *B. subtilis* 168, the naturally transformable strain, does not contain endogenous plasmids, plasmid-based cloning vectors were initially extracted from other Gram-positive bacteria, such as *Staphylococcus aureus* and lactic acid bacteria. A considerable number of such plasmids are able to replicate in *B. subtilis*. Some of them naturally contain genes encoding resistance to antibiotics, which provides them with a selectable marker. Some of these non-native plasmid replicons, such as pUB110, pC194 and pE194, are still used in *B. subtilis*. These plasmids were derived from *S. aureus* and carry useful resistance markers against, respectively, kanamycin, chloramphenicol and erythromycin. Extended reviews on these plasmid vectors are available (Bron, 1990; Jannière *et al.*, 1993; Bron *et al.*, 1999).

Mode of Replication and Plasmid Instability

The use of the plasmids mentioned above generally suffers from a significant problem: plasmid instability often occurs, in particular when inserts are present. There are two forms of instability. The first, segregational instability, relates to the loss of the entire plasmid. The second, structural instability, corresponds to the loss of part of the plasmid.

An important cause of both kinds of instabilities is the mode of plasmid replication. Most non-native plasmids, such as those described above, use the rolling-circle mode of replication, which is characterized by the uncoupled synthesis of the leading and lagging strands. As a consequence, single-stranded replication intermediates are produced which have to be converted to double-stranded molecules in a second replication step. For this conversion, specific sequences, the single-strand origins (SSOs), are important. Since SSOs are normally rather host-specific, undesirable replication intermediates can accumulate with non-native rolling-circle plasmids, which quite often results in increased levels of both segregational and structural instability (for reviews, see: Gruss and Ehrlich, 1989; Bron, 1990; Bron *et al.*, 1999).

It should be emphasised that, although inefficient rolling-circle replication is a major factor in plasmid instability, it is certainly not the only one. In addition to replication errors, problems resulting from imperfect homologous or non-homologous recombination can also occur. A discussion of these systems goes beyond the scope of this chapter. More detailed information can be found elsewhere (Gruss and Ehrlich, 1989; Bron, 1990; Jannière *et al.*, 1993; Bron *et al.*, 1999).

Although the majority of plasmids found in Gram-positive bacteria, in particular those smaller than 10 kb, seem to use the rolling-circle mode of replication, plasmids using the theta mode of replication have also been extracted from such bacteria. In theta plasmids, the synthesis of the leading and lagging strand are coupled and single-stranded replication intermediates are not observed. This generally results in improved stabilities.

Improved Cloning Vectors

Two main approaches have been followed to improve the stability of plasmid cloning vectors in *B. subtilis*. The first concerns the use of theta-type plasmids and the second the use of endogenous rolling-circle plasmids from *Bacillus* species.

Theta-type plasmids
Although theta-type plasmids have been extracted from *Bacillus* species, they have not often been used for vector development. A reason for that may be that theta-replicating plasmids of the pAMβ1 family, originally isolated from *Enterococcus faecalis*, replicate well in *B. subtilis* and show relatively low levels of instability. Several variants of this replicon are available, such as pIL252 and pHV1432 (for a review, see Bron, 1990; Jannière *et al.*, 1993). Another variant of this replicon, pAMS100 (Kiewiet *et al.*, 1993), has improved stability properties. In this plasmid a strong transcription terminator (the T_1T_2 terminator from *Escherichia coli*), together with a stability determinant from the parental pAMβ1 plasmid, were introduced.

Endogenous rolling-circle plasmids
Considering the potential instability problems often observed with non-native rolling-circle plasmids in *B. subtilis*, vectors have been constructed that are based on rolling-circle plasmids from *B. subtilis* strains other than strain 168 (which does not have any plasmid by itself). Several such plasmids have been described (Meijer *et al.*, 1998). One of the best known examples is pTA1060, which was used for the construction of a versatile cloning vector, pHB201 (Figure 3). The latter is a stable shuttle vector between *E. coli* and *B. subtilis* (Bron *et*

al., 1998). It carries the replication functions of pTA1060 and large inserts can stably be cloned in the extended polylinker site, which is positioned in the *lacZα* part of the plasmid. Expression of genes cloned in this polylinker site is driven by the strong constitutive P59 promoter from *Lactococcus lactis*. The copy number of the plasmid is about 10 to 15 per cell.

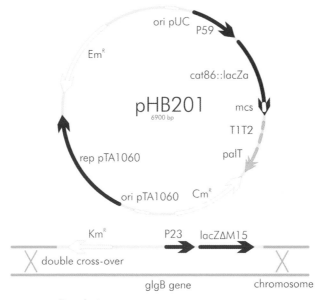

Conclusions
- versatile shuttle cloning system
- blue-white assay (X-gal)
- large inserts (up to 25kb)
- stable (>70% in 100 generations

Figure 3. Cloning system in *B. subtilis*. Plasmid pHB201 is a shuttle plasmid carrying the pUC (for *E. coli*), and pTA1060 (for *B. subtilis*) replication functions. The plasmid contains two selectable markers (Cm[R] and Em[R]) and a *lacZ*-alpha gene, the expression signals of which have been modified to allow the synthesis of the LacZ-alpha peptide in *B. subtilis*. The *lacZ*-alpha gene contains an extended MCS. The plasmid is stabilised through the highly active *palT* minus origin of replication of pTA1060 (a rolling circle-type plasmid), and the T_1T_2 transcription terminator. The plasmid can be used in conjunction with host strain 1012M15 (*Bacillus* Genetic Stock Centre, Ohio State University; strain 1A748), which is restriction-deficient and carries a *lacZΔ M15* gene in a non-essential part (*glgB* gene) of the chromosome (lower part of this Figure). This enables the blue-white assay for recombinant plasmids on X-gal plates by *lacZ*-alpha complementation.

Chromosomal Integration Systems

The use of chromosomal integration systems is a powerful tool for various applications and analyses in *B. subtilis*. The system is based on the natural transformation occuring in this bacterium, which renders the DNA that is taken up single-stranded. Since single-stranded DNA is highly recombinogenic, homologous fragments of internalised DNA recombine efficiently with the chromosome at the sites of homology. Two systems are frequently used: the first concerns chromosomal integrations involving single-crossover events while the second involves double-crossover events (Figure 4).

In standard applications, homologous DNA fragments are cloned in a plasmid that does not replicate autonomously in *B. subtilis*. Plasmids from *E. coli*, such as those based on the pUC and pSC101 families, are often used. The recombinant plasmids are propagated in *E. coli* and subsequently introduced in *B. subtilis* via natural transformation. The integrating plasmid should contain an antibiotic resistance marker that can be selected for in *B. subtilis*. Frequently used markers are

Figure 4. A schematic representation of chromosomal integration systems.
A: Campbell-type or single-crossover recombination. A typical vector is an *E. coli* plasmid, such as a pUC derivative, which contains replication functions, an antibiotic resistance determinant ("R"), and a fragment of *B. subtilis* DNA ("Y"). Single-crossover recombination at the homologous site in the chromosome (strain 1) results in the integration of the entire plasmid and, consequently, duplication of the cloned homologous fragment (strain 2). This may be followed by further amplifications, up to about 50 copies per chromosome (strain 3).
B: Double-crossover (replacement) recombination. The vector is usually an *E. coli* plasmid containing a region of homology with the *B. subtilis* chromosome (indicated as Y and Z), which is interrupted by other DNA sequences (here indicated as "R"). The latter can be as small as one basepair (mainly for the introduction of directed mutations in the chromosome). A double-crossover event in the flanking homologous regions (Y and Z) results in the replacement of the original part on the chromosome (W) by the new sequence(s), here called "R". Basically, any gene in the chromosome can be changed at any nucleotide position, either by a mutant basepair or by any other DNA sequence, or gene(s). In a variant of this application, the integration vector contains the 5'- and 3'-ends of the *amyE* gene (specifying α-amylase) at the Y and Z positions of the integration vector. This enables the insertion of any gene of interest at the *amyE* locus (see also explanation for Figure 6: pX plasmid and its use).

genes encoding resistance for chloramphenicol, erythromycine, kanamycine, or tetracycline.

Single-Crossover Recombination

As illustrated in the upper panel of Figure 4, a single-crossover event in the part of the plasmid that is homologous to a region of the chromosome will result in an integrated state in which the homologous part is duplicated. This process is also known as "Campbell-type"

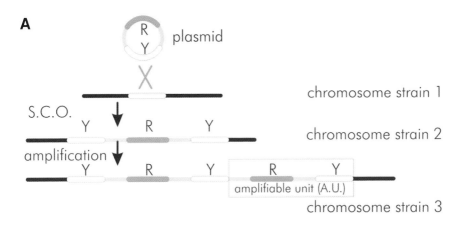

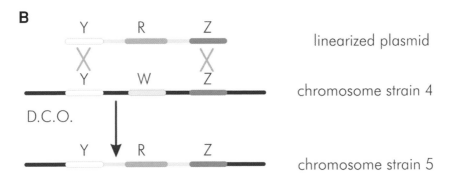

integration. The entire plasmid, including any cloned DNA fragments, is incorporated between the duplicated fragment. In many cases, further amplification of the integrated plasmid in the chromosome can occur. In particular, when antibiotic pressure can be applied, copy numbers of up to about 50 may be obtained (Meima *et al.*, 1995).

Double-Crossover Recombination

In this process, also called "replacement recombination", an even number (usually two) of recombination events is required (Figure 4, lower panel). Most often, the transforming plasmid is linearized at a site located outside of the cloned homologous fragment. This process results in the replacement of the chromosomal DNA fragment between the sites of cross-over by the "corresponding fragment" on the plasmid. The latter fragment can carry different gene(s), such as an antibiotic resistance marker, or (mutated) variants of the gene(s) to be studied. Integration by double-crossover recombination occurs at the single copy level.

Applications of Chromosomal Integration Systems

The chromosomal integration systems in *B. subtilis* have found many applications. An example of an integration vector is pMutin2 (Figure 5; Vagner *et al.*, 1998). This vector has been extensively used in Europe and Japan for gene function analyses (see the Genomics section). pMutin2 is particularly well suited for single-crossover integrations. It contains a promoter-less *lacZ* gene, as well as a ribosomal binding site active in *B. subtilis*. The *lacZ* gene is preceded by a polylinker site and the IPTG-inducible P_{spac} promoter.

For double-crossover recombination, the DNA fragment of interest is often cloned between the 5'- and 3'-ends of the *amyE* gene of *B. subtilis*, which specifies α-amylase. This enables the specific integration at the site of the α-amylase gene. A number of important applications of chromosomal integration systems are listed below.

Cloning

Single-crossover integration can be used to amplify (foreign) genes in the chromosome. The position of the integration will depend on the cloned fragment of homologous DNA. Such amplified regions are relatively stable, in particular when antibiotic pressure can be applied.

Gene expression

As illustrated with pMutin 2 (Figure 5), the cloning of an internal fragment of a gene (or operon) results in an integrant in which the promoter of the original gene controls the expression of the *lacZ* reporter. Promoter strength can thus be determined by measuring

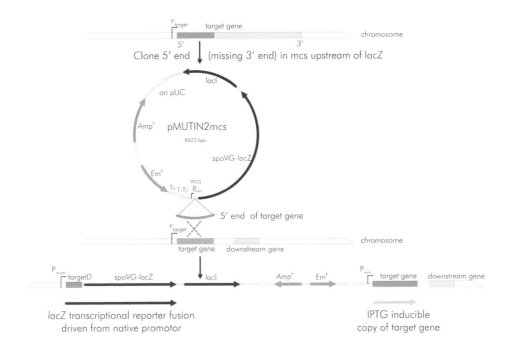

Figure 5. Plasmid pMutin2 is an integrational vector initially used by the European and Japanese *B. subtilis* Function Analysis (BSFA) groups to introduce, among others, loss-of-function mutations in genes of unknown function (Vagner *et al.*, 1998). The plasmid is based on the replication functions (ori) of pUC plasmids and carries two selectable markers (Ap[R] for use in *E. coli* and Em[R] for use in *B. subtilis* and *E. coli*). T_0 and $T_1 T_2$ indicate transcription terminators from phage lambda and the *E. coli* rRNA genes, respectively. P_{spac} is the LacI-controlled and IPTG-inducible promoter. The *lacZ* reporter gene lacks its own promoter, but is preceded by an efficient ribosome-binding site (from the *spoVG* gene) for *B. subtilis*. See text for further explanations.

β-galactosidase activities. Since the entire sequence of the *B. subtilis* genome is known, this can be done for any gene. Figure 5 also shows that in the same integration event, the second copy of the target gene is placed under control of the P$_{spac}$ promoter. If any intact gene lacking its own promoter is thus cloned, its transcription will be controlled by the regulatable P$_{spac}$ promoter.

Chromosome walking
Through the presence of the antibiotic resistance gene in the integrating plasmid, the site of integration resulting from a single-crossover is marked. The use of suitable restriction enzymes that introduce cuts in the DNA regions outside the integrated plasmid then enables excision of the entire plasmid with flanking DNA regions from the chromosome. The plasmid and flanking chromosomal DNA can be ligated and recovered after propagation in *E. coli*.

Gene function
Chromosomal integration systems can be used effectively in the analyses of gene function. The cloning of internal fragments of genes or operons, and their subsequent integration by single-crossover, results in the inactivation of the gene or operon. The effects of this inactivation can then be assessed. Using double-crossover integration, genes can be disrupted by introduction of other DNA sequences, or site-specific mutations can be introduced at virtually any nucleotide. This provides a powerful tool for gene function analyses (also see Genomics section).

Inducible Promoters

Inducible gene expression in *B. subtilis* usually involves the promoters of four different genes. The first is a modified promoter region of the *E. coli lac* operon. The second, third and fourth promoter regions involve *Bacillus* DNA sequences from the xylose-utilizing operon of *Bacillus megaterium* (*xylA*), the *citM* gene (encoding a secondary transporter for Mg-citrate complexes) of *B. subtilis*, and the *sacB* gene, encoding a *B. subtilis* extracellular levansucrase, respectively. Below, we briefly describe the characteristics of these systems.

P_{spac}, E. coli lac Operon

In *E.coli*, the *lac* operon is normally repressed by the product of the *lacI* repressor gene. Fast induction of the operon, including that of *lacZ* gene specifying β-galactosidase, occurs to high levels in the presence of β-galactosides. Isopropyl-β-D-thiogalactopyranoside (IPTG) resembles natural β-galactosides and acts as an efficient inducer for this system. The activity of the β-galactosidase gene can be easily monitored, by either using a blue-white color test on X-Gal-containing agar plates (a blue color develops when the enzyme is present), or by quantitative measurements of enzyme activity (Miller units).

Attempts were made to use this powerful *E. coli*-based system for controlled gene expression in *B. subtilis*. Unfortunately, the original promoter region of the *E.coli lac* operon is not active in *B. subtilis*. In order to make the system work, the 5'-sequences of a promoter from the *B. subtilis* phage SPO1 were fused to the 3'-sequences of the *E.coli lac* promoter, which still contains a functional operator region. The resulting hybrid promoter is active in *B. subtilis* and is controlled by the *lacI* gene product, which was modified to achieve constitutive expression in this organism (Yansura and Henner, 1984).

When P_{spac} is present in the same cell as *lacI*, genes located downstream of the promoter are induced up to about 50-fold with 1 mM IPTG. The promoter has been used successfully for the expression of numerous genes, both from plasmid and chromosomal locations. An example is pMutin2 which contains the promoter and its associated repressor gene (*lacI*). P_{spac} is fully functional in *E. coli*, allowing constructs to be tested in this bacterium before transfer to *B. subtilis*.

The main disadvantages of the P_{spac} promoter are: (i) ITPG is expensive and toxic, and therefore not suitable for large-scale fermentations or food production; (ii) promoter control is not very tight and small amounts of protein are synthesised even in the absence of IPTG, although this problem can be reduced by increasing the copy number of the *lacI* gene on a compatible multi-copy plasmid (*e.g.* pMap65; Dervyn and Ehrlich, 2001); and (iii) P_{spac} is not strong enough for very-large-scale protein production.

Promoters from Xylose Operons

In *Bacillus* species, the genes for the utilisation of xylose polymers are controlled at the transcriptional level through the activity of a repressor, the *xylR* gene product. XylR prevents transcription of target genes by binding to their operators in the absence of xylose. In the presence of xylose, repression is relieved and transcription of the xylose genes is activated.

The *B. subtilis* xylose-inducible promoter/operator elements have been used without modification to control expression from chromosomal and plasmid locations (Gartner *et al.*, 1992). A copy of the *xylR* gene is usually included on high-copy-number expression vectors to maintain a balance between the number of repressor molecules and operator sites. This is not strictly necessary when the target gene is integrated into the chromosome of *B. subtilis,* since the host would normally have its own copy of the repressor gene. Homology between the vector and chromosomal *xylR* genes is often used for integration. As with P_{spac}, *xylR*-controlled promoters are fully active in *E. coli,* where they also respond to the presence of xylose. Although genes in the xylose regulon are normally subject to catabolite repression, the catabolite responsive element is not included in the expression vectors and, thus, catabolite repression does not occur. A very tightly controlled xylose-dependent expression cassette for integration in the *amyE* locus, has been developed by Bhavsar *et al.* (2001). The cassette, which is located on plasmid pSWEET, contains the *xylR* gene, the *xylA* promoter and 5' sequences of *xylA* with an optimized catabolite-responsive element, as well as *xyl* operator sequences. An induction/repression ratio of 279 was reported for the pSWEET system.

In addition to the *B. subtilis*-based xylose-inducible systems described above, a versatile system derived from *Bacillus megaterium* was developed (Rygus and Hillen, 1991; Kim *et al.*, 1996). The system is based on the vector pX (Figure 6) and is designed for the integration of genes of interest at the chromosomal *amyE* locus of *B. subtilis.* Insertion in the *amyE* gene is easily detectable on starch-containing agar plates by the disappearance of halos (due to starch degradation) around recombinant colonies. The pX replicon is based on the *E. coli* pBR322 plasmid, which is unable to replicate in *B. subtilis*. Between the 5'- and 3'-ends of the *amyE* gene, providing the homology needed for integration, a cassette is placed that contains a selectable marker (chloramphenicol resistance), the *xylR* repressor gene, and the *xylA* promoter region. Expression from the latter region is controlled by the XylR repressor. Any gene of interest can be cloned in one of the available restriction sites that are present immediately downstream of the *xylA* promoter. Following its integration at *amyE*, the expression of the gene is then controllable by xylose addition. Induction of expression is normally performed by adding xylose from 0.1 to 2%,

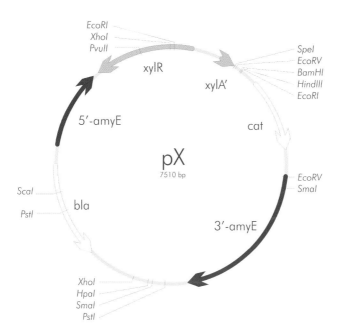

Figure 6. pX is an integrational vector designed for the insertion at the *amyE* locus (Kim *et al.*, 1996). *XylR* encodes the repressor for the *xylA* promoter/operator. This promoter is followed by a number of restriction sites that can be used for the cloning of inserts of interest. Selection of integrants in *B. subtilis* is possible through resistance to chloramphenicol (cat). pX is an *E. coli – B. subtilis* shuttle plasmid, enabling the use of *E. coli* as an intermediary host for cloning and the production of the recombinant integration plasmid. Selection in *E. coli* is based on resistance to ampicilline ("bla"). The 5'- and 3'-ends of the *amyE* gene, providing the homologous regions needed for integration, are indicated.

and increases in protein production up to 200-fold have been reported (Kim *et al.*, 1996). Although the system is not sensitive to general catabolite repression, it is sensitive to glucose repression. The initial cloning steps can be carried out in *E. coli* as an intermediate host. A further advantage of this system is that xylose is a cheap and readily available substrate that can be used for the induction of large-scale fermentations. A minor disadvantage is that, like systems involving $P_{spac,}$ small amounts of protein are synthesized even in the absence of xylose.

Additional variants of both the xylose-inducible and IPTG (P_{spac})-inducible systems that integrate in the *lacA* gene (specifying a β-galactosidase in *B. subtilis*) have been constructed (Härtl *et al.*, 2001). Using these vectors, the *amyE* and the *lacA* integration sites can now be used simultaneously, one for expression from the xylose-inducible promoter and the other from P_{spac}.

Citrate-Inducible Promoter

As shown in the group of Professor Sekiguchi (Department of Applied Biology, Shinshu University, Nagano, Japan), the *citM* gene of *B. subtilis* encodes a secondary transporter for Mg-citrate complexes and the expression of this gene requires the positive control of the CitST two-component system (Yamamoto *et al.*, 2000). The system is induced in the presence of 2 mM citrate. Expression from the *citM* promoter region is negatively controlled by glucose repression, but mutants have been constructed that lack the *cre*-site responsible for catabolite repression. An example is the pHYCM2 plasmid vector which replicates in *B. subtilis* and contains the *citM* promoter region without a *cre*-site. Genes of interest can be inserted downstream of this promoter and induced with citrate without being sensitive to catabolite repression. The system has been used for the synthesis of a number of proteins (Yamamoto *et al.*, 2000; Fukushima *et al.*, 2002).

SacR-Controlled Promoters

The inducible expression of *sacB*, the gene encoding extracellular levansucrase, by sucrose involves a number of regulatory mechanisms, both positive and negative, not all of which are fully understood. The *sacB* gene is positively controlled by sucrose in a process mediated by the *sacY* antiterminator. In addition, DegQ and SacU enhance transcription of this gene. Various expression cassettes have been based on the *sacB* promoter, which is particularly effective in *sacU^{hy}* (*degU^{hy}*) mutants (see, for instance, Zukowski and Miller, 1986; Wu *et al.*, 1991). Expression vectors based on the *sacB* promoter have the advantage that this promoter can be activated during exponential growth, when the expression of most *Bacillus* extracellular proteinases is repressed, and is not subject to catabolite repression. The level of induction is also controlled by the concentration of sucrose in the medium: the promoter is only weakly induced in the presence of 1 mM sucrose and fully induced at 30 mM.

Genomics

B. subtilis Genome Sequence

The worldwide interest in *B. subtilis* from both fundamental and applied standpoints formed the basis for attempts to sequence its genome as early as 1990. The model strain *B. subtilis* 168 was chosen for this purpose. Funding for this project was obtained in Europe (Commission of the EU) and Japan (Japanese government), and the project, involving about 25 research groups, was coordinated by F. Kunst (Institut Pasteur, Paris) and N. Ogasawara (Nara Institute of Science and Technology, Nara). Based on the genetic map, the chromosome was divided in regions of 140 kbp on average, and these were allocated to, and sequenced by the participating research groups. Several strategies were used for sequencing. The most important were the cloning of *B. subtilis* DNA fragments in *E. coli* vectors (both plasmids and lambda phage) and yeast artificial chromosomes (YACs), the use of chromosome walking techniques, and the use of long-range PCR techniques.

The entire sequence of *B. subtilis* 168 was completed in 1997 (Kunst *et al.*, 1997). This was the first sequence of a Gram-positive bacterium to become available in the public domain. The sequence information is accessible from the *SubtiList* database (http://genolist.pasteur.fr/SubtiList). Related information can be found on the BSORF-DB, in particular for ORF detection. A more detailed description of the information has been given by Moszer (2002).

The most important conclusions derived from the sequence are as follows. The genome of *B. subtilis* consists of 4,215 kbp and contains about 4,100 putative protein-encoding genes. No putative function could be ascribed to as many as 40% of all genes. The chromosome contains ten A/T-rich regions that represent about 10% of the chromosome. These regions represent prophages, prophage-like sequences or remnants thereof. The chromosome contains many gene duplications: only about 53% of all genes are unique and the others have one or more paralogous genes. Seventeen genes encoding sigma factors and 35 gene pairs encoding putative two-component regulator systems were detected. Transposons and IS-elements are absent from the *B. subtilis* 168 chromosome.

Gene Function Analysis and Genome Minimization

The genome sequencing project was followed by an extended project aimed at the analysis of the function of those genes for which no function could be ascribed or deduced. The project was called BSFA (*Bacillus subtilis* Function Analysis) and was coordinated by S.D. Ehrlich (INRA, Jouy en Josas, France) and N. Ogasawara (NAIST, Nara, Japan). About 25 research groups in Europe and Japan have participated in the project. The joint approach was to inactivate target genes using pMutin2 (Figure 5). This was achieved by cloning internal fragments of genes in pMutin2 and selecting chromosomal integrants resulting from single-crossovers. In cases where no viable integrants could be obtained, the target gene was placed under the control of the inducible P_{spac} promoter. Essential genes were considered as those that could not be inactivated, but which enabled cell growth when placed under the control of P_{spac} following IPTG induction.

The effects of the mutations on a considerable number of processes were analyzed. Included in the analyses were: metabolism of small molecules and inorganics, macromolecule metabolism, cell structure and mobility, stress and stationary phase, and cell processes (cell cycle, competence, sporulation/germination). The strength of the promoters of all genes tested was determined using *lacZ* fusions that were generated after integration of pMutin2 clones (Schumann *et al.*, 2001). Today, the collection of mutants generated with pMutin2, and derivatives thereof, comprises about 3,000 out of the 4,100 *B. subtilis* genes. This provides a vast resource for current and future research. The data is stored on the *Micado* web server (http://locus.jouy.inra.fr/cgi-bin/genmic/madbase_home.pl).

The most important conclusions and results from the BSFA project (Kobayashi *et al.*, 2003) are as follow. About 3,000 strains with mutations in specific genes are available.

The majority of these mutants have been characterized with respect to global phenotype and a global function assignment was possible for many of these. In particular, many mutants affected in stress responses were found. A set of promoters is now available enabling low-levels to high-levels of gene expression, in different media and at different growth stages. Only about 270 genes out of 4,100 seem to be essential for growth under the laboratory conditions used. The majority of these genes are involved in information processing, cell envelope biosynthesis, cell shape and division, and cell energetics. Most of these genes are non-paralogous genes and surprisingly few regulatory genes are present among the collection of essential ones.

In parallel with the BSFA project, another European project was carried out, which aimed at the minimization of the *B. subtilis* 168 chromosome. As a first approach, the regions corresponding to the prophage-like regions were deleted. In addition, three large regions specifying the non-ribosomal synthesis of two lipopeptides and a polyketide were deleted. All these deletions could be introduced separately without affecting any major physiological property of the cell. A multiple deletion strain has now been constructed which lacks six of the regions indicated above. The resulting strain lacks 320 kbp of DNA (7.7% of the chromosome) and, so far, no major effects on cell viability or key physiological and developmental processes have been observed (Westers *et al.*, 2003).

Recent Developments in *Bacillus* Genomics

Following the sequencing and gene function analysis projects, the European and Japanese *Bacillus* research groups have continued their joint efforts in what is often called "post-genomics". For this aim, the European groups have established the BACELL umbrella organisation. The aims of this organisation are to promote *Bacillus*-oriented research in Europe and to improve the interaction with the Commission of the EU and the *Bacillus* Industrial Platform (BACIP). All the EU-sponsored *Bacillus* projects are included under the BACELL umbrella. These projects include chromosome minimization, protein secretion and networks of gene regulation. The projects on protein secretion are coordinated by S. Bron and J.M. van Dijl (Groningen, Netherlands) and the current state of knowledge in this area of research is described in more detail in the Protein Secretion Pathways in Bacilli section.

In the BACELL NETWORK project (coordinated by C.R. Harwood, Newcastle, UK), the major aim is the analysis of networks of gene expression in *B. subtilis*, with the ultimate goal to obtain a holistic view of gene expression in the entire cell. Included in the analyses are the control of (i) metabolism of small molecules and large biomolecules, (ii) cell envelope synthesis, (iii) competence development, and (iv) protein secretion and intermediary metabolism. These analyses include several regulator systems, such as sigma factors, two-component systems, phospho-relays and genes involved in quorum sensing.

Several modern technologies are employed in recent *B. subtilis* research work in the various European and Japanese groups. These include: DNA macro- and microarrays for transcript profiling, proteome analyses, including mass spectometry (cytoplasmic, envelope and extracellular proteomes), fluorescent protein labelling for protein localization studies, studies aimed at understanding population heterogeneity and protein-protein interactions, and metabolome analyses.

Results of proteomic work can be found at the Sub2D database (Bernhardt *et al.*, 1999), and those on transcriptome analyses will be collected in the *SubScript* database, which is currently being set up (coodinated by I. Moszer; Institut Pasteur, Paris).

Conclusions and Outlook

The work summarized in the present chapter has significantly contributed to our fundamental understanding of the mechanisms controlling gene expression in *B. subtilis* and has led to the development of versatile cloning and expression systems for fundamental research and industrial applications alike. With respect to the latter, the detailed characterization of protein secretion pathways in *B. subtilis* and, perhaps most importantly, the identification of the challenges that translocated enzymes have to overcome *en route* to the cell's exterior environment, have been of major importance for the development of suitable hosts for high-level protein production.

Challenges that lie ahead include the elucidation of the specificity determinants and substrates of the Tat-dependent secretion route and, possibly, to tailor this pathway for secretion of heterologous or even cytoplasmic proteins. In line with this, modulation of the proteolytic environment at the cell wall-membrane interface, as well as the post-translocational quality control systems, will be instrumental for further improvement of *B. subtilis* and other bacilli as industrial hosts.

The development of even more tightly controlled promoter systems for gene expression without the need for specific and, in some cases, expensive inducers could be another important spin-off from the various genomics projects. It is evident that the current status of *B. subtilis* as being one of the best-characterized microbes has benefited tremendously from the establishment of its complete genome sequence. Similar sequencing efforts of other industrially important bacilli have already been initiated (*e.g.* the Genomik Netzwerk in Göttingen, Germany), and will undoubtedly result in new insights and principles in the very near future.

Finally, with growing concerns on certain aspects of human health, in particular in relation to combating pathogenic bacteria, *B. subtilis* could potentially become an important model organism for analyses of host-bacterium interactions. The idea behind this view is that through comparative genomics with other Gram-positive bacteria, medically relevant genes for host-bacterium interactions may be identified. Likewise, evidence is accumulating that *Bacillus* species play important roles as plant probiotics. With all the knowledge collected for this bacterium and the available experimental tools, it is now feasible to analyze possible mechanisms underlying these plant-microbe interactions. This is likely to be important for the sustainable production of plant crops.

Acknowledgements

We thank Albert Bolhuis, Harold Tjalsma, Jan D.H. Jongbloed, Wim J. Quax and other members of the Groningen and European *Bacillus* Secretion Groups for stimulating discussions, and Caroline Eschevins for providing Figure 1. J.M.v.D. and S.B. were supported by Grants QLK3-CT-1999-00413 and QLK3-CT-1999-00917 from the European Union.

References

Antelmann, H., Tjalsma, H., Voigt, B., Ohlmeier, S., Bron, S., van Dijl, J.M., and Hecker, M. 2001. A proteomic view on genome-based signal peptide predictions. Genome Research 11: 1484-1502.

Berks, B.C., Sargent, F., and Palmer, T. 2000. The Tat protein export pathway. Mol. Microbiol. 5: 260-274.

Bernhardt, J., Buttner, K., Scharf, C., and Hecker, M. 1999. Dual channel imaging of two-dimensional electropherograms in *Bacillus subtilis*. Electrophoresis 20: 2225-2240.

Bhavsar, A.P., Zhao, X., and Brown, E.D. 2001. Development and characterization of a xylose-dependent system for expression of cloned genes in *Bacillus subtilis*: conditional complementation of a teichoic acid mutant. Appl. Environ. Microbiol. 67: 403-410.

Bolhuis, A., Broekhuizen, C.P., Sorokin, A., van Roosmalen, M.L., Venema, G., Bron, S., Quax, W.J., and van Dijl, J.M. 1998. SecDF of *Bacillus subtilis*, a molecular Siamese twin required for the efficient secretion of proteins. J. Biol. Chem. 273: 21217-21224.

Bolhuis, A., Tjalsma, H., Smith, H.E., de Jong, A., Meima, R., Venema, G., Bron, S., and van Dijl, J.M. 1999a. Evaluation of bottlenecks in the late stages of protein secretion in *Bacillus subtilis*. Appl. Environ. Microbiol. 65: 2934-2941.

Bolhuis, A., Matzen, A., Hyyryläinen, H.L., Kontinen, V.P., Meima, R., Chapuis, J., Venema, G., Bron, S., Freudl, R., and van Dijl, J.M. 1999b. Signal peptide peptidase- and ClpP-like proteins of *Bacillus subtilis* are required for efficient translocation and processing of secretory proteins. J. Biol. Chem. 274: 24585-24592.

Bolhuis, A., Venema, G., Quax, W.J., Bron, S., and van Dijl, J.M. 1999c. Functional analysis of paralogous thiol-disulfide oxidoreductases in *Bacillus subtilis*. J. Biol. Chem. 274: 24531-24538.

Bron, S. 1990. Plasmids. In: Molecular Biology Methods for *Bacillus*. C.R. Harwood and S.M. Cutting, eds. Wiley and Sons Ltd., Chichester (UK). p. 75-175.

Bron, S., Bolhuis, A., Tjalsma, H., Holsappel, S., Venema, G., and van Dijl, J.M. 1998. Protein secretion and possible roles for multiple signal peptidases for precursor processing in Bacilli. J. Biotechnol. 64: 3-13.

Bron, S., Meima, R., van Dijl, J.M., Wipat, A., and Harwood, C.R. 1999. Molecular Biology and Genetics of *Bacillus* species. In: Manual of Industrial Microbiology and Biotechnology, 2nd ed. A.L. Demain and J.E. Davies, eds. ASM Press, Washington D.C. p. 392-416.

Bunai, K., Yamada, H., Hayashi, K., Nakamura, K., and Yamane, K.. 1999. Enhancing effect of *Bacillus subtilis* Ffh, a homologue of the SRP54 subunit of the mammalian signal recognition particle, on the binding of SecA to precursors of secretory proteins *in vitro*. J. Biochem. (Tokyo) 125: 151-159.

Chambert, R., Benyahia, F., and Petit-Glatron, M.F. 1990. Secretion of *Bacillus subtilis* levansucrase. Fe(III) could act as a cofactor in an efficient coupling of the folding and translocation processes. Biochem. J. 265: 375-382.

Chu, H.H., Hoang, V., Kreutzmann, P., Hofemeister, B., Melzer, M., and Hofemeister, J. 2002. Identification and properties of type I-signal peptidases of *Bacillus amyloliquefaciens*. Eur. J. Biochem. 269: 458-469.

Cristóbal, S., de Gier, J.W., Nielsen, H., and von Heijne, G.. 1999. Competition between Sec- and Tat-dependent protein translocation in *Escherichia coli*. EMBO J. 18: 2982-2990.

Dahl, M.K., Msadek, T., Kunst, F., and Rapoport, G. 1992. The phosphorylation state of the DegU response regulator acts as a molecular switch allowing either degradative enzyme synthesis or expression of genetic competence in *Bacillus subtilis*. J. Biol. Chem. 267: 14509-14514.

Dalbey, R. E., Lively, M.O., Bron, S., and van Dijl, J.M. 1997. The chemistry and enzymology of the type I signal peptidases. Prot. Science 6: 1129-1138.

Darmon, E., Noone, D., Masson, A., Bron, S., Kuipers, O.P., Devine, K.M. and van Dijl, J.M. 2002. A novel class of heat and secretion stress-responsive genes is controlled by the autoregulated CssRS two-component system of *Bacillus subtilis*. J. Bacteriol. 184: 5661-5671.

Dervyn, E. and Ehrlich, S.D. 2001. Analysis of essential genes. In: Functional analysis of bacterial genes, a practical manual. W. Schumann, S.D. Ehrlich, N. Ogasawara, eds. John Wiley and Sons, Ltd, Chichester, UK. p. 25-32.

de Gier, J.W., Scotti, P.A., Saaf, A., Valent, Q.A., Kuhn, A., Luirink, J., and von Heijne, G. 1998. Differential use of the signal recognition particle translocase targeting pathway for inner membrane protein assembly in *Escherichia coli*. Proc. Natl. Acad. Sci. U.S.A. 95: 14646-146451.

de Leeuw, E., te Kaat, K., Moser, C., Menestrina, G., Demel, R., de Kruijff, B., Oudega, B., Luirink, J., and Sinning, I. 2000. Anionic phospholipids are involved in membrane association of FtsY and stimulate its GTPase activity. EMBO J. 19: 531-541.

Deutscher, J., Galinier, A., and Martin-Verstraete, I. 2002. Carbohydrate uptake and metabolism. In: *Bacillus subtilis* and its closest relatives: from genes to cells. A.L. Sonenshein, J.A. Hoch and R. Losick, eds. ASM press, Washington D.C. p. 129-150

D'Souza, C., Nakano, M.M., and Zuber P. 1994. Identification of *comS*, a gene of the *srfA* operon that regulates the establishment of genetic competence in *Bacillus subtilis*. Proc. Natl. Acad. Sci. USA 91: 9397-9401.

Dubnau, D., and Lovett, C.M., Jr. 2002. Transformation and recombination. In: *Bacillus subtilis* and its closest relatives: from genes to cells. A.L. Sonenshein, J.A. Hoch and R. Losick, eds. ASM press, Washington D.C. p. 453-471.

Fekkes, P., and Driessen, A.J.M. 1999. Protein Targeting to the Bacterial Cytoplasmic Membrane. Microbiol. Mol. Biol. Rev. 63: 161-173.

Fukushima, T., Yamamoto, H., Atrih, A., Foster, S.J., and Sekiguchi, J. 2002. A polysaccharide deacetylase gene (*pdaA*) is required for germination and for production of muramic δ-lactam residues in the spore cortex of *Bacillus subtilis*. J. Bacteriol. 184: 6007-6015.

Gartner, D., Degenkolb, J., Ripperberger, J.A.E., Allmansberger, R., and Hillen, W. 1992. Regulation of the *Bacillus subtilis* W23 xylose utilization operon: interaction of the Xyl repressor with the *xyl* operator and the inducer xylose. Mol. Gen. Genet. 232: 415-422.

Gruss, A., and Ehrlich, S.D. 1989. The family of highly interrelated single-stranded deoxyribonucleic acid plasmids. Microbiol. Rev. 53: 231-241.

Haldenwang, W.G., and Losick, R. 1980. Novel RNA polymerase sigma factor from *Bacillus subtilis*. Proc. Natl. Acad. Sci. USA. 77: 7000-7004.

Haldenwang, W.G., Lang, N., and Losick, R. 1981. A sporulation-induced sigma-like regulatory protein from *B. subtilis*. Cell 23: 615-624.

Hamoen, L.W., Eshuis, H., Jongbloed, J., Venema, G., and van Sinderen, D. 1995. A small gene, designated *comS*, located within the coding region of the fourth amino acid-activation domain of *srfA*, is required for competence development in *Bacillus subtilis*. Mol. Microbiol. 15: 55-63.

Hamoen, L.W., van Werkhoven, A.F., Bijlsma, J.J., Dubnau, D., and Venema G. 1998. The competence transcription factor of *Bacillus subtilis* recognizes short A/T-rich sequences arranged in a unique, flexible pattern along the DNA helix. Genes Dev. 12: 1539-50.

Härtl, B., Wehrl, W., Wiegert, T., Homuth, G., and Schumann, W. 2001. Development of a new integration site within the *Bacillus subtilis* chromosme and construction of compatible expression cassettes. J. Bacteriol. 183: 2696-2699.

Harwood, C.R. 1992. *Bacillus subtilis* and its relatives: molecular biological and industrial workhorses. Trends Biotechnol. 10: 247-256.

Helmann, J.D. 1991. Alternate sigma factors and the regulation of flagellar gene expression. Mol. Microbiol. 5: 2875-2882.

Helmann, J.D. 1999. Anti-sigma factors. Curr. Opin. Microbiol. 2: 135-141.

Helmann, J.D., and Moran, C.P., Jr. 2002. RNA polymerase and sigma factors. In: *Bacillus subtilis* and its closest relatives: from genes to cells. A.L. Sonenshein, J.A. Hoch and R. Losick, eds. ASM press, Washington D.C. p. 289-312.

Henkin, T.M., Grundy, F.J., Nicholson, W.L., and Chamblis, G.H. 1991. Catabolite repression of α-amylase gene expression in *Bacillus subtilis* involves a trans-acting gene product homologous to *Escherichia coli lacI* and *galR* repressors. Mol. Microbiol. 5: 575-584.

Herbort, M., Klein, M., Manting, E.H., Driessen, A.J.M., and Freudl, R.J. 1999. Temporal expression of the *Bacillus subtilis secA* gene, encoding a central component of the preprotein translocase. J. Bacteriol. 181: 493-500.

Hirose, I., Sano, K., Shiosa, I., Kumano, M., Nakamura, K., and Yamane, K. 1999. Proteome analysis of *Bacillus subtilis* extracellular proteins: a two-dimensional protein electrophoretic study. Microbiol. 146: 65-75.

Hoch, J.A. 1993. The phosphorelay signal transduction pathway in the initiation of *Bacillus subtilis* sporulation. J. Cell. Biochem. 51: 55-61.

Honda, K., Nakamura, K., Nishiguchi, M., and Yamane, K. 1993. Cloning and characterization of a *Bacillus subtilis* gene encoding a homolog of the 54-kilodalton subunit of mammalian signal recognition particle and *Escherichia coli* Ffh. J. Bacteriol. 175: 4885-4894.

Horsburgh, M.J., and Moir, A. 1999. Sigma(M), an ECF RNA polymerase sigma factor of *Bacillus subtilis* 168, is essential for growth and survival in high concentrations of salt. Mol. Microbiol. 32: 41-50.

Huang, X.J., Fredrick, K.L., and Helmann, J.D. 1998. Promoter recognition by *Bacillus subtilis* sigma(W) - Autoregulation and partial overlap with the sigma(X) regulon. J. Bacteriol. 180: 3765-3770.

Hyyryläinen, H.L., Vitikainen, M., Thwaite, J., Wu, H., Sarvas, M., Harwood, C.R., Kontinen, V.P., and Stephenson, K. 2000. D-alanine substitution of teichoic acids as a modulator of protein folding and stability at the cytoplasmic membrane-cell wall interface of *Bacillus subtilis*. J. Biol. Chem. 275: 26696-703.

Jannière, L., Gruss, A., and Ehrlich, S.D. 1993. Plasmids. In: *Bacillus subtilis* and other Gram-positive Bacteria. A.L. Sonenshein, J.A. Hoch and R. Losick, eds. ASM press, Washington D.C. p. 625-644.

Jongbloed, J.D.H., Martin, U., Antelmann, H., Hecker, M., Tjalsma, H., Venema, G., Bron, S., van Dijl, J.M. and Müller, J. 2000. TatC is a specificity determinant for protein secretion via the twin-arginine translocation pathway. J. Biol. Chem. 275: 41350-41357.

Jongbloed, J.D.H., Antelmann, H., Hecker, M., Nijland, R., Bron, S., Airaksinen, U., Pries, F., Quax, W.J., van Dijl, J.M. and Braun, P.G. 2002. Selective contribution of the twin-arginine translocation pathway to protein secretion in *Bacillus subtilis*. J. Biol. Chem. 277: 44068-44078.

Kim, L., Mogk, A., and Schumann, W. 1996. A xylose-inducible *Bacillus subtilis* integration vector and its application. Gene 181: 71-76.

Kiewiet, R., Kok, J., Seegers, J.F.M.L., Venema, G., and Bron, S. 1993. The mode of replication is a major factor in segregational instability in *Lactococcus lactis*. Applied Environm. Microbiol. 59: 358-364.

Kobayashi, K., Ehrlich, S.D., Ogasawara, N., and many others. 2003. Essential *Bacillus subtilis* genes. Proc Natl Acad Sci USA 100: 4678-4683.

Kontinen V.P, and Sarvas, M. 1993. The PrsA lipoprotein is essential for protein secretion in *Bacillus subtilis* and sets a limit for high-level secretion. Mol. Microbiol. 8: 727-737.

Kunst, F., Debarbouille, M., Msadek, T., Young, M., Mauel, C., Karamata, D., Klier, A., Rapoport, G., and Dedonder, R. 1988. Deduced polypeptides encoded by the *Bacillus subtilis* sacU locus share homology with two-component sensor-regulator systems. J. Bacteriol.170: 5093-5101.

Kunst, F., Ogasawara, N., Moszer, I., and many others. 1997. The complete genome sequence of the Gram-positive bacterium *Bacillus subtilis*. Nature 390: 249-256.

Kusukawa, N., Yura, T., Ueguchi, C., Akiyama, Y., and Ito, K. 1989. Effects of mutations in heat-shock genes *groES* and *groEL* on protein export in *Escherichia coli*. EMBO J. 8: 3517-3512.

Leloup, L., Driessen, A.J.M., Freudl, R., Chambert, R., and Petit-Glatron, M. F.. 1999. Differential dependence of levansucrase and alpha-amylase secretion on SecA (Div) during the exponential phase of growth of *Bacillus subtilis*. J. Bacteriol. 181: 1820-1826.

Meens, J., Herbort, M., Klein, M., and Freudl, R. 1997. Use of the pre-pro part of *Staphylococcus hyicus* lipase as a carrier for secretion of *Escherichia coli* outer membrane protein A (OmpA) prevents proteolytic degradation of OmpA by cell-associated protease(s) in two different Gram-positive bacteria. Appl. Environ. Microbiol. 63: 2814-2820.

Meijer, W.J.J., de Jong, A., Wisman, G.B.A., Tjalsma, H., Venema, G., Bron, S., and van Dijl, J.M. 1995. The endogenous *Bacillus subtilis* (natto) plasmids pTA1015 and pTA1040 contain signal peptidase-encoding genes: identification of a new structural module on cryptic plasmids. Mol. Microbiol. 17: 621-631.

Meijer, W.J.J., Wisman, G.B.A., Terpstra, P., Thorsted, P.B., Thomas, C.M., Holsappel, S., Venema, G., and Bron, S. 1998. Rolling-circle plasmids from *Bacillus subtilis*: complete nucleotide sequences and analyses of genes of pTA1015, pTA1040, pTA1050, and pTA1060, and compatisons with related plasmids from Gram-positive bacteria. FEMS Microbiology Reviews 21: 337-368.

Meima, R., Haijema, B.J., Venema, G., and Bron, S. 1995. Overproduction of the ATP-dependent nuclease AddAB improves the structural stability of a model plasmid system in *Bacillus subtilis*. Mol. Gen. Genet. 248: 391-398.

Meima, R., Eschevins, C., Fillinger, S., Bolhuis, A., Hamoen, L.W., Dorenbos, R., Quax, W.J., van Dijl, J.M., Provvedi, R., Chen, I., Dubnau, D. and Bron, S. 2002. The *bdbDC* operon of *Bacillus subtilis* encodes thiol-disulphide oxidoreductases required for competence development. J. Biol. Chem. 277: 6994-7001.

Moran, C.P., Jr., Lang, N., LeGrice, S.F.J., Lee, G., Stevens, M., Sonenshein, A.L., Pero, J., and Losick, R. 1982. Nucleotide sequences that signal the initiation of transcription and translation in *Bacillus subtilis*. Mol. Gen. Genet. 186: 339-346.

Moszer, I. 2002. The *Bacillus subtilis* genome, genes, and functions. In: *Bacillus subtilis* and its closest relatives: from genes to cells. A.L. Sonenshein, J.A. Hoch and R. Losick, eds. ASM press, Washington D.C. p. 7-11.

Müller, J.P., Ozegowski, J., Vettermann, S., Swaving, J., van Wely, K.H., and Driessen, A.J.M. 2000. Interaction of *Bacillus subtilis* CsaA with SecA and precursor proteins. Biochem. J. 348: 367-73.

Nakamura, K., Yahagi, S., Yamazaki, T., and Yamane, K. 1999. *Bacillus subtilis* histone-like protein, HBsu, is an integral component of a SRP-like particle that can bind the Alu domain of small cytoplasmic RNA. J. Biol. Chem. 274: 13569-13576.

Nakano, M.M., and Zuber, P. 2002. Anaerobiosis. In: *Bacillus subtilis* and its closest relatives: from genes to cells. A.L. Sonenshein, J.A. Hoch and R. Losick, eds. ASM press, Washington D.C. p. 393-400

Nakayama, T., Irikura, M., Kurogi, Y., and Matsuo, H. 1981. Purification and properties of RNA polymerases from mother cells and forespores of sporulating cells of *Bacillus subtilis*. J Biochem (Tokyo) 89: 1681-1691.

Nielsen, H., Engelbrecht, J., Brunak, S., and von Heijne, G. 1997. Identification of prokaryotic and eukaryotic signal peptides and prediction of their cleavage sites. Protein Eng. 10: 1-6.

Noone, D., Howell, A., and Devine, K.M. 2000. Expression of *ykdA*, encoding a *Bacillus subtilis* homologue of HtrA, is heat shock inducible and negatively autoregulated. J. Bacteriol. 182: 1592-1599.

Nouwen, N. and Driessen, A.J.M. 2002. SecDFyajC forms a heterotetrameric complex with YidC. Mol. Microbiol. 44: 1397-1405.

Novak, P., and Dev, I.K. 1988. Degradation of a signal peptide by protease IV and oligopeptidase A. J. Bacteriol. 170: 5067-5075.

Perego, M. 2003. TPR-mediated interaction of Rap proteins with the ComA DNA-binding domain inhibits competence development in *B. subtilis*. Functional Genomics of Gram-positive Microorganisms, 12th International Conference on Bacilli, Baveno, Italy, June 23 2003.

Perego, M., Spiegelman, G.B., and Hoch, J.A. 1988. Structure of the gene for the transition state regulator *abrB*: regulator synthesis is controlled by the *spo0A* sporulation gene in *Bacillus subtilis*. Mol. Microbiol. 2: 689-699.

Perego, M., and Hoch, J.A. 2002. Two-component systems, phosphorelays, and regulation of their activities by phosphatases. In: *Bacillus subtilis* and its closest relatives: from genes to cells. A.L. Sonenshein, J.A. Hoch and R. Losick, eds. ASM press, Washington D.C. p. 473-481.

Price, C.W., Gitt, M.A., and Doi, R.H. 1983. Isolation and physical mapping of the gene encoding the major sigma factor of *Bacillus subtilis* RNA polymerase. Proc. Natl. Acad. Sci. USA. 80: 4074-4078.

Priest, F.G. 1989. Isolation and identification of aerobic endospore-forming bacteria. In: C.R. Harwood, ed., Biotechnology Handbooks 2: *Bacillus*. Plenum Publishing Corp., New York. p. 27-56.

Priest, F.G., and Harwood, C.R. 1994. *Bacillus* species. In: Y.H. Hui and G.G. Khachatourians, eds., Food Biotechnology.VCH Publishers Inc., New York. p. 377-421.

Pugsley, A.P. 1993. The complete general secretory pathway in Gram-negative bacteria. Microbiol. Rev. 57: 50-108.

Rocha, E.P.C., Danchin, A., and Viari, A. 1999. Translation in *Bacillus subtilis*: roles and trends of initiation and termination, insights from a genome analysis. Nucleic Acids Res. 27: 3567-3576.

Rygus, T., and Hillen, W. 1991. Inducible high-level expression of heterologous genes in *Bacillus megaterium* using the regulatory elements of the xylose-utilization operon. Appl. Microbiol. Biotechnol. 35: 594-599.

Schulz, A., Tzschaschel, B., and Schumann, W. 1995. Isolation and analysis of mutants of the *dnaK* operon of *Bacillus subtilis*. Mol. Microbiol. 15: 421-429.

Schumann, W., Ehrlich, S.D., and Ogasawara, N. 2001. Functional analysis of bacterial genes, a practical manual. John Wiley and Sons, Ltd, Chichester, UK.

Scotti, P.A., Urbanus, M.L., Brunner, J., de Gier, J.W., von Heijne, G., van der Does, C., Driessen, A.J.M., Oudega, B., and Luirink J. 2000. YidC, the *Escherichia coli* homologue of mitochondrial Oxa1p, is a component of the Sec translocase. EMBO J. 19: 542-549.

Serror, P., and Sonenshein, A.L. 1996. CodY is required for nutritional repression of *Bacillus subtilis* genetic competence. J. Bacteriol. 178: 5910-5915.

Spiess, C., Beil, A., and Ehrmann, M. 1999. A temperature-dependent switch from chaperone to protease in a widely conserved heat shock protein. Cell 97: 339-347.

Spizizen, J. 1958 Transformation of biochemically deficient strains of *Bacillus subtilis* by deoxyribonucleate. Proc Natl Acad Sci USA 44: 1072-1078.

Steinmetz, M., Kunst, F., and Dedonder, R. 1976. Mapping of mutations affecting synthesis of exocellular enzymes in *Bacillus subtilis*. Identity of the *sacUh*, *amyB*, and *pap* mutations. Mol. Gen. Genet. 148: 281-285.

Stephenson, K., and Harwood, C.R. 1998. Influence of a cell-wall-associated protease on production of alpha-amylase by *Bacillus subtilis*. Appl. Environ. Microbiol. 64: 2875-2881.

Stephenson, K., Carter, N.M., Harwood, C.R., Petit-Glatron, M.F., and Chambert, R. 1998. The influence of protein folding on late stages of the secretion of α-amylases from *Bacillus subtilis*. FEBS Lett. 430: 385-389.

Stöver, A.G., and Driks, A. 1999a. Secretion, localization, and antibacterial activity of TasA, a *Bacillus subtilis* spore-associated protein. J. Bacteriol. 181: 1664-1672.

Stöver, A.G., and Driks, A. 1999b. Control of synthesis and secretion of the *Bacillus subtilis* protein YqxM. J. Bacteriol. 181: 7065-7069.

Stragier, P., Kunkel, B., Kroos, L., and Losick, R. 1989. Chromosomal rearrangement generating a composite gene for a developmental transcription factor. Science 243: 507-512.

Stülke, J., and Hillen, W. 2000. Regulation of carbon catabolism in *Bacillus* species. Annu. Rev. Microbiol. 54: 849-880.

Sun, D.X., Cabrera-Martinez, R.-M., and Setlow, P. 1991. Control of transcription of the *Bacillus subtilis spoIIIG* gene, which codes for the forespore-specific transcription factor sigma G. J. Bacteriol. 173: 2977-2984.

Thomas, J.D., Daniel, R.A., Errington, J., and Robinson, C. 2001. Export of active green fluorescent protein to the periplasm by the twin-arginine translocase (Tat) pathway in *Escherichia coli*. Mol. Microbiol. 39: 47-53.

Tjalsma, H., Bolhuis, A., van Roosmalen, M.L., Wiegert, T., Schumann, W., Broekhuizen, C.P., Quax, W.J., Venema, G., Bron, S., and van

Dijl, J.M. 1998. Functional analysis of the secretory precursor processing machinery of *Bacillus subtilis*: identification of a eubacterial homolog of archaeal and eukaryotic signal peptidases. Genes Dev. 12: 2318-2331.

Tjalsma, H., Kontinen, V.P., Prágai, Z., Wu, H., Meima, R., Venema, G., Bron, S., Sarvas, M., and van Dijl, J.M. 1999a. The role of lipoprotein processing by signal peptidase II in the Gram-positive eubacterium *Bacillus subtilis*: signal peptidase II is required for the efficient secretion of α-amylase, a non-lipoprotein. J. Biol. Chem. 274: 1698-1707.

Tjalsma, H., van den Dolder, J., Meijer, W.J.J., Venema, G., Bron, S., and van Dijl, J.M. 1999b. The plasmid-encoded type I signal peptidase SipP can functionally replace the major signal peptidases SipS and SipT of *Bacillus subtilis*. J. Bacteriol. 181: 2448-2454.

Tjalsma, H., Bolhuis A., Jongbloed, J.D.H., Bron, S., and van Dijl, J.M. 2000a. Signal peptide-dependent protein transport in *Bacillus subtilis*: a Genome-based Survey of the Secretome. Microbiol. Mol. Biol. Rev. 64: 515-547.

Tjalsma, H., Stöver, A.G., Driks, A., Venema, G., Bron, S., and van Dijl, J.M. 2000b. Conserved Serine and Histidine Residues are Critical for Activity of the ER-type Signal Peptidase SipW of *Bacillus subtilis*. J. Biol. Chem. 275: 25102-25108.

Tjalsma, H., Bron, S. and van Dijl, J.M. 2003. Complementary impact of paralogous oxa1-like proteins of *Bacillus subtilis* on post-translocational stages in protein secretion J. Biol. Chem. 278: 15622-15632.

Turgay, K., Hamoen, L.W., Venema, G., and Dubnau, D. 1997. Biochemical characterization of a molecular switch involving the heat shock protein ClpC, which controls the activity of ComK, the competence transcription factor of *Bacillus subtilis*. Genes Dev 11: 119-128.

Vagner, V., Dervyn, E., and Ehrlich, S.D. 1998. A vector for systematic gene inactivation in *Bacillus subtilis*. Microbiology 144: 3097-3104.

Valent, Q.A., de Gier, J.W., von Heijne, G., Kendall, D.A., ten Hagen-Jongman, C.M., Oudega, B., and Luirink, J. 1997. Nascent membrane and presecretory proteins synthesized in *Escherichia coli* associate with signal recognition particle and trigger factor. Mol. Microbiol. 25: 53-64.

van Dijl, J.M., de Jong, A., Vehmaanperä, J., Venema, G., and Bron, S. 1992. Signal peptidase I of *Bacillus subtilis*: patterns of conserved amino acids in prokaryotic and eukaryotic type I signal peptidases. EMBO J. 11: 2819-2828.

van Dijl, J.M., de Jong, A., Venema, G., and Bron, S. 1995. Identification of the potential active site of the signal peptidase SipS of *Bacillus subtilis*: structural and functional similarities with LexA-like proteases. J. Biol. Chem. 270: 3611-3618.

van Dijl, J.M., Bolhuis, A., Tjalsma, H., Jongbloed, J.D.H., de Jong, A. and Bron, S. 2002. Protein transport pathways in *Bacillus subtilis*: a genome-based road map. In: *Bacillus subtilis* and its closest relatvies: from genes to cells. A.L. Sonenshein, J.A. Hoch, and R. Losick eds. ASM Press, USA, p. 337-355.

van Roosmalen, M.L., Jongbloed, J.D.H., Dubois, J.-Y. F., Venema, G., Bron, S. and van Dijl, J.M. 2001. Distinction between major and minor *Bacillus* signal peptidases based on phylogenetic and structural criteria. J. Biol. Chem. 276: 25230-25235.

van Sinderen, D., ten Berge, A., Haijema, B.J., Hamoen, L., and Venema, G. 1994. Molecular cloning and sequence of *comK*, a gene required for genetic competence in *Bacillus subtilis*. Mol. Microbiol. 11: 695-703.

Veltman, O.R., Vriend, G., Hardy, F., Mansfeld, J., van den Burg, B., Venema, G., and Eijsink, V.G.H.. 1997. Analysis of a calcium binding surface loop critical for the stability of the thermolysin-like protease of *Bacillus stearothermophilus*. Eur. J. Biochem. 248: 433-440.

Wandersman, C. 1989. Secretion, processing and activation of bacterial extracellular proteases. Mol. Microbiol. 3: 1825-1831.

Westers, H., Dorenbos, R., Van Dijl, J.M., Kabel, J., Flanagan, T., Devine, K.M., Jude, F., Seror, S.J., Beekman, A.C., Darmon, E., Eschevins, C., de Jong, A., Bron, S., Kuipers, O.P., Albertini, A.M., Antelmann, H., Hecker, M., Zamboni, N., Sauer, U., Bruand, C., Ehrlich, S.D., Alonso, J.C., Salas, M., and Quax, W.J. 2003. Genome engineering reveals large dispensable regions in *Bacillus subtilis*. Mol. Biol. Evol. in press.

Wild, J., Rossmeissl, P., Walter, W.A., and Gross, C.A. 1996. Involvement of the DnaK-DnaJ-GrpE chaperone team in protein secretion in *Escherichia coli*. J. Bacteriol. 178: 3608-3613.

Wisotzkey, J.D., Jurtshuck, P.J., Fox, G.E., Reinhard, G., and Poralla, K. 1992. Comparative sequence analysis of the 16S rRNA (rDNA) of *Bacillus acidocaldarius* and *Bacillus cycloheptanicus* and proposal for creation of a new genus *Alicyclobacillus* gen. nov. Internat. J. System. Bacteriol. 42: 236-269.

Wu, S.C., Ye, R., Wu, X.C., Ng, S.C., and Wong S.L. 1998. Enhanced secretory production of a single-chain antibody fragment from *Bacillus subtilis* by coproduction of molecular chaperones. J. Bacteriol. 180: 2830-2835.

Wu, S.C., Yeung, J.C., Duan, Y., Ye, R., Szarka, S.J., Habibi, H.R., and Wong, S.L. 2002. Functional production and characterization of a fibrin-specific single-chain antibody fragment from *Bacillus subtilis*: effects of molecular chaperones and a wall-bound protease on antibody fragment production. Appl Environ Microbiol 68: 3261-3269.

Wu, X., Lee, L., and Wong, S.-L. 1991. Engineering a *Bacillus subtilis* expression-secretion system with a strain deficient in six extracellular proteases. J. Bacteriol. 173: 4952-4958.

Yamamoto, H., Murata, M., and Sekiguchi, J. 2000. The CitST two-component system regulates the expression of the Mg-citrate transporter in *Bacillus subtilis*. Molec. Microbiol. 37: 898-912.

Yansura, D.G., and Henner, D.J. 1984. Development of an inducible promoter for controlled expression in *Bacillus subtilis*. P. 249-263. *In* Ganesan, A.T., and Hoch, J.A. (eds.), Genetics and Biochemistry of Bacilli. Academic Press, Orlando.

Yuan, G., and Wong, S.-L. 1995. Isolation and characterization of *Bacillus subtilis* regulatory mutants: evidence for *orf39* in the *dnaK* operon as repressor gene in regulating the expression of both *groE* and *dnaK*. J. Bacteriol 177: 6462-6468.

Zuber, U. and Schumann, W. 1994. CIRCE, a novel heat shock element involved in regulation of heat shock operon *dnaK* of *Bacillus subtilis*. J. Bacteriol. 176: 1359-1363.

Zukowski, M.M., and Miller, L. 1986. Hyperproduction of an intracellular heterologous protein in a *sacU^h* mutant of *Bacillus subtilis*. Gene 46: 247-255.

5

Saccharomyces cerevisiae Protein Expression: from Protein Production to Protein Engineering

Ronald T. Niebauer and Anne Skaja Robinson

Abstract

The yeast *Saccharomyces cerevisiae* has long been recognized and used as a host for protein expression since it can offer the processing steps of mammalian systems along with the ease of use of microbial systems. Here, some of the factors enabling effective protein expression are highlighted including vector design, expression strategies, potential problems encountered and methods to deal with these problems. Current and future challenges of the *Saccharomyces cerevisiae* system are also addressed. Emerging technologies based on functional studies, which can be applied to applications such as drug discovery, protein engineering for enhanced properties, and functional genomics are addressed. The versatility associated with using *Saccharomyces cerevisiae* as a tool to address these current and future challenges should allow a continued role in protein production, and in understanding and engineering proteins as well.

Introduction

The yeast *Saccharomyces cerevisiae* (Sc) has been used for hundreds of years in applications such as brewing and baking. Sc is a unicellular eukaryote that combines the benefits of bacterial and mammalian cells. Similar to bacterial cells, Sc is easily manipulated genetically and can be grown to high yield. Like mammalian cells, Sc has the capability of post-translational processing steps that are often necessary for producing authentic products. As an expression system Sc has a proven track record, is nonpathogenic and is a "generally regarded as safe" (GRAS) organism, meaning that it is not harmful to humans. The first recombinant protein expression in yeast was reported in 1981 (Hitzeman *et al.*, 1981). Today, several recombinant vaccines including the first commercialized recombinant vaccine (hepatitis B) are Sc derived (Valenzuela *et al.*, 1982). Table 1 includes a brief list of proteins that have been expressed in Sc highlighting the diversity in product source, protein complexity, and expression strategy.

Although Sc was the first and initially the most popular yeast used for protein production, alternate yeasts have been used to address some problems encountered with the Sc system. Yeasts and fungi have been found with greater promoter strength, secretion efficiency, and the ability to achieve higher cell densities (see Chapters 6 and 7). Even with these advances, Sc still remains a popular host primarily due to its proven history, the existence of various mutants and the availability of a fully sequenced and annotated genome (Goffeau *et al.*, 1996). Evidence of the continued importance of Sc has been shown through the production of medically important human proteins such as G-protein coupled receptors (Kapat *et al.*, 2000) and its versatility in expressing a wide range of proteins (Schuster *et al.*, 2000).

Future work with Sc will most likely include its traditional role in protein production for research and industrial purposes as well as an expanding future role geared toward functional studies. Examples of these types of applications, discussed in more detail later, include advances such as yeast surface display, which can be used to select for enhanced protein properties, and the yeast 2-hybrid system in which particular protein-protein interactions are identified. As we enter the

Table 1. Examples of proteins expressed in *Saccharomyces cerevisiae*.

Protein	Source	Secreted	Promoter	Signal sequence	Notes	Reference
Alcohol dehydrogenase	*Drosophila melanogaster*	No	Galactose, alcohol dehydrogenase	no	Functional dimers produced at 3.5% of extracted yeast protein	Atrian *et al.*, 1990
Alpha amylase	Rice	Yes	Yeast enolase	rice alpha amylase	Used in carbohydrate conversion; 45kDa	Kumagai *et al.*, 1990
Antibody (chimeric)	Mouse and human	Yes	Yeast phosphoglycerate kinase	yeast invertase	Functional antibody; 48 kDa	Horwitz *et al.*, 1988
Beta-glucosidase	*Pyrococcus furiosus*	Yes	Galactose	synthetic pre-pro	Tetrameric, hyperthermopholic enzyme; ~10 mg/L	Smith *et al.*, 2002
Beta2-adrenergic receptor	Human	Yes, plasma membrane	Galactose	alpha factor fusion	GPCR functionally expressed at 115 pmol/mg membrane protein	King *et al.*, 1990
Core antigen	Hepatitis B virus	No	Glyceraldehyde phosphate dehydrogenase (GPD)	no	22 kDa, expressed at 40% of soluble protein	Kniskern *et al.*, 1986
Erythropoietin	Human	Yes	Alpha factor, GPD	alpha factor	Glycosylated hormone	Elliot, *et al.*, 1989
Factor XIII	Human	No	Alcohol dehydrogenase	no	Blood clotting enzyme; 83 kDa	Bishop *et al.*, 1990
Major envelope	Epstein-Barr virus	Yes	Temperature, galactose	pre-pro	400kDa glycoprotein	Schultz *et al.*, 1987
Reverse transcriptase	HIV type1	No	Glyceraldehyde phosphate dehydrogenase	no	Viral enzyme; 117 kDa	Bathurst *et al.*, 1990

post-genomic era in which many genes from different organisms have been identified, one of the challenges that lies ahead is determining the function and interactions of these proteins. Sc will most likely play a pivotal role in tackling these problems, as it is a versatile host not only for protein production but also for protein engineering applications.

In this review we aim to provide an overview of Sc protein expression general enough for the beginner yet helpful to the specialist. The principles we discuss can be applied both to the laboratory and production scale for applications in basic research, commercial, medical, and food biotechnology. The reader is directed to other reviews for further reading (Romanos *et al.*, 1992; Hadfield *et al.*, 1993; Hinnen *et al.*, 1995; Mackay *et al.*, 1996). Specific reviews cover industrial production (Mendoza-Vega *et al.*, 1994; Akada, 2002), the Sc genome database (http://genome-www.stanford.edu/ Saccharomyces) and genetic nomenclature (Cherry *et al.*, 1998), and detailed methods related to Sc cell biology and genetics (Guthrie *et al.*, 2002). This review begins with the basics of protein expression in Sc including vector systems and an overview of expression strategies. The next section discusses some of the problems encountered when using this expression system along with practical approaches to overcome these problems. Finally, some specific examples show current trends for using this expression system.

Vector Systems

There are several key components of an Sc vector system. Depending on the goal, there are different types of plasmids to choose from with various levels of stability and gene copy number. Most systems used are shuttle vectors that contain, in addition to yeast origins of replication, bacterial origins of replication and selection markers for propagation in *Escherichia coli*. A yeast selection marker is needed to ensure the retention and propagation of the vector upon transformation and growth in yeast. A yeast signal sequence can be important for targeting the desired protein product to a desired location, for example extracellular secretion. Recently, a flexible system of yeast vectors with a variety of promoters and selection markers has been described (Funk *et al.*, 2002).

Transformation and Selection Markers

For *Saccharomyces cerevisiae* the common techniques for DNA transformation are the lithium acetate method (Ito *et al.*, 1983) and electroporation (Meilhoc *et al.*, 1990). A new technique known as *kar*-mediated plasmid transfer (Georgieva *et al.*, 2002) has been developed for plasmid transfer between yeast strains. Selection markers are of paramount importance for proper transformant selection. The two major categories of markers are auxotrophic and dominant markers. Auxotrophic markers are used in the transformation of mutant strains that lack the ability to synthesize a component needed for survival. To ensure proper selection, the strain is grown in a media lacking the metabolite that is required by the strain. The mutant yeast strains must contain a chromosomal mutation resulting in the loss of a marker. Common examples of auxotrophic selection markers include the amino acids leucine (*LEU2*) (Beggs, 1978), tryptophan (*TRP1*) (Tschumper *et al.*, 1980), histidine (*HIS3*) (Struhl *et al.*, 1980) and the nucleic acid base uracil (*URA3*) (Rose *et al.*, 1984). For a dominant selection marker, the vector contains genes that enable the cell to gain resistance to some type of antibiotic or potentially cytotoxic agent. Examples include the use of the antibiotic G418 (*G418^R*) (Webster *et al.*, 1983) and copper (*CUP1*) (Fogel *et al.*, 1982).

Autoselection has also been used for plasmid retention. The goal of autoselection is to ensure that the correct plasmid is maintained no matter what the media may be. The use of any media of choice may lead to better growth behavior and could lead to lower costs for large-scale operations. A specific example is the case involving the expression of the yeast killer toxin and corresponding immunity gene (Bussey *et al.*, 1985). Based on this system any cells with the plasmid of interest are viable, while those without the plasmid are not viable due to the expressed toxin.

Dominant markers are advantageous because various host strains can be used. In contrast, the host strain for auxotrophic markers needs to contain a particular mutation. Rich media can be used with a dominant selection marker, usually resulting in better growth characteristics than with minimal media that is used for auxotrophic markers. A concern when using a dominant marker is that spontaneous resistant clones can arise. In practice, the autoselection systems have not seen widespread use.

An interesting case study is the use of *LEU2-d* that contains a defective marker (Erhart *et al.*, 1983). The *LEU2* gene encodes beta-isopropylmalate dehydrogenase and is complementary to a *leu2* mutant. In the altered form (*LEU2-d*) the promoter is truncated and high copy numbers are required to produce enough gene products for the cell to survive. As a result the cells that grow typically have an increased plasmid copy number, the importance of which is discussed in detail below.

Types of Vectors

One way to classify yeast vectors is based upon the ability of the vector to replicate. A brief summary of the types of vectors is shown in Table 2. Yeast integrating plasmids (YIP) are integrated into the yeast chromosome and do not have their own origin of replication so they cannot replicate on their own. Systems that are able to replicate on their own are called yeast replicating plasmids (YRP), which often include an origin of replication by using a chromosomal autonomously replicating sequence (ARS). If a centromeric sequence, which allows for stable segregation of the plasmids during budding, has been added to a replicating vector the vector is known as a yeast centromeric plasmid (YCP). Yeast episomal plasmids (YEP) obtain their origin of replication from an endogenous yeast plasmid known as the 2μ plasmid.

For the YIPs, the DNA is usually integrated into the host by homologous recombination in which a specific site is targeted (Hinnen *et al.*, 1978). When an auxotrophic marker is targeted, the DNA is

Table 2. Summary of *Saccharomyces cerevisiae* vector systems.

Vector Type	Relative Stability	Copy number per cell	Reference
Integrative (YIP)	stable (+++++)	1	Hinnen et al., 1978
Multi-integrative (rDNA based)	fairly stable (++++)	1 to 100	Lopes et al., 1989, 1996
Multi-integrative (Ty based)	stable (+++++)	1 to 30	Parekh et al., 1996
Replicating (YRP)	unstable (+)	1 to 20	Murray et al., 1983
Centromeric (YCP)	stable (+++++)	1 to 2	Clarke et al., 1980
Episomal (YEP) 2 micron	fairly stable (+++)	25 to 100	Futcher et al., 1984
Episomal (YEP) 2 micron LEU2-d	fairly stable (+++)	100 to 200	Rose et al., 1990

integrated at one or two copies per cell. To achieve higher copies of the cloned gene, other sequences that appear more frequently can be used. For example, the ribosomal DNA (rDNA) cluster includes nearly 140 tandem repeats of a particular unit on chromosome XII (Petes, 1980). Using the rDNA cluster, transformants with 100-200 integrated copies have been achieved (Lopes et al., 1989, 1991). In most cases the number of integrated copies is not nearly as high (Fujii et al., 1990) or the strains are not as stable (Lopes et al., 1996). Another targeted site is the delta sequence that is normally present at about 100 copies per haploid genome (Shuster et al., 1990). The delta sequences can exist independently or part of a Ty element, which is a transposon, found in many yeast strains. Stable strains with up to 30 copies have been generated using this method (Parekh et al., 1996).

Other methods of integration include replacement (transplacement) and retrotransposition. In replacement, double homologous recombination is used to replace yeast DNA (Rothstein, 1983). The transformation is usually not efficient, but the resulting strain created is a stable single copy strain. Retrotransposition can be used with the Ty1 and Ty3 retrotransposons of Sc (Wang et al., 1996). This method uses the mechanisms of the Ty elements to carry out replication and integration and relies on a reverse transcription step that makes it different from the other methods.

Some of the trade-offs of the different systems include the copy number, which can range from 1 to approximately 200 per cell, and stability. Normally a high gene copy number is desired because this may result in high protein levels, although there are several examples in which the relationship does not hold (Parekh et al., 1995, 1997; Wittrup et al., 1994). YRPs can have reasonably high copy numbers (1-20), but are generally unstable and hence they are rarely used (Murray et al., 1983; Romaros et al., 1992). In order to ensure stability a centromeric sequence can be added (YCP), but the result is a low copy number of 1-2 per cell (Clarke et al., 1980). This plasmid system is ideal for expression studies when high yields are not needed. YEPs have been widely used for heterologous protein expression because they are fairly stable and can have relatively high copy numbers of more than 30 per cell (Futcher et al., 1984; Romanos et al., 1992). As mentioned before there are variants with partially defective selection markers (*LEU2-d*) that have very high copy numbers at approximately 100-200 per cell (Rose *et al.*, 1990). Although the 2μ plasmid is the most commonly used, it has certain limitations as detailed below.

The yeast 2μ plasmid (reviewed in Rose *et al.*, 1990) is naturally present in most yeast cells and is stably maintained even though it does not give an obvious advantage to the host. In contrast to the case where the native 2μ plasmid is stable by itself, an engineered 2μ vector inserted into a host can decrease plasmid stability. The engineered plasmid is only fairly stable under most conditions resulting in the need for the use of selective media for large-scale work (Futcher *et al.*, 1984; Schwartz *et al.*, 1988). Upon transformation there often is heterogeneity in the transformants with this plasmid. Additionally, the direct relationship between copy number and protein levels does not always hold, which in some cases can be attributed to the accumulation of unfolded protein (Parekh *et al.*, 1995, 1997). The actual design of some of the engineered 2μ plasmids may be to blame for some of the problems with this particular system because deletions or interruptions of functional regions of the plasmid have occurred in some cases. In comparison to alternate yeasts, the lower secretory capacity attributed to Sc may be partially due to these problems with the 2 μ plasmid.

Due to the limitations of the other systems the integrative plasmid is a possible alternative. The major advantage of the integrative plasmid is that it is stable so the cells can be grown in any optimal media after initial selection. A problem that may occur with high copy integration is faulty integration resulting in chromosomal rearrangement or deletion. Since copy numbers are normally low for the integrated plasmids, different techniques have been employed to try to increase the copy number for these plasmids. Progress in this area includes the aforementioned integrated, stable copy vector (Parekh *et al.*, 1996) as well as a novel vector that allows for the regulated insertion of cloned genes (Lee *et al.*, 1997).

Promoters and Terminators

Promoters have traditionally been divided into two categories: constitutive and regulated. The sources of actual promoter sequences include native and engineered systems. Commonly used constitutive promoters are derived from upstream sequences for genes encoding proteins present in high levels in the cell at nearly all times. Examples of these are genes encoding for enzymes such as the triose-phosphate isomerase (*TPI1*) (Smith *et al.*, 1985), glyceraldehyde-3-phosphate dehydrogenase (*GAP*) (Holland *et al.* 1980, Rosenberg *et al.*, 1990) and 3-phosphoglycerate kinase (*PGK1*) (Hitzeman *et al.*, 1983, Kingsman *et al.*, 1990). Nearly all promoters are regulated to some extent and promoters have various strengths. Regulated promoters can be controlled by the presence or absence of a particular media component such as galactose, or be controlled by another process variable such as temperature. Examples of regulated promoters include galactose inducible (*GAL*) (Johnston, 1987), acid phosphatase regulated (*PHO*) (Vogel *et al.*, 1990), and glucose repressed such as alcohol dehydrogenase (*ADH*) (Price *et al.*, 1990). For a detailed listing of promoters used in Sc systems the reader is referred to other sources (Das *et al.*, 1990; Romanos *et al.*, 1992).

Regulated promoters enable variable expression levels and regulated onset of gene expression. In cases where cell growth is affected by expression due to toxicity or heavy metabolic burden, regulated promoters are typically used. For large-scale operations constitutive expression is straightforward and enables constant process conditions for protein expression. Regardless of the promoter, the expression of a specific protein of interest has a major effect on the yield of protein, which can vary widely due to protein toxicity and folding limitations.

A terminator region 3' to the gene of interest will stop translation. The terminator ensures efficient messenger RNA 3' formation. If it is not present, other genes on the plasmid may be expressed. As a word of caution, some prokaryotic and higher eukaryotic terminators are not effective in yeast and termination and identification of precise signals is poorly understood. Commonly used terminator regions come from

MFalpha1 (Brake *et al.*, 1984), *CYC1* (Russo *et al.*, 1989), and *ADH1* (Urdea *et al.*, 1983).

Expression Strategies

A protein can be expressed intracellularly or be targeted for secretion. Cases when intracellular expression is adequate include proteins that are normally expressed in the cytoplasm or proteins that contain few disulfide bonds. An advantage of intracellular expression is that a homogenous product normally results, as modifications such as glycosylation do not occur. One advantage of secreting a protein is the ease of downstream purification since Sc secretes low levels of native proteins. Secretion also allows for post-translational modifications to occur. For example, disulfide bond formation and oligomerization often needed for function are more efficiently carried out in the secretory pathway. Another advantage of secretion is that it avoids the possible toxic build-up of a foreign protein. Cytoplasmically expressed proteins have reached levels of more than 50% of total cell protein (Hallewell *et al.*, 1987). Typically 20% of soluble proteins or on the order of grams per liter are achieved for cytoplasmic expression (Hadfield *et al.*, 1993), while secreted proteins levels often reach mg/L levels (Das *et al.*, 1990).

Intracellular Expression

Several approaches have been developed for the intracellular expression of protein. Direct intracellular expression simply involves placing an initiation codon at the 5' end of the DNA sequence. One method used to facilitate purification is to use an N- or C-terminal tag, but these tags can lead to aggregation. N- and C-terminal processing steps that occur intracellularly may be different than that of secreted proteins. For cytosolic proteins the removal of the N-terminal methionine residue and the acetylation of the N-terminal residue are common events (Moerschell *et al.*, 1990), but in many cases these modifications do not alter the functionality of the protein. Ubiquitin tags have been used to avoid the problem of a modified

N-terminus. In the cell, the ubiquitin is cleaved releasing the full-length protein as a product (Ecker *et al.*, 1989). Recently, a technique was described that involved the use of both ubiquitin and a purification tag that enabled the production and purification of a homogenous protein (Einhauer *et al.*, 2002). Protein stability can be a concern for intracellular expression due to the numerous proteases present in the cytosol. This problem has been overcome in certain situations by fusion to a known stable protein such as human superoxide dismutase (Cousens *et al.*, 1987).

Protein Secretion

The secretory pathway describes the series of compartments through which proteins are trafficked or transported to their final destination - secretion out of the cell, to the plasma membrane, or to the vacuole (Figure 1). Along the pathway various processing steps can occur, some of which ensure the functionality of the protein. Much of the understanding of the secretory pathway has come from analysis of secretory mutants (Novick *et al.*, 1980, 1981). Briefly, a protein is synthesized on the ribosome and then enters the endoplasmic reticulum (ER) either co- or post-translationally. Folding and some post-translational modifications may take place in the ER, the first compartment of the secretory pathway. The protein can then be transported to the Golgi complex where further processing can occur, after which the protein is sent to its final destination. Details of the complexity of the secretory pathway including quality control mechanisms and vesicle transport interactions are areas of active research.

In terms of vector design, key components are N-terminal leader or signaling sequences used to direct the nascent polypeptide into the secretory pathway. These signal sequences (presequences) are usually in the range of 15 to 30 amino acids and often include one or more charged residues followed by a sequence of hydrophobic amino acids and a sequence for cleavage in the ER by signal peptidase (Ngsee *et al.*, 1989).

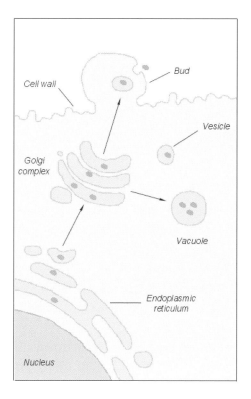

Figure 1. Schematic of the secretory pathway in yeast. Proteins destined for secretion are translocated into the endoplasmic reticulum, where proteins are folded and undergo some post-translational modifications, such as disulfide bond formation. Proteins are then transported in vesicles to the Golgi for further processing, and ultimately to the cell surface, primarily through the growing bud.

In addition to the signal sequence an additional sequence (prosequence) may be used to ensure appropriate trafficking and processing. The combined sequence (presequence and prosequence) is referred to as a secretion leader. The prosequence is present between the signal peptide and the final protein. Possible functions of the prosequence include helping or hindering folding, acting as an enzymatic inhibitor, or acting as a secondary signal sequence. The prosequences that have been used vary in size. One of the most widely studied examples is the secretion leader of the yeast mating alpha factor (Kurjan *et al.*, 1982), which includes a prosequence that can be glycosylated (Julius *et al.*, 1984).

The commonly used *MFalpha1* consists of a presequence of 19 amino acids and a prosequence of approximately 61 amino acids. The Kex2 protease cleaves the prosequence in the late Golgi.

Some of the first amino acid sequences used as leader sequences were those that were isolated from proteins that are normally secreted in Sc such as the alpha factor described above, invertase (*SUC2*) (Perlman *et al.*, 1982), and acid phosphatase (*PHO5*) (Arima *et al.*, 1983). The sequence can also come from the native protein that is being expressed or from an engineered construct. Although different sequences have been used, there seems to be no one optimal sequence that will work for all proteins, and the actual level of secreted protein may vary dramatically from protein to protein. For instance, the secretion of bovine pancreatic trypsin inhibitor in Sc was found to vary more than five-fold when 4 different leaders were investigated (Arnold *et al.*, 1998). In fact, some proteins appear to be secreted properly with no signal sequences (Blachly-Dyson *et al.*, 1987). If a protein is not being secreted, the solution usually involves empirically determining a different sequence that will work. For a thorough listing of signal sequences used, the reader is referred to previous reviews (Das *et al.*, 1990; Hadfield *et al.*, 1993).

Problems and Precautions

As with any protein expression system there are certain features that may be problematic. The exact causes of some of these problems are not well understood. For Sc, the actual yield of protein varies depending on the particular protein being expressed. Problems can be rooted at the genetic, transcriptional, translational or post-translational level. In comparison to other systems, the most common disadvantages of using Sc include problems involved with glycosylation, proteolysis, and bottlenecks in protein expression. Advances have addressed many of these problems and research is ongoing in many of the areas.

Host Strain Selection

The selection of a host strain is a key part to engineering Sc for protein production and should be one of the initial considerations in designing an expression system. It is important to note that there are different types of Sc cells (Dickinson *et al.*, 1999). The term haploid is used to describe cells that have one copy of each chromosome per cell. A diploid has two copies of each chromosome per cell. **A** and α cells are haploid cells that are different mating types - **a** cells secrete **a**-factor, while α cells secrete α factor. When **a** and α cells are cocultured a diploid is formed. In expression studies, the focus is not on mating or the cell cycle, so haploid cells of the same mating type are normally used.

Mutant strains are important in numerous applications. In order to ensure proper selection, the appropriate auxotrophic or nutritional mutation in the host strain needs to be present. Mutant strains have been isolated that reduce glycosylation and proteolysis which are discussed in more detail below. Super secreting strains were isolated by using a screen to identify mutants with improved secretion (Smith *et al.*, 1985). The mutants were believed to have increased secretory capacity by shifting the distribution of the product towards secretion, away from vacuolar degradation. Other mutations have been found that act on the transcriptional level (Sakai *et al.*, 1988).

At the genetic level, plasmid stability can be a key issue (Zhang *et al.*, 1996). Instabilities can arise due to segregational or structural problems. Segregational problems occur when the plasmids are not distributed properly to the daughter cells. Structural problems can result from rearrangements or loss of plasmid DNA. To address these problems, the DNA can be integrated into the chromosome for increased stability. For cases when a 2μ-based plasmid is used it has been found that certain partitioning components should be preserved to ensure stability (Mackay *et al.*, 1996).

Transcriptional Problems

Most yeast genes do not have introns, so mRNA with introns most likely will not be processed properly (Beggs *et al.*, 1980). Therefore, cDNA constructs or genes without introns should be used. The messenger RNA level is influenced by the rate of initiation, the elongation process, and the rate of turnover. However, there are few examples in which mRNA stability is the main factor in low protein yields.

mRNA levels can be reduced if the termination sequence is removed (Zaret *et al.*, 1982). Sometimes unmodified foreign genes are expressed in Sc, and examples of bacterial genes fortuitously expressed exist (Marczynski *et al.*, 1985), but in these cases the expression levels of the undesired fortuitously expressed protein tend to be low. Pausing or termination due to specific sequences can affect the elongation of the RNA. Typically these problems have occurred in rich AT regions and one potential solution is to use chemical synthesis of genes to increase the GC content while retaining the amino acid sequence (Romanos *et al.*, 1991).

Translational Problems

Similar to transcription, initiation and elongation can play a role in the efficiency of translation. Translation in yeast is sensitive to the 5' untranslated regions as well as the coding region of the mRNA where hairpin loops can form that may inhibit efficient translation (Baim *et al.*, 1988). The yeast consensus sequence at the initiation point is different from that of higher eukaryotes (Cigan *et al.*, 1987), hence it is recommended to avoid foreign non-coding sequences in yeast vectors. The codon usage is known to have an effect on translational elongation since the bias seems to correlate to the abundance of tRNAs, but is normally not a major problem unless there are very high levels of mRNA (Romanos *et al.*, 1992).

Post-Translational Issues

There are vast arrays of post-translational events that can occur in Sc. Some of these events are beneficial, while some are undesirable. Like mammalian systems, Sc can achieve most post-translational events including amino-terminal acetylation, phosphorylation, myristylation and isoprenylation. A few of the modifications that Sc is unable to perform include gamma-carboxylation and hydroxymethylation (Mackay *et al.*, 1996). This section highlights some of the typical post-translational events including glycosylation, proteolysis, and protein folding.

Glycosylation

Glycosylation (reviewed by Kukuruzinska *et al.*, 1987; Innis, 1989) is the most common and complex form of post-translational modification (Meynial-Salles *et al.*, 1996). Glycosylation is important for the functionality of many proteins and is believed to help in protein targeting, solubility, binding events, and stability (Wang *et al.*, 1996). For Sc, glycosylation can occur in the ER and Golgi. N-linked inner core glycosylation occurs in the ER. Potential sites on the protein for this to occur include asparagine/x/serine and asparagine/x/threonine, where x can be any amino acid except proline. These sites are necessary but not sufficient for glycosylation to occur. The carbohydrate complex that is added to the protein consists of two N-acetylglucosamine molecules, nine mannose sugars, and three glucose molecules, which are linked to the asparagine. Before exiting the ER, three of the glucose residues along with one mannose residue are removed. These initial core additions in Sc are similar to that in mammalian cells.

Outer core N-linked glycosylation and O-linked glycosylation (attached to serine or threonine) can occur in the Golgi. It is at this stage that Sc and other eukaryotes differ (Figure 2). In other eukaryotes, mannose residues are removed and additional sugars may be added resulting in shortened side branches that end with sugars such as galactose and sialic acid (Romanos *et al.*, 1992). Sc lacks the Golgi mannosidase that is present in higher eukaryotes and as a result

A

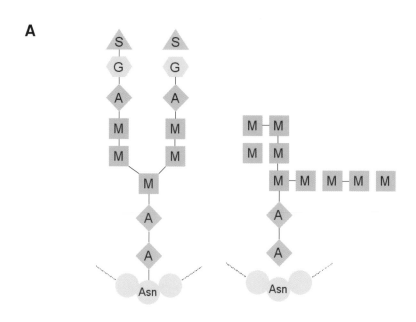

B

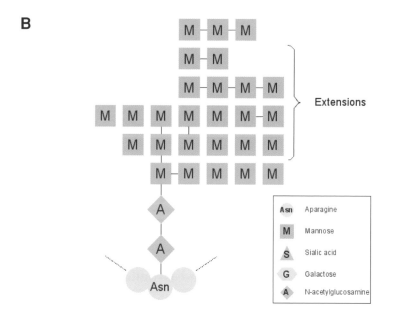

Figure 2. Outer core N-linked glycosylation for: A. Mammalian systems; B. *Saccharomyces cerevisiae*. Mannose addition for *S. cerevisiae* is typically heterogeneous. Asparagine (Asn) is the site of protein chain N-linked glycosylation for both systems.

elongation may occur. In fact, hyperglycosylation, which is the heterogeneous addition of 40 or more mannose residues, may occur. The modifications and extensions made in the Golgi can be different for each particular protein and may hinder purification or alter functionality. In addition, these modifications may render the protein immunogenic in humans. Due to the varying additions that occur, non-homogeneity in the product may result.

The problems associated with glycosylation can be controlled in various ways. The glycosylation sites can be mutated, but this may be undesirable because functionality may depend on certain levels of glycosylation. Glycosylation mutants have been identified that interfere with certain processing steps. In particular, mutants in mannan biosynthesis (*mnn*) that reduce the extensive outer chain additions have been used successfully (Tsai *et al.*, 1984).

Proteolysis

Even if a protein is being synthesized at a high rate, the degree of degradation must be taken into account when considering the final yield. Certain proteases that reside in the vacuole, including carboxypeptidases and aminopeptidases, have been found to lower the yields of native and heterologous proteins (reviewed by Van den Hazel *et al.*, 1996; Jones, 2002). In cases of secretory expression, vacuolar enzymes have been released into the medium (Stevens *et al.*, 1986). Proteins that are expressed intracellularly are especially prone to proteolysis following exposure to vacuolar proteases upon cell lysis. To avoid this problem, strains that are deficient in the major vacuolar proteases can be used. In addition, protease inhibitors are routinely added to lysates (Jones, 2002). Whereas the vacuolar degradation normally affects long-lived proteins, short-lived proteins can be degraded in the cytosol. This form of degradation involves ubiquitin, which is a 76 amino acid protein that may become linked to intracellular proteins. The ubiquitin pathway involves degradation by protease complexes, and is driven by ATP (reviewed by Dice, 1987).

Specific proteolytic cleavage is necessary for removing signal sequences to enable functionality of the protein. The Kex2 protease, which cleaves the mating factor signal prosequence, cleaves internal lysine-arginine or arginine-arginine pairs (Bitter *et al.*, 1984). To avoid cleavage within a protein of interest, a mutation to lysine-lysine is typically performed. In cases where the processing is not occurring properly the processing site can be modified, for example by adding an appropriate spacer (Kjeldsen *et al.*, 1996). Alternatively, protease expression can be increased (Egel-Mitani *et al.*, 1990).

Folding and Transport Bottlenecks

Often when there is a high gene copy number, unfolded protein accumulates in the ER, ultimately resulting in attenuation of protein production (reviewed by Sidrauski *et al.*, 1998). Various chaperones and foldases that are key in the ER folding machinery have been co-expressed to help relieve this bottleneck (Robinson *et al.*, 1994, Harmsen *et al.*, 1996, Robinson *et al.*, 1996, Shusta *et al.*, 1998). The most successful improvements in expression have relied on overexpression of protein disulfide isomerase (PDI), which plays a key role in disulfide bond formation in the ER (Robinson *et al.*, 1994, Shusta *et al.*, 1998). In some cases, the rate-limiting step in the process was shown to be disulfide bond formation. A key question that has yet to be fully answered is if there is an optimal synthesis level for protein expression.

Recent Advances in Protein Engineering and Functional Studies

Protein Engineering

Various techniques in genetic and cellular engineering have been used to tailor the expression system to produce more protein. For example, yeast strains are engineered to reduce the activity of proteases (reviewed by Jones, 2002). Cellular proteins such as chaperones and foldases, whose expression levels can be modified, ensure correct protein folding (see previous section "*Folding And Transport Bottlenecks*"). Protein engineering is being applied in many expression applications as well. In this section, techniques for designing proteins with enhanced properties and higher protein yields are discussed.

Sc Cell Surface Display

Screening techniques are used to identify and select proteins with new functions. The goal of cell-surface display is to express a particular protein on the surface of a cell, then screen for ligand binding affinity, stability, enzymatic activity, or to detect molecular interactions. A yeast surface display technique has been developed that offers the ability to express proteins with post-translational modifications, a higher quantitative precision than phage display, and more versatility and rapidity than mammalian systems (Wittrup, 2001).

The Sc surface display system uses a native cellular receptor, the α agglutinin receptor. The agglutinin receptor is composed of two subunits, Aga1p and Aga2p, which are held together by a disulfide bond. The protein of interest can be genetically fused to the carboxyl terminus of the *AGA2*. Since the Aga1p is secreted from the cell and covalently linked to the yeast cell wall, the protein of interest is then tethered to the surface of the yeast cell in its functional form and is accessible to ligands or other molecules without any steric interference. One example of the power of this system is a study in which an antifluorescein single chain antibody with improved binding characteristics was isolated from a library of mutated antibodies by using flow cytometric cell sorting. By screening libraries of 10^5-10^7 random mutants, mutants with dissociation kinetics four orders of magnitude slower than wild type were isolated (Boder *et al.*, 2000). Another application of this technique involved the isolation of single chain antibodies that had a higher affinity for a particular domain of a T-cell receptor (Kieke *et al.*, 1997).

Structure Based Design

Enhancement of protein expression has also been achieved by structure-based rational engineering. The production of insulin is currently required in large quantities for treatment of diabetes (Kjeldsen *et al.*, 2002). Based on the idea that a more stable precursor would allow for more efficient folding and transport through the secretory pathway, the yield of an insulin analogue was increased 5-fold (Kjeldsen *et al.*, 2002).

Analyzing Gene Function

With the large amount of information on protein sequence generated by the human genome project, much of the future work will focus on understanding function and protein interactions. One interesting observation is that many cellular structures, pathways, and signaling show a high level of conservation between species. In particular, many Sc genes have been found to have homologues in humans, making Sc an ideal model experimental system (Mushegian *et al.*, 1997). For example the human gene related to cystic fibrosis (cystic fibrosis transmembrane conductance regulator) has a counterpart in yeast named *YCF1* (Szczypka *et al.*, 1994). Specific information on the Sc genome and proteins can be obtained from several sources (reviewed by Lewitter, 2002). The following section highlights some of the new techniques and screening approaches using Sc expression technology.

One technique for analyzing gene function uses gene deletions. After deleting an open reading frame, the resulting Sc strains can be assayed for growth to identify the genes essential for viability (Winzeler *et al.*, 1999). Gene expression studies have also been described using serial analysis of gene expression (SAGE) that allows for the quantitative analysis of a large number of transcripts (Velculescu *et al.*, 1995).

One method to gain insight into *in vivo* protein function is to develop cell-based assays in Sc. There are numerous advantages associated with Sc including the ability to produce functional heterologous proteins and a thorough understanding of the Sc genetics allowing for the development of mutants or testing for complementarity. There are many different types of cell-based assays currently being used (Munder *et al.*, 1999). A substitution assay involves replacing a particular yeast protein with a similar protein from a different organism. Differential expression assays are used to determine what factor changes a given phenotype, with applications in identifying and characterizing the function of genes and gene products. With advances in high throughput screening techniques these cell-based assays are important in many diverse future applications, some of which are described below. More information can be found elsewhere on the potential use of Sc for the structure-function analysis of membrane proteins (reviewed by Bill, 2001), as well as the applications of yeast in drug discovery (reviewed by Ma, 2001).

Substitution Assay – Membrane Protein Screens

The G-protein coupled receptors (GPCRs) are a medically important class of proteins that have been associated with numerous diseases (Flower, 1999). Studies of membrane proteins lag considerably behind studies of soluble proteins, despite the fact that approximately one quarter to one third of most proteomes encodes membrane proteins. The major problem in understanding this class of proteins has been the difficulty in expressing enough functional proteins to obtain structural data (reviewed by Grisshammer *et al.*, 1995). Of the approximately 1000 GPCRs in humans, many with unknown ligands (orphans), no human GPCRs have been crystallized to date. Since GPCRs have been linked to cancer, heart disease, asthma, diabetes and several other diseases, effective methods of designing drugs to interact with these proteins are necessary.

The GPCRs are composed of a seven transmembrane spanning domain that is coupled to a heterotrimeric G-protein inside the cell. Upon ligand binding to a peptide, hormone, or small molecule, this class of proteins mediates a wide variety of responses within the cell. One downstream response is the activation of metabolic pathways within the cell (For example, see Figure 3). Cell-based assays for these GPCRs can provide much needed insight into this large superfamily of membrane proteins.

A

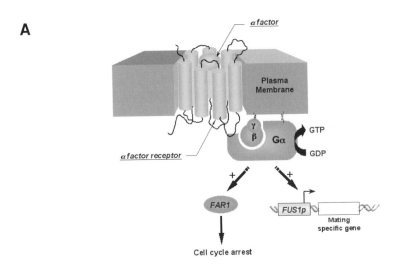

B

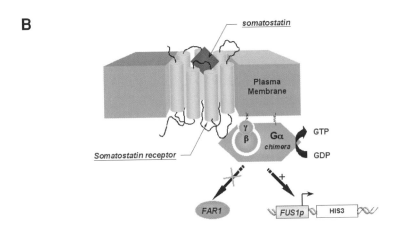

Figure 3. Schematic of GPCR signaling pathways. A. Native α-factor receptor signaling pathway. The signal transduction of **a**-mating type yeast cells is stimulated by binding of the small pheromone α factor to its receptor. Once activated, the α subunit of the G-protein (G$_\alpha$) dissociates from the βγ subunit complex, activating the MAP kinase pathway. This includes Far1 activation, resulting in cell-cycle arrest, and the expression of mating genes. B. Engineered yeast system for ligand screening of the somatostatin receptor. The somatostatin receptor (SST$_2$) is expressed in yeast, and the G$_\alpha$ protein is replaced by a yeast-mammalian G$_\alpha$ chimera. The FAR1 gene is deleted to prevent cell cycle arrest. Upon ligand binding and release of the G$_\alpha$-chimera, downstream activation of the Fus1-His3 reporter enables histidine biosynthesis. Screening on histidine-free media enables selection of ligand analogues. Modified from Pausch, 1997.

Recent advances have been made in designing high-throughput screens to find agonists (activators) and antagonists (inhibitors) for this class of proteins using Sc. In mating type **a** cells, an endogenous GPCR known as the α factor receptor is present. Upon activation by α factor the receptor changes conformation, the G-protein dissociates, activating mating-related cellular pathways and cell cycle arrest (Figure 3A). In order to design a screen for human GPCRs, several modifications were made to the wild type receptor system (Price *et al.*, 1995; Pausch, 1997). The mammalian somatostatin receptor was expressed recombinantly, and the wild-type pathway was modified so that the cells would only grow if the receptor was activated by ligand binding (Figure 3B). A chimeric yeast-human G_α-protein was created to enable interaction with the somatostatin receptor. *FAR1* was deleted so cell cycle arrest would not occur and the *FUS1* promoter was linked to the *HIS3* gene to enable ligand-binding directed expression of histidine. Selection of cells able to grow in the absence of histidine was then used to identify ligands that interacted with the GPCR and caused *FUS1* mediated *HIS3* expression. Somatostatin was shown to induce a dose-dependant increase in growth on media lacking histidine (Price *et al.*, 1995). This example shows the potential of using Sc as an experimental platform for expressing complex proteins and screening for particular downstream pathway activation.

Differential Assay – Cisplatin Resistance

Sc has been used to identify particular genes important for specific functions. For example, novel genes were identified that conferred resistance to cisplatin (Burger *et al.*, 2000). Cisplatin is an agent that has been shown to have anticancer activities. Although this is a promising agent, cellular resistance to cisplatin has been a major obstacle. Sc was chosen as a model organism to gain an understanding of the mechanism of cellular resistance as well as to identify particular genes that cause the resistance. This was carried out by transforming Sc cells with a yeast genomic DNA library, screening for resistance to cisplatin, and identifying the gene or genes responsible for resistance. A number of interesting yeast genes that were linked to cisplatin resistance were identified and should lead to insight about mammalian systems (Burger *et al.*, 2000).

Genome-Wide Analysis

In monitoring the effect of a particular variable on gene expression, there may be thousands of genes affected. One developing technique to help monitor widespread changes in expression is DNA microarrays (Gasch, 2002). The procedure involves printing thousands of gene sequences on an array or slide, which allows for large-scale genomic studies (DeRisi *et al.*, 1997). The arrays can be made of oligonucleotides or complementary DNA (reviews by Brown *et al.*, 1999; Lockhart *et al.*, 2000). For example, RNA from cell populations under different conditions is used to make fluorescently labeled cDNAs that can hybridize to the DNA microarray. Based on differential fluorescent signals, imaging of the array gives a snapshot of relative gene expression under different conditions. Analysis of the changes can give insight into the function of particular genes.

A developing approach involves using protein microarrays (Zhu *et al.*, 2001). This technique lies closer to directly determining function by looking at proteins themselves (proteome level) and not their precursors (transcriptome level). Although the capacity to express and purify the necessary proteins in a high-throughput fashion and ensure that they are functional when placed on a solid surface is still emerging, this technique shows promise. Another technique that looks specifically at proteins is two-dimensional gel electrophoresis coupled with peptide-mass fingerprinting (Shevchenko *et al.*, 1996) which allows for the large-scale identification of yeast proteins.

Advances in the use of reporter genes have enabled insight into the trafficking of proteins. The use of the green fluorescent protein (GFP) as a reporter protein has been revolutionary (Cubitt *et al.*, 1995). GFP can be fused to a particular protein of interest and the localization of the protein can be monitored *in vivo* using confocal microscopy. The advantages of using GFP include that functionality of its fusion partner is normally not perturbed, no additional substrate is needed, real-time events can be monitored, and mutants with enhanced properties are readily available (Tatchell *et al.*, 2002). By determining the location of native Sc proteins, information that leads to functional roles can be deduced. Monitoring the localization of heterologous proteins can be used to gain an understanding of the mechanisms of protein trafficking and the identification of any potential bottlenecks. New fluorescent proteins such as a red fluorescent protein from the *Discosoma* species (Bevis *et al.*, 2002; Campbell *et al.*, 2002) as well as color variants of GFP have enabled enhanced monitoring of protein trafficking through dual labeling. GFP fluorescence has also been used to quantitate protein levels and has found use as an on-line reporter of protein expression (Li *et al.*, 2000).

Protein-Protein Interactions

The characterization of protein interactions on a large scale is particularly challenging (reviewed by Auerbach *et al.*, 2002). As of 2001, six years after the yeast genome was available, there were still approximately 1900 yeast genes coding for proteins with unknown functions (Kumar *et al.*, 2001). An even larger problem is that an estimated 80% of the predicted human genes code for proteins of unknown function (Aach *et al.*, 2001). In what has been termed 'guilt by association', it is hoped that identifying a binding partner will give insight into a proteins' function. Described below are some of the techniques that are being used to determine interaction partners.

Yeast 2-hybrid System

The yeast two-hybrid system is a well-established technique that can be used to determine which protein from a library of proteins will interact with a protein of interest (Fields *et al.*, 1989). One potential application of this approach is the design of therapeutics to disrupt particular protein interactions. The 2-hybrid system is based on the fact that transcription factors usually have two discrete domains, a DNA binding domain and an activator domain (Figure 4A). One vector encodes the protein of interest (the bait) fused to a DNA binding domain. The second vector encodes a library of proteins (the prey) that are fused to an activation domain. A reporter gene is placed downstream of the DNA binding site. When there is protein-protein interaction, the moieties of the transcription factor are joined and the

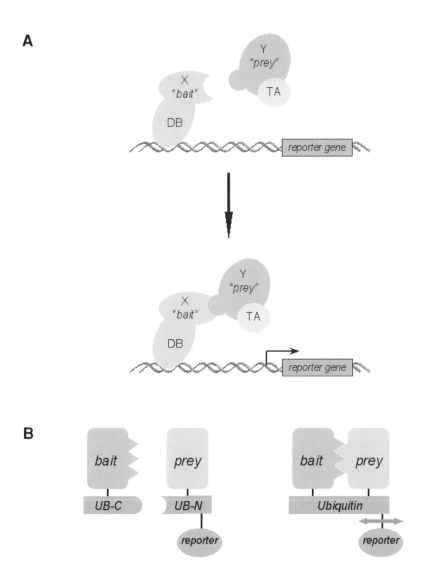

Figure 4. Identification of protein-protein interactions. A. Yeast two-hybrid system. The protein of interest ("bait") is genetically linked to the DNA binding domain (DB) of the transcription factor, while a library of possible interacting proteins ("prey") are linked to the activation domain of the transcription factor. Positive interactions result in transcription factor binding to the DNA, and reporter gene synthesis. B. Split-ubiquitin technique. The C-terminal half of ubiquitin (UB-C) is fused to a gene of interest ("bait") and the N-terminal half (UB-N) is fused to both the gene of a potential interaction partner ("prey") and a transcription factor. When the two proteins interact, the ubiquitin halves reassemble, which allows ubiquitin specific proteases to recognize and cleave the full-length ubiquitin at the C-terminal end to release the reporter.

reporter gene expression is activated. Numerous improvements have been developed to reduce false positives (Colas *et al*., 1998). Other similar techniques have been developed including a novel method based on the Ras pathway in yeast (Hubsman *et al*., 2001).

Split-Ubiquitin Technique

A split-ubiquitin system has been developed that is also useful for detecting protein interactions (Johnsson *et al*., 1994). In this technique ubiquitin is expressed as an N- and C-terminal half (Figure 4B). The C-terminal half is fused to a gene of interest ("bait") and the N-terminal half is fused to both the gene of a potential interaction partner ("prey") and a transcription factor. When the two proteins interact, the ubiquitin halves reassemble, which allows ubiquitin specific proteases to recognize and cleave the full-length ubiquitin at the C-terminal end. The cleavage releases the transcription factor that then activates reporter genes in the nucleus. While the two-hybrid technique actually monitors potential interactions that occur in the yeast nucleus, the split-ubiquitin technique has been designed to monitor interactions that may not occur in the nucleus such as that of integral membrane proteins (Stagljar *et al*., 1998).

Large Scale Techniques

Insight into particular cellular networks can potentially be achieved by determining which proteins interact in whole genomes. Information from these types of studies can be used to help identify the function of novel proteins by determining which known proteins they interact with. In a larger view, these types of studies can lead to a better understanding of entire protein-protein interaction networks within a cell - a large task since Sc has approximately 6,000 proteins. The two-hybrid technique has been adapted to high throughput applications in what are known as the matrix (array) approach and the library screening approach (reviewed by Legrain *et al*., 2000). For the matrix approach (Uetz *et al*., 2000) an array of prey clones come into contact with an array of bait clones and are allowed to mate resulting in a diploid strain in which the bait and prey can interact. Interactions are identified by expression of a reporter gene. In the library approach (Ito *et al.*, 2000, 2001), one bait protein is screened against a library and interactions are identified based on growth in selective media. When comparing two large-scale studies, a surprising result was that only about 20% of the interactions in the two datasets overlapped (Ito *et al.*, 2001) showing that the results of these experiments should be judged with both caution and optimism.

High-Throughput Assays on a Genomic Scale

Previously, yeast have been used in characterizing eukaryotic genomes through the use of yeast artificial chromosomes (YAC) which are specialized vectors that can contain large (100-1000kb) fragments of DNA (Burke *et al.*, 1987). No longer just a tool for mapping genomes, Sc is being used as a host to profile the human genes. Recently, a streamlined approach was described that showed how heterologous genes were assayed using yeast as a surrogate (Tugendreich *et al.*, 2001). This research is driven by the information that has been obtained from sequencing of the human genome and the challenge of determining the function of the proteins and how some of them are related to various diseases. There are gene families such as GPCRs and kinases that are known to be successful drug targets. Due to the fact that many of these gene families are rather large (~600 from kinase family) there is a need for a rapid method to test potential small molecule inhibitors. In Tugendreich *et al.*, (2001), heterologous cDNAs were transferred into yeast vectors and phenotypic profiling was used to screen for chemical inhibitors. The result is a process that uses Sc to assay heterologous genes in a high-throughput fashion.

Conclusion

Humankind has been using the yeast Sc for centuries in baking and brewing (reviewed by Mortimer, 2000). Today Sc is used in biological and biomedical research, health care, food technology, industrial fermentations, and environmental applications. The widespread use of Sc is the result of time and detailed research by many different groups and individuals (Evans *et al.*, 1991). By the end of the 17th century yeast cells were first observed under a microscope by van Leeuwenhoek. In the 1800s Pasteur showed that fermentation was carried out by living cells. The 19th and early 20th century saw advances in yeast cell biology and the use of yeast to study metabolism and fermentation energetics. In 1978 the first successful yeast transformation opened the doorway to many of the protein production applications discussed in this review. By the early 1980s, the pioneering work by Schekman led to a better understanding of the secretory pathway in yeast essential for the production of vaccines and enzymes (Novick *et al.*, 1980, 1981). Sc has been at the forefront of expression technology for a period of time, but problems and limitations of Sc led to research of non-Sc yeasts. Interest was revived in 1996 as the Sc genome project was completed. In particular, Sc will most likely be important as a model system in gaining an understanding of the function of the many proteins in the human body. Cell-based assays hold great potential in the development of medications for the treatment of human illnesses. A better understanding of the yeast machinery should lead to advances in protein expression technology.

Acknowledgements

The authors thank Damien Thevinin for generously creating the artwork for all of the figures. In addition, we thank Jason Smith and Dane Wittrup for helpful comments on the manuscript. This work was supported in part by funding from NSF BES 9984312.

References

Aach, J., Bulyk, M.L., Church, G.M., Comander, J., Derti, A., and Shendure, J. 2001. Computational comparison of two draft sequences of the human genome. Nature. 409: 856-859.

Akada, R. 2002. Genetically modified industrial yeast ready for application. J. Biosci. Bioeng. 94: 536-544.

Arima, K., Oshima, T., Kubota, I., Nakamura, N., Mizunaga, T., and Toh-e, A. 1983. The nucleotide sequence of the yeast PHO5 gene: a putative precursor of repressible acid phosphotase contains a signal peptide. Nucleic Acids Res. 11: 1657-1672.

Arnold, C.E., Parekh, R.N., Yang, W., and Wittrup, K.D. 1998. Leader peptide efficiency correlates with signal recognition particle dependence in *Saccharomyces cerevisiae*. Biotech. Bioeng. 59: 286-293.

Atrian, S., Gonzalez-Duarte, R., and Fothergill-Gilmore, L. 1990. Synthesis of *Drosophila melanogaster* alcohol dehydrogenase in yeast. Gene. 93: 205-212.

Auerbach, D., Thaminy, S., Hottiger, M.O., and Stagljar, I. 2002 The post-genomic era of interactive proteomics: facts and persectives. Proteomics. 2: 611-623.

Baim, S.B., and Sherman, F. 1988. mRNA structures influencing translation in the yeast *Saccharomyces cerevisiae*. Mol. Cell. Biol. 8: 1591-1601.

Bathurst, I.C., Moen, L.K., Lujan, M.A., Gibson, H.L., Feucht, P.H., Pichuantes, S., Craik, C.S., Santi, D.V., and Barr, P.J. 1990. Characterization of the human immunodeficiency virus type-1 reverse transcriptase enzyme produced in yeast. Biochem. Biophys. Res. Commun. 171: 589-595.

Beggs, J.D. 1978. Transformation of yeast by a replicating hybrid plasmid. Nature. 275: 104-109.

Beggs, J.D., van den Berg, J., Van Ooyen, A., and Weissman, C. 1980. Abnormal expression of chromosomal rabbit beta-globin gene in *Saccharomyces cerevisiae*. Nature. 283: 835-840.

Bevis, B.J., and Glick, B.S. 2002. Rapidly maturing variants of the *Discoma* red fluorescent protein (DsRed). Nat. Biotechnol. 20: 83-87.

Bill, R.M. 2001. Yeast – a panacea for the structure-function analysis of membrane proteins? Curr. Genet. 40: 157-171.

Bishop, P.D., Teller, D.C., Smith, R.A., Lasser, G.W., Gilbert, T., and Seale, R.L. 1990. Expression, purification, and characterization of human factor XIII in *Saccharomyces cerevisiae*. Biochemistry. 29: 1861-1869.

Bitter, G.A., Chen, K.K., Banks, A.R., Lai, P.H. 1984. Secretion of foreign proteins from *Saccharomyces cerevisiae* directed by α-factor gene fusions. Proc. Natl. Acad. Sci. USA. 81: 5330-5334.

Blachly-Dyson, E., and Stevens, T.H. 1987. Yeast carboxypeptidase Y can be translocated and glycosylated without its amino-terminal signal sequence. J. Cell Biol. 104: 1183-1191.

Boder, E.T., Midlefort, K.S., and Wittrup, K.D. 2000. Directed evolution of antibody fragments with monovalent femtomolar antigen-binding affinity. Proc. Natl. Acad. Sci. USA. 97: 10701-10705.

Brake, A.J., Merryweather, J.P., Coit, D.G., Heberlein, U.A., Masiarz, F.R., Mullenbach, G.T., Urdea, M.S., Valenzuela, P., and Barr, P.J. 1984. α-factor-directed synthesis and secretion of mature foreign proteins in *Saccharomyces cerevisiae*. Proc. Natl. Acad. Sci. USA. 81: 4642-4646.

Brown, P.O., and Bostein, D. 1999. Exploring the new world of the genome with DNA microarrays. Nat. Genet. 21: 33-37.

Burger, H., Capello, A., Schenk, P.W., Stoter, G., Brouwer, J., and Nooter, K. 2000. A genome-wide screening in *Saccharomyces cerevisiae* for genes that confer resistance to the anticancer agent cisplatin. Biochem. Biophys. Res. Comm. 269: 767-774.

Burke, D.T., Carle, G.F., and Olson, M.V. 1987. Cloning of large segments of exogenous DNA into yeast by means of artificial chromosome vectors. Science. 236: 806-812.

Bussey, H., and Meaden, P. 1985. Selection and stability of yeast transformants expressing cDNA of an M1 killer toxin-immunity gene. Curr. Genet. 9: 285-291.

Campbell, R.E., Tour, O., Palmer, A.E., Steinbach, P.A., Baird, G.S., Zacharias, D.A., Tsien, R.Y. 2002. A monomeric red fluorescent protein. Proc. Natl. Acad. Sci. USA. 99: 7877-7882.

Cherry, J.M., Adler, C., Ball, C., Chervitz, S.A., Dwight, S.S., Hester, E.T., Jia, Y., Juvik, G., Roe, T., Schroeder, M., Weng, S., and Botstein, D. 1998. SGD: *Saccharomyces genome* database. Nucleic Acids Res. 26: 73-79.

Cigan, A.M., and Donahue, T.F. 1987. Sequence and structural features associated with translational initiator regions in yeast - a review. Gene. 59: 1-18.

Clarke, L., and Carbon, J. 1980. Isolation of a yeast centromere and construction of functional small circular chromosomes. Nature. 257: 504-509.

Colas, P., and Brent, R. 1998. The impact of two-hybrid and related methods on biotechnology. Trends Biotechnol. 16: 355-363.

Cousens, L.S., Shuster, J.R., Gallegos, C., Ku, L.L., Stempien, M.M., Urdea, M.S., Sanchez-Pescador, R., Taylor, A., and Tekamp-Olson, P. 1987. High level expression of proinsulin in the yeast, *Saccharomyces cerevisiae*. Gene. 61: 265-275.

Cubitt, A.B., Heim, R., Adams, S.R., Boyd, A.E., Gross, L.A., and Tsien, R.Y. 1995. Understanding, improving and using green fluorescent proteins. Trends Biochem. Sci. 20: 448-455.

Das, R.C., and Campbell, D.A. 1990. Host cell control of heterologous protein production in *Saccharomyces cerevisiae*. In: Yeast Strain Selection. Chandra J. Panchal, ed. Marcel Dekker, New York, New York. p.311-342.

DeRisi, J.L., Iyer, V.R., and Brown, P.O. 1997. Exploring the metabolic and genetic control of gene expression on a genomic scale. Science. 278: 680-686.

Dice, J.F. 1987. Molecular determinants of protein half-lives in eukaryotic cells. FASEB J. 1: 349-357.

Dickinson, J.R., and Schweizer, M. 1999. The metabolism and molecular physiology of *Saccharomyces cerevisiae*. Taylor and Francis, London.

Ecker, D.J., Stadel, J.M., Butt, T.R., Marsh, J.A., Monia, B.P., Powers, D.A., Gorman, J.A., Clark, P.E., Warren, F., Shatzman, A., and Crooke, S.T. 1989. Increasing gene expression in yeast by fusion to ubiquitin. J. Biol. Chem. 264: 7715-7719.

Egel-Mitani, M., Flygenring, H.P., and Hansen, M.T. 1990. A novel aspartyl protease allowing KEX2-independent MF α propheromone processing in yeast. Yeast. 6: 127-137.

Einhauer, A., Schuster, M., Wasserbauer, E., and Jungbauer, A. 2002. Expression and purification of homogenous proteins in *Saccharomyces cerevisiae* based on ubiquitin-FLAG fusion. Protein Expr. Purif. 24: 497-504.

Elliott, S., Giffin, J., Suggs, S., Lau, E.P., and Banks, A. R. 1989. Secretion of glycosylated human erythropoietin from yeast directed by the α-factor leader region. Gene. 79: 167-180.

Erhart, E., and Hollenberg, C.P. 1983. The presence of a defective LEU2 gene in 2 μ DNA recombinant plasmids of *Saccharomyces cerevisiae* is responsible for curing and high copy number. J. Bacteriol. 156: 625-635.

Evans, I., and McAthey, P. 1991. Comparitive genetics of important yeasts. In: Genetically-engineered proteins and enzymes from yeast: production control. A. Wiseman, ed. Ellos Horwood, West Sussex, England. p. 11-74.

Fields, S., Song, O. 1989. A novel genetic system to detect protein-protein interactions. Nature. 340: 245-246.

Flower, D.R. 1999. Modelling G-protein-coupled receptors for drug design. Biochim. Biophys. Acta. 1422: 207-234.

Fogel, S., and Welch, J.W. 1982. Tandem gene amplification mediates copper resistance in yeast. Proc. Natl. Acad. Sci. USA. 79: 5342-5346.

Fujii, T., Kondo, K., Shimizu, F., Sone, H., Tanaka, J.I., and Inoue, T. 1990. Application of a ribosomal DNA integration vector in the construction of a brewer's yeast having α-acetolactate decarboxylase activity. Appl. Environ. Microbio. 56: 997-1003.

Funk, M., Niedenthal, R., Mumberg, D., Brinkmann, K., Ronicke, V., and Henkel, T. 2002. Vector systems for heterologous expression of proteins in *Saccharomyces cerevisiae*. Meth. Enzymol. 350: 248-257.

Futcher, A.B., and Cox, B.S. 1984. Copy number and the stability of 2-μ circle-based artificial plasmids of *Saccharomyces cerevisiae*. J. Bacteriol. 157: 283-290.

Gasch, A.P. 2002. Yeast genomic expression studies using DNA microarrays. Meth. Enzymol. 350: 393-414.

Georgieva, B., and Rothstein, R. 2002. Kar-mediated plasmid transfer between yeast strains: alternative to traditional transformation methods. Meth. Enzymol. 350: 278-289.

Goffeau, A., Barrell, B.G., Bussey, H., Davis, R.W., Dujon, B., Feldmann, H., Galibert, F., Hoheisel, J.D., Jacq, C., Johnston, M., Louis, E.J., Mewes, H.W., Murakami, Y., Philippsen, P., Tettelin, H., and Oliver, S.G. 1996. Life with 6000 genes. Science. 274: 546-551.

Grisshammer, R., and Tate, C.G. 1995. Overexpression of integral membrane proteins for structural studies. Q. Rev. Biophys. 28: 315-422.

Guthrie, C., and Fink, G.R. eds. 2002. Methods in Enzymology: Guide to yeast genetics and molecular and cell biology. Academic Press, San Diego, CA.

Hadfield, C., Rainna, K.K., Shashimenon, K., and Mount, R.C. 1993. The expression and performance of cloned genes in yeast. Mycol. Res. 97: 897-944.

Hallewell, R.A., Mills, R., Tekamp-Olson, P., Blacher, R., Rosenburg, S., Otting, F., Masiarz, F.R., and Scandella, C.J. 1987. Amino terminal acetylation of authentic human Cu,Zn superoxide-dismutase produced in yeast. Bio-tech. 5: 363-366.

Harmsen, M.M., Bruyne, M.I., Raue, H.A., Maat, J. 1996. Overexpression of binding protein and disruption of the PMR1 gene synergistically stimulate secretion of bovine prochymosin but not plant thaumatin in yeast. Appl. Microbiol. Biotech. 46: 365-370.

Hinnen, A., Buxton, F., Chaudhuri, B., Heim, J., Hottiger, T., Meyhack, B., and Pohlig, G. 1995. Gene expression in recombinant yeast. In: Gene expression in recombinant microorganisms. A. Smith, ed. Marcel Dekker, New York. p. 121-193.

Hinnen, A., Hicks, J.B., and Fink, G.R. 1978. Transformation of yeast. Proc. Natl. Acad. Sci. USA. 75: 1929-1933.

Hitzeman, R.A., Hagie, F.E., Levine H.L., Goeddel, D.V., Ammerer, G., and Hall, B.D. 1981. Expression of a human gene for interferon in yeast. Nature. 293: 717-722.

Hitzeman, R.A., Leung, D.W., Perry, L.J., Kohr, W.J., Levine, H.L., and Goeddel, D.V. 1983. Secretion of human interferons by yeast. Science. 219: 620-625.

Holland, J.P., and Holland, M.J. 1980. Structural comparison of two nontandemly repeated yeast glyceraldehyde-3-phosphate dehydrogenase genes. J. Biol. Chem. 255: 2596-2605.

Horwitz, A.H., Chang, C.P., Better, M., Hellstrom, K.E., and Robinson, R.R. 1988. Secretion of functional antibody and Fab fragment from yeast cells. Proc. Natl. Acad. Sci. USA. 85: 8678-8682.

Hubsman, M., Yudkovsky, G., and Aronheim, A. 2001. A novel approach for the identification of protein-protein interaction with integral membrane proteins. Nucleic Acids Res. 29:E18.

Innis, M.A. 1989. Glycosylation of heterologous proteins in *Saccharomyces cerevisiae*. In: Yeast genetic engineering. Barr, P. J., Brake, A. J., and Valenzuela, P., eds. Butterworth, Stoneham, MA. p. 233-246.

Ito, T., Chiba, T., Ozawa, R., Yoshida, M., Hattori, M., and Sakaki, Y. 2001. A comprehensive two-hybrid analysis to explore the yeast protein interactome. Proc. Natl. Acad. Sci. USA. 98: 4569-4574.

Ito, H., Fukada, Y., Murara, K., and Kimura, A. 1983. Transformation of intact yeast cells treated with alkali cations. J. Bacteriol. 153: 163-168.

Ito, T., Tashiro, K., Muta, S., Ozawa, R., Chiba, T., Nishizawa, M., Yamamoto, K., Kuhara, S., and Sakaki, Y. 2000. Toward a protein-protein interaction map of the budding yeast: A comprehensive system to examine two-hybrid interactions in all possible combinations between the yeast proteins. Proc. Natl. Acad. Sci. USA. 97: 1143-1147.

Johnsson, N., and Varshavsky, A. 1994. Split ubiquitin as a sensor of protein interactions *in vivo*. Proc. Natl. Acad. Sci. USA. 91: 10340-10344.

Johnston, M., 1987. A model fungal gene regulatory mechanism: the GAL genes of *Saccharomyces cerevisiae*. Microbiol. Rev. 51: 458-476.

Jones, E.W. 2002. Vacuolar proteases and proteolytic artifacts in *Saccharomyces cerevisiae*. Meth. Enzymol. 351: 127-150.

Julius, D., Schekman, R., and Thorner, J. 1984. Glycosylation and processing of prepro-α-factor through the yeast secretory pathway. Cell. 36: 309-318.

Kapat, A., Jaakola, V.P., Heimo, H., Liiti, S., Heikinheimo, P., Glumoff, T., and Goldman, A. 2000. Production and purification of recombinant human alpha 2C2 adrenergic receptor using *Saccharomyces cerevisiae*. Bioseperation. 9: 167-172.

Kieke, M.C., Cho, B.K., Boder, E.T., Kranz, D.M., and Wittrup, K.D. 1997. Isolation of anti-T cell receptor scFv mutants by yeast surface display. Protein Eng. 10: 1303-1310.

King, K., Dohlman, H.G., Thorner, J., Caron, M.G., and Lefkowitz, R.J. 1990. Control of yeast mating signal transduction by a mammalian beta2-adrenergic receptor and Gs_α subunit. Science. 250: 121-123.

Kingsman, S.M., Cousens, D., Stanway, C.A., Chambers, A., Wilson, M., and Kingsman, A.J. 1990. High-efficiency yeast expression vectors based on the promoter of the phosphoglycerate kinase gene. Meth. Enzymol. 185: 329-341.

Kjeldsen, T., Brandt, J., Andersen, A.S., Egel-Mitani, M., Hach, M., Petterson, A.F., and Vad, K. 1996. A removable spacer peptide in an alpha-factor-leader/insulin precursor fusion protein improves processing and concomitant yield of the insulin precursor in *Saccharomyces cerevisiae*. Gene. 170: 107-112.

Kjeldsen, T., Ludvigsen, S., Diers, I., Balschmidt, P., Sorensen, A.R., and Kaarsholm, N.C. 2002. Engineering-enhanced protein secretory expression in yeast with application to insulin. J. Biol. Chem. 277: 18245-18248.

Kniskern, P.J., Hagopian, A., Montgomery, D.L., Burke, P., Dunn, N.R., Hofmann, K.J., Miller, W.J., and Ellis, R.W. 1986. Unusually high-level expression of a foreign gene (hepatitis B virus core antigen) in *Saccharomyces cerevisiae*. Gene. 46: 135-141.

Kukuruzinska, M.A., Bergh, M.L.E., and Jackson, B.L. 1987. Protein glycosylation in yeast. Ann. Rev. Biochem. 56: 915-944.

Kumagai, M.H., Shah, M., Terashima, M., Vrkljan, Z., Whitaker, J.R., and Rodriguez, R.L. 1990. Expression and secretion of rice α-amylase by *Saccharomyces cerevisiae*. Gene. 94: 209-216.

Kumar, A., Snyder, M. 2001. Emerging technologies in yeast genomics. Nat. Rev. Genet. 2: 302-312.

Kurjan, J., and Herskowitz, I. 1982. Structure of a yeast pheromone gene (MF α): a putative α-factor precursor contains four tandem copies of mature α factor. Cell. 30: 933-943.

Lee, F.W.F., and Da Silva N.A. 1997. Sequential delta-integration for the regulated insertion of cloned genes in *Saccharomyces cerevisiae*. Biotechnol. Prog. 13: 368-373.

Legrain, P., and Selig, L. 2000. Genome-wide protein interaction maps using two-hybrid systems. FEBS Lett. 480: 32-36.

Lewitter, F. 2002. Database resources relevant to yeast biology. Meth. Enzymol. 350: 373-379.

Li, J., Wang, S., VanDusen, W.J., Schultz, L.D., George, H.A., Herber, W.K., Chae, H.J., Bentley, W.E., and Rao, G. 2000. Green fluorescent protein in *Saccharomyces cerevisiae*: Real-time studies of the GAL1 promoter. Biotech. Bioeng. 70: 187-196.

Lockhart, D.J., and Winzeler, E.A. 2000. Genomics, gene expression and DNA arrays. Nature. 405: 827-836.

Lopes, T.S., DeWijs, I.J., Steenhauer, S.I., Verbakel, J., and Planta, R.J. 1991. Factors affecting the mitotic stability of high-copy-number integration into the ribosomal DNA of *Saccharomyces cerevisiae*. Yeast. 12: 467-477.

Lopes, T.S., Hakaart, G.J., Koerts, B.L., Raue, H.A., and Planta, R.J. 1990. Mechanism of high-copy-number integration of pMIRY-type vectors into the ribosomal DNA of *Saccharomyces cerevisiae*. Gene. 105: 83-90.

Lopes, T.S., Klootwijk, J., Veenstra, A.E., van der Aar, P.C., van Heerikhuizen, H., Raue, H.A., and Planta, R.J. 1989. High-copy-number integration into the ribosomal DNA of *Saccharomyces cerevisiae*: A new vector for high-level expression. Gene. 79: 199-206.

Ma, D., 2001. Applications of yeast in drug discovery. Prog. Drug Res. 57: 117-162.

Mackay, V.L., and Kelleher, T. 1996. Methods for expressing recombinant proteins in yeast. In: Protein Engineering and Design. Paul R. Carey, ed. Academic Press, San Diego, California. p.105-153.

Marczynski, G.T., and Jaehning, J.A. 1985. A transcription map of a yeast centromere plasmid: unexpected transcripts and altered gene expression. Nucleics Acids Res. 13: 8487-8506.

Meilhoc, E., Masson, J.-M., and Tessie, J. 1990. High efficiency transformation of intact yeast cells by electroporation. Biotechnology (N.Y). 8: 223-227.

Mendoza-Vega, O., Sabatie, J., and Brown, S.W. 1994. Industrial production of heterologous proteins by fed-batch cultures of the yeast *Saccharomyces cerevisiae*. FEMS Microbiol. Rev. 15: 369-410.

Meynial-Salles, I., and Combes, D. 1996. *In vitro* glycosylation of proteins: an enzymatic approach. J. Biotechnol. 46: 1-14.

Moerschell, R.P., Hosokawa, Y., Tsunasawa, S., and Sherman, F. 1990. The specificities of yeast methionine aminopeptidase and acetylation of amino-terminal methionine *in vivo*. J. Biol. Chem. 265: 19638-19643.

Mortimer, R.K. 2000. Evolution and variation of the yeast (*Saccharomyces*) genome. Genome Res. 10: 403-409.

Munder, T., and Hinnen, A. 1999. Yeast cells as tools for target-oriented screening. Appl. Microbiol. Biotechnol. 52: 311-320.

Murray, A.W., and Szostak, J.W. 1983. Pedigree analysis of plasmid segregation in yeast. Cell. 34: 961-970.

Mushegian, A.R., Bassett, D.E., Boguski, M.S., Bork, P., and Koonin, E.V. 1997. Positionally cloned human disease genes: patterns of evolutionary conservation and functional motifs. Proc. Natl. Acad. Sci. USA. 94: 5831-5836.

Ngsee, J.K., Hansen, W., Walter, P., and Smith, M. 1989. Cassette mutagenic analysis of the yeast invertase signal peptide: Effects on protein translocation. Mol. Cell. Biol. 9: 3400-3410.

Novick, P., Field, C., and Schekman, R. 1980. Identification of 23 complementation groups required for posttranslational events in the yeast secretory pathway. Cell. 21: 205-215.

Novick, P., Ferro, S., and Schekman, R. 1981. Order events in the yeast secretory pathway. Cell. 25: 460-469.

Parekh, R.N., Forrester, K., and Wittrup, K.D. 1995. Multicopy overexpression of bovine pancreatic trypsin inhibitor saturates protein folding and secretory capacity of *Saccharomyces cerevisiae*. Protein Expr. Purif. 6: 537-545.

Parekh, R.N., Shaw, M.R., and Wittrup, K.D. 1996. An integrating vector for tunable, high copy, stable integration into the dispersed Ty delta sites of *Saccharomyces cerevisiae*. Biotechnol. Prog. 12: 16-21.

Parekh, R.N., and Wittrup, K.D. 1997. Expression level tuning for optimal heterologous protein secretion in *Saccharomyces cerevisiae*. Biotech. Progress. 13: 117-122.

Pausch, M.H. 1997. G-protein-coupled receptors in *Saccharomyces cerevisiae*: High-throughput screening assays for drug discovery. Trends Biotech. 15: 487-494.

Perlman, D., Halvorson, H.O. and Cannon, L.E. 1982. Presecretory and cytoplasmic invertase polypeptides encoded by distinct mRNAs derived from the same structural gene differ by a signal sequence. Proc. Natl. Acad. Sci. USA. 79: 781-785.

Petes, T.D. 1980. Unequal meiotic recombination within tandem arrays of yeast ribosomal DNA genes. Cell. 19: 765-774.

Price, L.A., Kajkowski, E.M., Hadcock, J.R., Ozenberger, B.A., and Pausch, M.H. 1995. Functional coupling of a mammalian somatostatin receptor to the yeast pheromone response pathway. Mol. Cell. Biol. 15: 6188-6195.

Price, V.L., Taylor, W.E., Clevenger, W., Worthington, M., and Young, E.T. 1990. Expression of heterologous proteins in *Saccharomyces cerevisiae* using ADH2 promoter. Meth. Enzymol. 185: 308-318.

Robinson, A.S., Bockhaus, J.S., Voegler, A.C., and Wittrup, K.D. 1996 Reduction of BiP levels decreases heterologous protein secretion in *Saccharomyces cerevisiae*. J. Biol. Chem. 271: 10017-10022.

Robinson, A.S., Hines, V., Wittrup, K.D. 1994. Protein disulfide isomerase overexpression increases secretion of foreign proteins in *Saccharomyces cerevisiae*. Bio/technol. 12: 381-384.

Romanos, M.A., Makoff, A.J., Fairweather, N.F., Beesley, K.M., Slater, D.E., Rayment, F.B., Payne, M.M., and Clare, J.J. 1991. Expression of tetanus toxin fragment C in yeast: gene synthesis is required to eliminate fortuitous polyadenylation sites in AT-rich DNA. Nucleic Acids Res. 19: 1461-1467.

Romanos, M.A., Scoere, C.A., and Clare, J.J. 1992. Foreign gene expression in yeast. Yeast. 8: 423-488.

Rose, A.B., and Broach, J.R. 1990. Propagation and expression of cloned genes in yeast 2 μ circle-based vectors. Meth. Enzymol. 185: 234-279.

Rose, M., Grisafi, P., and Botstein, D. 1984. Structure and function of the yeast URA3 gene: expression in *Escherichia coli*. Gene. 29: 113-124.

Rosenberg, S., Coit, D. and Tekamp-Olson, P. 1990. Glyceraldehyde-3-phosphate dehydrogenase-derived expression cassettes for constitutive synthesis of heterologous proteins. Meth. Enzymol. 185: 341-351.

Rothstein, R.J. 1983. One-step gene disruption in yeast. Meth. Enzymol. 101: 202-211.

Russo, P., and Sherman, F. 1989. Transcription terminates near the poly(A) site in the CYC1 gene of the yeast *Saccharomyces cerevisiae*. Proc. Natl. Acad. Sci. USA. 86: 8348-8352.

Sakai, A., Shimizu, Y., and Hishinuma, F. 1988. Isolation and characterization of mutants which show an oversecretion phenotype in *Saccharomyces cerevisiae*. Genetics. 119: 499-506.

Schultz, L.D., Tanner, J., Hofmann, K.J., Emini, E.A., Condra, J.H., Jones, R.E., Kieff, E., and Ellis, R.W. 1987. Expression and secretion in yeast of a 400-kDa envelope glycoprotein derived from Epstein-Barr virus. Gene. 54: 113-123.

Schuster, M., Einhauer, A., Wasserbauer, E., Subenbacher, F., Ortner, C., Paumann, M., Werner., and Jungbauer A. 2000. Protein expression in yeast; comparison of two expression strategies regarding protein maturation. Biotech. 84: 237-248.

Schwartz, L.S., Jansen, N.B., Ho, N.W.Y., and Tsao, G.T. 1988. Plasmid instability kinetics of the yeast S288C pUCKm8 in non-selective and selective media. Biotechnol. Bioeng. 332: 733-740.

Shevchenko, A., Jensen, O.N., Podtelejnikov, A.V., Sagliocco, F., Wilm, M., Vorm, O., Mortensen, P., Shevchenko, A., Boucherie, H., and Mann, M. 1996. Linking genome and proteome by mass-spectrometry: large-scale identification of yeast proteins from two-dimensional gels. Proc. Natl. Acad. Sci. USA. 93: 14440-14445.

Shusta, E.V., Raines, R.T., Pluckthun, A., and Wittrup, K.D. 1998. Increasing the secretory capacity of *Saccharomyces cerevisiae* for production of single-chain antibody fragments. Nat. Biotechnol. 16: 773-777.

Shuster, J.R., Lee, H., and Moyer, D.L. 1990. Integration and amplification of DNA at yeast delta sequences. Yeast. 6:S79.

Sidrauski, C., Chapman, R., and Walter, P. 1998. The unfolded protein response: an intracellular signaling pathway with many surprising features. Trends Cell Biol. 8: 245-249.

Smith, J.D., and Robinson, A.S. 2002. Overexpression of an archaeal protein in yeast: secretion bottleneck at the ER. Biotech. Bioeng. 79: 713-723.

Smith, R.A., Duncan, J.M., and Moir, D.T. 1985. Heterologous protein secretion from yeast. Science. 229: 1219-1224.

Stagljar, I., Korostensky, C., Johnsson, N., teHeesen S. 1998. A genetic system based on split-ubiquitin for the analysis of interactions between membrane proteins *in vivo*. Proc. Natl. Acad. Sci. USA. 95: 5187-5192.

Stevens, T.H., Rothman, J.H., Payne, G.S., and Schekman, R. 1986. Gene dosage-dependent secretion of yeast vacuolar carboxypeptidase Y. J. Cell. Biol. 102: 1551-1557.

Struhl, K., and Davis, R.W. 1980. A physical, genetic, and transcriptional map of the yeast his3 gene of *Saccharomyces cerevisiae*. J. Mol. Biol. 136: 309-332.

Szczypka, M.S., Wemmie, J.A., Moye-Rowley, W.S., and Thiele, D.J. 1994. A yeast metal resistance protein similar to human cystic fibrosis transmembrane conductance regulator (CFTR) and multi-drug resistance-associated protein. J. Biol. Chem. 269: 22853-22857.

Tatchell, K., and Robinson, L.C. 2002. Use of green fluorescent protein in living yeast cells. Meth. Enzymol. 351: 661-683.

Tsai, P.K., Frevert, J., and Ballou, C.E. 1984. Carbohydrate structure of *Saccharomyces cerevisiae mnn9* mannoprotein. J. Biol. Chem. 259: 3805-3811.

Tschumper, G., and Carbon, J. 1980. Sequence of a yeast DNA fragment containing a chromosomal replicator and the TRP1 gene. Gene. 10: 157-166.

Tugendreich, S., Perkins, E., Couto, J., Barthmaier, P., Sun, D., Tang, S., Tulac, S., Nguyen, A., Yeh, E., Mays, A., Wallace, E., Lila, T., Shivak, D., Prichard, M., Andrejka, L., Kim, R., and Melese, T. 2001. A streamlined process to phenotypically profile heterologous cDNAs in parallel using yeast cell-based assays. Genome Res. 11: 1899-1912.

Uetz, P., Giot, L., Cagney, G., Mansfield, T.A., Judson, R.S., Knight, J.R., Lockshon, D., Narayan, V., Srinivasan, M., Pochart, P., Qureshi-Emili, A., Li, Y., Godwin, B., Conover, D., Kalbfleisch, T., Vijayadamodar, G., Yang, M., Hohnston, M., Fields, S., and Rothberg, J.M. 2000. A comprehensive analysis of protein-protein interactions in *Saccharomyces cerevisiae*. Nature. 403: 623-627.

Urdea, M.S., Merryweather, J.P., Mullenbach, D.C., Corr, D., Heberlein, U., Valenzuela, P., and Barr, P.J. 1983. Chemical synthesis of a gene for human epidermal growth factor urogastrone and its expression in yeast. Proc. Natl. Acad. Sci. USA. 80: 7461-7465.

Valenzuela, P., Medina, A., Rutter, W.J., Ammerer, G., and Hall, B.D. 1982. Synthesis and assembly of hepatitis B virus surface antigen particles in yeast. Nature. 298: 347-350.

Van den Hazel, H.B., Kielland-Brandt, M.C., and Winther, J.R. 1996. Review: biosynthesis and function of yeast vacuolar proteases. Yeast. 12: 1-16.

Velculescu, A.E., Zhang, L., Vogelstein, B., and Kinzler, K.W. 1995. Serial analysis of gene expression. Science. 270: 484-487.

Vogel, K., and Hinnen, A. 1990. The yeast phosphatase system. Mol. Microbiol. 4: 2013-2017.

Wang, C., Eufemi, M., Turano, C., and Giartosio, A. 1996. Influence of carbohydrate moiety on the stability of glycoproteins. Biochemistry. 35: 7299-7307.

Wang, X., and Da Silva, N.A. 1996. Site-specific integratin of heterologous genes in yeast via Ty3 retrotransposition. Biotechnol. Bioeng. 51: 703-712.

Webster, T.D., and Dickson, R.C. 1983. Direct selection of *Saccharomyces cerevisiae* resistant to the antibiotic G418 following transformation with a DNA vector carrying the kanamycin resistance gene of Tn*903*. Gene. 26: 243-252.

Winzeler, E.A., (50 others) and Davis, R.W. 1999. Functional characterization of the S. cerevisiae genome by gene deletion and parallel analysis. Science. 285: 901-906.

Wittrup, K. D., Robinson, A. S., Parekh, R. N., and Forrester, K. J. 1994. Existence of an optimum expression level for secretion of foreign proteins in yeast. Ann. N. Y. Acad. Sci. 745: 321-30.

Wittrup, K.D. 2001. Protein engineering by cell-surface display. Curr. Op. Biotech. 12: 395-399.

Zaret, K.S., and Sherman, F. 1982. DNA sequence required for efficient transcription termination in yeast. Cell. 28: 563-573.

Zhang, Z., Moo-Young, M., and Chisti, Y. 1996. Plasmid stability in recombinant *Saccharomyces cerevisiae*. Biotechnol. Adv. 14: 401-435.

Zhu, H., Bilgin, M., Bangham, R., Hall, D., Casamayor, A., Bertone, P., Lan, N., Jansen, R., Bidlingmaier, S., Houfek, T., Mitchell, T., Miller, P., Dean, R., Gerstein, M., and Snyder, M. 2001. Global analysis of protein activities using proteome chips. Science. 293: 2101-2105.

6

Production of Recombinant Proteins Using the Methylotrophic Yeasts *Pichia pastoris* and *Pichia methanolica*

Bruce L. Zamost, Gary J. Lesnicki
and Sanjay Jain

Abstract

Pichia expression systems have been used to produce a wide variety of mammalian, microbial, and plant proteins. The methylotrophic yeasts *Pichia pastoris* and *Pichia methanolica* are capable of high level expression when heterologous genes are placed under transcriptional control of the methanol-inducible alcohol oxidase (*AOX1 or AUG1*) promoters. Additional inducible and constitutive promoters have also been used for recombinant protein synthesis. *Pichia* systems have a number of desirable features: (1) the yeast host strains are readily amenable to genetic manipulations; (2) genetic stability can be ensured by integrating expression cassette(s) into the host chromosome; (3) cultures readily adapt to high-biomass fermentations in low-cost, defined medium; (4) high level expression with disulfide bond formation is possible; and (5) *Pichia* can perform specific post-translational modifications.

Introduction

The recent application of molecular biology in the production of biopharmaceuticals and proteins of industrial significance has led to the development of heterologous systems suitable for expressing foreign proteins at high levels. Methylotrophic yeasts are one such system. These yeasts can utilize methanol as their sole carbon and energy source, and have been used with great success for the synthesis of many recombinant proteins (Dale *et al.*, 1999; Cereghino and Cregg, 2000; Gellissen,2000).

The *Pichia* expression systems have a number of desirable features. The yeast host strains are readily amenable to genetic manipulations. Auxotrophic mutant strains facilitate the introduction of protein expression cassettes and the generation of genetically stable expression hosts (Hollenberg and Gellissen, 1997; Cereghino and Cregg, 2000). Stability is maintained by integration of the cassette(s) into the host chromosome (Romanos *et al.*, 1992; Hollenberg and Gellissen, 1997;Cereghino and Cregg, 2000). Cultures readily adapt to high-biomass fermentations in a low-cost, defined medium (Cregg, 2000) and these yeasts are able to introduce eukaryotic specific post-translational modifications including O- and N-linked glycosylation, lipid addition and disulfide bridge formation (Faber *et al.*, 1995; Eckart and Bussineau, 1996; Guarna *et al.*,1997; Hollenberg and Gellissen, 1997; Cereghino and Cregg, 2000).

An additional advantage of *Pichia* expression systems is that properly folded proteins can be efficiently produced in the growth medium by fusing the coding region of target proteins to the signal sequences of yeast or mammalian secretory proteins (Tschopp *et al.*, 1987a; Digan *et al.*, 1989). A specific advantage of protein secretion in *Pichia pastoris* is that this organism only secretes low levels of native proteins, resulting in easy recovery of active product from the medium (Eckart and Bussineau, 1996; Sreekrishna *et al.*, 1997;Cereghino and Cregg, 2000). As a result, methylotrophic yeasts have proven useful for the production of numerous eukaryotic heterologous proteins (Eckart and Bussineau,1996; Sreekrishna *et al.*, 1997; Cereghino and Cregg, 2000).

The methylotrophic yeasts *Pichia pastoris* and *Pichia methanolica* share the same methanol utilization pathway (Koutz *et al.*, 1989; Rodriguez *et al.*, 1997; Nakagawa *et al.*, 1999). Initially, methanol is brought into the peroxisome, where it is oxidized to formaldehyde by alcohol oxidase(s). Formaldehyde then diffuses out into the cytoplasm where it is processed by the xylulose monophosphate pathway to generate energy and biosynthetic intermediates.

Both species have two alcohol oxidase genes: *AOX1* and *AOX2* in *P. pastoris* (Koutz *et al.*, 1989; Cereghino and Cregg, 2000) and *AUG1* and *AUG2* in *P. methanolica* (Raymond *et al.*, 1998; Nakagawa *et al.*, 1999). Transcription of these genes is completely repressed at moderate levels of glucose or ethanol, but is derepressed by carbon starvation and powerfully induced by addition of methanol (Guarna *et al.*, 1997; Raymond *et al.*, 1998; Cereghino and Cregg., 2000). With these regulatory triggers, alcohol oxidase promoters are particularly well suited for driving heterologous gene expression

In *P. pastoris,* gene expression systems have been developed in which the heterologous gene of interest, under the control of the *AOX1* promoter, is introduced on an integrative plasmid into strains containing two functional *AOX* genes (Mut[+] phenotype) as well as strains in which the stronger *AOX1* gene function has been disrupted (Mut[s] phenotype). Disruption of both *AOX* genes results in a Mut[-] phenotype (Higgins *et al.*, 1998; Cereghino and Cregg, 2000). The *AOX1* promoter is maximally induced at low methanol concentrations, whereas the *AOX2* promoter is activated at higher methanol concentrations (Nakagawa *et al.*, 1999).

Similar systems are now available in *P. methanolica* (Raymond *et al.*, 1998). *AUG1* gene disruptions result in a slow methanol utilization phenotype (Mut[s]), disruption of *AUG2* does not markedly influence methanol-dependent growth, and *aug1Δ aug2Δ* (Mut⁻) double mutants are totally deficient in methanol utilization (Raymond *et al.*, 1998). A number of studies indicate that Mut[s] strains produce higher levels of recombinant proteins than wild type (Mut[+]) strains while utilizing substantially less methanol (Cereghino and Cregg, 2000; reviewed by: Faber *et al.*, 1995; Higgins *et al.*, 1998). Finally, *P. methanolica* and *P. pastoris* produce equivalent levels of recombinant proteins when transformed with similar heterologous gene expression cassettes (Raymond *et al.*, 1998).

Methanol Metabolism in Yeast

Methanol is assimilated by *P. methanolica* and other methylotophic yeasts through the action of alcohol oxidases and a variety of other enzymes. These enzymes are stored in the yeast peroxisome. Glucose-grown cells only contain a small volume of peroxisomes, but the volume of peroxisomes greatly increases upon methanol growth (Faber *et al.*, 1995). When methanol-grown cells are shifted to glucose, peroxisomes are quickly degraded by vacuolar proteases. A number of *P. methanolica* and *P. pastoris* mutants defective in peroxisome degradation or autophagy have been isolated (Kulackovsky *et al.*, 1998). These strains may allow higher growth during mixed feeding of glucose and methanol in fed-batch fermentations.

In *Pichia*, peroxisome-resident FAD-dependent alcohol oxidase oxidizes methanol to formaldehyde. Hydrogen peroxide is also produced and quickly degraded by catalase. Formaldehyde is converted to dihydroxy acetone (DHA) and glyceraldehyde-3-phosphate (GAP) by dihydroxy acetone synthase in the peroxisome. Phosphorylated DHA and GAP condense to form fructose-1,6-bisphosphate, which is converted to fructose-6-phosphate (F6P) by fructose-1,6-bisphosphatase. Two molecules of F6P and one molecule of dihydroxy acetone phosphate form 3 molecules of xylulose-5-phosphate (X5P). The X5P is recycled in the peroxisome to form the xylulose monophosphate cycle (Sibirny *et al.*, 1988, Sibirny, 1996).

Dissimilation of formaldehyde takes place in the yeast cytosol. Formaldehyde is oxidized to formate by GSH dependent formaldehyde dehydrogenase and S-formyl glutathione hydrolase. The formate is next oxidized to CO_2 by formate dehydrogenase. A GSH independent pathway that leads to the formation of methyl formate from formaldehyde and methanol by methyl formate synthase has also been described (Sakai *et al.*, 1995). This activity was induced in *Pichia methanolica* AKU 4262 by formaldehyde addition during growth on methanol. Intact cells were then used to convert methanol at a 2M concentration into 135 mM methylformate (Murdanato *et al.*,1997).

Catabolite repression of enzymes involved in methanol catabolism has been demonstrated in methylotrophic yeasts. Production of methanol oxidizing enzymes is repressed by growth on ethanol or glucose and these pathways are separately regulated (Sakai et.al., 1987, Sibirny, et.al., 1986). Wild type *Pichia pinus* MH4 exhibits diauxic growth on methanol and glucose, as methanol utilization does not occur until glucose is fully utilized (Alamae and Simisker, 1994). Mutant strains of *P. pinus* MH4 were obtained after selection on plating media containing the 2-deoxyglucose (a non metabolizable glucose analog which selects for mutants refractile to glucose repression). One of these strains (X6) synthesizes alcohol oxidase and catalase during growth on glucose. Resistance to glucose repression was not linked to an altered hexokinase function, but possibly to changes in the low affinity glucose transport system (Alamae and Simisker, 1994).

Alcohol Utilization Genes

Two genes encoding alcohol oxidases have been identified in *P. methanolica* (*AUG1* or *MOD1* and *AUG2* or *MOD2*) and in *P. pastoris* (*AOX1* and *AOX2*) (Raymond *et al.*, 1998, Ellis et.al., 1985). The alcohol oxidases Aug1p and Aug2p share 84% identity and Aug1p is 69% identical to Aox1p from *P. pastoris* (Raymond et.al., 1998). In *P. pastoris* and *P. methanolica*, *AOX1* and *AUG1* encode the majority of the alcohol oxidase activity present in the cell. Disruption of *AOX1* or *AUG1* results in strains that grow poorly on methanol in comparison to the wild type or *AUG2*/*AOX2* disrupted strains (Cregg *et al.*, 1989,

Raymond *et al.*, 1999). At present, the physiological function and importance of the *AUG2* and *AOX2* is not understood.

Alcohol oxidase (AOD) is a homo-octameric flavoprotein consisting of eight ≈ 70-kDa subunits that each contain a flavin adenine dinucleotide molecule (Nakagawa *et al.*, 1999). Cell free extracts of *P. methanolica* PMAD11 have nine alcohol oxidase activities based on activity staining in native polyacrylamide gels. The AODs exhibiting the smallest and largest electrophoretic mobility were termed MOD-A and MOD-B respectively. These oxidases have different enzymatic activity, substrate specificity and content of modified FAD. The enzymes also contain different N-terminal sequences indicating that two different subunits of AOD are produced in *P. methanolica* (Nakagawa *et al.*, 1996).

It was theorized that the different AOD forms originated as a single gene product that underwent post-translational modification (Gruzman *et al.*, 1996). A second hypothesis was that the nine AOD bands represent a mixture of two gene products (Nakagawa *et al.*, 1999). To determine which hypothesis was correct, the *MOD1* and *MOD2* genes form *P. methanolica* were cloned and expressed in an alcohol oxidase depleted strain of *Candida boidini* (Nakagawa *et al.*, 1999). Expression of *MOD1* and *MOD2* allowed the *aod1Δ* host strain to grow on methanol, and that of either *MOD1* or *MOD2* gave a single AOD band on zymograms, which corresponded to MOD-B and MOD-A, respectively. The Mod1p and Mod2p proteins were purified, mixed in various activity ratios and subject to zymogram analysis. Results showed a ladder of nine bands whose profile coincided with that observed in methanol grown cells. Co-expression of *MOD1* and *MOD2* in the *aod1Δ* host strain also resulted in multiple band formation. *In vitro* and *in vivo* ladder formation only occurred when Mod1p and Mod2p were present. The nine bands correspond to two homo-oligomers and 7 hetero octamers, supporting the model proposed by Nakagawa (Nakagawa *et al.*, 1999).

The expression of *MOD1* and *MOD2* appears to be controlled differently depending on methanol concentration: Mod1p is produced at low methanol concentrations while Mod2p is produced at high methanol concentrations. The regulation of MOD1 and MOD2 expression was investigated in *MOD⁻* and *MOD2* disrupted strains of *P. methanolica* (Nakagawa, et.al 2001). Cell free extracts of the wild type strain PMAD11, and the disrupted strains PMAD12 (*mod1Δ*), PMAD13 (*mod2Δ*) and PMAD14 (*mod1Δ mod2Δ*) were examined by zymography. When grown on methanol PMAD12 and PMAD13 exhibited AOD activity and a single band on zymograms. PMAD14 did not produce AOD activity or a band on the zymogram. Cell free extracts of the wild type strain grown on glycerol also gave a single band when analyzed by zymograhy. The *mod1Δ*, *mod2Δ*, and *mod1Δ mod2Δ* strains grew as well as the wild type on glycerol, but only the *mod2Δ* strain produced a band on the zymogram. This band was in the same location as the AOD band from the wild type strain suggesting that only *MOD1* is expressed during growth on glycerol. The purified AOD from glycerol-grown cells is the homo-octameric form of Mod1p. Thus, *MOD2* is only induced by methanol while *MOD1* is induced by both methanol and glycerol, and the two genes are controlled by distinct regulatory mechanism (Nakagawa, et.al. 2001).

Pichia pastoris

Introduction to *Pichia pastoris*

During the 1960s there was interest in using the ability of bacteria to grow on methanol for the production of single cell protein (SCP). Several companies (ICI and Hoechst) even developed SCP processes employing methylotrophic bacteria. This led to the discovery of yeasts capable of growth on methanol for SCP production. In 1969, Koichi Ogata and coworkers first described the ability of certain yeast organisms to utilize methanol as a sole source of carbon and energy (Ogata *et al.*, 1969). During the 1970s, the Phillips Petroleum Company developed methods for growing *Pichia pastoris* using methanol as carbon source in a high cell density, continuous fermentation process, yielding 125-150 g/L dry cell weight (Wegner, 1983, 1990). This was mainly developed for the application of SCP as high protein animal feed. In 1988, the process was scaled up to 25,000 L scale by Provesta Corporation (a subsidiary of Phillips Petroleum Company).

Later, Phillips Petroleum contracted with the Salk Institute Biotechnology/Industrial Associates, Inc. (SIBIA, La Jolla, CA) to develop *P. pastoris* as an expression system for production of heterologous proteins. SIBIA scientists isolated the alcohol oxidase promoter, and constructed expression vectors and host strains for heterologous protein expression. The high cell density fermentation process initially developed for SCP production combined with the strong, regulated alcohol oxidase promoter resulted in high levels of expression of heterologous proteins. In 1993, Phillips Petroleum sold the *P. pastoris* expression system technology to Research Corporation Technologies (Tuczon, AZ), and licensed Invitrogen Corporation (Carlsbad, CA) to sell *Pichia pastoris* expression kit consisting of expression vectors and host strains for protein expression.

Host Strains

Methanol Utilization Phenotype

Most *P. pastoris* host strains are derived from the wild type strain NRRL-Y11430 (Northern Regional Research Laboratories, Peoria, IL) and grow on methanol exhibiting the methanol utilization plus phenotype (Mut⁺). They have one or more auxotrophic mutations, which allows transformation of expression vectors with appropriate selection markers. For example, GS115 (*his4*) is the most commonly used host strain for heterologous protein expression. Mut⁺ strains have both alcohol oxidase genes (*AOX1* and *AOX2*) present. Strains with *AOX* mutations, or deletions do not metabolize methanol as a carbon source or do so slowly, which sometimes result in higher production of foreign proteins. KM71 (*his4 arg4 aox1Δ::SARG4*) has a partially deleted *AOX1* gene replaced with the *S. cerevisiae ARG4* gene (Cregg and Madden, 1987). This strain grows slowly on methanol and exhibits the methanol utilization slow phenotype (Mut^s). MC100-3 (*his4 arg4 aox1Δ::SARG4 aox2Δ::Phis4*) has both *AOX* genes deleted and is therefore unable to grow on methanol (methanol utilization minus (Mut⁻) phenotype) (Cregg *et al.*, 1989). All strains are able to induce protein expression from the *AOX1* promoter upon methanol addition (Chiruvolu *et al.*, 1997).

Protease-Deficient Strains

Optimizing the expression of secreted heterologous protein is often challenging due to their susceptibility to proteases produced by the host organism. Protease-deficient host strains have been useful to improve expression in *S. cerevisiae* (Sander *et al.*, 1994, Jones, 1991) and *E. coli* (Meerman and Georgiou, 1994). The protease activities of *S. cerevisiae* have been characterized in detail elsewhere (Van Den Hazel *et al.*, 1996) and seem to be similar to those of *Pichia pastoris*. Most proteolytic activities in yeast cells are located within membrane-bound vacuoles. Proteinase A, a vacuolar aspartyl protease encoded by the *S. cerevisiae PEP4* gene, is capable of self-activation, as well that of other vacuolar proteases including carboxypeptidase Y and proteinase B. Carboxypeptidase Y is completely inactive prior to proteinase A-mediated proteolytic activation, and proteinase B (encoded by the *PRB1* gene in *S. cerevisiae*) is 50% bioactive in its precursor form prior to activation by proteinase A. Therefore, strains deficient in proteinase A are also deficient in carboxypeptidase Y acitivity and partially deficient in proteinase B activity.

Several protease deficient strains of *P. pastoris* (SMD1163: *his4 pep4 prb1*; SMD1165: *his4 prb1*; and SMD1168: *his4 pep4*) have been constructed to improve the expression of proteins susceptible to proteolytic degradation (Brierley, 1998; White *et al.*, 1995). More recently, a protease deficient host strain named SMD1168 (Δ*pep4::URA3* Δ*kex1::SUC2 his4 ura3*) was developed to inhibit the proteolysis of murine and human endostatin since the Kex1 protease can cleave carboxy-terminal lysine and arginine residues in endostatin (Boehm *et al.*, 1999).

Two additional strategies have been successfully used to reduce the degradation of heterologous proteins secreted in the growth medium of *P. pastoris*: supplementation of the fermentation medium with complex protein hydrolysates such as casamino acids or YP (Siegel *et al.*, 1990, Clare *et al.*, 1991b, Barr *et al.*1992, Sreekrishna 1993) and adjusting the pH of the medium during induction phase to a value where degradation is reduced (*e.g.* pH 3). These strategies can be combined with the use of protease deficient strains. For example, no

IGF-I could be detected in the growth medium of *P. pastoris* unless the pH was adjusted to 3, and further improvements were achieved by using a *pep4* strain in a medium buffered to pH 3 during induction (Brierley *et al.*, 1994).

Expression Vectors

P. pastoris has been widely used as an expression system for the production of heterologous proteins for several reasons: its *AOX1* promoter is strong and tightly controlled; genetically stable integration vectors are available; it is capable of secretion and correct protein folding; it can be fermented to high cell density on simple defined medium; fermentation scale-up is straightforward and high levels of recombinant protein are typically achieved.

All expression vectors are shuttle vectors with an origin of replication for maintenance in *E.coli* and selection markers functional in one or both organisms. Most expression vectors contain an expression cassette consisting of a 5' *AOX1* promoter sequence and 3' *AOX1* transcription termination sequence (Koutz *et al.*, 1989). Vectors are also available with the secretion signals of *S. cerevisiae* α-mating factor (MF) or native *P. pastoris* acid phosphatase (*PHO1*) for the secretion of foreign proteins. Although heterologous proteins are generally expressed from the methanol-inducible *AOX1* promoter, there could be a regulatory issue due to methanol being fire hazard and its use in large-scale fermentations. Therefore, other promoters systems have been developed and are briefly described in the following paragraphs.

AOX1 Promoter

As previously mentioned, there are two alcohol oxidase genes in *P. pastoris* (*AOX1* and *AOX2*) and AOX1 is responsible for the majority of the alcohol oxidase activity in the cell (Cregg *et al.*, 1989; Ellis *et al.*, 1985, Tschopp *et al.*, 1987b). *AOX2* encodes a protein that is 97% identical to and has approximately the same specific activity as that encoded by *AOX1* (Cregg *et al.*, 1989). Expression of *AOX1* is controlled at the transcriptional level. The regulation of the *AOX1* gene appears to be under the control of both carbon catabolite repression/derepression and a carbon-source specific induction mechanism similar to that implicated in the regulation of the *S. cerevisiae GAL1* gene (see Alcohol Utilization Genes).

GAP Promoter

The *P. pastoris* glyceraldehyde 3-phosphate dehydrogenase (*GAP*) gene promoter allows strong constitutive expression on glucose and its strength is comparable to that of the *AOX1* promoter (Waterham *et al.*, 1997). In fact, expression of β-lactamase under GAP promoter control in glucose-grown cells was significantly higher than under *AOX1* promoter control in methanol-grown cells. Recently, Goodrick *et al.* (2001) demonstrated high-level expression and stabilization of recombinant human chitinase in a continuous constitutive expression system using the GAP promoter and growth on either glycerol or glucose.

FLD1 Promoter

The *FLD1* gene encodes a glutathione-dependent formaldehyde dehydrogenase, a key enzyme required for the metabolism of methanol as a carbon source and certain alkylated amines such as methylamine as nitrogen sources. The FLD1 protein of *P. pastoris* shares 71% identity with FLD from the n-alkane-assimilating yeast *Candida maltosa*, and exhibits 61-65% identity with dehydrogenase class III enzymes from humans and other higher eukaryotes. Using β-lactamase as a reporter, Shen *et al.* (1998) have shown that the *FLD1* promoter is strongly and independently induced by either methanol as a sole carbon source (with ammonium sulfate as nitrogen source) or methylamine as a sole nitrogen source (with glucose as carbon source). The expression levels are comparable to those achieved with the *AOX1* promoter.

CUP1 Promoter

The *CUP1* promoter from *S. cerevisiae* is a copper-inducible promoter that functions in *P. pastoris* (Koller, *et al.*, 2000). The level of induction is dependent on the concentration of copper in the medium. The *CUP1* promoter was cloned into a pPIC3K vector, replacing the *AOX1* promoter, and used to induce β-galactosidase production from the *lacZ* gene. Induction with 100 µM copper led to expression levels 26-fold over basal levels.

PEX8 and YPT1 Promoters

P. pastoris PEX8 (a peroxisomal matrix protein essential for peroxisome biogenesis, Liu *et al.*, 1995) and YPT1 (a GTPase involved in secretion, Segev *et al.*, 1988) are expressed from moderate promoters compared to those driving the transcription of the *AOX1*, *GAP* and *FLD1* genes. Hence, the *PEX8* and *YPT1* promoters may be useful for the expression of toxic proteins, or when proteins have a tendency to misfold, are incorrectly processed or become mislocalized.

When placed under *PEX8* promoter control, heterologous proteins are expressed at a low level on glucose and induced when cells are shifted to methanol. Sears *et al.* (1998) compared the expression of bacterial β-glucuronidase (GUS) under the control of the strong constitutive *GAP* promoter, the moderate *YPT1* promoter, and the strong inducible *AOX1* promoter. *AOX1*-control led to very high GUS expression in methanol induction conditions, compared to the other two systems. However, both the *GAP* and *YPT1* promoters were constitutively active on various carbon sources (glucose, glycerol, mannitol, and methanol).

Selection Markers

Expression vectors can be directed to integrate into the *Pichia* genome depending on the restriction site of plasmid DNA digestion. Selection markers have been limited to the biosynthetic pathway genes *HIS4* (histidinol dehydrogenase gene) from either *P. pastoris* or *S. cerevisiae*

(Cregg *et al.*, 1985), *ARG4* (argininosuccinate lyase gene) from *S. cerevisiae* (Cregg and Madden, 1989), the kanamycin resistance gene which confers G418 resistance (Scorer *et al.*, 1993), the *Sh ble* gene from *Streptoalloteichus hindustanus* (Higgins *et al.*, 1998) which confers resistance to Zeocin (Zeo^r), and blasticidin S deaminase gene from *Aspergillus terreus* which confers resistance to blasticidin (Kimura *et al.*, 1994). In some cases multiple selection markers are required either for protein expression or for improving protein expression yields. Stable expression of human type III collagen requires coexpression of prolyl-4-hydroxylase in the synthesis and assembly of trimeric collagen. Three markers (*ARG4*, *HIS4*, and Zeo^r) were required to coexpress three polypeptides in the same strain. More recently, new biosynthetic markers have been reported (Cereghino *et al.*, 2001): the *P. pastoris ADE1* (PR-amidoimidazole-succinocarboxamide synthase), *ARG4* (argininosuccinate lyase), and *URA3* (orotidine 5'-phosphate decarboxylase) genes. Several auxotrophic host strains have also been developed with various combinations of auxotrophic mutations like *ade1*, *arg4*, *his4*, and *ura3* (Cereghino, *et al.*, 2001).

There is a definite advantage in using *P. pastoris* marker genes, because one can integrate into the genome at specific gene locus. The *ADE1* and *URA3* selection markers also have certain advantages. *P. pastoris ade1* mutants display a strong pink color phenotype which reverts to wild-type white upon transformation to ADE$^+$. This is helpful for the initial screening of integrated transformants of *ADE1* vectors. The *URA3* marker offers the ability to select either for or against a URA$^+$ phenotype (Boeke *et al.*, 1984). In addition to select for strains transformed to URA$^+$ phenotype with URA3 vectors, one can also select for URA$^-$ strains on media containing the drug FOA (plus uracil). The later technique is often useful if one needs to use production strains for further improvements or gene disruption (pop-in/pop-out method as described by Scherer and Davis, 1979). This strategy was successfully used for generating protease knockout mutants of an IGF-1 production strain (Brierley, 1998).

Secretion Signal Sequences

Although several secretion signal sequences from different sources have been successfully used in *P. pastoris*, results are often variable. Several heterologous proteins have been secreted using expression vectors carrying native signal sequences, or *S. cerevisiae* α-factor prepro peptide, or *P. pastoris* acid phosphatase (*PHOI*) signal sequence. Expression of human serum albumin (HSA) using the native signal peptide resulted in expression levels up to 10 g/L in a strain containing three copies of the gene (Barr *et al.*, 1992). The *S. cerevisiae SUC2* signal sequence improved the secretion of bacterial α-amylase 2- to 3-fold (Paifer *et al.*, 1994). A synthetic hybrid signal peptide based on the *PHOI* sequence with an additional 19 residues, including a Kex2 cleavage site, improved secretion of tick anticoagulant protein (Laroche *et al.*, 1994). The secretion and yield of recombinant human insulin was improved using a synthetic leader and spacer sequence (Kjeldsen *et al.*, 1999).

The *S. cerevisiae* mating factor, α -mating factor (MF) prepro peptide has been used with the most success (Cereghino and Cregg, 2000). Several combinations of native and heterologous signal sequences were studied for the secretion of recombinant human insulin-like growth factor (IGF-1), as summarized in Table 1. This signal sequence consists

Table 1. Effect of secretion signal sequence on secretion of recombinant human IGF-I (rhIGF-I) expression in *Pichia pastoris*.

	Secretion Signal Sequence	Relative rhIGF-I Levels
AMF	*Saccharomyces cerevisiae* α-mating factor prepro	100%
INV	*Saccharomyces cerevisiae* invertase pre signal	16%
INVS	Invertase pre+ SLDKR pro	26%
HAS	Human serum albumin pre-pro	33%
PSS	*Pichia pastoris* acid phosphatase pre	9%
AM5	PSS+ SLDKR pro	3%
PKV	PSS pre+ *Kluyveromyces lactis* killer toxin pro	4%
PPI	PSS+ 25 AA mature protein+ EKR (KEX-2 cleavage site)	84%

of a 19-amino acid signal (Pre) sequence followed by a 66-residue (pro) sequence containing three consensus N-linked glycosylation sites and a dibasic Kex2 endopeptidase processing site (Kurjan and Herskowitz, 1982). Processing of this signal sequence involves three steps. First the pre-sequence is removed by signal peptidase in the ER. Further processing of the pro-sequence involves the action of an endopeptidase encoded by the *KEX2* gene, which cleaves C-terminally to a specific Lys-Arg sequence, and a dipeptidyl aminopeptidase encoded by the *STE13* gene, which removes N-trminal Glu-Ala repeats (Brake, 1990). The correct processing of the α-factor preprosequence is very much protein specific and several cases have been reported in both *Saccharomyces* as well as *Pichia* where proteins have failed to give the correct processing to yield mature protein.

Glycosylation of Proteins

Glycosylation is the most common and complex form of post-translational modification. *P. pastoris* has the ability to add both O- and N-linked carbohydrate moieties to secreted proteins (Goochee *et al.*, 1991). O-linked oligosaccharides in mammals are composed of a variety of sugars, such as galactose, N-acetylgalatosamine, and sialic acid. However, lower eukaryotes such as *P. pastoris* add O-linked oligosaccharides composed solely of mannose residues. In addition, different host strains may add O-linked sugars on different residues in the same protein, sometimes even if the protein is not naturally glycosylated by its native host. For example, Insulin-like growth factor-I (IGF-I) is not glycosylated in humans, however, the *P. pastoris* expressed product had O-linked mannose added to 15% of the product (Brierley, 1998).

The structure of carbohydrates added to the secreted proteins is organism specific. N-asparagine-linked oligosaccharides on proteins secreted from *S. cerevisiae* have been shown to be antigenic when injected into mammals. Grinna and Tschopp studied the oligosaccharide chains of glycoproteins from *P. pastoris* and discovered that the average chain length was only Man8-14 compared to Man>40 in *S. cerevisiae*. The length of oligosaccharide chains added to

P. pastoris secreted invertase is much shorter than those from *S. cerevisiae* (Grinna and Tschopp, 1989). *P. pastoris* secreted invertase has majority of chains in the size range Man8-14GlcNAc2 (Man- mannose, GlcNAc- N-acetylglucosamine). However, only 20% of *S. cerevisiae* chains are of this size. *S. cerevisiae* secreted proteins are generally hyperglycosylated (50-150 mannose residues), but the longest chains contain only about 30 mannose residues in case of *P. pastoris* secreted invertase and did not contain immunogenic terminal α-1,3-linked mannose residues (Tschopp *et al.*, 1987a; Trimble *et al.*, 1991). As a result, *P. pastoris* invertase is relatively homogeneous with a molecular mass of 85-90 kDa, compared to *S. cerevisiae* invertase, which is much larger and more heterogeneous with a mass of 100-140 kDa.

Two strategies have been used to address the issue of protein glycosylation in *P. pastoris*. One is to eliminate glycosylation sites from the sequence of the secreted protein (Asami *et al.*, 2000) and the other is to engineer *P. pastoris* strains with more human-like glycosylation properties. For example *Trichoderma reesei* 1,2-α-D-mannosidase was expressed in *P. pastoris* (Callewaert *et al.*, 2001) to convert yeast type Man8GlcNAc2 structures to human type Man5GlNAc2 which are not substrates for hyperglycosylation in the Golgi. The genetically engineered strain was shown to alter the glycosylation patterns of two different reporter glycoproteins, influenza virus haemagglutinin and *Trypanosoma cruzi* trans-sialidase. This suggests that N-glycan engineering can be effectively accomplished in *P. pastoris*.

Copy Number Effects

Heterologous protein expression in *Pichia pastoris* can be improved by progressively increasing the number of copies of the gene expression cassette using in vitro multimerization. Constructs containing up to six direct repeats of an IGF-1 expression cassette were created, resulting in a 5-fold increase in IGF-1 protein expression as shown in Table 2. (Brierley *et al.*, 1994). Several authors have reported increased protein expression by utilizing multicopy transformants in *Pichia*

Table 2. Effect of expression cassette copy number on recombinant human IGF-I (rhIGF-I) expression in *Pichia pastoris*.

Cassette copy number	Phenotype	pH	RP-HPLC (mg/L)*	Cell density (wet g/L)
Six	Mut⁺	5.0	3	325
Six	Mut⁺	3.0	121	385
Four	Mut⁺	3.0	103	350
Two	Mut⁺	3.0	39	430
One	Mut⁺	3.0	14	415

*HPLC values reported as authentic IGF-I values, which represent approximately 20% of the total IGF-I forms present in HPLC chromatogram.

pastoris. For example, by integrating more than 20 copies of an expression cassette encoding the gene for tumor necrosis factor (TNF), a 200-fold increase in TNF expression was observed (Sreekrishna *et al.*, 1989). Other examples of high level expression resulting from multicopy integrants include tetanus toxin fragment C (14-copies and 6-fold increase), murine epidermal growth factor (19-copies and 13-fold increase), aprotinin (5-copies and 7-fold increase), and Hepatitis B surface antigen (8-copies and 12-fold increase) (Clare *et al.*, 1991a ; Vassileva *et al.*, 2001).

However, increasing gene copy number does not always have a positive effect on protein expression. Thill *et al.* (1990) observed that the amount of bovine lysozyme secreted per expression cassette was approximately 5-fold lower in a 3-copy strain than in a 1-copy strain. However, Northern analysis indicated that the increase in copy number correlated with an increase in steady state levels of mRNA encoding bovine lysozyme. It seems that the 3-copy strain has lower secretion efficiency (46% versus 75% for the one copy strain), leading to a greater percentage of the synthesized protein retained inside the cell.

If the level of secreted protein per expression cassette remains constant as the number of cassettes increases, the protein secretion rate is efficient relative to the rate protein synthesis. If the rate of synthesis becomes greater than the rate of secretion in multi-copy strains, the product may accumulate intracellularly, resulting in decreased levels of secreted product per cassette. Consequently, the relatively constant level of increase of secreted product with copy number also indicates that the secretory pathway has not reached its maximal capacity to secrete proteins. Our observations with IGF-1 also showed increase in protein expression using a 6-copy number expression plasmid. However, when copy number was further increased using zeocin selection there was not significant increase in expression and production levels dropped with very high copy number strains (data not shown) as described in the following section.

Transformation and Screening for High Copy number Integrants

Most commonly, the GS115 *HIS4* auxotrophic strain derived from the wild type *P. pastoris* host strain has been used for expression of heterologous proteins. For methanol utilization positive (Mut$^+$) strains, the expression vector is linearized with *Stu*I restriction enzyme whose recognition sequence is located within the *HIS4* gene on the vector. This directs the integration of the linear vector into the *HIS4* locus of the GS115 genome via an additive homologous recombination event. For methanol utilization slow (Mut^s) strains, the expression vector should be linearized with *Bgl*II whose recognition sequence is located within the *AOX1* structural gene on the vector. In these strains, the *AOX1* chromosomal gene is disrupted by integration of the vector at the *AOX1* locus. Plasmid transformation is performed by electroporation or spheroplasting, and transformants are screened for histidine prototrophy and growth on methanol. Integration of the plasmid in the genome can be verified by Southern hybridization. Shake flask screening is generally carried out in phosphate buffered YNB (yeast nitrogen base) medium supplemented with 2% glycerol. After 24 h, cells are transferred to fresh medium containing 1% methanol, and samples are taken every day for expression analysis.

Zeocin resistance is a dominant selectable marker, which allows for selection of transformants in both *E.coli* and *Pichia pastoris*. An additional advantage of using this marker is the ease of isolating multicopy transformants by selecting for higher zeocin resistance. In the case of the IGF-I production strain improvements, the idea was to integrate extra copies of expression cassette in the genome of the production strain SMD1120, using pPICZ based expression vectors. PPICZ vectors are ~3.3 kb- 3.6 kb in size with unique MCS region, zeocin resistance, and *AOX1* promoter and terminator sequences. *Bam*HI and *Bgl*II restriction sites were used to digest pPICZ as well as pIGF201 plasmids, which resulted into pZIA-1. Using pZIA-1 which is one-copy plasmid, two-copy (pZIA-2), 4-copy (pZIA-4) and six-copy (pZIA-6) plasmids were constructed. Similar constructs were also made using *Pichia pastoris* acid phosphatase (PPI) secretion signal sequence instead of the *S. cerevisiae* α-mating factor signal sequence, to see if the secretory pathway is saturated or different signal sequence may have different effects.

Initially, the SMD1120 IGF-I production strain was transformed with these new zeocin vectors to amplify *IGF-I* gene copy number and confirm if there were additional copy number effects. But since this strain already carries 6-8 copies of the *IGF-I* gene, it was decided to use wild type *Pichia* strain NRRL11430 (Mut⁺) as well as Mut⁻ strains, transformed with zeocin vectors. Transformation was carried out with lithium chloride treatment or with electroporation. MIC (minimal inhibitory concentration) of the different strains against varying concentrations of zeocin antibiotic on YPD plates was carried out prior to transformation, in order to select initial selection conditions. Further screening of transformants was carried out on increasing concentration of zeocin (300, 600, 1000, 2000 µg/mL), to isolate transformants having varying copies of expression plasmid integrated in the genome. After fermentation, lower expression levels were found with high copy number plasmids and hyper zeocin resistance strains. Improved expression was seen with low copy number plasmids indicating that integration of few more IGF-1 copies in the genome might further increase expression levels, and is very close to saturation.

Fermentation

Fermentation protocols for the production of yeast single cell protein (SCP) using *Pichia pastoris* were initially developed by the Provesta corporation. The process was scaled-up to 25,000 L scale as a continuous fermentation process. Later on, protocols for the recombinant protein production were developed by SIBIA, San Diego (Brierley *et al.* 1990b). The high-density fed-batch fermentation process for IGF-1 production utilizes basal defined medium supplemented with glycerol in the batch medium (Brierley, 1998). There are typically three different phases of fermentation. Cells are initially grown in batch mode using glycerol as the sole carbon source. Once glycerol is exhausted, a limiting glycerol feed is initiated in fed-batch mode. This ensures that there is no glycerol accumulation during the time where cell mass continues to increase. The culture is then shifted to a methanol feed, where cell mass continues to accumulate and the recombinant protein is produced. Often, the level of target protein expression is proportional to the volume of methanol fed during the fermentation process.

The fermentation medium is comprised of basal salts supplemented with glycerol (4%) as the carbon source. The initial media volume should be approximately 50% of the total working volume of the fermentor vessel. After autoclaving and cooling, a trace metals solution is added. The fermentor is then inoculated with 5-10% (v/v) based on the initial fermentation volume. The batch culture continues until glycerol is completely consumed (12-24 h), which is indicated by a sharp spike in dissolved oxygen (DO) levels. During this phase 90-150 g/L of wet cell weight is expected. Temperature is usually controlled at 30° C, pH at 5.0 and foaming by automatic addition of antifoam. Dissolved oxygen levels are maintained at higher than 20-30% saturation using DO cascade by increasing agitation, air flow rate and pure oxygen enrichment.

At the end of the batch phase, a 50% (w/v) glycerol feed is initiated using a feed rate of approximately 20 mL/h per liter of initial fermentation volume. Glycerol feeding continues for approximately 4 h or until approximately 80 ml/L of initial fermentation volume of the glycerol solution has been fed to the fermentor. DO spikes can be performed at regular intervals to verify that the culture is growing under carbon limitation. This results into 180-220 g/L of wet cell weight.

At the onset of the methanol fed-batch induction phase, the glycerol feed is stopped and a 100% methanol feed is initiated to induce the culture at very slow feed rate. The feed rate is increased slowly in three stages as the biomass accumulates and the cells become adapted to the methanol carbon source. The pH can be decreased at this time for proteins sensitive to proteases.

During first 2-3 h, the methanol feed rate is approximately 3.5 mL/h per liter of initial volume. Methanol will tend to accumulate in the fermentor as the cells become adapted to methanol at which time the methanol should decrease to limiting amounts. The feed rate is then doubled to 7 mL/h per liter initial volume. DO spikes should be performed to ensure that the culture is limited on methanol. After about 2-3 h, the feed rate is increased to 11 mL/h per liter initial volume for at least 24 h. The methanol feed can then be increased to 13 mL/h/L for the remainder of the fermentation. If methanol becomes non-limiting, the methanol feed rate should be decreased to avoid toxic levels of methanol accumulation. The duration of the methanol fed-batch induction phase is approximately 72 h with a total of 740 mL methanol fed per liter of initial volume. The cell density at the end of induction phase can reach up to 450 ± 100 g/L wet cell weight (see Figure 1).

A direct correlation between gene copy number and IGF-1 expression in Mut$^+$ strains was observed. As the cassette copy number decreased, IGF-1 expression levels decreased. Likewise, a 6-copy Mut^s strain appeared to produce lower levels of IGF-1 than a 6-copy Mut$^+$ strain. Also, there seemed to be slightly more degradation in 6-copy compared to 4-copy number strains as determined by RP-HPLC. The *PEP4* protease gene in the 6-copy Mut$^+$ strain was disrupted to help prevent proteolytic degradation of IGF-1 secreted in the fermentation medium. The IGF-1 production process was scaled up to 1,500 L for clinical trial material with a linear increase in fermentation yields

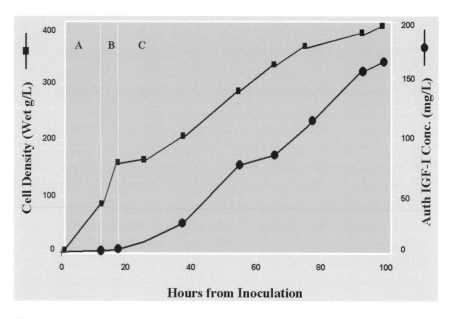

Figure 1. Fermentation profile for a protease deficient *Pichia pastoris* strain expressing recombinant human IGF-I. The strain contained a multicopy vector containing 6 copies of the AOX1- rhIGF-1 gene. A: Glycerol batch phase, B: Glycerol fed-batch phase, and C: Methanol fed-batch induction phase).

(140 ± 40 mg/L authentic IGF-1). Commercial scale production was further carried out at 10,000 L scale.

The single feed fermentation method was utilized to study the effects of specific growth rate on production of recombinant ovine interferon (r-oIFN) and protease by a Mut$^+$ strain of *Pichia pastoris* X-33 (Sinha *et al.*, 2003). Growth models were made using methanol as the carbon source. A methanol control loop was established to control methanol feeding. The maximum expression levels of r-oIFN were observed with a growth rate of $\mu = 0.025$ h^{-1} and a biomass wet weight of 386 g/L. When higher specific growth rates were achieved by increased methanol feeding, the specific production rate of r-oIFN fell sharply. This was correlated with a large increase in specific protease production rate that peaked at a specific growth rate of $\mu = 0.035$ h^{-1}. These data indicate that by controlling the specific growth rate and using methanol as the substrate, recombinant expression can be optimized while protease expression can be minimized. It should be noted that the expression and growth rates are protein specific and different effects are seen with different proteins.

Mixed feed fed-batch fermentations where nutrients are fed into the fermentor is the prevalent mode of operation in recombinant processes involving *P. pastoris*. By controlling the specific growth rate, the accumulation of undesirable by-products and deleterious effects such as catabolite repression or substrate inhibition can be minimized, thus leading to high cell densities and product concentrations. Methanol is fed during Pichia fermentations to induce protein expression, since the *AOX1* promoter is regulated by a combination of repression/ derepression by glycerol and induction by methanol. Mixed substrate feeding has been shown to improve overall productivity especially for the strains exhibiting a Mut^s phenotype (Brierley *et al.*, 1990a; Brierley *et al.*, 1990b; Loewen *et al.*, 1997). Glycerol or sorbitol are added with methanol during the induction phase. Sorbitol appears to be less repressive to the *AOX1* promoter compared to glycerol, but the cell yield on sorbitol is lower than that on glycerol (Thorpe *et al.*, 1999). However, the specific rate of product formation is higher on sorbitol/methanol compared to glycerol/methanol.

Table 3. Recombinant human IGF-I (rhIGF-I) expression in MUT+ and MUT- strains of *Pichia pastoris* under mixed-feed fermentation conditions.

Mixed Feed	Mut$^+$/ pep4$^-$ pIGF206	Mut$^+$/ Pep4$^+$ pIGF206	Mut$^-$/ Pep4$^+$ pIGF206	Mut$^-$/ Pep4$^+$ pZIA1
50% Glycerol	14%	No data	12%	No data
50% Glycerol + 2% Methanol	40%	13%	22%	19%
50% Glycerol + 5% Methanol	50%	18%	36%	31%
50% Glycerol + 10% Methanol	52%	30%	42%	No data

4% glycerol in batch medium, and 0.5% methanol injected at 4 h feed to initiate residual methanol. Feed rates similar to Mut$^+$ feed rates (11-13 ml/hr/L) for 3 days. Results are shown as relative expression levels using 100% methanol.

A mixed feed fed-batch fermentation process (glycerol/methanol) was also developed for IGF-1 production for all three phenotypes, Mut+, Mut^s and Mut ‾. Based on preliminary fermentation results with the high-copy number strains, a mixed feed fermentation protocol was tried in order to slow down the induction and to see if further increases in final IGF-1 expression could be made. Mixed feeding was optimized using the IGF-1 production strain (SMD1120 strain) and different ratios of glycerol and methanol. The same strategy was applied to new zeocin strains, which resulted in yield improvements of IGF-1 expression by 20-30% as shown in Table 3.

The effect of feeding mixed carbon sources on the expression of angiostatin was investigated using a Mut^s strain of *P. pastorsis* (GS115; Xie *et al.*, 2003). Methanol and glycerol were co-fed to improve cell growth and angiostatin production levels over those obtained using single feed fermentation. Methanol feeding was automatically controlled while glycerol feeding was adjusted to produce oscillations in the dissolved oxygen concentrations based on a cycle of glycerol overfeeding and limitation. The methanol was maintained at 5 g/L throughout the induction phase of the fermentation. Low glycerol feed rates resulted in low growth rates and low angiostatin expression (35 mg/L). Adjusting the glycerol feed rates to give oscillations in the DO levels (maintained above 20% saturation) greater amounts of glycerol were fed. This kept the fermentation under glycerol-limited condition and improved cell density and angiostatin expression. The fermentations were shown to be nitrogen limited, and when ammonia was added to the new fermentation protocol, the angiostatin yields were increased approximately 3-fold to 108 mg/L. The ratio of consumed glycerol to methanol during the expression phase was 1.5:1 (w/w) and indicates that methanol plays a key role in growth.

Growth models based on a glycerol and methanol mixed feeding strategy were studied during the expression of botulinum neurotoxin type C (BONT/C). This work (Zhang, *et al.*, 2003) combined previous studies on the use of mixed feeds with controlling specific growth rates by the rate of methanol addition. Growth models were made that describe the relationship between the specific growth rate (μ) and the specific glycerol/methanol consumption rate. A Mut+ strain of GS115 was used in this work. The maximum specific growth rate during methanol feeding was $\mu_{MeoH} = 0.020\ h^{-1}$, while the maximum specific growth rate on glycerol was $\mu_{Gly} = 0.177\ h^{-1}$. The highest expression levels were obtained when $\mu_{MeOH} = 0.015\ h^{-1}$, and was used to study growth interactions during mixed feeding. The optimal desired μ_{Glky}/μ_{MeOH} is around 2 for highest expression of BONT/C, while optimal feed rate ratio of $F_{Gly}:F_{MeOH}$ was derived to be 0.889.

An efficient production process has also been developed for the production of recombinant HSA (human serum albumin) using *P. pastoris* strains (Barr *et al.*, 1992). The process yields approximately 10 g/L by optimization of methanol feed and minimization of proteolytic degradation (Kobayashi *et al.*, 2000). A continuous fermentation process has been developed using the constitutive GAP promoter for the production of recombinant human chitinase (Goodrick *et al.*, 2001). Expression levels of about 200-400 mg/L of rh-chitinase were demonstrated in fed-batch fermentations, but there was extensive proteolytic degradation under these conditions. The continuous fermentation process was carried out for 30 days with a productivity level of approximately 360 mg/L per day, and no proteolytic degradation was observed.

Pichia methanolica

Introduction to *Pichia methanolica*

The methylotrophic yeast *Pichia methanolica* was first described by Kato and coworkers in 1974 (Kato *et al*, 1974). *Pichia methaolica* is a homothallic haploid that can be induced to mate by nutritional starvation (Tolstorukov, 1982). The first thorough genetic studies with *P. methanolica* were performed in the laboratory of I.I. Tolstorukov in the Scientific Center for Microbial Genetics and Bioengineering in Moscow (Tolstorukov *et al*, 1977, 1982). These studies were performed using *Pichia pinus* MH4. This strain later became synonymous with *P. methanolica* (Sibirny 1996, Tolstorukov, 1994).

The first transformations of *P. methanolica* were performed using the *LEU2* gene of *Saccharomyces cerevisiae* to complement a *leu1* mutation in *P. methanolica* (Tarutina and Tolstorukov, 1991). The cloning and sequencing of the *P. methanolica ADE1* gene led to a suitable selective marker for homologous transformation (Heip et. al 1991). Mutations in the *ADE1* gene leads to accumulation of a red pigment on adenine limited medium (Smirnov *et al.*, 1967). This allows for the easy selection of strains transformed with the *ADE1* gene. Homologous transformation using the *P. methanolica ADE1* gene and heterologous transformation using the *S. cerevisiae ADE2* gene was accomplished using an *ade1* mutant host (Heip et.al. 1993). Only linearized plasmids or DNA fragments were effective in transformation. The introduced DNA fragments circularized causing unstable transformants or integrated at different sites in the genome.

The *ADE2* gene of *P. methaolica* was cloned by cross species complementation in an *S. cerevisiae ade2* mutant (Raymond et. al. 1998). Transforming mutants from different complementation groups with the *ADE2* gene resulted in the isolation of an *ade2* mutant strain of *P. methanolica* CBS6515. This *ade2-11* mutant strain, PMAD11, was used to develop an efficient transformation and expression system in *P. methanolica* (Raymond et.al. 1998).

Host Strains

All of the commercially available strains for cloning in *P. methanolica* were derived from PMAD11 (*ade2-11*). Deletion mutations in the two major yeast protease genes *PEP4* and *PRB1* (Jones 1991) were introduced using loop in-loop out mutagenesis (Raymond and Vanaja, 1999). The resulting strains PMAD15 (*ade2-11 pep4Δ*) and PMAD16 (*ade2-11 pep4Δ prb1Δ*) showed a decrease in protease activity. Using the APNE overlay assay (Jones 1991), where protease proficient colonies become red upon addition of the overlay, the PMAD16 strain remained white and showed no indication of vacuolar protease activity.

The wild type PMAD11 strain carries two genes (*AUG1* and *AUG2*) for the production of alcohol oxidase. Deletion mutations in PMAD11 were introduced using loop in-loop out mutagenesis. Isogenic strains PMAD12 (*ade2-11 aug1Δ*), PMAD13 (*ade2-11 aug2Δ*) and PMAD14 (*ade2-11 aug1Δ aug2Δ*) were obtained. The PMAD12 strain grew slowly in minimal methanol medium, while the *AUG2* disrupted strain PMAD13 grew as well as the wild type PMAD11. The double mutant PMAD14 did not grow at all in minimal methanol medium, which indicated that there were only 2 alcohol utilization genes (Raymond et. al.1998).

Additional strains of *P. methanolica* were prepared using the PMAD16 (Mut$^+$) host strain, which is deficient in PEP4 and PRB1 protease activities. Using loop in-loop out mutagenesis, the alcohol utilization genes were disrupted. The methanol utilization minus (Mut$^-$) strain PMAD17 (*ade2-11 pep4Δ prb1Δ aug1Δ aug2Δ*) and the slow methanol utilizing strain (Mut^s) PMAD18 (*ade2-11 pep4Δ prb1Δ aug2Δ*) were obtained (Raymond 1999).

AUG1 Expression Vector

The first expression vectors made for *P. methanolica* used *AUG1* transcriptional control elements and the *ADE2* selectable marker assembled into a pUC19 vector. To make this vector, pUC19 was digested with *Eco*RI and *Hin*dIII and a linker encoding *Not*I-*Sfi*I-*Bam*HI-*Not*I restriction sites was inserted. The 1,350 base pair *AUG1* promoter was subcloned as a *Bg*/II- *Bam*HI fragment into the *Bam*HI site of the modified pUC19 vector. The 1,600 bp *AUG1* terminator sequence was subcloned as a *Bam*HI-*Bg*/II fragment into the *Bam*HI-digested derivative. The *ADE2* gene was subcloned into a unique *Xba*I site in the *AUG1* terminator to yield expression vector pCZR134. Protein coding sequences have been cloned into the unique *Eco*RI, *Bam*HI or *Spe*I sites (Raymond et.al. 1998).

For transformation into *P. methanolica* PMAD11, the pUC19 backbone of pCZR134 is digested with either *Not*I or *Sfi*I to liberate the linear DNA fragment of the expression cassette. Efficient transformation of *P. methanolica* is performed using electroporation (Raymond et. al. 1998). It was shown that transformation with 1 µg of circular pCZR134 only produced 1-2 ADE$^+$ transformants while transformation with 1 μg of linearized plasmid yielded 10-20 ADE$^+$ transformants. Transformation with linearized plasmid cut with *Spe*I yielded about 10,000 ADE$^+$ transformants. Transformed linear fragments were shown to undergo repair and circularize becoming unstable episomes. The unstable transformants would turn pink, allowing for the selection of stable white colonies. Screening the transformants on ADE$^-$ agar plates containing 1.2 M sorbitol enhanced the isolation of stable ADE$^+$ transformants (Raymond, 1999).

Expression vectors for *P. methanolica* are now prepared in *Saccharomyces cerevsiae* using homologous recombination. Homologous recombination allows *S. cerevisiae* to efficiently join double stranded DNA fragments. Three DNA components are required for plasmid assembly. These are the cDNA donor fragment, the acceptor vector (expression vector) into which the cDNA is to be sublconed and recombinatorial linkers. The linkers share sequence overlap with the vector on one end and the cDNA to be cloned on the other. The expression vector carries a CEN-ARS sequence along with the *URA3* gene for replication and selection in yeast. Amp^r and *ori* sequences are included for shuttling in *E. coli*. This method was used to sublcone the human leptin gene (Zhang et. al. 1994) into a pCZR182 vector containing the *S. cerevisiae* α-factor preprotein coding region (Raymond *et al* 1999b, Kurjan and Herskowitz, 1982).

P. methanolica PMAD11 was transformed with the vector containing the α factor prepro sequence fused to the N-terminal of the leptin gene. The protein was expressed and secreted but the Kex2 cleavage site (Lys-Arg-Val-Pro) was not cleaved by *P. methanolica* or *S. cerevisiae*. Expression of a modified leader fused to the N-terminal of a human cytokine receptor gene in *P. methanolica* PMAD16 resulted in a secreted protein without the α factor prepro sequence. This Kex2 celavage site was changed to Lys-Arg-Glu-Glu-Glu and was cleaved by an endogenous Kex2 like protease in *P. methanolica* (Raymond, 1999). The Kex2 cleavage site Asp-Lys-Arg-Glu-Ala was shown to work efficiently in *P. methanolica* (data not shown). Altering the P3 position of the α factor prepro sequence to Tyr-Lys-Arg – native protein enhances Kex2 cleavage without carrying over an amino acid to the native protein (Rockwell *et al.* 1997).

Recently, a *P. methanolica* expression kit has been developed and marketed by Invitrogen Life Technologies (San Diego, CA). The kit is supplied with either the PMAD11 or PMAD16 strains of *P. methanolica*. There are two expression vectors available and each is supplied as three different versions. The pMET vector is supplied in three different versions (A, B, C) to facilitate in-frame cloning with the C-terminal peptide encoding a V5 epitope and a hexahistidine tag. The V5 epitope allows for detection of expressed proteins with antibodies, while the polyhistidine tag aids in purification of the expressed protein. The pMETα vector is similar to pMET but contains the α factor signal sequence to allow for secretion of expressed protein. All vectors contain multiple cloning sites for the introduction of heterologous genes (Invitrogen Manual, 2003)

The *P. methanolica* expression kit was used to clone and express human Hepatitis B Virus polymerase (HHBV-P) and duck Hepatitis B Virus polymerase (DHBV-P). The HHBV-P gene was cloned into the *Sal*I site of the pMET-A vector and pMETαA vector. The DHBV gene was cloned into the *Eco*RI and *Spe*I sites of the pMET-A vector and pMETαA vector. The resulting plasmids were introduced into *E. coli* and amplified. The expression cassette was excised and used to transform electrocompetent cells of PMAD11 or PMAD16. Using antibodies to the V5 epitope, expression of the cloned proteins was detected by Western analysis in cells transformed with the pMET-A vector but not with the pMETαA vector. Proteins were purifed using Ni-NTA resin to bind to the hexahistidine tag at the C-terminus of the proteins. The proteins were shown to be active, stable and not aggregated (Choi et. al. 2002)

Screening Integrants

After transformation, about 100 stable transformants are obtained from 1 μg of transforming DNA. In order to screen a larger number of *Pichia methanolica* ADE[+] transformants, a colony lift technique has been used. *P. methanolica* vectors do not have zeocin selectable markers, although *P. methanolica* can be inhibited by zeocin. The alternative method utilizes antibodies to the foreign protein being expressed, the V5 epitope or the α factor secretion leader followed by detection of expressing colonies on X-ray film. Stable transformants are picked from sorbitol ADE⁻ plates (described earlier) and plated onto a minimal methanol agar plates using a grid pattern. The agar surface is overlayed with a nitrocellulose filter, and grown 2-3 days. Colonies adhering to the filter are lysed and the filters are probed with an antibody using Western blotting (Wuestehube et. al 1996). For expression of secreted protein, the nitrocellulose filters can be blotted directly.

Expression vectors integrate preferentially into the chromosome of *P. methanolica* by non-homologous recombination. Integration of the expression cassette at a variety of locations results in the Mut⁺ phenotype. Occasionally (<10%) homologous recombination at the *AUG1* site occurs and leads to strains with Mutˢ phenotype (Raymond et. al. 1998). Mutˢ strains do not produce the AUG1 methanol oxidase and grow much slower on methanol plates than Mut⁺ cells. This slow growth can be used to distinguish Mut⁺ from Mutˢ strains. By replica plating onto minimal plates plus dextrose and minimal plates plus methanol, slow growing colonies on the methanol plates can be selected.

Shake Flask Media for Screening Transformants of *Pichia methanolica*

In most microbial expression/screening groups, strains are not handed off for fermentation until expression levels can be confirmed and compared between strains. The usual method for comparing strains is to grow the strains in shake flask cultures in one or two media optimized for expression with the host strain used, or in one medium using different conditions (*e.g.* different pH or growth temperature). In shake flask cultures, it is common to simply add discreet doses of methanol to 0.5% at predetermined intervals. This continuously exposes cells to fluctuating levels of methanol and likely numerous periods in which the methanol has been depleted and is no longer at inducing levels (Guarna, *et al.*, 1997). When entering a fermentor system, methanol feeding strategies can become much more sophisticated, allowing for more consistent induction and growth characteristics.

A number of different carbon sources and media recipes have been evaluated for growth and protein expression by *P. methanolica* PMAD16. *P. methanolica* is capable of growth in a defined medium on glucose, fructose and mannose, but does not grow on the disaccharides maltose or sucrose. Growth on methanol as the sole carbon source yields only half the biomass obtained on glucose or mannose. No growth is obtained in a minimal defined medium with glycerol as the sole carbon source. Glycerol is used as the main carbon source in *P. pastoris* fermentations. Cell free extracts of *P. methanolica* PMAD16 grown on glycerol were used to study *AUG1* and *AUG2* deletions as well as AOD activity (Nakagawa *et al.*, 2001). Good growth of *P. methanolica* PMAD 16 can be obtained in complex medium containing glycerol, but the glycerol is not utilized. This may be the reason for this discrepancy

A recurring problem with screening *P. methanolica* strains in shake flasks has been the lack of a good recipe that provided detectable expression yields. Part of the problem in developing a good shake flask medium is obtaining good cell biomass, while keeping the pH of the medium from becoming too acidic. The BMMY medium used for growing *P. pastoris* utilizes phosphate to buffer the medium. A drawback with this recipe is that this medium does not have good buffering capacity at pH values lower than 5.5. A new modified version of BMMY (10 g/L yeast extract, 20 g/L tryptone, 13.4 g/L yeast nitrogen base, prepared in 100 mM citrate phosphate buffer) has been used for *P. methanolica*. The use of citrate-phosphate buffer allows the medium to be prepared in a range from pH 3.5 to pH 6.5. This is important since not all proteins are expressed well at pH 6.0 used in BMMY.

Expression of leptin by *P. methanolica* in shake flasks was studied using the citrate-phosphate buffered medium. This was investigated using PMAD16 grown in BMMY plus 2% methanol. The pH of the BMMY medium was varied in different shake flasks from 4.0 to 6.0. Cultures were started on 1% methanol and after 24 h growth, each flask received additional methanol to bring the final amount added to 2% (v/v). Cells were grown at 30°C with agitation at 250 rpm. Samples taken from each flask were analyzed for cell growth, pH and leptin expression. A maximum dry cell weight (DCW) of 12 g/L was obtained at pH 6.0. The DCW decreased as the initial pH of the BMMY medium decreased, with only 9.2 g/L obtained at pH 4.0. The pH of the medium increased during growth and the final pH of the shake flasks ranged between 5.0 and 6.2 for all the pH ranges tested. Leptin expression was approximately the same in all flasks tested.

Fermentation

Pichia fermentations generate large amounts of heat due to the high cell densities and the inefficiencies involved in methanol oxidation. Therefore, a robust, high-capacity temperature regulation system is required. Typically, optimum growth of *P. methanolica* occurs at 30°C (Nakagawa *et al.*, 2001). It is necessary, however, to investigate this parameter for each target protein as protein stability is greatly influenced by culture temperature. Recent studies indicate that, in some cases, low temperatures markedly increase protein production (Jahic *et al.*, 2003) Similarly, the demand for oxygen can be very high. This requires operation at elevated pressures or supplementation of the air sparge stream with pure oxygen.

During the fermentation of *P. methanolica,* various feeding protocols for the fed-batch and induction phases have been employed to maximize the expression of recombinant proteins in different phenotypes. Fermentation protocols for all three strains typically involve growth on excess glucose during the batch phase followed by a glucose fed-batch growth limiting phase and transition to a methanol induction phase. Most methanol feeding strategies involve continuously feeding methanol at a preprogrammed, exponentially increasing flow in order to control the culture growth rate at some predetermined value (Rodríguez, *et al.*, 1997; Trinh *et al.*, 2003; Zhang *et al.*, 2000). Other methanol feeding strategies have been developed utilizing on-line methanol sensor systems, allowing for maintenance of a constant methanol concentration in the fermentor (Guarna *et al.*, 1997; Hellwig *et al.*, 2001; Hong *et al.*, 2002; Mayson *et al.*, 2003; Mining *et al.*, 2001; Trinh *et al.*, 2003; Wagner *et al.*, 1997). Each protein produced will likely have its own expression profile in relation to methanol concentration.

Media for the initial batch phase of the fermentation consists of minimal basal salts media, trace salts, magnesium sulfate, vitamin stock, glucose, and chemical antifoaming agents (*e.g.* Antifoam 289 or 204). Growth during the initial batch phase ranges between 10-14 g/L DCW depending on the amount of glucose used (Zamost, 2001). Following batch growth on glucose (approximately 10-h), continuous glucose feeding can be commenced. A limited glucose feed profile is adopted to allow for sufficient growth to occur with all phenotypic strains. Because the *AUG1* gene is disrupted in the Mut[s] and Mut[-] strains, the potential for cellular growth utilizing methanol as the sole carbon and energy source is minimal. Therefore, in order to achieve reasonable cellular yields and promote cellular growth, glucose feeding is continued at a limiting rate. Glucose feeding should proceed until the culture reaches a wet cell weight of approximately 180 to 200 g/L (20 to 25 h total elapsed fermentation time) at which time the culture should be induced with methanol.

A number of different feeding strategies have been used to study the expression of secreted, recombinant human leptin by *Pichia methanolica* PMAD16 using the *AUG1* promoter (Zamost and Raymond, 1999, Zamost, 2001). Strategies investigated using fed-batch fermentations included: glucose and methanol co-feeding; various combinations of glucose and methanol feed rates; glucose feeding followed by methanol feeding; fed-batch with glucose and ethanol feeding; glucose only feeding; methanol feeding using a methanol controller and a number of other combinations. Fermentations run with glucose dosing for 30 h after the batch phase, followed by methanol co-feeding for 48 h resulted in leptin yields of 160 mg/L with a dry cell weight of 80 g/L (Figure 2). Fermentations run with glucose feeding followed by methanol only feeding resulted in leptin yields of 140 mg/L with a biomass of 75 g/L. It is interesting to note that good expression of leptin was obtained using glucose and ethanol co-feeding in place of methanol. Later experiments showed that good leptin expression yields (30 mg/L) were obtained during carbon limited feeding of glucose alone (Zamost, 2001). This indicates that the *AUG1* promoter can be de-repressed to allow expression during growth on glucose.

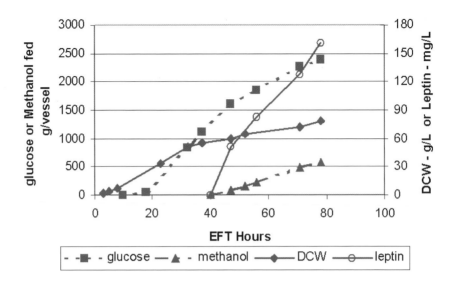

Figure 2. Production of rh Leptin in a mixed feed fermentation of *Pichia methanolica* PMAD16. The culture was grown in a batch mode on glucose, followed by glucose feeding starting at 10 hours elapsed fermentation time (EFT). At 40 hours EFT, the glucose feed rate was decreased and methanol feeding was started. The feeds were both continued at a fixed rate until the end of the fermentation. The glucose (60% v/v) and methanol (100 %) were fed into a 6.6 L vessel containing an initial starting volume of 2.5 L.

Because of the different methanol metabolizing abilities of the different phenotypes, the methanol induction phases differ significantly. Mut[+] strains have the ability to utilize methanol as their sole carbon and energy source, therefore methanol is sufficient to both support growth and drive protein expression. Mut[s] phenotypes have only the functional *AUG2* chromosomal gene and an *AUG1* driven gene expression cassette integrated into the chromosome. For Mut[s] strains, mixed glucose/methanol induction is necessary, as growth on methanol due to *AUG2* activity is limited, especially at low methanol concentrations. Although a contentious issue, Mut[s] strains may produce higher levels of recombinant proteins than wild type Mut[+] strains when transformed with the same expression cassette, while at the same time utilizing significantly less methanol (Cereghino and Cregg, 2000, Faber *et al.,* 1995). Because of the decreased ability of Mut[s] strains to metabolize methanol, higher methanol concentrations can be tolerated without the corresponding increase in inhibitory byproducts that is seen with Mut[+] strains.

Both preprogrammed continuous feeding strategies and controlled methanol concentrations have been investigated. The preprogrammed methanol feeding strategy for Mut[s] strains differs slightly from Mut[+] strains. The lower methanol utilization rate may cause methanol to build up to slightly higher levels, however this has not been seen to be problematic up to 1.5% methanol. Mut[s] strains are not as sensitive to high methanol concentrations and higher methanol concentrations have been shown to increase protein production (Mayson *et al.*, 2003). When methanol induction is initiated, glucose feed is continued to promote cellular growth. It is important that glucose does not build up in the medium and a limited glucose feed rate should be adopted. The methanol feed rate is initially set low in order to allow the culture to adapt to growth on methanol. After an acclimatization period of approximately 1-3 h, the methanol feed rate can be increased to promote cellular growth and protein expression (Mayson *et al.*, 2003)

Preprogrammed continuous methanol feeding is a complicated and laborious process. Methanol sensor systems have been developed that allow on-line monitoring and control of methanol concentrations in the fermentation medium. These systems may measure the amount of methanol in the medium indirectly through measurement of off-gas methanol content, or directly through the use of an in situ probe. By controlling methanol concentration, protein production can be significantly increased (Guarna *et al.*, 1997; Mayson *et al.*, 2003, Hellwig *et al.*, 2001). The usefulness of on-line methanol monitoring and control is very apparent with Mut[s] strains. The inability to estimate methanol concentrations via the DO spike method makes continuous feeding strategies extremely challenging. It is not uncommon for fermentations to be overfed methanol to toxic levels or underfed methanol resulting in poor protein expression.

Optimization of methanol concentration in Mut[s] cultures of *P. methanolica* PMAD18 has been analyzed in detail (Mayson *et al.*, 2003). Various methanol concentrations, controlled by an on-line monitoring and control system, were investigated during expression of the human transferrin N lobe. A mixed glucose and methanol feeding scheme was used during these investigations. Optimum cellular growth of 100 g/L DCW was obtained at 1.0% (v/v) methanol concentrations. Optimum protein productivity was obtained in cultures with methanol concentration controlled at 0.7% (v/v). Cultures controlled at 0.7% methanol, resulted in a 4-fold increase in protein production (450 mg/L) versus cultures controlled at 0.3% (Figure 3). Production of transferrin using a programmed methanol feed of 3 gh^{-1} during glucose feeding yielded only half the maximum expression level (217 mg /L) with a decreased biomass of 70 g/L DCW. Methanol utilization rates peak at 0.15 g g^{-1} DCW for controlled methanol concentrations of 0.7% and 1%. This optimum will likely be different for various recombinant protein constructs, however this suggests that higher methanol concentrations should be investigated for Mut^s strains versus Mut^+ strains.

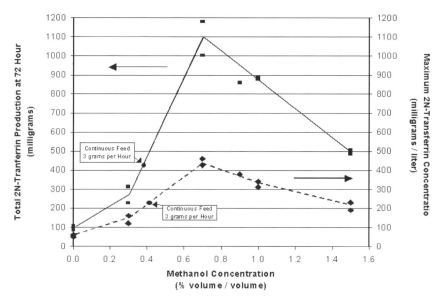

Figure 3. Effects of methanol concentration on 2N-transferrin production in fed-batch cultures of *Pichia methanolica* Mut^s PMAD18-p1. Maximum volumetric production (milligrams / liter) for each controlled methanol concentration is shown (◆). Total 2N-transferrin (milligrams) that could be recovered at 72 hours is also shown (■). All cultures were induced at a wet cell weight of 189-204 grams / liter with methanol and subsequently controlled at a constant concentration utilizing. A continuous methanol feeding strategy of 3 grams / hour is also shown (●). Note that culture volumes increased with increasing methanol addition.

Concluding Remarks

The *Pichia pastoris* and *Pichia methanolica* expression systems have been widely used for the expression of a variety of recombinant proteins. The commercial availability of both systems makes them easy to obtain and employ. This has led to improvements in both systems, including new promoters and signal sequences. There are currently 6 promoters available for driving protein expression in *P. pastoris*. Two new constitutive promoters for protein expression in *P. methanolica*, *GAP1* And *GAP2*, are also available (Raymond, 2002). The *GAP1* promoter was used successfully to express rh-leptin in fed-batch and continuous cultures of *P. methanolica* grown on glucose or fructose (Zamost, 2001).

The *P. pastoris* expression system has been used successfully to obtain very high levels of protein expression. A *P. pastoris* transformant containing 15 copies of a vector for gelatin expression was able to produce secreted gelatin at 14.8 g/L (Werten *et al.*, 1999). Most of the high expression yields obtained in *P. pastoris* are due to the integration of multi-copy expression vectors and the ability to select for these transformants. Methanol induced cultures of *P. methanolica* and *P. pastoris*, each containing single copies of the genes for glutamic acid decarboxylase (*GAD*) and transferrin N lobe, have been shown to produce equivalent amounts of the proteins (Raymond *et al.*, 1998, Guarna *et al* 1997, and Mayson *et al*, 2003). Multi-copy integration events do occur in *P. methanolica* and were shown to improve expression yields of *GAD*. The ability of transformed *P. methanolica* to form ARS-independent mini plasmids that integrate into the genome may also make it easier to obtain high copy number integration without the need for site-specific homologous integration (Tarutina and Tolstorukov, 2002). A multi-copy integration system of *P. methanolica* is currently under development.

Acknowledgements

The authors would like to thank Brian Mayson, Doug Kilburn, Noel Fong, Russell Brierley and Susan Holderman for their contributions and comments to this paper.

References

Alamae. T., and Simisker, J. 1994. Isolation and preliminary characterization of *Pichia pinus* mutants insensitive to glucose repression. Yeast 10:1459-1466.

Asami, Y., Nagano, H., Ikematsu, S., and Murasugi, A. 2000. An approach to the removal of yeast specific O-linked oligo-mannose from human midkine expressed in *Pichia pastoris* using site-specific mutagenesis. J. Biochem. 128: 823-826.

Ashin, V.V. and Trotsenko, Y.A, 1998. Formation and distribution of modified FAD between isozymes of alcohol oxidase in the methylotrophic yeast *Pichia methanolica*. Biochemistry (Moscow) 63:1407-1413.

Ashin, V.V. and Trotsenko, Y.A, 2000. Alcohol oxidase of the methylotrophic yeast: new findings. J. molecular Catalysis B: Enzymatic 10:295-303.

Barr, K.A., Hopkins, S.A., and Sreekrishna, K. 1992. Protocol for efficient secretion of HSA developed from *Pichia pastoris*. Pharm Eng. 12: 48-51.

Boehm, T., Pirie-Shepard, S., Trinh, L.B., Shiloach, J., and Folkman, J. 1999. Disruption of KEX1 gene in *Pichia pastoris* allows expression of full-length murine and human endostatin. Yeast. 15: 563-567.

Boeke, J.D., LaCroute, F., and Fink, G.R. 1984. A positive selection for mutants lacking orotidine5'phosphate decarboxylase activity in yeast: 5-fluoro-orotic acid resistance. Mol. Gen. Genet. 197: 345-346.

Brake, A.J., Merryweather, J.P., Coit, D.G., Heberlein, U.A., Masiarz, F.R., Mullenbach, G.T., Urdea, M.S., Valenzuela, P., and Barr, P.J. 1984. Alpha-factor directed synthesis and secretion of mature foreign proteins in *Saccharomyces cerevisiae*. Proc. Natl. Acad. Sci. USA. 81: 4642-4646.

Brake, A.J. 1990. Alpha-factor leader directed secretion of heterologous proteins from yeast. Methods Enzymol. 185: 408-421.

Brierley, R.A. 1998. Secretion of recombinant human insulin-like growth factor I (IGF-I). Methods Mol. Biol. 103: 149-177.

Brierley, R.A., Siegel, R.S., Bussineau, C., Craig, W.S., Holtz, G.C., Davis, G.R., and Buckholz, R.G. 1990a. Mixed feed yeast fermentation. International Patent. WO 90/03431.

Brierley, R.A., Bussineau, C., Kosson, R., Melton, A., and Siegel, R.S. 1990b. Fermentation development of recombinant *Pichia pastoris* expressing the heterologous gene: bovine lysozyme. Ann. N.Y. Acad. Sci. 589: 350-362.

Brierley, R.A., Davis, G.R., and Holtz, G.C. 1994. Production of insulin-like growth factor-1 in methylotrophic yeast cells. US Patent No.5,324,639.

Callewaert, N., Laroy, W., Cadirgi, H., Geysens, S., Saelens, X., Min, J.W., and Contreras, R. 2001. Use of HDEL-tagged *Trichoderma reesei* mannosyl oligosaccharide 1,2-alpha-D-mannosidase for N-glycan engineering in *Pichia pastoris*. FEBS Lett. 503(2-3):173-178.

Cereghino, G.P.L., Cereghino, J.L., Sunga, A.J., Johnson, M.A., Lim, M., Gleeson, M., and Cregg, J.M. 2001. New selectable marker/auxotrophic host strain combinations for molecular genetic manipulation of *Pichia pastoris*. Gene. 263: 159-169.

Cereghino, J. L., and Cregg, J.M. 2000. Heterologous protein expression in the methylotrophic yeast *Pichia pastoris* FEMS Microbiol Rev. 24:46-66

Chiruvolu, V., Cregg, J.M., and Meagher, M.M. 1997. Recombinant protein production in an alcohol oxidase-defective strain of Pichia pastoris in fed-batch fermentations. Enzyme Microb. Technol. 21 : 277-283.

Choi, J., Kim, E.E., Park, Y.I, and Han, Y.S. 2002. Expression of the active human and duck hepatitis B virus polymerases in heterologous system of *Pichia methanolica*. Antiviral Research 55: 279-290.

Clare, J.J., Rayment, F.B., Ballantine, S.P., Sreekrishna, K., and Romanos, M.A. 1991a. High level expression of tetanus toxin fragment C in *Pichia pastoris* strains containing multiple tandem integration of the gene. Bio/Technology. 9: 455-460.

Clare, J.J., Romanos, M.A., Rayment, F.B., Rowedder, J.E., Smith, M.A., Payne, M.M., Sreekrishna, K., and Henwood, C.A. 1991b. Production of mouse epidermal growth factor in yeast: high level secretion using *Pichia pastoris* strains containing multiple gene copies. Gene. 105: 205-212.

Cregg, J.M., Madden, K.R., Barringer, K.J., Thill, G.P., and Stillman, C.A. 1989. Functional characterization of the two alcohol oxidase genes from the yeast *Pichia pastoris*. Mol. Cell Biol. 9:1316-1323.

Cregg, J.M., Barringer, K.J., Hessler, A.Y., and Madden, K.R. 1985. *Pichia pastoris* as a host system for transformation. Mol. Cell. Biol. 5: 3376-3385.

Cregg, J.M., and Madden, K.R. 1987. Development of yeast transformation systems and construction of methanol-utilization-defective mutants of *Pichia pastoris* by gene disruption. In: Biological Research on Industrial Yeasts. G.G. Stewart, I. Russell, R.D. Klein, and R.R. Hiebsch, eds. CRC Press, Boca Raton, FL. vol.2. p. 1-18.

Cregg, J.M., and Madden, K.R. 1989. Use of site specific recombination to regenerate selectable markers. Mol. Gen. Genet. 219: 320-323.

Dale, C., Allen, A., and Fogerty, S. 1999. *Pichia pastoris*: a eukaryotic system for the large scale production of biopharmaceuticals. BioPharm Nov. 36-32.

Digan, M.E., Lair, S.V., Brierley, R.A., Siegel, R.S., Williams, M.E., Ellis, S.B., Kellaris, P.A., Provow, S.A., Craig, W.S., Velicelebi, G., Harpold, M.M., and Thill, G.P. 1989. Continuous production of a novel lysozyme via secretion from the yeast, *Pichia pastoris*. Bio/Technology. 7: 160-164.

Eckart,M.R., and Bussineau, C.M. 1996. Quality and authenticity of heterologous proteins synthesized in yeast. Curr. Opin. Biotechnol 7:525-530.

Ellis, S.B., Brust, B.F., Koutz, P.J., Waters, A.F., Harpold, M.M., and Gingeras, T.R. 1985. Isolation of alcohol oxidase and two other methanol regulatable genes from the yeast *Pichia patoris*. Mol. Cell. Biol. 5:1111-1121.

Faber: K.N., Harder, W., Abb, G., and Veenhius, M. 1995. Review: methylotrophic yeasts as factories for the production of foreign proteins. Yeast 11 1331-1344.

Gellissen, G. 2000. Heterologous protein production in methylotrophic yeast. Appl. Microbiol. Biotech 54:741-750.

Gleeson, M.A., and Howard, B.D. 1994. Genes which influence Pichia proteolytic activity, and uses therefor. US Patent. 5,324,660.

Goochee, C.F., Gramer, M.J., Andersen, D.C., Bahr, J.B., and Rasmussen, J.R. 1991. The oligosaccharides of glycoproteins: bioprocess factors affecting oligosaccharide structure and their effect on glycoprotein properties. Bio/Technology. 9: 1347-1355.

Goodrick, J.C., Xu, M., Finnegan, R., Schilling, B.M., Schiavi, S., Hoppe, H., and Wan, N.C. 2001. High-level expression and stabilization of recombinant human chitinase produced in a continuous constitutive *Pichia pastoris* expression system. Biotechnol. Bioeng. 74: 492-497.

Grinna, L.S., and Tschopp, J.F. 1989. Size distribution and general structural features of N-linked oligosaccharides from the methylotrophic yeast, *Pichia pastoris*. Yeast. 5: 107-115.

Gruzman, M.B., Titorenko, V.I., Ashin, V.V., Lusta, K.A., and Trotsenko, Y.A. 1996. Multiple molecular forms of alcohol oxidase from the methylotrophic yeast Pichia methanolica. Biochemistry (Moscow), 61:1537- 1544.

Guarna, M.M., Lesnicki, G.J., Tam, B.M., Robinson, J., Radziminski, C.Z., Hasenwinkle, D., Boraston, A., Jervis, E., MacGillivray, R.T.A , Turner, R.F.B., and Kilburn, D.G. 1997. On-line monitoring and control of methanol concentration in shake flask cultures of *Pichia pastoris*. Biotechnol. Bioeng. 56:279-286

Hellwig ,S., Emde ,F., Raven N. P.G., Henke, M., van der Logt , P., and Fischer, R. 2001. Analysis of single-chain antibody production in *Pichia pastoris* using on-line methanol control in fed-batch and mixed-feed fermentations. Biotechnol. Bioeng. 74: 344-352.

Heip, T.T., Kukilov, V.N., and Pavlov, Y.I. 1991. The structure of the ADE1 gene from the methylotrophic yeast *Pichia pinus*: comparison with the AIR carboxylase gene from other species. Abstr. 15th Intl. Spec. Symp. On Yeasts, Riga p55.

Heip, T.T., Noskov, V.N., and Pavlov, Y.I. 1993. Transformation in the methylotrophic yeast *Pichia methanolica* utilizing homologous ADE1 and heterlogous *Saccharomyces* ADE2 and LEU2 genes as genetic markers. Yeast 9:1189-1197.

Higgins, D.R., Busser, K., Cominsky, J., Whittier, P.S., Purcell, T.J., and Hoefler, J.P. 1998. Small vectors for expression based on dominant drug resistance with direct multicopy selection. Methods Mol. Biol. 103: 41-53.

Hollenberg, C.P., and Gellissen, G. 1997. Production of recombinant proteins by methylotrophic yeasts. Curr. Opin. Biiotechnol. 8:554-560.

Hong, F., Meinander, N.Q, and Jönsson, L.J. 2002. Fermentation strategies for improved heterologous expression of laccase in *Pichia pastoris*. Biotech. Bioeng. 79: 438-449.

Invitrogen Life Technologies. 2003. A manual of methods for expression of recombinant proteins in *Pichia methanolica*. Version C 062102.

Jahic, M., Gustavsson, M., Jansen AK, Martinelle, M., and Enfors S.O. 2003. Analysis and control of proteolysis of a fusion protein in *Pichia pastoris* fed-batch processes. J. Biotechnol. 102: 45-53.

Jones, E.W., 1991. Tackling the protease problem in *Saccharomyces cerevisae*. Methods Enzymol. 194: 428-453.

Kato, K., Kurimura, Y., Makiguchi, N., and Asi, Y. 1974. Determination of strongly methanol assimilating yeasts. J. Gen. Appl. Microbiol . 20:123-27

Kimura, M., Kamakura, T., Tao, Q.Z., Kaneko, I., and Yamaguchi, I. 1994. Cloning of the blasticidin S deaminase gene (BSD) from *Aspergillus terreus* and its use as a selectable marker for *Schizosaccharomyces pombe* and *Pyriculuria oryzae*. Mol. Gen. Genet. 242: 121-129.

Kjeldsen, T., Pettersson, A.F., and Hach, M. 1999. Secretory expression and characterization of insulin in *Pichia pastoris*. Biotechnol. Appl. Biochem. 29:79086.

Kobayashi, K., Kuwae, S., Ohya, T., Ohda, T., Ohyama, M., Ohi, H., Tomomitsu, K., and Ohmura, T. 2000. High-level expression of recombinant human serum albumin from the methylotrophic yeast *Pichia pastoris* with minimal protease production and activation. J Biosci Bioeng. 89: 55-61.

Koller, A., Valesco, J., and Subramani, S. 2000. The CUP1 promoter of *Saccharomyces cerevisae* is inducible by copper in *Pichia pastoris*. Yeast 16:651-656.

Koutz, P., Davis, G.R., Stillman, C., Barringer, K., Cregg, J.M., and Thill, G. 1989. Structural comparison of the *Pichia pastoris* alcohol oxidase genes. Yeast. 5: 167-177.

Kulackovsky, A.R., Stasyk, O.V. *et al.* 1998. Nutrition and ultrastructure of the mutants of methylotrophic yeasts defective in biogenesis and degradation of peroxisomes. Food. Technol. Biotechnol. 36: 19-26.

Kurjan, J. and Herskowitz, I. 1982. Structure of a yeast pheromone gene (MFα): a putative α- factor precursor contains four tandem copies of mature α- factor. Cell 39: 993-943.

Laroche, Y., Storme, V., De Mutter, J., Messens, J., and Lauwereys, M. 1994. High level secretion and very efficient isotopic labeling of tick anticoagulant peptide (TAP) expressed in the methylotrophic yeast, *Pichia pastoris*. Bio/Technology. 12:1119-1124

Liu, H., Tan, X., Russell, K.A., Veenhuis, M., and Cregg, J.M. 1995. *PER3*, a gene required for peroxisome biogenesis in Pichia pastoris, encodes a peroxisomal membrane protein involved in protein import. J. Biol. Chem. 270: 10940-10951.

Loewen, M.C., Liu, X., Davies, P.L., and Daugulis, A.J. 1997. Biosynthetic production of type II antifreeze protein: fermentation by *Pichia pastoris*. Appl. Microbiol. Biotechnol. 48: 480-486.

Mason A.B., Woodworth, R.C. et. al. 1996. Production and isolation of the recombinant N-lobe of human serum tranferrin from the methylotrophic yeast *Pichia pastoris*. Protein Expr. Purif. 8:119-125.

Mayson, B.E., Kilburn, D.G., Zamost, B. L., Raymond, C.K., and Lesnicki, G.J. 2003. Effects of methanol concentration on expression levels of recombinant protein in fed batch cultures of *Pichia methanolica*. Biotechnol. Bioeng. 81: 291-298.

Meerman, H.J., and Georgiou, G. 1994. Construction and characterization of a set of *E. coli* strains deficient in all known loci affecting the proteolytic stability of secreted recombinant proteins. Bio/Technology. 12: 1107-1110.

Minning,S., Serrano, A., Solà, C., Schmid, R.D., and Valero, F. 2001. Optimization of the high-level production of *Rhizopus oryzae* lipase in *Pichia pastoris*. J. Biotechnol. 86: 59-70.

Murdanoto, A.P., Sakai, Y., Sembiring, L., Tani, Y., and Kato, N. 1997. Ester synthesis by NAD^+ dependent dehydrogenation of hemiacetal: Production of methyl formate by cells of methylotrophic yeats. Biosci. Biotech. Biochem. 61:1391-1393.

Nakagawa, T., Uchimura, T., and Komagata, K. 1996. Isozymes of methanol oxidase in a methanol utilizing yeast Pichia methanolica IAM 12901. J. Ferment. Biotechnol. 81:498-503.

Nakagawa, T.,Mukaiyama, H., Yurimoto, H., Sakai, Y., and Kato, N. 1999. Alcohol oxidase hybrid oligomers formed *in vivo* and *in vitro*. Yeast 15: 1223-1230.

Nakagawa, T., Sakai, Y Mukaiyama, H., Mizumura, T., Miyaji, T., Yurimoto, H., Kato, N., and Tomizuka, N. 2001. Analysis of alcohol oxidase isozymes in gene-disrupted strains of methylotrophic yeast *Pichia methanolica*. J. Biosci. Bioeng.

Ogata, K., Nishikawa, H., and Ohsugi, M. 1969. A yeast capable of utilizing methanol. Agric. Biol. Chem. 33: 1519-1520.

Paifer, E., Margolles, E., Cremata, J., Montesino, R., Herrera, L., and Delgado, J.M. 1994. Efficient expression and secretion of recombinant alpha amylase in *Pichia pastoris* using two different signal sequences. Yeast. 10: 1415-1459.

Raymond, C.K., Bukowski T., Holderman, S.D. Ching, A., Vanaja, E., and Stamm, M.R. 1998. Development of the methylotrophic yeast *Picha methanolica* for the expression of the 65 kilodalton isoform of human glutamate decarboxylase. Yeast 14: 11-23

Raymond C.K. 1999. Composition and methods for producing heterologous polypeptides in *Pichia methanolica*. US patent 5,888, 768

Raymond, C.K., 1999b. Recombinant protein expression in *Pichia methanolica*. Gene Expression Systems, Academic Press, San Diego, CA. pp 193-209.

Raymond, C.K., and Vanaja, E. 1999. Generation of protease-deficient strains of *Pichia methanolica*" US Patent 6,153,424.

Raymond, C.K., Pownder, T.A., and Sexson, S.L. 1999. General methods for plasmid construction using homologous recombination. BioTechniques 26: 134-141

Raymond, C.K. 2002. *Pichia methanolica* glycerldehyde –3-phosphate dehydrogenase 2 promoter and terminator. US patent 6,440,720.

Rockwell, N. C., Wang, G.T., and Fuller R.S. 1997. Internally consistent libraries of fluorogenic substrates demosnstrate that Kex2 protease specificity is generated by multiple mechanisms. Biochemistry, 36: 1912-1917.

Rodríguez, J., Sánchez, K., Roca, H., and Delgado, J.M. 1997. Different methanol feeding strategies to recombinant *Pichia pastoris* cultures producing high level of dextranase. Biotechnol. Tech. 11: 461-466.

Romanos, M.A., Scorer, C.A., and Clare, J.J. 1992. Foreign gene expression: a review. Yeast. 8: 423-488.

Sakai, Y., Sawai, T. and Tani, Y. 1987. Isolation and characterization of a catabolite repression insensitive mutant of a methanol yeast, *Candida boidini* A5 producing alcohol oxidase in glucose containing medium. Appl. Environ. Microbiol. 53: 1812-1818.

Sakai, Y., Murdanoto, A.G., Sembiring, L., Tani, Y., and Kato, N. 1995. A novel formaldehyde oxidation pathway in methylotrophic yeasts: Methylformate as a possible intermediate. FEMS Microbiol. Lett. 127: 229-234.

Sander, P., Grunewald, S., Bach, M., Haase, W., Reilander, H., and Michel, H. 1994. Heterologous expression of the human D2S dopamine receptor in protease-deficient *Saccharomyces cerevisiae* strains. Eur. J. Biochem. 226: 697-705.

Scherer, S., and Davis, R.W. 1979. Replacement of chromosome segments with altered DNA sequences constructed *in vitro*. Proc. Natl. Acad. Sci. USA. 76: 4951-4955.

Scorer, C.A., Buckholz, R.G., Clare, J.J., and Romanos, M.A. 1993. The intracellular production and secretion of HIV-I envelope protein in the methylotrophic yeast *Pichia pastoris*. Gene. 136: 111-119.

Sears, I.B., O'Connor, J., Rossanese, O.W., and Glick, B.S. 1998. A versatile set of vectors for constitutive and regulated gene expression in Pic*hia pastoris*. Yeast. 14: 783-790.

Segev, N., Mulholland, J., and Botstein, D. 1988. The yeast GTP-binding YPT1 protein and a mammalian counterpart are associated with the secretion machinery. Cell. 52: 915-924.

Shen, S., Sulter, G., Jeffries, T.W., and Cregg, J.M. 1998. A strong nitrogen source-regulated promoter for controlled expression of foreign genes in the yeast *Pichia pastoris*. Gene. 216: 93-102.

Sibirny, A., Titorenko V.I., Benevolensky, S.V., and Tolsturokov, I.I. 1986. On the differences in the mechanism of ethanol and glucose induced catabolite repression in the yeast *Pichia pinus*. Genetika 22: 584-592.

Sibirny,.A., Titorenko V.I., Gonchar M.V., Ubiyvovk, V.M. and Vitvitskaya, O.P. 1988. Genetic control of methanol utilization in yeasts J. Basic. Microbiol. 28: 293-319.

Sibirny, A, 1996. *Pichia methanolica* (*Pichia pinus* MH4). Nonconventional yeasts in biotechnology. Klaus Wolf, ed. Springer, Berlin

Siegel, R.S., Buckholz, R.G., Thill, G.P., and Wondrack, L.M. 1990. Production of epidermal growth factor in methylotrophic yeasts. International Patent Application, WO 90/10697.

Sinha, J., Plantz, P.A., Zhang, W., Guothro, M., Schlegel, V., Liu, C., and Meagher, M.M. 2003. Improved production of recombinant ovine interferon by Mut + strain of *Pichia pastoris* using an optimized methanol feed profile. Biotechnol. Progress 19: 794-802.

Smirnov, M.N., Sirnov,V.N., and Inget-Vechtomov, S.G. 1967. Red pigment of *Saccharomyces cerevisiae* adenine mutant. Mol Biol (Moscow) 1: 639-647.

Sreekrishna, K., Nelles, L., Potenz, R., Cruze, J., Mazzaferro, P., Fish, W., Fuke, W., Holden, K., Phelps, D., Wood, P, and Parker, K. 1989. High level expression, purification and characterization of recombinant human tumor necrosis factor synthesized in the methylotrophic yeast *Pichia pastoris*. Biochemistry. 28: 4117-4125.

Sreekrishna, K. 1993. Strategies for optimizing protein expression and secretion in the methylotrophic yeast *Pichia pastoris*. In: Industrial Microorganisms: Basic and Applied Molecular Genetics. R.H. Baltz, G.D. Hegeman, and P.L. Skatrud, eds. American Society of Microbiology, Washington, D.C. p. 119-126.

Sreekrishna, K., Brankamp R.G., Kropp K.E., Blankenship D.T., Tsay J.T., Smith P.L., Wierschke J.D., Subramaniam, A., and Birkenberger, L.A. 1997. Strategies for optimal synthesis and secretion of heterologous proteins in the methylotrophic yeast *Pichia pastoris*. Gene 190: 55-62.

Tarutina, M.G., and Tolstorukov, I.I. 1991. Transformation of the methylotrophic yeast *Pichia (Pichia methaolica) pinus* MH4 with the LEU2 gene of the yeast Saccharomyces cerevisiae. Abst. of the 15th International Specialized Symposium on Yeasts Riga (U.S.S.R.) Septemeber 30- October 6 p 141.

Tarutina, M.G., and Tolstorukov, I.I. 2002. Formation of ARS-independent miniplasmids upon transformation of yeast *Pichia methanolica* with DNA molecules containing transforming and non transforming genes. genetika. 38: 1451-1462.

Thill, G.P., Davis, G.R., Stillman, C., Holtz, G., Brierley, R., Engel, M., Buckholtz, R., Kinney, J., Provow, S., Vedvick, T., and Siegel, R.S. 1990. Positive and negative effects of multi-copy integrated expression vectors on protein expression in *Pichia pastoris*. In: Proceedings of the 6th International Symposium on Genetics of Microorganisms. Heslot, H., Davies, J., Florent, J., Bobichon, L., Durand, G., and Penasse, L. Societe Francaise de Microbiologie, Paris. Vol.2. p. 477-490.

Thorpe, E.D., d'Anjou, M.C., and Daugulis, A.J. 1999. Sorbitol as a non-repression carbon source for fed-batch fermentations of recombinant *Pichia pastoris*. Biotechnol. Lett. 21: 669-672.

Tolstorukov, I.I., Dutova T.A., Benevolensky S.V, and. Soom, Y.O 1977. Hybridization and genetic analysis of methanol oxidizing yeast *Pichia pinus* MH4. Genetika 13: 322-329

Tolstorukov, I.I., Benevolensky, S.V., and Efremov, B.D. 1982. Genetic control of cell type and complex organization of mating type locus in the yeast *Pichia pinus*. Curr. Genet 5: 137-142.

Tolstorukov, I.I. 1994. Genome structure and re-identification of the taxonomic status of *Pichia pinus* MH4 genetic lines. Genetika 30: 635-640.

Trimble, R.B., Atkinson, P.H., Tschopp, J.F., Townsend, R.R., and Maley, F. 1991. Structure of oligosaccharides on Saccharomyces SUC2 invertase secreted by the methylotrophic yeast *Pichia pastoris*. J. Biol. Chem. 266: 22807-22817.

Trinh, L.B., Phue, J.N., and Shiloach, N. 2003. Effect of methanol feeding strategies on production and yield of recombinant mouse endostatin from *Pichia pastoris*. Biotechnol Bioeng 82: 438-444.

Tschopp, J.F., Sverlow, G., Kosson, R., Craig, W., and Grinna, L. 1987a. High-level secretion of glycosylated invertase in the methylotrophic yeast, *Pichia pastoris*. Bio/Technology. 5: 1305-1308.

Tschopp, J.F., Brust, P.F., Cregg, J.M., Stillman, C.A., and Gingeras, T.R. 1987b. Expression of LacZ gene from two methanol-regulated promoters in *Pichia pastoris*. Nucleic Acids Res. 15: 3859-3876.

Van Den Hazel, B.H., Kielland-Brandt, M.C., and Winther, J.B. 1996. Review: Biosynthesis and function of yeast vacuolar proteases. Yeast. 12: 1-16.

Vassileva, A., Chugh, D.A., Swaminathan, S., and Khanna, N. 2001. Effect of copy number on the expression levels of Hepatitis B surface antigen in the methylotrophic yeast *Pichia pastoris*. Protein Expr.Purif. 21: 71-80.

Wagner, L.W., Matheson, N.H., Heisey R..F., and Schneider, K. 1997. Use of a silicone tubing sensor to control methanol concentration during fed batch fermentation of *Pichia pastoris*. Biotechnology Tech. 11: 791-795.

Waterham, H.R., Digan, M.E., Koutz, P.J., Lair, S.V., and Cregg, J.M. 1997. Isolation of the *Pichia pastoris* glyceraldehydes-3-phosphate dehydrogenase gene and regulation of its promoter. Gene. 186: 37-44.

Wegner, G.H. 1983. Biochemical conversions by yeast fermentation at high cell densities. US Patent. 4,414,329.

Wegner, G.H. 1990. Emerging applications of the methylotrophic yeasts. FEMS Microbiol. Rev. 7: 279-283.

Werten, M.W., Van Den Bosch, T.J. Wind, R.D., Mooibroek, H., and De Wolf, F.A. 1999. High yield secretion of recombinant gelatins by *Picha pastoris*. Yeast 15:1087-1096.

White, C.E., Hunter, M.J., Meininger, D.P., White, L.R., and Komives, E.A. 1995. Large scale expression, purification and characterization of small fragments of thrombomodulin: the roles of the sixth domain and of methionine 388. Protein Eng. 8: 1177-1187.

Wuestehube, L.J., Duden R.A., Eun, A., Hamamoto, S., Koen, P., Ran, R., and Schekman, R. 1996. New mutants of *Saccharomyces cerevisiae* affected in the transport of proteins from the endoplasmic reticulum to the golgi complex. Genetics 142: 393-406.

Xie, J., Zhang, L., Ye, Q., Zhou, Q., Xin, L., Du, P., and Gan, R. 2003. Angiostatin production in cultivation of recombinant *Pichia pastoris* fed with mixed carbon sources. Biotechnol. Lett. 25:173-177.

Zamost B. L., and Raymonk, C.K. 1999. Production of leptin using a new yeast expression system: *Pichia methanolica*. Abstract 075. Book of abstracts, 217th National Meeting of the American Chemical Society. March 21-25, 1999.

Zamost, B.L. 2001. Method for producing proteins in transformed *Pichia*. US patent 6,258,559.

Zhang, Y., Proneca, R., Maffei, M., Barone, M., Leoplod, L., and Friedman, J.M. 1994. Positional cloning of the mouse obese gene and its human homologue. Nature 372: 425-427.

Zhang, W., Bevins M.A., Plantz B.A., Smith L.A., and Meagher M.M. 2000. Modeling *Pichia pastoris* growth on methanol and optimizing the production of recombinant protein, the heavy-chain fragment C of Botulinium neurotoxin, serotype A. Biotechnol Bioeng 70: 1-8.

Zhang, W., Potter, K.J., Plantz, B.A., Schlegel, V.L., Smith, L.A., and Meagher, M.M. 2003. *Pichia pastoris*: fermentation with mixed feeds of glycerol and methanol: growth kinetics and product improvements. J. Ind. Microbiol. 30: 210-215.

7

Advances in Protein Expression in Filamentous Fungi

Hendrik J. Meerman, Aaron S. Kelley, and Michael Ward

Abstract

Filamentous fungi have a long history of safe industrial cultivation for the efficient production of a diverse set of compounds, including proteins, organic acids and secondary metabolites such as antibiotics, carotenoids, and vitamins. In particular, both endogenous and exogenous enzymes are currently manufactured commercially using fungal host systems. The advantages of recombinant protein expression in filamentous fungi have been detailed extensively over the years, such as in several recent excellent reviews (Punt *et al.*, 2002; Bergquist *et al.*, 2002; Archer, 2000). In this chapter we will recapitulate some of the highlights featured in those reviews and emphasize some recent advances that have significantly improved the understanding and utilization of fungal hosts. Specifically, we will give examples of the different strategies used to produce various proteins in fungi at Genencor, and a perspective on improvements that are in progress in other laboratories. We will also address the regulation of expression and secretion of foreign proteins, and discuss the effect of fermentation technology on protein production in fungi.

Introduction

As exemplified in this book, many hosts can be utilized for the expression of recombinant proteins, ranging from prokaryotes, such as *Escherichia coli* and *Bacillus subtilis*, to eukaryotes, such as yeast (*e.g. Saccharomyces cereviseae* and *Pichia pastoris*), baculovirus-insect cells, cultured mammalian cells and even transgenic animals and plants. The choice of a host depends on factors such as the native source of the desired protein, the need for secretion versus intracellular accumulation, the susceptibility of the desired protein to host proteases, the final application for the protein product, and the familiarity of the laboratory with the expression system. For a particular protein of interest, it is often not possible to predict in advance which host is most appropriate, although some judgment can be made. Often, a host system that is closely related to the source organism of the desired protein may be most appropriate and, certainly, filamentous fungi are the first choice for production of filamentous fungal proteins. However, it is not possible to use a closely related host and source organism if, for example, one is trying to take advantage of the lower cost of production offered by microbial systems for the manufacture of mammalian proteins.

Filamentous fungi have a long history of safe industrial cultivation for the efficient production of a diverse set of compounds, including proteins, organic acids and secondary metabolites such as antibiotics, carotenoids, and vitamins. Both endogenous (native) and exogenous (foreign) enzymes are currently manufactured commercially using fungal host systems. In addition to food ingredients, fungal biomass itself, *e.g.* various mushrooms and Quorn™ (*Fusarium venenatum*), is sold directly for consumption by the food industry.

Some of the relevant features and perceived advantages of using filamentous fungal species as commercial production hosts are as follows. They can be cultivated in simple, inexpensive media, and naturally secrete a variety of native proteins into the medium in very large amounts. Industrial-scale fermentation and downstream recovery processes have been developed. Several proteins produced in filamentous fungi have been designated as GRAS (generally recognized as safe) and are considered appropriate for human consumption. Molecular biology tools are available for the rapid construction of production strains. The secretory system of fungi has similarities with that of higher eukaryotes. Homologs of many structural proteins, chaperones and foldases are present in fungi. As with higher eukaryotes, fungi glycosylate most secreted proteins, although the structure of the glycan is of the high-mannose type rather than the complex type produced by mammalian or plant cells.

There have been a number of recent excellent reviews covering various aspects of the expression and secretion of recombinant proteins in filamentous fungi (Punt *et al.*, 2002; Bergquist *et al.*, 2002; Archer, 2000; Maras *et al.*, 1999b; Archer and Peberdy, 1997; Gouka *et al.*, 1997a). Punt *et al.* (2002) discuss the heterologous expression of the human cytokine interleukin 6 and a fungal manganese peroxidase in *Aspergillus niger*, while Bergquist *et al.* (2002) focus on the aspects affecting expression of thermophilic xylanases in *Trichoderma reesei*. In contrast, Archer (2000) concentrates on the issues and recent advances in understanding the factors affecting the production of fungal enzymes and non-protein ingredients in food. Maras *et al.* (1999b) focus on what is known about glycosylation in filamentous fungi; and Archer and Peberdy (1997) and Gouka *et al.* (1997a) on the molecular basis for control of protein expression, secretion and structure.

In this chapter, we give examples of the different strategies used at Genencor to produce various proteins in fungi and comment on improvements that are in progress in other laboratories. To appreciate the highly efficient, natural secretion machines that filamentous fungi are, we first discuss the current status of our ability to manipulate their genetics and review the current understanding of the regulation of protein production at the molecular level. We recapitulate some of the highlights featured in the aforementioned reviews and emphasize some recent advances that have significantly improved the understanding and utilization of fungal hosts. We also discuss the effect of fermentation technology on protein production in fungi. The potential of using filamentous fungi to produce therapeutic proteins is also addressed. Finally, we will highlight how advances in the areas of genomics, metabolomics and proteomics have further improved our understanding of foreign protein production in filamentous fungi.

Genetic Manipulation

Work on secretion of recombinant proteins by *Aspergillus nidulans* began immediately after DNA-mediated transformation methods were developed for this model organism. Soon afterwards, it became possible to perform similar studies with those species (*e.g. A. niger, A. oryzae* and *Trichoderma reesei*) used in industrial fermentation processes to produce high titers of native secreted proteins (*e.g.* glucoamylase, alpha-amylases and cellulases). More recently, several other species have been tested as alternatives including *Penicillium* (Graessle *et al.*, 1997; Belshaw *et al.*, 2002), *Fusarium* (Royer *et al.*, 1995), and those reviewed in Radzio and Kück (1997). Research into the basic biology of fungi is generally pursued with the better-characterized species *A. nidulans* and *Neurospora crassa*. Sophisticated genetic tools are available in these species but this is often not the case for industrial fungal strains. Considerable research on secretion and production of foreign proteins has also been conducted with wild-type isolates of *A. niger*. In contrast, the strains employed by industrial laboratories are typically proprietary ones that have been developed over many years and with multiple rounds of mutagenesis and screening to increase productivity of secreted proteins in submerged culture. These may provide an advantage in terms of secreted protein titers compared to strains used in academic settings.

Transformation Techniques

Integrating vectors are commonly used for transformation of filamentous fungi, although autonomously replicating vectors have been developed for some species. For example, vectors that incorporate the AMA1 sequence from *A. nidulans* will replicate autonomously in a variety of *Aspergilli* and closely related species (Aleksenko and Clutterbuck, 1997). Replicating vectors generally allow a much higher frequency of transformation than integrating vectors but are not stably maintained and are lost during growth in the absence of selective pressure. Integrating vectors may recombine at either homologous or non-homologous sites in the genome (reviewed in Timberlake, 1991), with the relative frequency dependent on the design of the plasmid and the fungal species. In general, non-homologous integration of a vector that contains a region of homology with the genome is more prevalent than homologous integration. However, the fact that homologous integration occurs at some observable frequency enables directed gene disruption and modification to be performed.

T. reesei can be transformed with foreign or native genes by a variety of techniques. Perhaps the oldest method involves polyethylene glycol (PEG) transformation of protoplasts (Case *et al.*, 1979). At Genencor we have successfully used this technique for many years to generate *T. reesei* production strains. However, PEG transformation has some drawbacks that make it less attractive compared to newly developed techniques. For example, the initial step of creating protoplasts typically requires optimization of the treatment conditions for each different host to be transformed. In addition, successfully transformed protoplasts can be multi-nucleate, and require purification of the uni-nucleate spores.

One newer fungal transformation technique involves using *Agrobacterium tumefaciens* to introduce DNA into fungal hyphae or spores (Gouka *et al.*, 1999). This method, which uses *A. tumefaciens* as an intermediate plasmid carrier, is simpler and yields many more stable transformants of *T. reesei* compared to PEG transformation (E. Bodie, unpublished). The only drawback of this technique is that some *A. tumefaciens* flanking DNA is introduced upon integration. In addition to introducing genes to be expressed, *A. tumefaciens* transformation may be useful for other applications that require high transformation efficiencies, such as insertional mutagenesis.

Another transformation technique involves bombarding fungal spores with tungsten particles coated with the desired DNA for insertion (Hazell *et al.*, 2000; Lorito *et al.*, 1993). This method, called biolistic transformation, has helped to improve the throughput of fungal transformations at Genencor. The ability to transform uni-nucleate spores with this method is an advantage relative to PEG transformation. Biolistic transformation also produces a high percentage of stable clones with the desired gene insertion (B. Bower, unpublished).

For research purposes, DNA-mediated transformation frequencies of some fungal species like *A. nidulans* and *A. niger* with replicating vectors can be sufficiently high to allow screening of expression libraries (*e.g.* cDNA libraries or genomic libraries) for gene cloning (Gems *et al.*, 1994; Verdoes *et al.*, 1994). It is also possible to create libraries of mutations in a gene encoding a secreted protein of interest. When coupled with growth of the fungi and protein production in microtiter plates, this allows high-throughput screening to be performed. However, it is important to take care that post-translational modifications do not confound these studies. For example, it is likely that the glycosylation pattern and extent will depend on the host. This is true even if a protein from one fungus is produced in another closely related fungus, and it is important to check that these differences do not significantly affect the properties of the desired protein (glycosylation and product quality are discussed in more detail in a subsequent section). Although handling filamentous fungi in a microtiter format is not as easy as with unicellular organisms such as bacteria or yeasts, it is certainly an appropriate approach when suitable assays are available. For example, Genencor has produced laccases from a variety of fungi (both basidiomycetes and ascomycetes) in *A. niger* and screened mutant libraries to identify variants with improved properties (H. Wang, unpublished).

Genetic Markers

There are several different selectable markers available for fungal transformation (Mach and Zeilinger, 1998). Acetamidase (*amdS* from *A. nidulans*) and bacterial hygromycin B phosphotransferase (*hph*) are two of the most common dominant selectable markers. The *pyr4* or *pyrG* genes (which encode orotidine monophosphate decarboxylase in the uridine biosynthetic pathway in *Trichoderma* and *Aspergillus*, respectively) represent a commonly used marker for complementation of uridine auxotrophs. The dominant selectable markers have the benefit of not requiring an auxotrophic mutant strain as recipient. Conveniently, both the *amdS* and *pyr4/G* genes can also be selected against (by selecting for resistance to fluoroacetate or fluoroorotic acid, respectively) and eliminated from a transformed strain so that they can be used again in subsequent transformations.

Promoter Selection

A number of promoters, either constitutive or regulated, have been widely used to drive expression of genes of interest in filamentous fungi. These promoters are suitable for many types of research projects in *A. nidulans*, *N. crassa* and *A. niger*. For constitutive expression a promoter regulating a gene in the glycolysis pathway is commonly employed. An example is the *Aspergillus gpdA* (glyceraldehyde phosphate dehydrogenase) promoter that is incorporated into the frequently used vector pAN52 and its derivatives (Punt *et al.*, 1991b).

An example of a regulated promoter is the *A. nidulans alcA* (alcohol dehydrogenase) promoter. This promoter has found favor because of its high level of expression and well-characterized and relatively simple mode of regulation (applications for this promoter are reviewed in Nikolaev *et al.*, 2002). Two regulatory proteins, CreA and AlcR, bind to the *alcA* promoter. CreA binding is involved in repression during growth on readily utilized carbon sources such as glucose, while AlcR binding is implicated in induction in the presence of ethanol or other compounds (*e.g.* threonine and ethylamine) that are metabolized through the co-inducer acetaldehyde. Expression from the *alcA* promoter is often induced after pre-growth on glucose by addition of ethanol (or another inducer) in the presence of low amounts of fructose. The induced level of expression from the *alcA* promoter can be increased by over-production or constitutive production of AlcR and derepression in the presence of glucose through removal of CreA binding sites in the promoters (Mathieu and Felenbok, 1994; Panozzo *et al.*, 1998). The *alcA/alcR* system has been adapted for *A. niger*, which does not have a native, inducible *alc* system (Nikolaev *et al.*, 2002), and even transferred to higher plants (Caddick *et al.*, 1998; Roslan *et al.*, 2001). Other commonly used promoters with well-characterized regulation are the *A. nidulans niaD-niiA* bi-directional promoter region and the *A. nidulans amdS* promoter (Punt *et al.*, 1991a; Davis *et al.*, 1993).

While not transferable to yeast, promoters from one filamentous ascomycete species will generally function in another. However, optimal expression is often not achieved. For large volume (commercial) production of a secreted protein in filamentous fungi, one must normally combine the use of an industrial species with a native promoter from a gene encoding a protein naturally secreted at high levels. Thus, combinations of *A. niger* and the *glaA* (glucoamylase) promoter, *T. reesei* and the *cbh1* (cellobiohydrolase I) promoter, or *A. oryzae* and one of the alpha-amylase promoters (*e.g. amyB*) are typically used. The regulation of these promoters is described in more detail below.

Regulation of Native Protein Expression

The cellulase enzymes secreted by *T. reesei* are composed of a complement of endo- and exoglucanases that act synergistically to degrade cellulose (a highly ordered polymer of β-1,4 linked glucose residues) to cellobiose (glycosyl-β-1,4-glucose). The most abundant component is the exoglucanase CBHI and this single protein accounts for as much as 50% of the total protein secreted by the organism under inducing conditions. Industrial strains of *T. reesei* can produce in excess of 35 g/L total secreted protein under optimized fermentation conditions (Durand *et al.*, 1988). Similarly, industrial strains of *A. niger* and *A. oryzae* are renowned for prolific secretion of glucoamylase and α-amylases.

Common themes have emerged in studies of the regulation of secreted cellulases, hemicellulases and amylases in *Trichoderma* and

Aspergillus strains (Tsukagoshi *et al.*, 2001). Production of the enzymes is subject to carbon catabolite repression, at least partly mediated by the *cre1*/*creA* gene. Low levels of extracellular enzymes may be produced either constitutively or when available soluble carbon is at a low level. As a result, soluble sugars are released from available insoluble substrates (*e.g.* starch or plant cell wall polymers). The presence of soluble sugars, or of transglycosylation products derived by the action of (for example) β-glucosidases or α-glucosidases, is responsible for the induction of the extracellular enzyme genes. For instance, transcription of the major cellulase genes of *T. reesei* is induced by the presence of cellulose, as well as that of a variety of disaccharides including cellobiose, lactose (galactosyl-β-1,4-glucose), and sophorose (glycosyl-β-1,2-glucose) (Ilmén *et al.*, 1997). Sophorose is by far the most potent inducer (Mandels *et al.*, 1962) and can be derived from the action of β-glucosidases on cellobiose or glucose (Vaheri *et al.*, 1979). Similarly, starch or maltose, isomaltose or other disaccharides released from it, directly or via transglycosylation, have been shown to induce the production of glucoamylases and α-amylases in *Aspergillus* strains.

In practice, protein production in liquid culture of *T. reesei* is accomplished by first growing the production strain in the presence of repressing levels of glucose followed by transfer to inducing conditions such as cellulose or low levels of lactose as sole carbon sources. In contrast, protein production using *A. niger* may be continuous throughout the culture using high levels of maltose or maltodextrin as carbon source. Alternatively, it is possible to minimize expression from the *glaA* promoter during initial growth of *A. niger* by using xylose as a sole carbon source and then inducing expression by switching to maltose (Fowler *et al.*, 1990).

Transcriptional regulation of cellulolytic gene expression in *T. reesei*, or of amylase expression in *Aspergillus* strains involves multiple different binding proteins and is only partially understood. In *Aspergilli*, many of these promoters include CCAAT elements to which a complex of at least three proteins (HapB, HapC and HapE) binds (Brakhage *et al.*, 1999). This may contribute to high expression levels by maintaining the chromatin in a transcriptionally active state (Narendja *et al.*, 1999).

In addition, there are binding sites for transcriptional activators and repressors that are more specific for a subset of the genes. For example, Ace1 and Ace2 are factors that bind to the *T. reesei cbh1* promoter and act, respectively, to repress or enhance production of cellulases and xylanases (Aro *et al.*, 2001; Aro *et al.*, 2003). Activators of amylase and xylanase expression in *Aspergillus* (AmyR and XynR, respectively) or of xylanase expression in *Trichoderma* (Xyr1) have also been described (reviewed in Mach and Zeilinger, 2003; Tsukagoshi *et al.*, 2001).

Fermentation Technology

While for research purposes sufficient quantities of a desired protein can often be obtained in flask cultures, commercial manufacture normally requires a significant increase in production scale. For example, a commercial scale vessel for the production of industrial enzymes is typically hundreds of thousands of liters. Issues such as mixing, heat and mass transfer, and the high liquid head pressure at that scale provide production challenges that cannot be easily simulated in typical laboratory scale vessels (scale-up considerations are reviewed in detail in Reisman, 1993). Due to the cost of production equipment at the commercial scale, volumetric productivity (the amount of product produced per unit volume per hour) is a critical determinant of the economics of a manufacturing process. To maximize production of a desired protein product, it is important to optimize the concentration of viable, productive biomass.

Filamentous fungi are typically characterized by their long, branched mycelia in submerged culture, but cultivation conditions can affect their morphology (the spatial structure of filamentous organisms) and dictate the productivity of the culture (*e.g.* the ability to produce a protein product). Therefore, it is critical to characterize the effect of morphology on fungal physiology during fermentation (for a detailed review of the parameters affecting fungal morphology see Pazouki and Panda, 2000). Morphology of filamentous organisms can be defined on both a macroscopic and a microscopic scale (McIntyre *et al.*, 2001). Macro-morphology reflects the overall broth morphology (pellets, clumps and/or freely dispersed mycelia) and directly affects production by influencing mixing and oxygen transfer in a fermentor. Micro-morphology is characterized by the overall branching frequency of the fungus, the thickness of the cell wall and other factors that influence protein or small molecule secretion.

In this section we discuss the effect of fungal morphology on fermentation and scale-up, and highlight some recent advances in fermentation technology, including developments in biomass concentration quantification and the use of a unique carbon substrate in cellulase manufacture. We also discuss the viability of solid-state culture for the commercial manufacture of proteins.

Scale-Up and Fungal Morphology

One of the most challenging issues concerning fungal fermentations is the high broth viscosity resulting from fungal morphology. In most cases, a more diffuse morphology is better for production, but in turn leads to higher viscosity, decreased oxygen transfer and mixing. Viscosity issues become even more important as engineers begin to scale a process from the laboratory into full-scale manufacture. Figure 1 shows how viscosity might vary versus shear rate for a typical fungal fermentation (similar to Ryu and Humphrey, 1972. Because the average shear rate, the physical stress experienced by the fungal mycelia due to the force exerted by rotation of the agitator blades, is dependent on fermentor configuration, the apparent viscosity (as measured at the average shear rate) varies with vessel size. This change in apparent viscosity also changes the $K_L a$ (oxygen mass transfer coefficient). Thus, as the size of the fermentor is increased, the apparent viscosity increases and the ability to aerate the broth is diminished. To avoid oxygen transfer limitations at manufacturing scale, it is advisable to operate laboratory equipment below its capabilities in a range that is more reflective of what can be achieved at larger scale.

Since oxygen limitation in fungal fermentations is a common issue, the ability to control fungal morphology can be advantageous. Attempts are also being made to explore the relationship between fungal

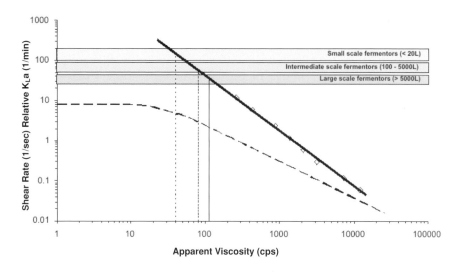

Figure 1. Dependence of shear rate and K_La on viscosity for a typical fungal fermentation. The solid line shows a linear relationship for viscosity vs. shear rate on a log-log plot. The dashed curve represents how K_La, and therefore oxygen transfer, decrease with increasing viscosity beyond a certain threshold. As the size of the fermentation vessel increases, the apparent viscosity increases due to decreasing shear rate as represented by the horizontal band. This leads to a lower maximal oxygen transfer rate, which can cause difficulties in scaling fungal fermentations.

morphology and production (see Schügerl *et al.* (1998) for an excellent review on the influence of process parameters on morphology and enzyme production in *Aspergillus*). Recently, Bhargava *et al.* (2003) observed a decrease in broth viscosity of an *A. oryzae* culture when glucose was pulsed rather than fed continuously. Feeding the same total amount of glucose either continuously or by pulsing in 300 second cycles resulted in no apparent change in biomass concentration or cellular metabolism as measured by oxygen uptake rate or base consumption for pH control. However, the size (average projected area) of the pulse-fed mycelia was half that observed when feeding continuously leading to a concomitant reduction in viscosity. There was no significant difference in the amount of extracellular protein secreted by either method, although there was some increase in titer variability in the pulse-fed fermentations. In theory, this finding may allow one to increase the quantity of productive biomass in a pulse-fed relative to a continuously fed fermentation, as lower viscosity should increase the oxygen transfer capability of the broth. Assuming equivalent specific productivity (the amount of production per unit biomass per hour), an increased biomass concentration would result in an overall increase in the volumetric productivity of the fermentation.

Strain morphology can also be manipulated through genetic modification. For example, Muller *et al.* (2002) altered the micro-morphology of *A. oryzae* by manipulating chitin synthesis. Disruption of chitin synthase B (encoded by the *chsB* gene) decreased the tip extension rate 88% and increased the branching frequency 188%, whereas the absence of chitin synthesis myosin A (encoded by *csmA*) had a more moderate effect with 25% decrease in both tip extension and branching. Similarly, Bocking *et al.* (1999) demonstrated a linear relationship between the amount of branching and culture viscosity in mutated strains of *A. oryzae*. This led to an overall increase of 12% in volumetric amylase productivity because the lower viscosity of the mutant allowed for higher cell density fermentation.

Biomass Quantitation

To optimize the amount of productive biomass in a protein production process, one must quantitate it. Traditional biomass quantitation techniques such as dry cell weight (DCW) or turbidity provide no information on the quality, activity or viability of the microorganisms, and therefore cannot be used alone to accurately characterize the specific performance (productivity) of the biomass in a bioreactor. Determination of fungal viability by culturability (dilution and incubation) is particularly difficult due to the inherent multicellular morphology of filamentous fungi. Genencor recently evaluated a capacitance probe (Aber Instruments Ltd., Aberystwyth, UK) for online measurement of cell concentration and viability. This probe only measures viable cells and can give some indication of problems if active biomass begins to decline. Because the probe detects the capacitance built across an intact cell membrane and not optically, it can be used in media with suspended solids, which have become more common as agricultural waste products are added to the fermentation media as nutrient sources.

In another effort to effectively measure the concentration of active biomass, Hassan *et al.* (2002) demonstrated that fluorescence-based image analysis could be successfully applied for the quantitative, rapid and accurate assessment of filamentous fungal viability. This study correlated culturability and growth rate of *Trichoderma harzianum* with the percentage of active biomass as measured by the percent of fluorescein diacetate (FDA) stained hyphal area. Results showed that estimated viable biomass in the fermentor was strongly influenced by process conditions (*e.g.* whether the fermentation was run in batch or fed-batch mode). This relatively simple method should allow a fermentation development engineer to assess the effect of process conditions on filamentous fungal viability and design the process to maximize the amount of viable (and productive) biomass in the fermentor.

Substrate Utlilization

In large-scale commercial protein manufacture one must consider the cost and availability of the raw materials, especially that of the carbon source. This issue is exemplified by the current commercial manufacture of cellulase in *T. reesei*, which commonly utilizes lactose as both the carbon substrate and the inducer of cellulase expression. While lactose is generally available, the current supply is insufficient to meet the projected demand for the manufacture of cellulase needed to convert biomass to inexpensive sugars for the production of fuel ethanol (Hettenhaus and Glassner, 1997; see: http://ceassist.com/producti.htm). In contrast, glucose is plentiful, costs less than half as much per kg of dry weight, but normally represses cellulase production. While cellulose is technically another alternative to lactose, the practical handling considerations of dealing with a heterogeneous feed make it less desirable. At Genencor we have recently developed an economical method that allows the use of glucose in combination with the potent inducer sophorose for the production of whole cellulase by *T. reesei*.

Currently, only chemically synthesized sophorose is commercially available in very small quantities costing thousands of dollars per gram.

To assess whether the strong cellulase induction by sophorose could overcome the repression by glucose when both are combined as the carbon feed source for commercial cellulase production, we developed an inexpensive and efficient way to produce a glucose solution containing a significant concentration of sophorose. The addition of a proprietary whole cellulase mixture produced by *T. reesei* to a concentrated glucose solution utilizes the natural transglycosylation activity of the organism's native β-glucosidases for the conversion of glucose to sophorose. In a 60% glucose solution, the addition of 10 g/L whole cellulase produces about 15 g/L sophorose when incubated at 65°C for 3 days (Kelley *et al.*, 2002, 2003). There is no need to purify the β-glucosidase from the whole cellulase product for the conversion reaction, nor the sophorose from the glucose for its utilization in cellulase production. We have found that when fed at the optimum carbon limiting feed flow rate, the combined glucose/sophorose solution resulting from the above reaction can generate at least equivalent specific cellulase productivity as the current industry standard lactose. In addition to the more abundant supply, better availability and cheaper cost of glucose relative to lactose, the higher feed concentration results in less dilution in the fermentor, which can translate to an increased volumetric productivity. Indeed, this transglycosylated glucose feed has been demonstrated to give at least equivalent volumetric productivity of a protein-engineered xylanase variant over-expressed using the *cbh1* promoter in *T. reesei* (H.J. Meerman, unpublished). It is expected that other heterologous genes expressed using the *cbh1* promoter will also be successfully produced in *T. reesei* using this combined carbon source.

Solid-State Fermentation

While the previous discussion in this chapter has been geared towards submerged fermentation, many of the techniques to improve protein expression could be readily applied to solid-state fermentation (SSF) as well. The benefits of SSF for fungal hosts are frequently debated. Many authors claim that the higher product titers in SSF make it clearly superior to submerged culturing (Robinson *et al.*, 2001). While this may be true concerning product titers, the authors making these assertions often neglect the need to extract the product from the solid growth media which usually dilutes it to concentrations at or below those produced in submerged culture. In addition, mass transfer constraints limit the growth and volumetric productivity achievable in SSF relative to high cell density submerged cultivation (Schügerl *et al.*, 1998). A recent review gives a very objective and comprehensive analysis of the current state of SSF (Mitchell *et al.*, 2000).

SSF, although utilized in the production of enzymes and fermented foods for many years, is still in the early stages of development and has some substantial hurdles to overcome. Control of operational variables such as temperature, humidity and pH remain very difficult in SSF (Mitchell *et al.*, 2000). Mixing is needed to avoid development of temperature and substrate gradients, but may be inadvisable since some microorganisms are sensitive to this type of shear (te Biesebeke *et al.*, 2002; Nagel *et al.*, 2001). Water evaporation, which is used to remove excess heat, must be balanced to avoid drying the solid media too much, as this is detrimental to cell viability. Changes in pH can be made relatively easily by addition of NH_3 gas, but measuring the pH *in situ* can be problematic. It is often measured offline by dilution of the solid media in water. Measurement of biomass concentration in SSF is another difficult task. Attempts are usually made either to separate the biomass from the solid substrate and measure its metabolic activity, or to infer its concentration by measuring a specific biomass component (Mitchell *et al.*, 2000).

As mentioned earlier, recovery of the desired product from SSF is a poorly studied process. While different laboratory techniques have been used, most of these would not economically scale for an industrial process. Both the percentage of the product that is extracted as well as its concentration after extraction will affect the economics. Recovery of enzyme products by direct pressing of the solids shows the most promise for recovery of SSF material (Roussos *et al.*, 1992). With a press/wash/press protocol, Roussos *et al.*, (1992) were able to recover 85-95% of a cellulase with a concentration ratio (fresh weight solids/ final leachate weight) of 1.2. The effectiveness of this method may vary depending upon how tightly the solid media binds water. Despite the effectiveness shown by the authors, no further scale-up development has occurred for this technique.

SSF may find an initial niche with processes that do not need further product purification, such as delignified wheat straw used as a feed supplement or using SSF to remove toxic compounds from feed (Mitchell *et al.*, 2000). One could even imagine a process where a produced enzyme is immobilized to the culture's growth substrate and sold as a product. Even with the current reluctance to adopt SSF there will be a future need for solid-state technology, since some products are not produced well in submerged cultures.

Recombinant Protein Production Strategies

Many published papers describe the production of foreign proteins in filamentous fungi for different purposes. The level of protein expression required is dictated by the use to which the protein will be put. Protein may be produced for research purposes (*e.g.* for functional or crystallographic studies) and mutated forms of a protein may be expressed in protein engineering projects and to investigate structure-function relationships (*e.g.* Cao *et al.*, 2000). At the other extreme, recombinant proteins can be produced for commercial manufacture. For example, several enzymes native to other fungi have been produced in *A. niger*, *A. oryzae* or *T. reesei* for commercial use in the textile, pulp and paper, detergent or animal feed industries.

As previously mentioned, it is desirable to use a strong native promoter to maximize the production of a recombinant protein. Codon optimization has been shown to enhance the translation of some foreign proteins in fungi (*e.g.* the production of a xylanase from a thermophilic bacterium in *T. reesei* by Te'o *et al.*, 2000). To facilitate secretion, the minimal requirement is an N-terminal signal sequence that is removed as the protein passes through the membrane of the endoplasmic reticulum (ER). Although this may suffice for the efficient secretion of some proteins, a strategy in which the target protein is fused to the C-terminus of an endogenous secreted polypeptide may be more appropriate for other proteins. To minimize degradation of the extracellular protein, some of the proteases naturally produced by the host strain can be removed by gene manipulation or otherwise inactivated (van den Hombergh *et al.*, 1997b). The exact details of these strategies depend on the protein of interest and the identity of the host and examples are provided below.

Proteins derived from closely related fungal species are generally produced efficiently in *Aspergillus* and *Trichoderma*. In these cases, high titers of secreted protein can be obtained, for example, by fusion of the relevant gene (including the region encoding its native secretion signal sequence) to the *glaA* or *cbh1* promoters and expression in *A. niger* or *T. reesei* respectively. For example, at Genencor we have produced several different fungal endoglucanases in *A. niger* (Goedegebuur *et al.*, 2002) or laccases from various related fungi in *T. reesei* (E. Bodie, H.J. Meerman, and H. Wang, unpublished). Other examples include production of a *Rhizomucor miehei* aspartic protease in *A. niger* (Ward and Kodama, 1991), a *Penicillium* aspartic protease in *A. niger* (Cao *et al.*, 2000) and a glucoamylase from *Hormoconis resinae* in *T. reesei* (Joutsjoki *et al.*, 1993).

It is possible to integrate an expression cassette for the gene of interest in a region of the genome that is known to be highly transcribed (*e.g.* the *cbh1* locus of *T. reesei*; Joutsjoki, 1994). However, transformants that produce high titers of the desired protein can be isolated readily by allowing random integration of the expression vector and screening transformants to identify the best producers. Unlike those from yeast or non-fungal genes, the introns present in genes from other filamentous fungi can generally be processed in *Aspergillus* or *Trichoderma* species (*e.g.* Joutsjoki, 1994).

The titer of non-fungal (*e.g.* mammalian or bacterial) proteins secreted by filamentous fungi is generally significantly lower than for endogenous proteins or for proteins from other fungi, but they can be improved, often by more than ten fold, if produced as a fusion with a native secreted protein. Examples of native fungal proteins used as fusion partners are glucoamylase in *A. niger* and cellobiohydrolase I in *T. reesei*. Both of these proteins have two separately folded domains, an N-terminal catalytic domain and a C-terminal substrate-binding domain, connected by an extended linker region. Expression vectors can be constructed to replace the coding region for the binding domain with that of the protein of interest (see Figure 2). To produce an

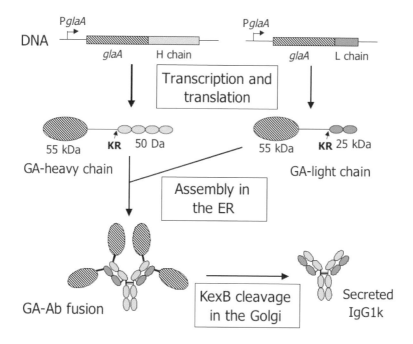

Figure 2. Schematic of our antibody (Ab) expression strategy in *A. niger*. Each antibody light (L) chain and heavy (H) chain is expressed as a separate protein fusion with *A. niger* glucoamylase (GA) using its native *glaA* promoter. The four-chain antibody molecule is expected to fold and assemble in the lumen of the endoplasmic reticulum (ER). Both L and H chains must fold correctly with appropriate intrachain (2 and 4, respectively in each L and H chain) and interchain (4) disulfide formation. KexB cleavage sites were included to allow release of the antibody chains from GA as it traverses the Golgi apparatus.

authentic N-terminus on a desired protein, this production strategy requires a means to cleave the fusion protein at the correct position. Often this is achieved by inclusion of a protease cleavage site immediately before the N-terminus of the desired protein. The KexB protease in *A. niger* (known as Kex2 in *T. reesei*) is a kexin-like serine proteinase thought to reside in the Golgi apparatus (Jalving *et al.*, 2000). It cleaves proteins preferentially at the C-terminal side of the dibasic amino acid sequence Lys-Arg in regions of the folded protein that are exposed or accessible. There are many published examples of foreign proteins produced in this way (reviewed in Gouka *et al.*, 1997a). It is suggested that fusion to the C-terminus of an endogenous secreted protein may overcome bottlenecks in translation or translocation of the non-fungal protein and may improve protein folding in the ER, thus reducing degradation (Nyyssönen *et al.*, 1995; Gouka *et al.*, 1997a, 1997b). The extent and fidelity of KexB cleavage seems to depend on the sequence and structure of the N-terminal part of the desired protein. In fact, Spencer *et al.* (1998) demonstrated that the fidelity of cleavage at the KexB site in *A. niger* could be improved by changing the adjacent amino acid sequence.

Even if folding and secretion are efficient, production of a candidate protein can be limited by host proteolysis. In addition to the aforementioned KexB protease involved in proteolytic processing during secretion, there are many other known proteases in *A. niger* (reviewed by van den Hombergh *et al.*, 1997b). One general method to limit proteolysis is to utilize host strains deficient in one or more proteases. Berka *et al.* (1990) demonstrated that deletion of the *Aspergillus awamori* (since renamed *A. niger* var *awamori*) *pepA* gene encoding the major secreted aspartic proteinase reduced the total secreted proteinase activity significantly. van den Hombergh *et al.* (1997a) constructed a set of *A. niger* strains combining multiple protease gene disruptions. More recently, Moralejo *et al.* (2002) demonstrated the use of either antisense RNA or targeted gene disruption to increase the production of thaumatin in *A. awamori*. Media and cultivation conditions can also modulate protease expression.

As an example, we developed strains of *A. niger* var *awamori* in our laboratories for the commercial scale production of bovine chymosin for cheese manufacture. To obtain commercially sufficient expression, a strain was used that had a deletion in the gene encoding the major secreted aspartic proteinase (Berka *et al.*, 1990). The protein was produced as a secreted glucoamylase-prochymosin fusion protein that was cleaved by the proteolytic action of chymosin itself to release authentic mature chymosin (Ward *et al.*, 1990). Once a production strain was obtained, it was subjected to several rounds of mutagenesis and screening to increase productivity (Dunn-Coleman *et al.*, 1991). Ultimately, secreted titers of over 1 g/l were obtained and the purified chymosin was approved for use in human food. Subsequently, the improved strain was cured of the expression vector and retransformation showed that this strain (dgr246 P2) was also improved for the production of other foreign proteins (Ward *et al.*, 1993).

At Genencor we have successfully expressed several other heterologous genes using this fusion protein method, including the bacterial thermophilic endoglucanases E5 (cel5A) from *Thermobifida fusca* and cel12A from the extremophile *Actinomycete sp.* 11AG8 (M. Ward, E. Nyyssönen and H. Olivares, unpublished).

Therapeutic Proteins in Filamentous Fungi

Mammalian proteins, some of which could be potential therapeutics in humans, have been produced in filamentous fungi. Despite these initial studies, there has been no example to date of an approved therapeutic protein manufactured by expression in filamentous fungi. However, there is precedence for using fungal hosts for drug production, as several non-protein drugs are produced in different species of filamentous fungi (*e.g.* lovastatin from *A. terreus*, cyclosporin A from *Tolypocladium inflatum*, and antibiotics from *Penicillium chrysogenum* and *Acremonium chrysogenum*). In fact, some companies are manufacturing proteins in fungi for use in the pharmaceutical industry. Neose Technologies, Inc. has constructed a production facility to produce clinical and commercial-scale quantities of enzymes from *Aspergillus* under FDA current Good Manufacturing Practices (cGMP) guidelines. Agennix, in collaboration with DSM Pharma Products, has announced the opening of a manufacturing facility that includes a 35,000 L fermentor for the production of lactoferrin under cGMP guidelines for further clinical trials and commercialization.

Human lactoferrin has been produced in *A. niger* at titers of more than 5 g/L (Ward *et al.*, 1995; Sun *et al.*, 1999) using the glucoamylase fusion strategy with cleavage by KexB and employing several rounds of mutagenesis and screening for increased strain productivity. This protein is a broad spectrum antibiotic and is involved in regulating various immuno-stimulatory cytokines. Clinical trials are underway for indications in oncology, asthma and wound healing.

In a collaboration between Cytel Corporation (now part of Neose) and Genencor, the potential to produce mammalian glycosyltransferases in *A. niger* was explored. These enzymes can be used to synthesize complex carbohydrates or to modify the glycans attached to proteins after secretion. Glycosyltransferases are normally localized in the ER or Golgi apparatus of mammalian cells and possess an N-terminal secretion signal and membrane-spanning anchor region. To produce soluble secreted forms of the enzymes, an expression vector was designed to synthesize a truncated form of the transferase lacking signal sequence and membrane anchor region. This was fused to the C-terminus of the glucoamylase catalytic core and linker region after a KexB cleavage site. Although several host species have been tested, it has proven difficult to produce secreted glycosyltransferases at high titers. *A. niger* was shown to be a suitable host for production of several of these enzymes. For example, rat alpha 2,3 sialyltransferase (Wen *et al.*, 1992) was produced at titers of more than 1,000 U/L (>50 mg/L), which is a significant improvement over the titers obtained using mammalian cell culture or the baculovirus system. Several different mammalian fucosyltransferases have also been successfully produced in this way (for example, see Murray *et al.*, 1996).

Antibodies have many potential biotechnological applications (*e.g.* as research tools, in diagnostics or as affinity ligands in purification). There are also numerous monoclonal antibodies in development as potential therapeutics and it is desirable to reduce the cost of manufacture of these proteins as much as possible. For some therapies full-length antibodies are required, whereas antibody fragments may be suitable in other cases. Typically, full-length antibodies are manufactured in mammalian cell culture whereas antibody fragments (*e.g.* ScFv or Fab fragments) can be produced in microbial systems such as *E. coli* or *Pichia*. A number of alternative hosts are being investigated and two groups have tested production of antibody fragments in filamentous fungi. Frenken *et al.* (1998) produced a single chain antibody fragment as a glucoamylase fusion protein in *A. niger*. Nyyssönen *et al.* (1993) demonstrated that an antibody Fab fragment could assemble and be secreted in *T. reesei*. In the latter case, the truncated heavy chain of the antibody was produced as a fusion with the CBHI catalytic core and linker region whereas the light chain was produced in a non-fused form by direct expression from the *cbh1* promoter. A Kex2 cleavage site was designed to allow separation of the heavy chain from CBHI. However, analysis of the N-terminus of the released heavy chain demonstrated that cleavage had not occurred at the expected location, but rather at a position towards the N-terminus within the linker region of CBHI. Thus, a native proteinase may have cleaved rather than Kex2. Titers of secreted and assembled Fab of up to 150 mg/L were obtained in *T. reesei*.

Production of assembled antibodies involves an additional level of complexity compared to the production of a monomeric protein. Each antibody light (L) chain and heavy (H) chain must fold correctly with appropriate intrachain disulphide formation (two disulfides in each L chain and four in each H chain). In addition, two light chains and two heavy chains must assemble and form four interchain disulfides. Until now, there were no examples of production of full-length antibodies in filamentous fungi. While there are a few published examples of expression in other microbial systems, the amounts of antibody secreted have been low (Horwitz *et al.*, 1988), although Simmons *et al.* (2002) recently reported accumulation of 150 mg/L assembled antibody in the periplasm of *E. coli*. We recently performed a study to determine the feasibility of producing full-length IgG1 in *A. niger*. The schematic of the expression strategy is shown in Figure 2. We reasoned that maximum titers of secreted antibody would require expression of both L and H chains as fusion proteins with the *A. niger* glucoamylase. We included KexB cleavage sites to allow release of the antibody chains. The four-chain antibody molecule is expected to assemble in the lumen of the ER immediately after translocation across the ER membrane. However, cleavage by KexB would not be expected until later in the secretion process as the assembled antibody traversed the Golgi apparatus. This means that assembly of the antibody is even more complicated because glucoamylase is attached to the N-terminus of the L and H chains during the assembly process and the total size of the assembled antibody plus glucoamylase would be approximately 350-kDa. This assembly does occur efficiently and titers of up to 0.9 g/L of secreted antibody have been achieved with a humanized IgG1 (Ward *et al.*, 2003). The antibody has been purified from culture supernatant and has been shown to have similar affinity as antibody produced by mammalian cells. As has been seen previously for some other fusion proteins (Spencer *et al.*, 1998), cleavage by KexB was not 100% efficient. This may be due to the secondary structure of the L and H chain proximal to the cleavage site interfering with the KexB protease activity. The result was that some secreted assembled antibody retained a fused glucoamylase molecule on one or more chains. In addition, when cleavage occurred, the resulting N-terminus of the antibody chains was heterogeneous because cleavage did not always occur at the expected site. It was possible to achieve close to 100% cleavage at the desired position (immediately after the KR residues) by addition of three glycine residues at the N-terminus of both antibody chains. The three glycines remained attached to the N-terminus of the H and L chains of the antibody, but the affinity of the resulting molecule was unaffected.

There is a single N-linked glycosylation site on each H chain of IgG1. Analysis has shown that some of the H chains had a glycan occupying this site after production in *A. niger* while some did not. Thus, it is likely that some assembled antibody would contain one non-glycosylated and one glycosylated heavy chain and some antibodies may be completely non-glycosylated. Analysis of the glycan attached to the H chains in *A. niger* showed it to be of the high mannose-type but with some galactose, possibly in the form of terminal galactofuranose residues similar to that reported previously for other fungal proteins (Maras *et al.*, 1999a; Wallis *et al.*, 2001). There was heterogeneity in the size of the glycans, but the dominant form had two N-acetylglucosamine and nine hexose (mannose or galactose) residues (Figure 3).

For pharmaceutical use it would be beneficial to produce proteins with a reproducible pattern of glycosylation and a minimal amount of heterogeneity in size and sugar composition. Glycosylation can affect protein function and in some cases it is desirable to produce a protein with a human-like complex glycosylation pattern as opposed to the high-mannose type of fungal glycosylation. Significant progress is being made to modify fungal glycans and make them more human-like. Both *in vitro* and *in vivo* methods for modifying glycans are being tested. It has been shown that some of the glycan on the native CBHI

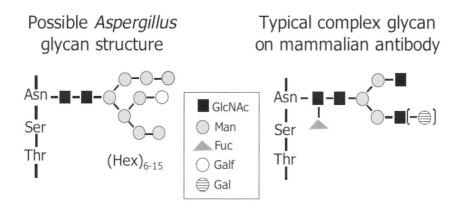

Possible *Aspergillus* glycan structure

Typical complex glycan on mammalian antibody

(Hex)$_{6-15}$

- ■ GlcNAc
- ○ Man
- ▲ Fuc
- ◯ Galf
- ⊜ Gal

Figure 3. Schematic representation of the possible glycan structure on Ab produced in *A. niger* compared to the typical glycan structure on an Ab produced in mammalian cell culture. Legend: Hex, hexose; GlcNAc, N-acetylglucosamine; Man, mannose; Fuc, fucose; Galf, galactofuranose; Gal, galactose.

enzyme secreted by *T. reesei* is in a form (Man$_5$GlcNAc$_2$) that can act as a substrate for mammalian glycosyltransferases. This enabled a hybrid mammalian-type glycan chain to be built *in vitro* using appropriate N-acetylglucosaminyltransferase, galactosyltransferase and sialyltransferase enzymes (Maras *et al.*, 1997). Expression of a human N-acetylglucosaminyltransferase in *T. reesei* allowed the first step in building the hybrid glycan to occur *in vivo* (Maras *et al.*, 1999a). However, many of the glycans on *T. reesei* proteins are larger than this core Man$_5$GlcNAc$_2$ form and are consequently not amenable to this modification. Therefore, it is desirable for the fungus to produce predominantly the Man$_5$GlcNAc$_2$ glycan that can be further modified *in vitro* or *in vivo*.

For *Pichia pastoris*, expression of a *T. reesei* α-1,2-mannosidase and localization to the ER causes a shift in the glycosylation pattern so that the majority of glycans are in the shorter Man$_5$GlcNAc$_2$ form (Callewaert *et al.*, 2001). A similar strategy is being tested in filamentous fungi (Maras *et al.*, 1999b).

Future Perspectives

Historically, the utilization of filamentous fungi for the production of foreign proteins has been driven by the characteristics of the proteins and their compatibility with a particular host. With advances in the areas of genomics, proteomics and systems biology, the use of filamentous fungi for protein expression is expanding rapidly. This section will highlight two areas of study that have further improved our understanding of foreign protein secretion in fungi.

Chaperone Proteins

It is tantalizing that native proteins can reach reported titers of more than 40 g/L in culture supernatants when mammalian proteins are produced at levels circa 1 g/L at best. While the limitation to production will vary from protein to protein, quality control in the ER is likely a major factor in preventing secretion of misfolded or improperly glycosylated proteins and directing them back into the cytoplasm for degradation by the proteosome. It is thought that foreign proteins are preferentially recognized by this system. Indeed, foreign protein secretion can elicit the Unfolded Protein Response (UPR), a specific response to the presence of misfolded proteins in the ER which leads to upregulation of genes encoding ER chaperones (*e.g.* BiP, calnexin) and foldases (*e.g.* protein disulfide isomerases and peptidyl-prolyl isomerases).

In light of the role played by chaperones in protein secretion, there have been a number of attempts to improve production of foreign secreted protein by overexpression of individual native chaperones or foldases (reviewed by Conesa *et al.*, 2002). Despite encouraging results from similar work in *Saccharomyces cerevisiae*, these attempts did not meet with much success in fungi, although improved secretion was obtained in a few cases. For example, calnexin overexpression improved production of *Phanerochaete* manganese peroxidase in *A. niger,* whereas BiP overexpression had the opposite result (Conesa *et al.*, 2002). Moralejo *et al.* (2001) demonstrated that increasing the levels of PdiA in the ER by 2 to 4 fold improved plant thaumatin secretion by *A. awamori*. However, higher levels of PdiA were detrimental. It is now recognized that the results of these experiments are often protein and chaperone or foldase dependent. That is, for this strategy to work for a particular foreign protein, one needs to identify the correct foldase or chaperone to overexpress in combination with the target protein.

In yeast, a transcriptional activator, Hac1p, has been identified that is responsible for increased expression of ER foldases and chaperones involved in the folding of secreted proteins. *hac1* mRNA contains an unconventional intron that must be spliced for translation to produce active Hac1p. A membrane protein, Ire1p, which senses the level of unfolded protein within the ER is thought to excise this intron. Travers *et al.* (2000) used microarrays to identify the yeast genes for which mRNA levels varied in response to ER stress (such as treatment with 1,4-dithiothreitol [DTT]). More recently, Saloheimo *et al.* (2003) have demonstrated that homologs of Hac1 and Ire1 play a similar role in *Trichoderma* and *Aspergillus*. Genes encoding many of the ER foldases and chaperones have also been cloned from filamentous fungi (reviewed in Conesa *et al.*, 2001). The transcript levels of some of these genes have been shown to increase as a result of ER stress caused by chemicals and, in some cases, by the secretion of a foreign protein (Wang *et al.*, 2003; Saloheimo *et al.*, 1999; Wiebe *et al.*, 2001; Wang and Ward, 2000, Conesa *et al.*, 2002).

Overproduction of HacA in *A. niger* gives rise to a constitutive UPR and secretion of foreign proteins can be improved as a result (Valkonen *et al.*, 2003). Production levels of both a *Trametes versicolor* laccase and bovine preprochymosin could be improved by this method although it is very interesting to note that simultaneous production of native secreted proteins like α-amylase and β-glucosidase was significantly reduced in these strains. As we understand more about the secretion process and its regulation in filamentous fungi we hope to be able to manipulate strains in rational ways so that foreign proteins can be produced more efficiently.

Fungal Genomics

Among the fungi, *Saccharomyces cerevisiae* was the first free-living organism for which a complete genome sequence was available (Goffeau *et al.*, 1996). Functional genomics methods, (*e.g.* microarray technology for transcript profiling and high throughput gene deletion) have since been developed. The genomes of *Neurospora crassa* (Galagan *et al.*, 2003), *Magnaporthe grisea, Phanerochaete chrysosporium* and *Aspergillus nidulans* have been sequenced to high coverage and are publicly available, while sequencing of *Aspergillus fumigatus* is in progress. Of the industrially important species, the *A. niger* genome has been sequenced but is not currently publicly available. The *A. oryzae* genome is being sequenced by a consortium including the National Institute of Technology and Evaluation in Japan and is expected to become publicly available in 2003. A public effort to sequence *T. reesei* is planned in the U.S. by the Joint Genomes Institute funded by Department of Energy. Over the course of 2003, it is expected that the sequences of several more filamentous fungal genomes will become available, mainly as a result of the commitment by the Whitehead Institute at MIT to sequence a series of academically or economically important and taxonomically diverse fungi (for details on the Fungal Genome Initiative, see: http://www-genome.wi.mit.edu/seq/fgi/).

All of the above sequence data will provide an extremely valuable resource. Obviously, cloning individual genes that can be recognized on the basis of their sequence will be facilitated. In addition, comparative genomics will highlight similarities and differences in gene content and organization among different fungi that can be related to lifestyle and taxonomy. Microarrays can be designed to monitor transcript levels corresponding to all genes identified in the genome.

Even in the absence of complete genome sequences, it has been possible to initiate functional genomics studies with fungi of industrial importance. As an example, we will describe work with *T. reesei* that has been performed by Genencor in collaboration with the research group of Prof. Ralph Dean at North Carolina State University (NCSU). cDNA libraries were constructed from RNA isolated after growth under a wide variety of conditions, including cellulase and hemicellulase inducing conditions. At NCSU, 18,000 individual cDNA clones were sequenced with a single read from the 5' end. These expressed sequence tags (ESTs) were assembled into overlapping sequences to give 2,101 contigs and a remaining set of 3,030 singleton reads which did not overlap with any other sequences. This sampling of the expressed sequences of the genome could represent close to 50% of the predicted 10,000 genes. Searching this database has revealed several sequences encoding previously unrecognized *T. reesei* cellulase and hemicellulase genes despite the fact that this fungus has been studied for several decades as a cellulase-producing organism (Foreman *et al.*, 2003).

Sixty base pair unique oligonucleotides were designed based on EST sequences and synthesized in microarray format on glass slides (Hughes *et al.*, 2001). The microarrays were used to monitor changes in expression of all the corresponding genes. In this way, for example, genes that are co-regulated with cellulases under different conditions were identified (Foreman *et al.*, 2003). These genes may encode proteins directly involved in cellulose and hemicellulose degradation or in the synthesis and secretion of these enzymes.

In a similar study, Chambergo *et al.* (2002) generated an EST data set of 343 contigs and 808 singletons by sequencing the 5' ends of cDNA clones from a library made from RNA extracted after growth of *T. reesei* on glycerol. Under these growth conditions one would not expect to detect mRNA related to the secreted cellulases and hemicellulases, but instead a good representation of the central metabolism genes. Inserts from the library were amplified by PCR and spotted onto glass slides to produce microarrays. The purpose of the study was to monitor the gene expression profile during glucose exhaustion. These studies demonstrated that, at high glucose levels, gene expression is regulated to favor the tricarboxylic acid cycle over fermentation of pyruvate to, for example, ethanol as in yeast.

Methods are being developed that can be applied to monitor the proteome and metabolome of fungi. Metabolomics, the non-biased determination and analysis of the complete metabolite profile of an organism, is an emerging powerful tool to better understand microbial physiology and facilitate metabolic pathway engineering to increase the production of specific proteins and fine chemicals in industrial organisms. To assess whether metabolomics could be readily applied to improve protein secretion in filamentous fungi, Genencor collaborated with TNO Nutrition and Food Research (Zeist, Netherlands) to elucidate the metabolites involved in the regulation of cellulase production in *T. reesei*. The goal was to apply pattern recognition analysis to identify the metabolites correlating with cellulase induction in hyper-secreting variants of *T. reesei*. Samples of mycelia collected under inducing and non-inducing fermentation conditions were quenched to halt cellular metabolism. LC-MS and GC-MS methods were used to measure more than 260 metabolites, including all intracellular sugars, sugar phosphates, NAD(P)(H) and ADP/ATP in the cell (M.J. van der Werf, personal communication). Transcript profiling and proteomics analyses were done on selected samples to complement the metabolome data. The information generated from these studies will ultimately be combined with genomic and functional genomic data to provide an enormous body of knowledge covering molecular genetics, physiology and cell biology of the fungal cell. This, in turn, will undoubtedly be applied to improve protein production in fungi.

Concluding Remarks

Filamentous fungi have long been successfully used as commercial production hosts for a number of industrial proteins. The availability of tools for genetic manipulation, their high productivity, demonstrated safety, and ease of cultivation and recovery make fungi a natural choice to evaluate for production of a desired protein. As knowledge about their genetics and physiology increased, either empirically or through directed studies, their use has been more extensive. With more comprehensive understanding made possible through the availability of fungal genomes, we can further exploit their natural protein secretion capacity, particularly for generation of therapeutic proteins. Given the anticipated capacity constraints for therapeutic proteins currently in the pipeline, we believe that filamentous fungi should be aggressively pursued for the manufacture of many products.

Acknowledgements

The authors wish to thank Roopa Ghirnikar, Pamela Foreman and Tim Dodge for reading and giving feedback on the manuscript. This work was supported in part by a subcontract from The Office of Biomass Program, within the DOE Office of Energy Efficiency and Renewable Energy.

References

Aleksenko, A. and Clutterbuck, A.J. 1997. Autonomous plasmid replication in *Aspergillus nidulans*: AMA1 and MATE elements. Fungal Genet. Biol. 21: 373-387.

Archer, D.B. 2000. Filamentous fungi as microbial cell factories for food use. Curr. Opin. Biotechnol. 11: 478-483.

Archer, D.B., and Peberdy, J.F. 1997. The molecular biology of secreted enzyme production by fungi. Crit. Rev. Biotechnol. 17: 273-306.

Aro, N., Ilmén, M., Saloheimo, A., and Penttilä, M. 2003. ACEI of *Trichoderma reesei* is a repressor of cellulase and xylanase expression. Appl. Environ. Microbiol. 69: 56-65.

Aro, N., Saloheimo, A., Ilmén, M., and Penttilä, M. 2001. ACEII, a novel transcriptional activator involved in regulation of cellulase and xylanase genes of *Trichoderma reesei*. J. Biol. Chem. 276: 24309-24314.

Belshaw, N.J., Haigh, N.P., Fish, N.M., Archer, D.B., and Alcocer, M.J. 2002. Use of a histone H4 promoter to drive the expression of homologous and heterologous proteins by *Penicillium funiculosum*. Appl. Microbiol. Biotechnol. 60: 455-460.

Bergquist, P., Te'o, V., Gibbs, M., Cziferszky, A., de Faria, F.P., Azevedo, M., and Nevalainen, H. 2002. Expression of xylanase enzymes from thermophilic microorganisms in fungal hosts. Extremophiles. 6: 177-184.

Berka, R.M., Ward, M., Wilson, L.J., Hayenga, K.J., Kodama, K.H., Carlomagno, L.P., and Thompson, S.A. 1990. Molecular cloning and deletion of the gene encoding aspergillopepsin A from *Aspergillus awamori*. Gene. 86: 153-162.

Bhargava, S., Nandakumar, M.P., Roy, A., Wenger, K.S., and Marten, M.R. 2003. Pulsed feeding during fed-batch fungal fermentation leads to reduced viscosity without detrimentally affecting protein expression. Biotechnol. Bioeng. 81: 341-347.

Blinkovsky, A.M., Byun, T., Brown, K.M., and Golightly, E.J. 1999. Purification, characterization, and heterologous expression in *Fusarium venenatum* of a novel serine carboxypeptidase from *Aspergillus oryzae*. Appl. Environ. Microbiol. 65: 3298-3303.

Bocking, S.P., Wiebe, M.G., Robson, G.D., Hansen, K., Christiansen, L.H., and Trinci A.P. 1999. Effect of branch frequency in *Aspergillus oryzae* on protein secretion and culture viscosity. Biotechnol. Bioeng. 65: 638-648.

Brakhage, A.A., Andrianopoulos, A., Kato, M., Steidl, S., Davis, M.A., Tsukagoshi, N., and Hynes, M.J. 1999. HAP-Like CCAAT-binding complexes in filamentous fungi: implications for biotechnology. Fungal Genet. Biol. 27: 243-252.

Caddick, M.X., Greenland, A.J., Jepson, I., Krause, K.P., Qu, N., Riddell, K.V., Salter, M.G., Schuch, W., Sonnewald, U., and Tomsett, A.B. 1998. An ethanol inducible gene switch for plants used to manipulate carbon metabolism. Nat. Biotechnol. 16: 177-180.

Callewaert, N., Laroy, W., Cadirgi, H., Geysens, S., Saelens, X., Min Jou, W., and Contreras, R. 2001. Use of HDEL-tagged *Trichoderma reesei* mannosyl oligosaccharide 1,2-alpha-D-mannosidase for N-glycan engineering in *Pichia pastoris*. FEBS Lett. 503: 173-178.

Cao, Q., Stubbs, M., Ngo, K.Q.P., Ward, M., Cunningham, A., Pai, E.F., Tu, G., and Hofmann, T. 2000. Penicillopepsin-JT2, a recombinant enzyme from *Penicillium janthinellum* and the contribution of a hydrogen bond in subsite S3 to k(cat). Protein Science. 9: 991-1001.

Case, M.E., Schweizer, M., Kushner, S.R., and Giles, N.H. 1979. Efficient transformation of *Neurospora crassa* by utilizing hybrid plasmid DNA. Proc. Natl. Acad. Sci. USA. 76: 5259-5263.

Chambergo, F.S., Bonaccorsi, E.D., Ferreira, A.J., Ramos, A.S., Ferreira Junior, J.R., Abrahao-Neto, J., Farah, J.P., and El-Dorry, H. 2002. Elucidation of the metabolic fate of glucose in the filamentous fungus *Trichoderma reesei* using expressed sequence tag (EST) analysis and cDNA microarrays. J. Biol. Chem. 277: 13983-13988.

Conesa, A., Jeenes, D., Archer, D.B., van den Hondel, C.A.M.J.J., and Punt, P.J. 2002. Calnexin overexpression increases manganese peroxidase production in *Aspergillus niger*. Appl. Environ. Microbiol. 68: 846-851.

Conesa, A., Punt, P.J., van Lu_k, N., and van den Hondel, C.A.M.J.J. 2001. The secretion pathway in filamentous fungi: a biotechnological view. Fungal Genet. Biol. 33: 155-171.

Davis, M.A., Kelly, J.M., and Hynes, M.J. 1993. Fungal catabolic gene regulation: molecular genetic analysis of the amdS gene of *Aspergillus nidulans*. Genetica. 90: 133-145.

Dunn-Coleman, N.S., Bloebaum, P., Berka, R.M., Bodie, E., Robinson, N., Armstrong, G., Ward, M., Przetak, M., Carter, G.L., LaCost, R., Wilson, L.J., Kodama, K.H., Baliu, E.F., Bower, B., Lamsa, M., and Heinsohn, H. 1991. Commercial levels of chymosin production by *Aspergillus*. Biotechnology (N.Y.). 9: 976-981.

Durand, H., Clanet, M., and Tiraby, G. 1988. Genetic improvement of *Trichoderma reesei* for large scale cellulolase production. Enz. Microb. Technol. 10: 341-346.

Foreman, P.K., Brown, D., Dankmeyer, L., Dean, R., Diener, S., Dunn-Coleman, N.S., Goedegebuur, F., Houfek, T.D., England, G.J., Kelley, A., Meerman, H.J., Mitchell, T., Mitchinson, C., Olivares, H.A., Teunissen, P., Yao, J., and M. Ward. 2003. Synthesis and regulation of biomass-degrading enzymes in the filamentous fungus *Trichoderma reesei*. J. Biol. Chem. 278: 31988-31997.

Fowler, T., Berka, R.M., and Ward, M. 1990. Regulation of the *glaA* gene of *Aspergillus niger*. Curr. Genet. 18: 537-545.

Frenken, L.G., Hessing, J.G., van den Hondel, C.A.M.J.J., and Verrips, C.T. 1998. Recent advances in the large-scale production of antibody fragments using lower eukaryotic microorganisms. Res. Immunol. 149: 589-599.

Galagan, J.E., Calvo, S.E., Borkovich, K.A., Selker, E.U., Read, N.D., Jaffe, D., Fitzhugh, W., Ma, L.J., Smirnov, S., Purcell, S., Rehman, B., Elkins, T., Engels, R., Wang, S., Nielsen, C.B., Butler, J., Endrizzi, M., Qui, D., Ianakiev, P., Bell-Pedersen, D., Nelson, M.A., Werner-Washburne, M., Selitrennikoff, C.P., Kinsey, J.A., Braun, E.L., Zelter, A., Schulte, U., Kothe, G.O., Jedd, G., Mewes, W., Staben, C., Marcotte, E., Greenberg, D., Roy, A., Foley, K., Naylor, J., Stange-Thomann, N., Barrett, R., Gnerre, S., Kamal, M., Kamvysselis, M.,

Mauceli, E., Bielke, C., Rudd, S., Frishmann, D., Krystofova, S., Rasmussen, C., Metzenberd, R.L., Perkins, D.D., Kroken, S., Cogoni, C., Macini, G., Catcheside, D., Li, W., Pratt, R.J., Osmani, S.A., DeSouza, C.P., Glass, L., Orbach, M.J., Berglund, J.A., Voelker, R., Yarden, O., Plamann, M., Seiler, S., Dunlap, J., Radford, A., Aramayo, R., Natvig, D.O., Alex, L.A., Mannhaupt, G., Ebbole, D.J., Freitag, M., Paulsen, I., Sachs, M.S., Lander, E.S., Nusbaum, C., and Birren, B. 2003. The genome sequence of the filamentous fungus *Neurospora crassa*. Nature. 422: 859-868.

Gems, D., Aleksenko, A., Belenky, L., Robertson, S., Ramsden, M., Vinetski, Y., and Clutterbuck, A.J. 1994. An 'instant gene bank' method for gene cloning by mutant complementation. Mol. Gen. Genet. 242: 467-471.

Goedegebuur, F., Fowler, T., Phillips, J., van der Kley, P., van Solingen, P., Dankmeyer, L., and Power, S.D. 2002. Cloning and relational analysis of 15 novel fungal endoglucanases from family 12 glycosyl hydrolase. Curr. Genet. 41: 89-98.

Goffeau, A., Barrell, B.G., Bussey, H., Davis, R.W., Dujon, B., Feldmann, H., Galibert, F., Hoheisel, J.D., Jacq, C., Johnston, M., Louis, E.J., Mewes, H.W., Murakami, Y., Philippsen, P., Tettelin, H., and Oliver, S.G. 1996. Life with 6000 genes. Science. 274: 563-567.

Gouka, R.J., Punt, P.J., and van den Hondel, C.A.M.J.J. 1997a. Efficient production of secreted proteins by *aspergillus*: progress, limitations and prospects. Appl. Microbiol. Biotechnol. 47: 1-11.

Gouka, R.J., Punt, P.J., and van den Hondel, C.A.M.J.J. 1997b. Glucoamylase gene fusions alleviate limitations for protein production in *Aspergillus awamori* at the transcriptional and (post)translational levels. Appl. Environ. Microbiol. 63: 488-497.

Gouka, R.J., Gerk, C., Hooykaas, P.J.J., Bundock, P., Musters, W., Verrips, C.T., and de Groot, M.J.A. 1999. Transformation of *Aspergillus awamori* by *Agrobacterium tumefaciens*-mediated homologous recombination. Nat. Biotechnol. 6: 598-601.

Graessle, S., Haas, H., Friedlin, E., Kürnsteiner, H., Stöffler, G., and Redl, B. 1997. Regulated system for heterologous gene expression in *Penicillium chrysogenum*. Appl. Environ. Microbiol. 63: 753-756.

Hassan, M., Corkidi, G., Galindo, E., Flores, C., and Serrano-Carreon, L. 2002. Accurate and rapid viability assessment of *Trichoderma harzianum* using fluorescence-based digital image analysis. Biotechnol. Bioeng. 80: 677-684.

Hazell, B.W., Te'o, V.S.J., Bradner, J.R., Bergquist, P.L., and Nevalainen, K.M.H. 2000. Rapid transformation of high cellulase-producing mutant strains of *Trichoderma reesei* by microprojectile bombardment. Lett. Appl. Microbiol. 30: 282-286.

Hettenhaus, J., and Glassner, D. 1997. Cellulase assessment for biomass hydrolysis. Report sponsored by the U.S. D.O.E. Office of Energy Efficiency and Renewable Energy Ethanol Program, < http://ceassist.com/assessment.htm>.

Horwitz, A.H., Chang, C.P., Better, M., Hellstrom, K.E., and Robinson, R.R. 1988. Secretion of functional antibody and Fab fragment from yeast cells. Proc. Natl. Acad. Sci. USA. 85: 8678-8682.

Hughes, T.R., Mao, M., Jones, A.R., Burchard, J., Marton, M.J., Shannon, K.W., Lefkowitz, S.M., Ziman, M., Schelter, J.M., Meyer, M.R., Kobayashi, S., Davis, C., Dai, H., He, Y.D., Stephaniants, S.B., Cavet, G., Walker, W.L., West, A., Coffey, E., Shoemaker, D.D., Stoughton, R., Blanchard, A.P., Friend, S.H. and Linsley, P.S. 2001. Expression profiling using microarrays fabricated by an ink-jet oligonucleotide synthesizer. Nat. Biotechnol. 19: 342-347.

Ilmén, M., Salheimo, A., Onnela, M.L., and Penttilä, M.E. 1997. Regulation of cellulase gene expression in the filamentous fungus *Trichoderma reesei*. Appl. Environ. Microbiol. 63: 1298-1306.

Jalving, R., van de Vondervoort, P.J., Visser, J., and Schaap, P.J. 2000. Characterization of the kexin-like maturase of *Aspergillus niger*. Appl. Environ. Microbiol. 66: 363-368.

Joutsjoki, V.V., Torkkeli, T.K., and Nevalainen, K.M. 1993. Transformation of *Trichoderma reesei* with the *Hormoconis resinae* glucoamylase P (gamP) gene: production of a heterologous glucoamylase by *Trichoderma reesei*. Curr. Genet. 24: 223-228.

Joutsjoki, V.V. 1994. Construction by one-step gene replacement of *Trichoderma reesei* strains that produce the glucoamylase P of *Hormoconis resinae*. Curr. Genet. 26: 422-429.

Kelley, A.S., England, G.E., and Pepsin, M. 2002. Induction of gene expression from cellulase gene promoter sequences using a high concentration sugar mixture. U.S. PTO# 60/409,466, filed Sep 10, 2002.

Kelley, A.S., England, G.E., and Mitchinson, C. 2003. Cost effective sophorose cellulase production by *Trichoderma reesei*. Manuscript in preparation.

Lorito, M., Hayes, C.K., Di Pietro, A., and Harman, G.E. 1993. Biolistic transformation of *Trichoderma harzianum* and *Gliocladium virens* using plasmid and genomic DNA. Curr. Genet. 24: 349-356.

Mach, R.L., and Zeilinger, S. 2003. Regulation of gene expression in industrial fungi: *Trichoderma*. Appl. Microbiol. Biotechnol. 60: 515-522.

Mach, R.L., and Zeilinger, S. 1998. Genetic transformation of *Trichoderma* and *Gliocladium*. In: *Trichoderma* and *Gliocladium*, Vol 1. Kubicek, C.P., and Harman, G.E., eds. Taylor and Francis Ltd., London. p. 225-241.

Mandels, M., Parish, F.W., and Reese, E.T. 1962. Sophorose as an inducer of cellulase in *Trichoderma viride*. J. Bacteriol. 83: 400-408.

Maras, M., Saelens, X., Laroy, W., Piens, K., Claeyssens, M., Fiers, W., and Contreras, R. 1997. *In vitro* conversion of the carbohydrate moiety of fungal glycoproteins to mammalian-type oligosaccharides—evidence for N-acetylglucosaminyltransferase-I-accepting glycans from *Trichoderma reesei*. Eur. J. Biochem. 249: 701-707.

Maras, M., De Bruyn, A., Vervecken, W., Uusitalo, J., Penttilä, M., Busson, R., Herdewijn, P., and Contreras, R. 1999a. *In vivo* synthesis of complex N-glycans by expression of human N-acetylglucosaminyltransferase I in the filamentous fungus *Trichoderma reesei*. FEBS Lett. 452: 365-370.

Maras, M., van Die, I., Contreras, R., and van den Hondel, C.A.M.J.J. 1999b. Filamentous fungi as production organisms for glycoproteins of bio-medical interest. Glycoconj. J. 16: 99-107.

Mathieu, M., and Felenbok, B. 1994. The *Aspergillus nidulans* CREA protein mediates glucose repression of the ethanol regulon at various levels through competition with the ALCR-specific transactivator. EMBO J. 13: 4022-4027.

McIntyre, M., Muller, C., Dynesen, J., and Nielsen, J. 2001. Metabolic engineering of the morphology of *Aspergillus*. Adv. Biochem. Eng. Biotechnol. 73: 103-128.

Mitchell, D.A., Berovic, M., and Krieger, N. 2000. Biochemical engineering aspects of solid state bioprocessing. Adv. Biochem. Eng. Biotechnol. 68: 61-138.

Moralejo, F.J., Cardoza, R.E., Gutierrez, S., Lombraña, M., Fierro, F., and Martín, J.F. 2002. Silencing of the aspergillopepsin B (*pepB*) gene of *Aspergillus awamori* by antisense RNA expression or protease removal by gene disruption results in a large increase in thaumatin production. Appl. Environ. Microbiol. 68: 3550-3559.

Moralejo, F.J., Watson, A.J., Jeenes, D.J., Archer, D.B., and Martin, J.F. 2001. A defined level of protein disulfide isomerase expression is required for optimal secretion of thaumatin by *Aspergillus awamori*. Mol. Genet. Genomics. 266: 246-253.

Muller C, McIntyre M, Hansen K, and Nielsen J. 2002. Metabolic engineering of the morphology of *Aspergillus oryzae* by altering chitin synthesis. Appl. Environ. Microbiol. 68: 1827-1836.

Murray, B.W., Takayama, S., Schultz, J., and Wong, C.H. 1996. Mechanism and specificity of human alpha-1,3-fucosyltransferase V. Biochemistry. 35: 11183-11195.

Nagel, F.J., Tramper, J., Bakker M.S., and Rinzema, A. 2001. Temperature control in a continuously mixed bioreactor for solid-state fermentation. Biotechnol. Bioeng. 72: 219-230.

Narendja, F.M., Davis, M.A., and Hynes, M.J. 1999. AnCF, the CCAAT binding complex of *Aspergillus nidulans*, is essential for the formation of a DNase I-hypersensitive site in the 5' region of the *amdS* gene. Mol. Cell. Biol. 19: 6523-6531.

Nikolaev, I., Mathieu, M., van de Vondervoort, P., Visser, J., and Felenbok, B. 2002. Heterologous expression of the *Aspergillus nidulans* alcR-alcA system in *Aspergillus niger*. Fungal Genet. Biol. 37: 89-97.

Nyyssönen, E., Penttilä, M., Harkki, A., Saloheimo, A., Knowles, J.K., and Keränen, S. 1993. Efficient production of antibody fragments by the filamentous fungus *Trichoderma reesei*. Biotechnology (N.Y.). 11: 591-595.

Nyyssönen, E., and Keränen, S. 1995. Multiple roles of the cellulase CBHI in enhancing production of fusion antibodies by the filamentous fungus *Trichoderma reesei*. Curr. Genet. 28: 71-79.

Panozzo, C., Cornillot, E., and Felenbok, B. 1998. The CreA repressor is the sole DNA-binding protein responsible for carbon catabolite repression of the *alcA* gene in *Aspergillus nidulans* via its binding to a couple of specific sites. J. Biol. Chem. 273: 6367-6372.

Pazouki, M., and Panda, T. 2000. Understanding the morphology of fungi. Bioprocess Engineering. 22: 127-143.

Punt, P.J., Greaves, P.A., Kuyvenhoven, A., van Deutekom, J.C., Kinghorn, J.R., Pouwels, P.H., and van den Hondel, C.A.M.J.J. 1991a. A twin-reporter vector for simultaneous analysis of expression signals of divergently transcribed, contiguous genes in filamentous fungi. Gene. 104: 119-122.

Punt, P.J., van Biezen, N., Conesa, A., Albers, A., Mangnus, J., and van den Hondel, C.A.M.J.J. 2002. Filamentous fungi as cell factories for heterologous protein production. Trends Biotechnol. 20: 200-206.

Punt, P.J., Zegers, N.D., Busscher, M., Pouwels, P.H., and van den Hondel, C.A.M.J.J. 1991b. Intracellular and extracellular production of proteins in *Aspergillus* under the control of expression signals of the highly expressed *Aspergillus nidulans gpdA* gene. J. Biotechnol. 17: 19-33.

Radzio, R., and Kück, U. 1997. Synthesis of biotechnologically relevant heterologous proteins in filamentous fungi. Process Biochem. 32: 529-539.

Reisman, H.B. 1993. Problems in scale-up of biotechnology production processes. Crit. Rev. Biotechnol. 13: 195-253.

Robinson, T., Singh, D., and Nigam, P. 2001. Solid-state fermentation: a promising microbial technology for secondary metabolite production. Appl. Microbiol. Biotechnol 55: 284-289.

Roslan, H.A., Salter, M.G., Wood, C.D., White, M.R., Croft, K.P., Robson, F., Coupland, G., Doonan, J., Laufs, P., Tomsett, A.B., and Caddick M.X. 2001. Characterization of the ethanol-inducible *alc* gene-expression system in *Arabidopsis thaliana*. Plant J. 28: 225-235.

Royer, J.C., Moyer, D.L., Reiwitch, S.G., Madden, M.S., Jensen, E.B., Brown, S.H., Yonker, C.C., Johnston, J.A., Golightly, E.J., Yoder, W.T., and Shuster, J.R. 1995. *Fusarium graminearum* A 3/5 as a novel host for heterologous protein production. Biotechnology (N.Y.) 13: 1479-1483.

Roussos, S., Raimbault, M., Saucedo-Castaneda, G., and Lonsane, B.K. 1992. Efficient leaching of cellulases produced by *Trichoderma harzianum* in solid state fermentation. Biotechnol. Techniques 6: 429-432.

Ryu, D.Y. and Humphrey, A.E. 1972. A reassessment of oxygen-transfer rates in antibiotics fermentations. J. Ferment. Technol. 50: 424-431.

Saloheimo, M., Lund, M., and Penttilä, M.E. 1999. The protein disulphide isomerase gene of the fungus *Trichoderma reesei* is induced by endoplasmic reticulum stress and regulated by the carbon source. Mol. Gen. Genet. 262: 35-45.

Saloheimo, M., Valkonen, M., and Penttilä, M.E. 2003. Activation mechanisms of the HAC1-mediated unfolded protein response in filamentous fungi. Mol. Microbiol. 47: 1149-1161.

Schügerl, K.; Gerlach, S.R.; and Siedenberg, D. 1998. Influence of the process parameters on the morphology and enzyme production of *Aspergilli*. In: Advances in Biochemical Engineering/ Biotechnology, Vol. 60 (Relation between Morphology and Process Performance), Scheper, T., ed. Springer-Verlag, Berlin. p. 195-266.

Simmons, L.C., Reilly, D., Klimowski, L., Raju, T.S., Meng, G., Sims, P., Hong, K., Shields, R.L., Damico, L.A., Rancatore, P., and Yansura, D.G. 2002. Expression of full-length immunoglobulins in *Escherichia coli*: rapid and efficient production of aglycosylated antibodies. J. Immunol. Methods. 263: 133-147.

Spencer, J.A., Jeenes, D.J., MacKenzie, D.A., Haynie, D.T., and Archer, D.B. 1998. Determinants of the fidelity of processing glucoamylase-lysozyme fusions by *Aspergillus niger*. Eur. J. Biochem. 258: 107-112.

Sun, X.L., Baker, H.M., Shewry, S.C., Jameson, G.B., and Baker, E.N. 1999. Structure of recombinant human lactoferrin expressed in *Aspergillus awamori*. Acta Crystallogr. D. Biol. Crystallogr. 55: 403-407.

te Biesebeke, R., Ruijter, G., Rahardjo, Y.S., Hoogschagen, M.J., Heerikhuisen, M., Levin, A., van Driel, K.G., Schutyser, M.A., Dijksterhuis, J., Zhu, Y., Weber, F.J., de Vos, W.M., van den Hondel, C.A.M.J.J., Rinzema, A., and Punt, P.J. 2002. *Aspergillus oryzae* in solid-state and submerged fermentations. Progress report on a multi-disciplinary project. FEM Yeast Res. 2: 245-248.

Te'o, V.S.J., Cziferszky, A.E., Bergquist, P.L., and Nevalainen, K.M.H. 2000. Codon optimization of xylanase gene *xynB* from the thermophilic bacterium *Dictyoglomus thermophilum* for expression in the filamentous fungus *Trichoderma reesei*. FEMS Microbiol Lett. 190: 13-19.

Timberlake, W.E. (1991). Cloning and analysis of fungal genes. In: More Gene Manipulations in Fungi. Bennett, J.W., and Lasure, L.L., eds. Academic Press, Inc., New York. p. 56-61.

Travers, K.J., Patil, C.K., Wodicka, L., Lockhart, D.J., Weissman, J.S., and Walter, P. 2000. Functional and genomic analyses reveal an essential coordination between the unfolded protein response and ER-associated degradation. Cell. 101: 249-258.

Tsukagoshi, N., Kobayashi, T., and Kato, M. 2001. Regulation of the amylolytic and (hemi-) cellulolytic genes in *aspergilli*. J. Gen. Appl. Microbiol. 47: 1-19.

Vaheri, M.P., Leisola, M., and Kaupinnen, V. 1979. Transglycosylation products of the cellulase system of *Trichoderma reesei*. Biotechnol. Lett. 1: 41-46.

Valkonen, M., Ward, M., Wang, H., Penttilä, M., and Saloheimo, M. 2003. Improvement of foreign protein production in *Aspergillus niger var. awamori* by constitutive induction of the unfolded-protein response. Appl. Environ. Microbiol. In press.

van den Hombergh, J.P.T.W., Sollewijn Gelpke, M.D., van de Vondervoort, P.J.I., Buxton, F.P., and Visser, J. 1997a. Disruption of three acid proteases in *Aspergillus niger* - effects on protease spectrum, intracellular proteolysis, and degradation of target proteins. Eur. J. Biochem. 247: 605-613.

van den Hombergh, J.P.T.W., van de Vondervoort, P.J.I., Fraissinet-Tachet, L., and Visser, J. 1997b. *Aspergillus* as a host for heterologous protein production: the problem of proteases. Trends Biotechnol. 15: 256-263.

Verdoes, J.C., Punt, P.J., van der Berg, P., Debets, F., Stouthamer, A.H., and van den Hondel, C.A.M.J.J. 1994. Characterization of an efficient gene cloning strategy for *Aspergillus niger* based on an autonomously replicating plasmid: cloning of the *nicB* gene of *A. niger*. Gene. 146: 159-165.

Wallis, G.L., Hemming, F.W., and Peberdy, J.F. 2001. Beta-galactofuranoside glycoconjugates on conidia and conidiophores of *Aspergillus niger*. FEMS Microbiol. Lett. 201: 21-27.

Wang, H., Entwistle, J., Morlon, E., Archer, D.B., Peberdy, J.F., Ward, M., and Jeenes, D.J. 2003. Isolation and characterisation of a calnexin homologue, *clxA*, from *Aspergillus niger*. Mol. Genet. Genomics. 268: 684-691.

Wang, H., and Ward, M. 2000. Molecular characterization of a PDI-related gene prpA in *Aspergillus niger var. awamori*. Curr. Genet. 37: 57-64.

Ward, M., and Kodama, K.H. 1991. Introduction to fungal proteinases and expression in fungal systems. In: Structure and Function of Aspartic Proteinases: Genetics, Structure, Mechanisms. Dunn, B.M., ed. Plenum Press, New York. p. 149-160.

Ward, M., Lin, C., Victoria, D.C., Fox, B.P., Fox, J.A., Wong, D.L., Meerman, H.J., Pucci, J.P., Fong, R.B., Heng, M.H., Tsurushita, N., Gieswein, C., Park, M., and Wang, H. 2003. Production of antibodies and antibody fragments in *Aspergillus niger.* Appl. Environ. Microbiol. Submitted.

Ward, M., Wilson, L.J., and Kodama, K.H. 1993. Use of *Aspergillus* overproducing mutants, cured for integrated plasmid, to overproduce heterologous proteins. Appl. Microbiol. Biotechnol. 39: 738-743.

Ward, M., Wilson, L.J., Kodama. K.H., Rey, M.W., and Berka, R.M. 1990. Improved production of chymosin in *Aspergillus* by expression as a glucoamylase-chymosin fusion. Biotechnology (N.Y.). 8: 435-440.

Ward, P.P., Piddington, C.S., Cunningham, G.A., Zhou, X., Wyatt, R.D., and Conneely, O.M. 1995. A system for production of commercial quantities of human lactoferrin: a broad spectrum natural antibiotic. Biotechnology (N.Y.). 13: 498-503.

Wen, D.X., Livingston, B.D., Medzihradszky, K.F., Kelm, S., Burlingame, A.L., and Paulson, J.C. 1992. Primary structure of Gal beta 1,3(4)GlcNAc alpha 2,3-sialyltransferase determined by mass spectrometry sequence analysis and molecular cloning. Evidence for a protein motif in the sialyltransferase gene family. J. Biol. Chem. 267: 21011-21019.

Wiebe, M.G., Karandikar, A., Robson, G.D., Trinci, A.P., Candia, J.L., Trappe, S., Wallis, G., Rinas, U., Derkx, P.M., Madrid, S.M., Sisniega, H., Faus, I., Montijn, R., van den Hondel, C.A.M.J.J., and Punt, P.J. 2001. Production of tissue plasminogen activator (t-PA) in *Aspergillus niger.* Biotechnol. Bioeng. 76: 164-174.

8

Insect Cell
Expression Systems

Karthik Viswanathan, Yu-chan Chao,
John T. A. Hsu, and Michael J. Betenbaugh

Abstract

Insect cells in combination with infectious baculoviruses have been used for the production of numerous heterologous proteins. This expression system relies on the generation of recombinant baculoviruses in which viral genes active during the late phases of infection, but not essential for viral replication in cell culture, are replaced by DNA sequences of interest. Viral deletions and specific molecular markers ensure efficient production and selection of recombinant baculoviruses. The high strength of the polyhedrin and p10 promoters enables high level expression of target proteins starting approximately 20 hours post-infection and lasting until cell lysis, three to five days thereafter. Direct incorporation of foreign genes into the host chromosome and the development of non-lethal viruses can eliminate cell death that accompanies traditional baculovirus vector systems. Insect cell lines are being isolated and engineered to include complex post-translational processing events similar to those of mammalian cells in order to produce heterologous glycoproteins that are virtually identical to those generated by mammalian hosts. Insect larvae are also being utilized as a more economical method for production of recombinant proteins of interest. The convenience and flexibility of the baculovirus system has led to its expanding use in other areas. Display on the baculovirus surface is being used to evolve proteins with altered properties and as a tool to present antigens. Baculovirus vectors are also being evaluated as gene delivery vehicles to express foreign genes in mammalian cells and other non-permissive hosts. These developments will ensure that insect cells continue to be a versatile and widely used expression system for the foreseeable future.

Introduction

An ideal expression system should exhibit a number of properties of interest to the biotechnology community. First, it should be capable of yielding large amounts of heterologous proteins. Second, the expressed protein should be synthesized in a form that closely resembles that found in nature. Third, the genetic manipulations required for protein expression should be relatively straightforward and completed easily in a short time frame, especially if a large number of polypeptides are to be expressed.

Insect cells have gained wide popularity as an expression system because they meet many of these demands when combined with the baculovirus cloning technology. Baculoviruses contain a very strong polyhedrin promoter that enables high level expression of heterologous proteins to levels as high as 50% of the total cellular protein. Adaptations made to the baculovirus vector enable the generation of recombinant viruses in a few weeks using commercial kits that allow for their rapid isolation. Finally, since insect cells are eukaryotes, they can perform numerous post-translational modifications of expressed polypeptides. As a result, heterologous mammalian proteins synthesized in insect cells often resemble the authentic polypeptide.

Like other expression systems, insect cells also have limitations. First, the system requires infection of insect cell lines with a recombinant baculovirus encoding the gene of interest. The process is typically lethal and eventually leads to the death and lysis of the infected cell. Thus, production occurs only over the limited time period in which cells remain viable and capable of expressing the heterologous protein. Second, although the yields of intracellular proteins can be very high in insect cells, production of secreted proteins is usually lower due to saturation of the secretory apparatus. In addition, post-translational processing events, especially glycosylation, are often different in insect and mammalian cells which may limit the range of applications. Fortunately, research is under way to address many of these limitations. The creation of less lethal baculoviruses as well as the introduction of stable expression vectors in insect cells can eliminate the transient nature of expression while maintaining satisfactory yields. The engineering of insect cells to include genes associated with more complex glycoyslation patterns and the evaluation of alternative insect hosts may one day lead to the generation of cell lines capable of performing post-translational modifications identical to those of mammalian cells. Such adaptations to the original insect-baculovirus expression system will ensure that insect cells remain a popular expression system. In the following sections, we review the characteristics of the insect cell-baculovirus expression system and some of the modifications that are under way to adapt it for future applications in biotechnology.

Insects and Insect Cells

Insects and other invertebrates represent more than half of the animal species identified on the planet. While a few insects are useful to the clothing industry where they are used in the production of silks, most of the agricultural industry views insects as pests to crops and other plant life. As a result, the study of insects and the control of their proliferation has been a major consideration for the agricultural and forest service industry for many years. With the advent of the baculovirus technology, insects and insect cell lines have also been used to express heterologous proteins.

In an effort to characterize better insect traits, cell lines have been established from around 100 insect species from seven different orders of Arthropoda. These have been isolated from eggs or embryonic, larval, and adult tissues including hemocytes, ovaries, imaginal discs, testes, midgut, and the fat body. Cells from lysed tissue are cultured in rich medium until division is observed and subcultured until a new line is established. Such procedures have led to the establishment of a number of widely used cell lines such as *Spodoptera frugiperda* (Sf-21). A particular clonal isolate of *Spodoptera frugiperda* called Sf-9 is now the most widely used insect cell line (Figure 1). The screening of established cell lines has also led to the use of another insect cell line, *Trichoplusia ni* (TN-5B1-4), obtained from cabbage looper. These Lepidopteran cell lines are used primarily because of their ability to replicate baculovirus vectors and produce baculovirus-encoded recombinant proteins in culture. Additional lines have been

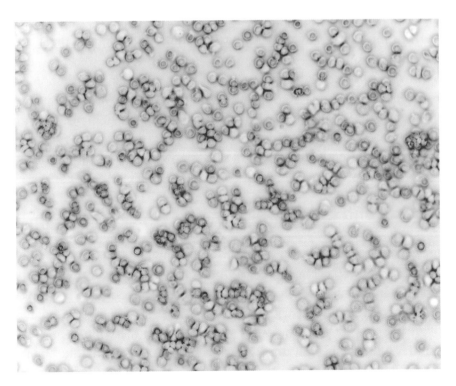

Figure 1. Sf-9 cells grown in insect cell culture medium. Photograph taken at 10X magnification.

established from *Bombyx mori* (silkworm) and *Lymantria dispar* (gypsy moth) and many other insect hosts. In addition to Lepidopteran species, cell lines are available from other orders of the insecta class. *Drosophila melanogaster*, the fruitfly from the Diptera order, is the most extensively studied insect species because of its importance in developmental genetics. The entire genome of *Drosophila* has been sequenced, further enhancing its position as a model for genomics investigation. However, *Drosophila* lines such as Schneider S2 cannot be infected by many of the widely used baculoviruses. Instead, foreign genes are typically incorporated into S2 cells using chromosomal integration techniques. The genome sequence of another invertebrate, *Anopholes gambi* (mosquito), has also been published and mosquito cell lines are available. The genomes of Lepidopteran species such as *Spodoptera frugiperda*, *Trichoplusia ni*, and *Lymantria dispar* have yet to be sequenced though some of these may be sequenced in the near future. Unfortunately, the lack of genomic information has hampered genetic investigations into these cell lines. When one considers the vast number of invertebrate species in the animal kingdom, it is clear that many more sequencing projects will be necessary in order to characterize invertebrate biology.

Many insect cell lines have been adapted for growth in serum free medium and can be cultivated as adherent cultures or in an anchorage-independent fashion in shaker and spinner flasks. Insect cells are typically 10-20 μm in diameter and grow between 25 and 30 °C without any requirement for CO_2 supplementation.

Baculovirus

A number of viruses that selectively infect and replicate in particular insect species have been identified. The best known insect viruses are baculoviruses and the most extensively studied is the *Autographica californica* multiple nuclear polyhedrosis virus (AcMNPV). However, a number of other baculoviruses such as the *Bombyx mori* nuclear polyhedrosis virus (BmNPV) and the *Lymantria dispar* nuclear polyhedrosis virus (LdNPV) have also been examined by biologists and agricultural scientists. "Baculo" refers to the characteristic rod-

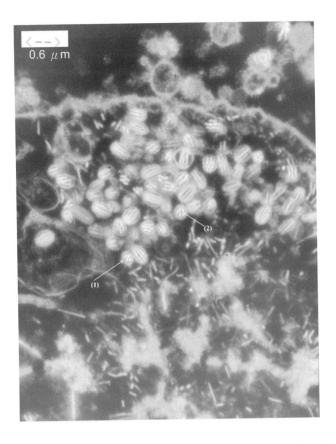

Figure 2. Transmission electron microscope image of Sf-9 cells infected with polyhedrin deleted AcMNPV. (1) Envelope; (2) Nucleocapsid.

shaped capsid of these viruses, which are typically 40-50 nm in diameter and 200-400 nm in length (Figure 2).

Baculovirus infection of a host insect cell proceeds by the replication cycle shown in Figure 3. Baculoviruses are deposited in two forms: budded and occluded (O'Reilly *et al.*, 1992; Shuler *et al.*, 1995). At the initial stages of infection (10-14 hours), membrane-enveloped viruses bud from the plasma membrane of an infected host in a form commonly referred to as extra-cellular virus (EV), non-occluded virus (NOV), or budded virus (BV). A second form of the virus, the occluded virus (OV), forms in the nucleus of infected cells beginning 20-24 hours after infection. This intracellular virus is enveloped and encased within a viral protein matrix called polyhedrin. These crystals of

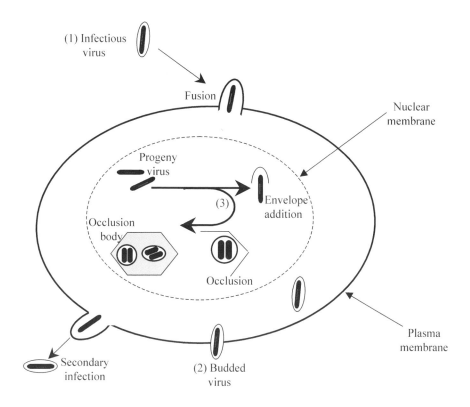

Figure 3. Baculovirus replication cycle in insect cells. (1) Virus infection; (2) Budded virus production; (3) Occlusion body formation.

protein, called occlusion bodies, polyhedra, or polyhedral inclusion bodies (PIBs), can grow to 5 μm in diameter and contain many virions. The OV encased within the polyhedrin protein matrix may contain many nucleocapsids within a single envelope for multiple nuclear polyhedrosis viruses (MNPVs) or a single nucleocapsid per envelope for single nuclear polyhedrosis viruses (SNPVs).

These polyhedra can accumulate in the cells of an infected insect host. Following the death and decomposition of the infected insect, polyhedra are deposited on the vegetation. The presence of the polyhedrin coat likely protects the encased OV from degradation in the environment during periods between transmission from one insect to another. Eventually, polyhedra are ingested by insect larvae feeding on the vegetation and transported with food to the midgut. Polyhedra dissolve in this alkaline environment, leading to the release of virions into the lumen of the midgut. The viruses fuse to the epithelial cells and cause an infection in the midgut tissue. Infected cells then release BV, which are responsible for secondary infection of other tissues by spreading through the hemolymph and trachea. After feeding for 5 to 7 days, the infected larva eventually becomes listless and its muscular structure dissolves in a process called wilting. The larva eventually ruptures, releasing about 25% of its dry weight in the form of PIBs that are deposited on vegetation for consumption by other insect larvae and re-initiation of the infection cycle.

The formation of OV during the late phase of baculovirus infection is accompanied by the abundant expression of two genes, the polyhedrin and p10 genes, starting at 20 hours and terminating 3 to 5 days following the onset of infection. The product of the polyhedrin gene (*polh*) is a 29-kDa protein that forms the matrix protecting the occluded viral nucleocapsids from degradation. The product of the p10 gene is a 10-kDa protein that forms large fibrous arrays in the nucleus and occasionally in the cytoplasm. In 1983, it was found that deletions in the polyhedrin gene were stable and that the mutant viruses could replicate in culture (Smith *et al.*, 1983a). Cells infected with these viruses lacked the occluded particles needed for insect-to-insect transmission but BV production was unaffected. BV could thus be obtained from the culture medium and used to infect other cells in culture in order to passage the altered virus. Subsequently, researchers took advantage of the fact that the polyhedrin was non-essential for BV cell-to-cell transmission by constructing recombinant viruses (Smith *et al.*, 1983b, Pennock *et al.*, 1984). Early efforts showed that the gene for interferon-β or another foreign gene could be substituted for the polyhedrin gene and that insect cells infected with this recombinant virus expressed the foreign gene product upon promoter activation 20-24 h post-infection. In these original experiments, recombinant cells were identified by the absence of occlusion bodies that are characteristic of wild-type viruses. Subsequently the p10 gene was also found to be non-essential for the generation of BV and recombinant vectors were established in which heterologous genes were inserted downstream of the p10 promoter (Vlak *et al.*, 1990). Because the polyhedrin and p10 promoters are very strong, foreign genes placed under their transcriptional control can be expressed at levels as high as 50% of the total cell protein, although this level is rarely achieved for recombinant proteins.

Methods for Introducing Foreign Genes into Baculovirus

The genome of baculovirus AcMNPV is approximately 130 kbp, which makes it difficult to find unique restriction sites for direct insertion of a heterologous gene. As a consequence, alternative strategies have been developed to introduce foreign genes into the viral genome. Although direct ligation via introduction of a unique restriction site is possible (Yang and Miller, 1998), homologous recombination is the most frequently used technique (Smith *et al.*, 1983b) (Figure 4). Originally, the wild type viral genome was used for this purpose (Smith *et al.*, 1983b). However, the recombination success rate was low (usually around or below 1%) and intensive screening was required to identify recombinant viruses.

Later, a unique restriction site was engineered into the baculovirus genome, allowing its linearization. The linearized DNA was 15- to 150-fold less infectious than the corresponding circular form when transfected into insect cells. However, this manipulation improved the rate of recombination 10-fold compared to circular DNA and 30% of the virus progeny were recombinant (Kitts *et al.*, 1990). An essential gene downstream of the polyhedrin expression locus was next removed from the linear viral DNA leading to nearly 100% recombination rates

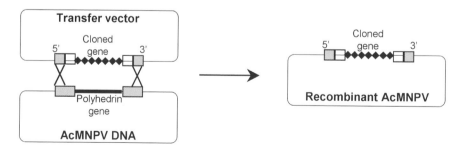

Figure 4. Homologous recombination between the loci of the transfer vector and baculovirus DNA. The recombination results in replacement of the baculovirus gene flanked by the homologous regions with the foreign gene.

(Kitts and Possee, 1993). In addition, the *p35* gene has been used as an effective dominant selectable marker to increase the recombination rate of AcMNPV (Lerch and Friesen, 1993).

Generation of infectious recombinant baculoviruses by transposon-mediated insertion of the foreign gene is another frequently used method (Luckow, *et al.*, 1993). This technology uses a baculovirus shuttle vector termed bacmid, that can replicate in *Escherichia coli* as a plasmid. The purified plasmid is used to infect susceptible lepidopteran insect cells. This approach reduces the recombination process from weeks to days. The Cre-lox system has also been tested (Peakman *et al.*, 1992). Although not widely used, this method may prove useful for the construction of baculovirus libraries. Recombinant baculovirus can also be constructed by homologous recombination with PCR-amplified DNA fragments (Gritsun *et al.*, 1997). This direct cloning method circumvents the use of bacterial transfer vectors.

Promoter Selection for Foreign Gene Expression

The baculovirus AcMNPV contains 154 putative open reading frames (ORFs) of 150 nucleotides or greater (Ayres *et al.*, 1994), suggesting that the virus is equipped with many promoters. These promoters are all under temporal regulation and their strengths differ. The time course of virus infection can be divided into at least three phases: early, late, and very late. Promoters from all phases have been used for foreign gene expression in the baculovirus expression vector system. As previously mentioned, recombinant proteins can be expressed at very high levels when the genes encoding these proteins are placed under transcriptional control of the two very late phase polyhedrin and *p10* promoters (Hasnain *et al.*, 1997; Lopez-Ferber *et al.*, 1995).

The most widely used promoter is polyhderin. It is a short (Possee and Howard, 1987; Rankin *et al.*, 1988) yet strong (Smith *et al.*, 1983a) promoter. Recombinant viruses in which a foreign gene is inserted upstream of the polyhedrin start codon are expressed at lower levels than when portions of the coding sequence for the polyhedrin gene are fused in frame with the target gene (Luckow *et al.*, 1988). Recently,

5' upstream elements have been found to activate gene expression from this promoter. Acharya and Gopinathan (2001) identified a 293 bp enhancer-like element located 1.0 kbp upstream of the initiation codon of the polyhedrin gene from the *Bombyx mori* nuclear polyhedrosis virus that leads to a 10-fold increase in expression from the minimal polyhedrin promoter. Two additional enhancer regions within 4 kbp of the polyhedrin promoter have also been described (Ramachandran *et al.*, 2001).

Although the polyhedrin promoter usually gives the highest yields of recombinant proteins, post-translational processing generally deteriorates during late and very late phases of baculovirus infection. Therefore, the use of early promoters can improve protein quality (Jarvis *et al.*, 1990; and Jarvis *et al.*, 1996; Lu *et al.*, 1997). The *ie1* promoter is one of the most frequently used early promoters for foreign gene expression. Although it is weak, *ie1* promoter activity can be greatly enhanced with the assistance of the homologous region (*hr*), a set of enhancers present in the baculovirus genome (Guarino *et al.*, 1986). The *ie1* gene product can in turn assist the enhancer for better expression from other promoters (Guarino *et al.*, 1991). Lu *et al.* (1997) showed that incorporation of the *hr* sequence from BmNPV into an actin promoter-based expression cassette results in a two orders of magnitude increase in transgene expression in transfected cells relative to a control cassette containing only the actin promoter. Furthermore, supplementation of transfected cells with the BmIE1 trans-activator resulted in a more than 1,000-fold increase in recombinant protein expression driven by the actin gene promoter. This expression casette was incorporated in a virus-free non-lytic insect cell expression system for continuous high-level production of recombinant proteins. Such a promoter may provide yields comparable to traditional baculovirus-based expression systems with the advantage of maintaining cell integrity for proper post-translational modifications of recombinant proteins.

A sequence upstream of the AcMNPV polyhedrin gene has also been found to activate strongly several full or minimal promoters obtained exogenously or derived endogenously from this baculovirus (Lo *et al.*, 2002). This sequence, named polyhedrin upstream (*pu*) sequence,

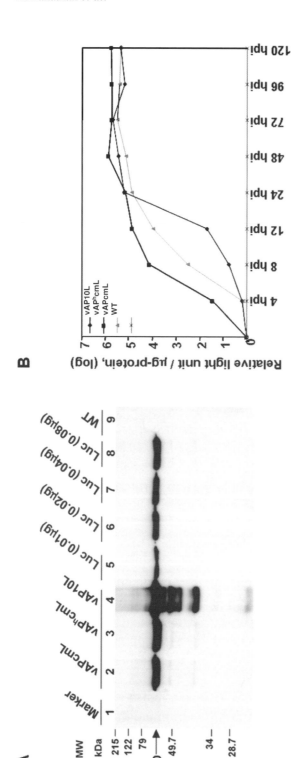

Figure 5. Comparative luciferase expression and degradation by the infection of different recombinant baculoviruses. (A) Western blot analysis of the luciferase expressed by different recombinant viruses. Lanes 5-8 includes chromatographically purified luciferase (Sigma) standards for the calibration of yields by different recombinant viruses. (B) Time course of luciferase expressions by different recombinant viruses. The luciferase expressed by recombinant viruses vAPcmL, vAP^hcmL, and vAP10L are driven by the CMVm, *hr*1-enhanced CMVm, and *p10* promoters, respectively. WT: wild type AcMNPV is used as a negative control. Reprinted by permission of Journal of Biological Chemistry (Copyright by the American Society for Biochemistry and Molecular Biology.)

contains three open reading frames, ORF4, ORF5 and *lef*-2. The homologous region of AcMNPV enhanced expression from these promoters in a synergistic manner. In the presence of *pu* and/or *hr*, luciferase was expressed much earlier for one of the tested promoters (the minimal CMV or CMVm promoter from human cytomegalovirus). Although less protein was obtained than with the *p10* promoter, superior enzymatic activity and no significant protein degradation was observed with the CMVm promoter (Figure 5). These results demonstrate that a synthetic promoter sequence containing *pu*, *hr*, and various full or minimal promoters, can provide early and strong expression of foreign proteins from insect cells.

Molecular Markers for the Identification of Recombinant Baculoviruses

The absence of occlusion bodies in the parental baculovirus was the earliest molecular marker used to identify successful recombinants (Smith *et al.*, 1983b). Although the cells are lysed by the end of baculovirus infection, identification of the recombinant plaques among the parental virus was very dificult in early baculovirus systems. Since occlusion bodies are a very late gene product, it usually takes more than 5 days after viral infection to confirm their absence or formation. Therefore the selection of recombinants was a particularly difficult task if one considers that only about 1% of the recombinations were successful after transfection. The screening method typically required selecting occlusion-minus (occ⁻) plaques that lacked production of OV particles as a result of the deletion of the polyhedrin gene. Occlusion-positive (occ⁺) plaques are opaque and refractile to light and distinguishable from occ⁻ plaques under the microscope. However, picking an occ⁻ plaque could be a formidable task for inexperienced researchers since young wild type plaques often show no sign of occlusion.

As discussed earlier, this problem has been largely resolved by using linear viral DNA to generate recombinant viruses. Even so, it is useful to verify that the recombinant viral stock is not contaminated with parental virus. For this purpose, and to facilitate genetic manipulations, various molecular markers have been developed.

*lac*Z was one of the first markers used to identify recombinant baculoviruses (Vlak *et al.*, 1990). A cassette containing the *lac*Z gene under the control of a heat-shock promoter from *Drosophila melanogaster* in a transfer vector allowed convenient identification of putative recombinants by virtue of β-galactosidase expression. The *lac*Z gene has also been used as a marker to detect contamination by parental viruses and is present in commercial baculovirus expression systems. Luciferase was later used as a sensitive reporter for rapid-screening of recombinant baculovirus following infection of insect cells and larvae (Richardson *et al.*, 1992). The green fluorescent protein (GFP) is a more sensitive and convenient marker for detecting recombinant viruses (Chao *et al.*, 1996; Wilson *et al.*, 1997). These studies also showed that GFP could serve as an efficient marker to detect the spreading of viruses in insect populations. The incorporation of toxic or insect growth regulatory genes into baculovirus vectors could be a useful tool to control insect pests in the fields. However, the spread of such engineered viruses is a potential threat to the biosafety of the environment, and the efficient tracing and confinement of recombinant viruses in the field using markers such as GFP is crucial for future pest management programs.

Baculovirus-Mediated Gene Expression in Non-Natural Hosts

Baculoviruses infect and replicate only in insect hosts. However, recent findings indicate that they can also transduce a variety of mammalian cells to mediate gene expression in dividing and non-dividing cells provided that the gene of interest is under the control of a mammalian promoter. For instance, AcMNPV was first reported to be capable of transducing human hepatocytes (Hofmann *et al.*, 1995). Since then, numerous studies have shown that AcMNPV can transduce other types of mammalian cells, including CHO, CV-1, HeLa, 293, BHK, Cos7, neural glial cells, mesenchymal cells, keratinocytes, and pancreatic islet cells (Boyce and Bucher, 1996; Condreay *et al.*, 1999; Dwarakanath *et al.*, 2001; Ma *et al.*, 2000; Sarkis *et al.*, 2000; Shoji *et al.*, 1997). Baculoviruses have also been utilized to deliver genes *in vivo* (Airenne *et al.*, 2000; Lehtolainen *et al.*, 2002; Pieroni *et al.*, 2001; Sarkis *et al.*, 2000).

Baculovirus transduction into mammalian cells is not toxic to the host and the virus does not replicate inside the transduced mammalian cells. These advantages have led to attempts to develop baculovirus vectors carrying mammalian expression cassettes for *in vitro* (Merrihew *et al.*, 2001) and *in vivo* (Airenne *et al.*, 2000; Pieroni *et al.*, 2001) gene therapy studies. Because baculoviruses are generally regarded as safe and nonpathogenic for vertebrates including humans, they represent a promising new class of gene delivery carriers. Furthermore, the tissue tropism of baculovirus-mediated gene delivery has been found to be different from that of other gene delivery vehicles (Lehtolainen *et al.*, 2002).

AcMNPV is also an efficient vector for gene transfer and expression in nonpermissive *Drosophila* S2 cells (Lee *et al.*, 2000). Nearly 100% of the S2 cells showed evidence of gene expression after transduction with recombinant baculoviruses carrying gene expression cassettes containing *Drosophila*-derived promoters (*e.g.* heat shock protein 70, actin 5C, or metallothionein promoters). In this study, S2 cells were shown to permit repetitive baculovirus transductions of different genes over time. Thus, baculovirus is an excellent tool for efficient gene transfer into multiple eukaryotic hosts. Applications of baculovirus vectors in this area are likely to expand as gene therapy increases and baculoviruses are designed with specific transduction properties.

Viral Surface Display

Baculoviruses infect insect cells by first attaching to a cellular receptor. Attached virions are then internalized by receptor-mediated endocytosis. Viral attachment to host cells is mainly due to the presence of the viral glycoprotein gp64 ($M_r \approx 64$-kDa) present on the virus envelope (Monsma *et al.*, 1996; Oomens *et al.*, 1995). The *gp64* gene encodes an N-terminal signal sequence and a mature domain that includes a transmembrane region. Upon expression in infected insect cells, gp64 is directed to the cell membrane by its signal peptide (Jarvis and Garcia, 1994). The mature protein oligomerizes and is presented on the surface of infected cells as homotrimers. In newly synthesized progeny virions, gp64 is incorporated into the budding virus and displayed on the viral peplomer.

This feature of gp64 display on the viral surface has been exploited in a manner analogous to bacterial phage display systems (Boublik *et al.*, 1995; Grabherr and Ernst, 2001; Possee, 1986). Surface display on baculoviruses has the advantage of potential post-translational modifications of displayed epitopes that are lacking in bacterial-based system. In the study of Boublik and coworkers, the gene encoding human immunodeficiency virus (HIV) gp120 was inserted into the mature domain of gp64 and expressed as a second copy in addition to endogenous gp64. The fusion protein co-oligomerized with wild-type gp64 and displayed gp120 on the viral surface (Boublik *et al.*, 1995). Using the same strategy, the HIV envelope protein gp41 (Grabherr *et al.*, 1997), green fluorescent protein and rubella virus envelope protein (Mottershead *et al.*, 1997) have been expressed and displayed on the baculovirus surface.

Displaying specific ligand-binding moieties on viral vectors is of increasing interest in the area of targeted gene therapy. Ojala *et al.* (2001) designed a baculovirus vector displaying a functional single chain antibody fragment (scFv) specific for the carcinoembryonic antigen (CEA) fused to gp64. The baculovirus with anti-CEA scFv displayed on its surface was shown to bind specifically and deliver genes to CEA expressing cells (PC-3).

Baculovirus surface display technology has also been successfully used to generate monoclonal antibodies (MAbs). Tanaka *et al.* (2002) produced MAbs recognizing human peroxisome proliferator-activated receptors (PPARs) using baculovirus particles displaying surface glycoprotein gp64-PPAR fusion proteins as the immunizing agent. The resulting antibodies recognized the gp64-PPARs fusion protein as well as mature, expressed PPARs proteins as demonstrated by immunohistochemistry, immunoblotting, and electrophoretic mobility shift assays (EMSAs). Thus, baculoviruses can be used to display antigens and produce functional monoclonal antibodies without requiring prior purification of the antigen.

Non-Lytic Protein Expression Systems

In 1983, the baculovirus expression vector system (BEVS) was developed based on a lytic viral infection (Smith *et al.* 1983b). Although popular in industry and molecular biological laboratories, this system has drawbacks associated with the infectious cycle. One disadvantage is that production of foreign proteins occurs during later stages of viral infection at a time when host cell functions are deteriorating. By using the human plasminogen activator gene as a model, baculovirus infection has been shown to have an adverse effect on the function of the Sf9 cellular secretory pathway (Jarvis, 1993; Jarvis *et al.*, 1996). Expression of chaperones and folding assistance factors has been utilized as one strategy to maintain expression of functional proteins during the later stages of the viral infection (Ailor and Betenbaugh, 1999). Eventually, however, the insect cell host succumbs to infection and thus the process is inherently transient in nature. In addition, a new batch of insect cells or larvae must be infected each time a recombinant protein is produced. Furthermore, this approach requires two separate components, a virus and a cell line, for expression of recombinant proteins. This approach differs from expression systems used in many other cell lines in which the host gene is stably integrated into the cell line of interest. Lu *et al.* (1997) noted many of the limitations of conventional lytic baculovirus expression systems, including the reduction in proper posttranslational modifications, the general inability to process complex, intron-containing transcripts during the late stages of infection, and the release of proteases that can degrade expressed heterologous products.

An expression system in which cells remain intact during the production phase would eliminate these problems. In an effort to avoid difficulties associated with insect cell death due to virus infection, a number of stable integration vectors have been developed that allow continuous protein expression of foreign genes in insect cell hosts. Stable expression of heterologous proteins using integrated vectors has been achieved in a variety of insect cell lines including Sf-9, *Trichoplusia ni* (*T. ni*), *Bombyx mori*, *Drosophila* S2, and *Aedes albopictus* (Patterson *et al.*, 1995). Jarvis *et al.* (1990) established an expression system with stable transfection of foreign genes driven by the immediate early *ie*1 promoter (see Promoter Selection for Foreign Gene Expression). Although the intensity of the *ie*1 promoter was significantly lower (10-50 times less) than the stronger polyhedrin promoter, the stably-transfected system produced similar amounts of secretory glycoproteins as obtained with recombinant viruses using the polyhedrin promoter (Jarvis *et al.*, 1996). While conventional baculovirus vectors produced greater amounts of enzymatic activity from two different prokaryotic genes, immediate-early vectors produced enzymatic activity levels that were as high or higher than those from the late phase promoter for two different eukaryotic genes encoding secretory pathway proteins. Comparable productivity between the stable and transient expression systems was likely due to the low yield of secretory proteins obtained using the polyhedrin promoter. Lu *et al.* (1997) showed that the promoter of the silkmoth cytoplasmic actin gene along with the *ie1* trans-activator gene and the *hr3* enhancer region of BmNPV could drive high level expression of foreign gene sequences stably integrated into insect cells. This strategy was later applied to the expression of a variety of recombinant proteins including juvenile hormone esterase and tissue plasminogen activator from a variety of insect cells at levels exceeding those obtained from comparable mammalian hosts (Farrell *et al.*, 1999; Keith *et al.*, 1999; Farrell *et al.*, 1998).

Although stable transfection techniques are promising, the establishment of nonlytic baculovirus expression system is another potential alternative to lethal viral vectors. The construction of a recombinant viruses is easier than establishing stable transfectants. The creation of cell lines typically takes about four weeks while a baculovirus vector can be generated from commercial systems in a week or two. Furthermore, development, storage and handling of viruses is easier than retaining numerous stably transfected cell lines when a large number of genes are to be evaluated.

In recent years, a nonlytic baculovirus infection system was established based on deletion of the *p35* gene of AcMNPV. Lee at al. (1998) found that the infection of Sf-9 cells with AcMNPV carrying a mutation or deletion in the apoptotic suppressor gene *p35* allowed the isolation of surviving insect cells that harbored persistent viral genomes. A persistent infection allowed only early gene expression and not expression from the very late polyhedrin gene. Nevertheless, this study provides an initial model for nonlytic baculovirus expression systems in the future. These stable expression methodologies are likely to become more widely used in insect cells provided that high expression levels and rapid generation of vectors can be achieved.

Larvae and Other Baculoviruses

A number of insect larvae such at *Trichoplusia ni* have been used to express proteins using either AcMNPV or other baculoviral vectors. The low cost of feeding and growing larvae makes this system potentially desirable for the economic production of high levels of recombinant proteins of interest. Purification of proteins from larvae may be slightly more difficult but inclusion of affinity tags can significantly ease the purification process (Cha *et al.*, 2002). The addition of GFP markers in the baculovirus of interest allows users to monitor expression levels in order to harvest under optimal conditions (Cha *et al.*, 1999; Kramer *et al.*, 2003).

A number of different viral vectors and larvae types have been considered for larval production. *Bombyx mori* (silkworm) larvae have been used to express many recombinant proteins in conjunction with the *Bombyx mori* nuclear polyhedrosis virus (BmNPV) containing a foreign gene of interest (Higashihashi *et al.*, 1991; Ho *et al.*, 1998; Murakami *et al.*, 2001; Okano *et al.*, 2000; Yang *et al.*, 2002). The silkworm can produce levels up to 400-600 μg/ml of proteins such as hepatitis B virus surface antigens (HBsAg) (Higashihashi *et al.*, 1991). Bovine interferon-gamma (βIFN-γ) produced form silkworm larvae was shown to have high-level biological activity (Murakami *et al.*, 2001). Recently, a method has been developed using a *piggyBac* transposon-derived vector to produce transgenic larvae (Handler, 2002; Tamura *et al.*, 2000). Employing the *piggyBac* vector, Tomita *et al.* (2003) demonstrated that transgenic silkworms could produce large quantities of recombinant human type III procollagen in cocoons. They estimated that 0.8 kg of recombinant collagen could be produced out of 100 kg of cocoon material. Transgenic silkworm production

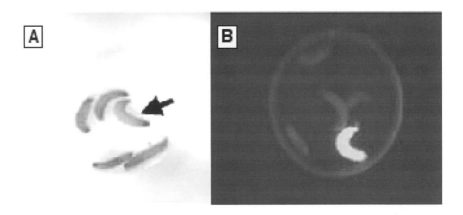

Figure 6. *Galleria mellonella* larvae infected (injected) with recombinant AcMNPV which codes GFP under the control of polyhedrin promoter. The larvae is ~2cm in length. (A) The infected larvae is noted by an arrow. (B) The GFP in the larvae can be observed by using a fluorescence device.

machinery may one day become an attractive route to produce large quantities of recombinant proteins.

The presence of sialic acid in the prothoracic glands of *Galleria mellonella* (Karacali, 1997) suggests that this insect may be an interesting host for the production of heterologous recombinant glycoproteins. Although this host is typically infected with the GmMNPV virus (McIntosh *et al.*, 1985), it has been shown that *G. mellonella* larvae can also be infected by recombinant AcMNPV. An AcMNPV vector, encoding enhanced green fluorescent protein (GFP) under the control of polyhedrin promoter, was injected into *G. mellonella* larvae. After 1 week, the worm was examined under a fluorescent light source in order to excite GFP expressed in the larvae. As shown in Figure 6, one out of five larvae was successfully infected and expressed GFP.

Post-Translational Modifications

The baculovirus expression system, like most other eukaryote hosts, retains the capacity to perform post-translational modifications (PTMs) including chemical covalent modifications such as phosphorylation, prenylation, glycosylation, acylation, methylation, sulfation, palmitoylation, and myristoylation along with other processing events such as proteolytic cleavage and disulfide bond formation. These alterations can affect the structure, function, targeting and activity of the expressed proteins. However, some PTMs produced in BEVS are found to be different and less efficient than those generated in the native mammalian hosts. These variant modifications constitute a major hindrance to the use of insect cell-derived proteins in structure-function studies and for the production of mammalian therapeutics. Phosphorylation is one such PTM that sometimes differs in insect cells. While serine and threonine phosphorylation are common in insect-derived proteins, tyrosine phosphorylation, which may require more specific kinases, is less commonly observed.

Glycosylation is a PTM that involves the attachment of oligosaccharides to polypeptides synthesized in the secretory compartment and the subsequent processing of these carbohydrate structures. While the baculovirus system is capable of carrying out both N- and O-glycosylation of proteins, the nature of these processing steps is typically not the same as in mammalian cells. Given the importance of glycosylation to protein activity and *in vivo* clearance rates, insect glycosylation has been studied extensively and insect cells have been subjected to genetic engineering in order to alter glyosylation profiles.

Glycosylation

The majority of current and potential recombinant therapeutics for human application are glycoproteins. Glycoproteins are typically modified by addition and processing of N- and O-glycans during transit through the secretory apparatus. Insect cells are capable of adding both N- and O-linked attachments but the subsequent modifications made to these proteins are often different compared to mammalian cells. In some cases, the occupancy of the glycosylation sites may also be problematic in insect cells (Ailor et al., 2000). Differences in the attached glycan structures of the recombinant protein can result in different biological activities, pharmacokinetic behaviors, and allergenic reactions in humans when used as a therapeutic.

Insect cells are capable of adding *N*-linked oligosaccharide attachments at Asn sites similar to those in mammalian cells but the subsequent modifications made to these proteins are often different in insect cells compared to mammalian cells. Typically, *N*-linked oligosaccharides from insect cells are truncated and terminate in mannose (Man) or occasionally *N*-acetylglucosamine (GlcNAc) instead of sialic acid (SA, Neu5Ac) or galactose termini of mammalian glycoproteins (Altmann *et al.*, 1999; Hsu *et al.*, 1997; Jarvis *et al.*, 1998; Kulakosky *et al.*, 1998). Lepidopteran insect cells contain enzymes involved in the earlier processing steps of *N*-glycosylation but typically lack the glycosyltransferases needed to convert these glycans to become complex-type (Altmann and Marz, 1995; Jarvis and Finn, 1996; Ren *et al.*, 1995; Velardo *et al.*, 1993). The inability of most lepidopteran cells to produce mammalian-type *N*-glycans is mainly attributed to (i) the lack of significant glycosyltransferases involving the addition of GlcNAc, Gal, and SA (Breitbach and Jarvis, 2001), (ii) the inability to produce the precursor nucleotide sugar CMP-*N*-acetylneuraminic acid (Lawrence *et al.*, 2001), and (iii) the presence of a unique β-*N*-acetylglucosaminidase which removes GlcNAc from the Manα(1,3) branch (Altmann *et al.*, 1999). In addition, many insect cell lines such at *Trichoplusia ni* (Tn-5B1-4) include an α(1,3)-Fucose attached to the innermost GlcNAc that may result in potentially allergenic attachments on some heterologous proteins (Hsu *et al.*, 1997). Moreover, baculovirus infection has been found to decrease transferase activities (van Die *et al.*, 1996). Overall, these properties preclude most insect cells from producing glycoproteins with advanced and desirable glycosylational modifications.

N-glycan processing in insect cells, like most other eukaryotes, starts with the transfer of a $Glc_3Man_9GlcNAc_2$ oligosaccharide from the donor dolichol-linked oligosaccharide to Asn on an Asn-X-Ser/Thr acceptor site of the protein. The attached oligosaccharide is subsequently processed by a pathway similar, but not identical, to mammalian cells. The stepwise processing is shown in Figure 7. The action of α-glucosidase I, II, and α-mannosidase removes all the

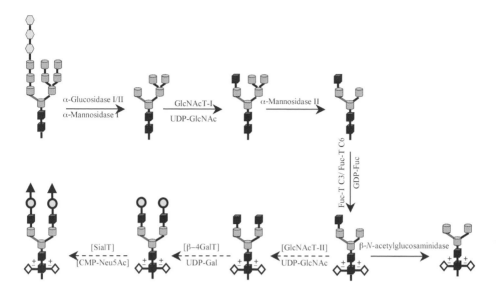

Figure 7. Typical *N*-Glycan processing pathway in lepidopteran insect cells. Broken arrows (---->) represent steps of the pathway and brackets [ ] represent the enzymes and sugar-nucleotides which are absent or limiting in many insect cell lines. Symbols: ⬡ , Glucose; ▨, Mannose; ◼,GlcNAc; ◇ , Fucose; ◯, Galactose; ▲, Sialic acid.

Glucose (Glc) attachments and the outer Man attachments. α-Glucosidase I cleaves the terminal glucose residue and one inner glucose residue, followed by the action of α-glucosidase II which cleaves the glucose residue attached to mannose. Then a number of mannose residues are removed by α-mannosidase I to yield a Man$_5$GlcNAc$_2$ structure. Some glycoproteins produced by lepidopteran insect cells have significant levels of high-mannose type Man$_{5-8}$GlcNAc$_2$ *N*-glycans. The presence of these high mannose *N*-glycans may be due to secretion of incompletely processed glycoproteins or as a result of the release of intracellular high-mannose *N*-glycans during the cell lysis (Hsu *et al.*, 1997). Processed Man$_5$GlcNAc$_2$, if not secreted, is subsequently acted upon by *N*-acetylglucos-aminyltransferase I (GlcNAcT-I) to add a GlcNAc to Manα(1,3) branch. Thereafter two Man residues are removed by α-mannosidase II to yield GlcNAcMan$_3$GlcNAc$_2$ structure. The enzymes involved in these steps have been shown to be active in a variety of insect cell lines including Sf-9, Sf-21 and Bm-N. At this point mammalian cells add GlcNAc to the other branch in a processing step carried out by GlcNacT-II. However, only low GlcNAcT-II activity has been found in most Lepidoptera cell lines studied to date. In addition, insect cell lines also have a β-*N*-acetylglucosaminidase activity which removes GlcNAc from the Manα(1,3) branch (Licari *et al.*, 1993; van Die *et al.*, 1996). Also at this stage the *N*-Glycan could be further modified by core α-1,3-fucosyltransferase (Fuc-T C3) and core α-1,6-fucosyltransferase (Fuc-T C6) to yield fucosylated forms. In some cases, additional Man groups will be removed by a subsequent mannosidase activity. Thus insect cells typically generate numerous paucimannosidic *N*-glycans containing $Man_{1-3}GlcNAc_2$ ($\pm Fuc_{1-2}$) structures along with a few structures including GlcNAc at the Man-α(1,3) or occasionally Man-α(1,6) branch. The following steps that occur in mammalian *N*-glycan processing are the addition of galactose and sialic acid by galactosyltransferase and sialyltransferase, respectively. Most insect cells lack these activities, although a few recent studies have suggested some cell lines may include more advanced processing (Palomares *et al.*, 2003, Joshi *et al.*, 2001). In addition, many insect cells do not have the capacity to generate the essential donor nucleotide, CMP-Neu5Ac, required for the transfer of a Neu5Ac sialic acid onto the growing polypeptide chain. Insect cells also lack UDP-GlcNAc-2-epimerase/ ManNAc kinase which is responsible for the conversion of UDP-GlcNAc to ManNAc-6-phosphate as well as the sialic acid synthase and CMP-sialic acid synthase activities for synthesis of the donor nucleotide CMP-Neu5Ac (Lawrence *et al.*, 2001).

Since structural and functional properties of a glycoprotein as well as *in vivo* clearance rates depend on the presence of proper glycosylation, establishing a platform technology that is able to produce mammalian-type glycoproteins in insect cells is highly desirable. As a result, metabolic engineering approaches have been undertaken in several laboratories to overcome the limitations in processing observed in insect cells. The incorporation of glycosyltransferases into insect cells has enabled the production of fully galactosylated and some sialylated glycoproteins from insect cell lines (Tomiya *et al.*, 2003; Seo *et al.*, 2001; Breitbach and Jarvis, 2001). In addition, the incorporation of genes responsible for the metabolic pathways has enabled insect cells to generate all the necessary nucleotide sugar precursors including CMP-Neu5Ac required for full sialylation (Lawrence *et al.*, 2001).

O-glycosylation is the attachment of sugars to specific hydroxylated amino acids like serine and threonine. *O*-linked glycans are added post-translationally to the fully folded proteins and may involve addition of either one sugar residue or the generation of an oligosaccharide built by transfer of multiple sugars from sugar-nucleotide donors. Insect cells are capable of carrying out *O*-glycosylation, but the nature of the *O*-glycan may not be the same as in mammalian cells. Specifically insect cells are not capable of synthesizing *N*-acetylneuraminic acid (sialic acid), so the *O*-glycans are not capped with sialic acids. In addition, *O*-glycans from insect cells tend to be shorter due to less complete additions of sugar residues onto the growing oligosaccharide chain. Pseudorabies virus (PRV) produced from mammalian cells contained primarily Galβ1-3GalNAc attachments with or without sialic acid residues. However, PRV from insect cells included the monosaccharide GalNAc attachment and lower amounts of the Galβ1-3GalNAc disaccharide with no sialic acid (Thomsen *et al.*, 1990).

Prenylation

Another type PTM that may also differ in insect cells compared to mammalian cells is the prenylation of a cysteine residue within the CAAX motif at the carboxyl terminus of a protein. The addition of the farnesyl group (15-carbons) or the geranyl-geranyl group (20-carbons) on cysteine residues of a protein is an important PTM called prenylation or isoprenylation. This PTM is catalyzed either by farnesyltransferase (FTase) or geranylgeranyltransferase (GGTase). These modifications were shown to be important for the intracellular localization and activity of many proteins involved in signal transduction pathways (Dietrich *et al.*, 1994; Kisselev *et al.*, 1994; Ray *et al.*, 1995).

Many proteins, such as Ras oncogene and heterotrimeric G proteins require the attachment of prenyl groups to become biologically active, and proteins with prenylation modifications can comprise 0.5 to 2% of total cellular protein (Casey *et al.*, 1991; Epstein *et al.*, 1991). The study of the Ras family has revealed that the farnesyl group is required for the transforming activity of Ras proteins (Kato *et al.*, 1992; Wittinghofer and Waldmann, 2000), and the Ras-family contains 141 members in humans alone (Venter *et al.*, 2001). Other than enabling membrane targeting, farnesylation is also important in protein-protein interactions.

The farnesylation process involves a three step modification of proteins containing CAAX motifs (C is a cysteine, A is usually an aliphatic amino acid, and X any amino acid). First, the cysteine is modified with either a farnesyl or a geranylgeranyl isoprenyl lipid. Second, the AAX tripeptide is trimmed through proteolysis to generate a C-terminus on the protein. Third, the now C-terminal prenylcysteine is methylated at the carboxyl group (Zhang and Casey, 1996).

Heterotrimeric G proteins (GTP hydrolases) are positioned in the inner face of the plasma membrane and represent an important family of signal transduction elements in cells. All G_γ subunits of G proteins have CAAX motifs at their carboxyl termini and prenylation of G_γ is important for their interaction with other G protein components, G protein coupled receptors (GPCRs), and subsequent effectors (Dietrich *et al.*, 1996; Kisselev *et al.*, 1994).

The baculovirus expression system has proved to be an effective tool for the study of GPCRs and heterotrimeric G proteins (Brys *et al.*, 2000; Butkerait *et al.*, 1995; Wenzel-Seifert *et al.*, 1998). However, prenylation of some recombinant proteins was found to be insufficient and severely heterogeneous (Bruel *et al.*, 2000; Cha *et al.*, 1997; Kalman *et al.*, 1995). As with glycosylation, insufficient prenylation was reported as a limitation to the use of the baculovirus expression system for the study of G protein activities. In contrast, Lindorfer *et al.* (1996) reported that insect cells could accurately and specifically add prenyl groups to recombinant proteins. Only a small percentage of improperly prenylated proteins were observed, and these might have been formed as a result of the overwhelmed prenyltransferase activity in baculovirus-infected insect cells.

In vertebrate cells, farnesyltransferase (FTase) and geranylgeranyl-transferase (GGTase) I- α subunit are cleaved by caspase-3 during apoptosis. In the baculovirus expression system, the endogenous FTase and GGTase may be labile following baculovirus infection. The coexpression of heterologous FTase or GGTase in BEVS might be one approach to obtain more complete prenylation for gene products of interest. An increase in the expression of proteins with homogeneously prenylated modifications may have a significant positive impact on functional and structural studies of prenylated proteins in insect cells.

Concluding Remarks

Insect cells and the baculovirus expression system have found widespread utility for the production of thousands of heterologous proteins from mammalian and other sources. This expression system is particularly useful for eukaryotic proteins that are not correctly folded in prokaryotes and include significant post-translation modifications. The insect cell-baculovirus system will continue to maintain its position as a host for the production of heterologous proteins as genomic investigations require rapid expression of large numbers of newly discovered genes. In addition, the use of insect cells is likely to expand in the future with the development of stable expression systems and modifications of the post-translational processing apparatus to mimic mammalian cell activities. Baculoviruses are also likely to find increasing roles as vectors for gene therapy and for gene transduction into mammalian cells and other non-permissive hosts. The use of the baculovirus display systems should provide another avenue in which insect cells are used to produce engineered proteins with altered properties. In summary, the exploitation of insect cells and insects for biotechnology purposes will continue to grow in the future due to research advances that will expand the versatility and capabilities of this important recombinant expression system.

References

Acharya, A., and Gopinathan, K.P. 2001. Identification of an enhancer-like element in the polyhedrin gene upstream region of *Bombyx mori* nucleopolyhedrovirus. J. Gen. Virol. 82: 2811-2819.

Ailor, E., and Betenbaugh, M.J. 1999. Modifying secretion and post-translational processing in insect cells. Curr. Opin. Biotechnol. 10: 142-145.

Ailor, E., Takahashi, N., Tsukamoto, Y., Masuda, K., Rahman, B.A., Jarvis, D.L., Lee, Y.C., and Betenbaugh, M.J. 2000. N-glycan patterns of human transferrin produced in *Trichoplusia ni* insect cells: effects of mammalian galactosyltransferase. Glycobiology. 10: 837-847.

Airenne, K.J., Hiltunen, M.O., Turunen, M.P., Turunen, A.M., Laitinen, O.H., Kulomaa, M.S., and Yla-Herttuala, S. 2000a. Baculovirus-mediated periadventitial gene transfer to rabbit carotid artery. Gene. Ther. 7: 1499-1504.

Altmann, F., Staudacher, E., Wilson, I.B., and Marz, L. 1999. Insect cells as hosts for the expression of recombinant glycoproteins. Glycoconj. J. 16: 109-123.

Altmann, F., and Marz, L. 1995. Processing of asparagine-linked oligosaccharides in insect cells: evidence for alpha-mannosidase II. Glycoconj. J. 12: 150-155.

Ayres, M.D., Howard, S.C., Kuzio, J., Lopez-Ferber, M., and Possee, R.D. 1994. The complete DNA sequence of *Autographa californica* nuclear polyhedrosis virus. Virology. 202: 586-605.

Boublik, Y., Di Bonito, P., and Jones, I.M. 1995. Eukaryotic virus display: engineering the major surface glycoprotein of the *Autographa californica* nuclear polyhedrosis virus (AcNPV) for the presentation of foreign proteins on the virus surface. Biotechnology. 13: 1079-1084.

Boyce, F.M., and Bucher, N.L.R. 1996. Baculovirus-mediated gene transfer into mammalian cells. Proc. Natl. Acad. Sci. USA. 93: 2348-2352.

Breitbach, K., and Jarvis, D.L. 2001. Improved glycosylation of foreign proteins by Tn-5B1-4 cells engineered to express mammalian glycosyltransferases. Biotechnol. Bioeng. 74: 230-239.

Bruel, C., Cha, K., Reeves, P.J., Getmanova, E., and Khorana, H.G. 2000. Rhodopsin kinase: expression in mammalian cells and a two-step purification. Proc. Natl. Acad. Sci. USA. 97: 3004-3009.

Brys, R., Josson, K., Castelli, M.P., Jurzak, M., Lijnen, P., Gommeren, W., and Leysen, J.E. 2000. Reconstitution of the human 5-HT(1D) receptor-G-protein coupling: evidence for constitutive activity and multiple receptor conformations. Mol. Pharmacol. 57: 1132-1141.

Butkerait, P., Zheng, Y., Hallak, H., Graham, T.E., Miller, H.A., Burris, K.D., Molinoff, P.B., and Manning, D.R. 1995. Expression of the human 5-hydroxytryptamine1A receptor in Sf9 cells. Reconstitution of a coupled phenotype by co-expression of mammalian G protein subunits. J. Biol. Chem. 270: 18691-18699.

Casey, P.J., Thissen, J.A., and Moomaw, J.F. 1991. Enzymatic modification of proteins with a geranylgeranyl isoprenoid. Proc. Natl. Acad. Sci. USA. 88: 8631-8635.

Cha HJ, Dalal NG, Pham MQ, Kramer SF, Vakharia VN, Bentley WE. 2002. Monitoring foreign protein expression under baculovirus p10 and polh promoters in insect larvae. Biotechniques. 32: 986-990.

Cha, H.J., Dalal, N.G., Pham, M.Q., Vakharia, V.N., Rao, G., and Bentley, W.E. 1999. Insect larval expression process is optimized by generating fusions with green fluorescent protein. Biotechnol. Bioeng. 65:316-324

Cha, K., Bruel, C., Inglese, J., and Khorana, H.G. 1997. Rhodopsin kinase: expression in baculovirus-infected insect cells, and characterization of post-translational modifications. Proc. Natl. Acad. Sci. USA. 94: 10577-10582.

Chao, Y.C., Chen, S.L., and Li, C.F. 1996. Pest control by fluorescence. Nature. 380: 396-397.

Condreay, J.P., Witherspoon, S.M., Clay, W.C., and Kost, T.A. 1999. Transient and stable gene expression in mammalian cells transduced with a recombinant baculovirus vector. Proc. Natl. Acad. Sci. USA. 96: 127-132.

Dietrich, A., Brazil, D., Jensen, O.N., Meister, M., Schrader, M., Moomaw, J.F., Mann, M., Illenberger, D., and Gierschik, P. 1996. Isoprenylation of the G protein gamma subunit is both necessary and sufficient for beta gamma dimer-mediated stimulation of phospholipase C. Biochemistry. 35: 15174-15182.

Dietrich, A., Meister, M., Brazil, D., Camps, M., and Gierschik, P. 1994. Stimulation of phospholipase C-beta 2 by recombinant guanine-nucleotide- binding protein beta gamma dimers produced in a baculovirus/insect cell expression system. Requirement of gamma-subunit isoprenylation for stimulation of phospholipase C. Eur. J. Biochem. 219: 171-178.

Dwarakanath, R.S., Clark, C.L., McElroy, A.K. and Spector, D.H. 2001. The use of recombinant baculoviruses for sustained expression of human cytomegalovirus immediate early proteins in fibroblasts. Virology. 284: 297-307.

Epstein, W.W., Lever, D., Leining, L.M., Bruenger, E., and Rilling, H.C. 1991. Quantitation of prenylcysteines by a selective cleavage reaction. Proc. Natl. Acad. Sci. USA. 88: 9668-9670.

Farrell, P.J., Behie, L.A., and Iatrou, K. 1999. Transformed Lepidopteran insect cells: new sources of recombinant human tissue plasminogen activator. Biotechnol. Bioeng. 64: 426-433.

Farrell, P.J., Lu, M., Prevost, J., Brown, C., Behie, L., and Iatrou, K. 1998. High-level expression of secreted glycoproteins in transformed lepidopteran insect cells using a novel expression vector. 60: 656-663.

Grabherr, R., and Ernst, W. 2001. The baculovirus expression system as a tool for generating diversity by viral surface display. Comb. Chem. High Throughput Screen. 4: 185-192.

Grabherr, R., Ernst, W., Doblhoff-Dier, O., Sara, M., and Katinger, H. 1997. Expression of foreign proteins on the surface of *Autographa californica* nuclear polyhedrosis virus. Biotechniques. 22: 730-735.

Gritsun, T.S., Mikhailov, M.V., Roy, P., and Gould, E.A. 1997. A new, rapid and simple procedure for direct cloning of PCR products into baculoviruses. Nucleic Acids Res. 25: 1864-1865.

Guarino, L.A., and Dong, W. 1991. Expression of an enhancer-binding protein in insect cells transfected with the *Autographa californica* nuclear polyhedrosis virus IE1 gene. J. Virol. 65: 3676-3680.

Handler, A.M. 2002. Use of the piggyBac transposon for germ-line transformation of insects. Insect Biochem. Mol. Biol. 32: 1211-1220.

Hasnain, S.E., Jain, A., Habib, S., Ghosh, S., Chaterjee, U., Ramachandran, A., Das,P., Venkaiah, B., Pandey, S., Liang, B., Ranjan, A., Natarajan, K., and Azim, C.A. 1997. Involvement of host factors in transcription from baculovirus very late promoters. Gene. 190: 113-118.

Higashihashi, N., Arai, Y., Enjo, T., Horiuchi, T., Saeki, Y., Sakano, K., Sato, Y., Takeda, K., Takashina, S., and Takahashi, T. 1991. High-level expression and characterization of hepatitis B virus surface antigen in silkworm using a baculovirus vector. J. Virol. Methods. 35: 159-167.

Ho, W.K., Meng, Z.Q., Lin, H.R., Poon, C.T., Leung, Y.K., Yan, K.T., Dias, N., Che, A.P., Liu, J., Zheng, W.M., Sun, Y., and Wong, A.O. 1998. Expression of grass carp growth hormone by baculovirus in silkworm larvae. Biochim. Biophys. Acta. 1381: 331-339.

Hofmann, C., Sandig, V., Jennings, G., Rudolph, M., Schlag, P. and Strauss, M. 1995. Efficient gene-transfer into human hepatocytes by baculovirus vectors. Proc. Nat. Acad. Sci. USA. 92: 10099-10103.

Hsu, T.A., Takahashi, N., Tsukamoto, Y., Kato, K., Shimada, I., Masuda, K., Whiteley, E.M., Fan, J.Q., Lee, Y.C., and Betenbaugh, M.J. 1997. Differential N-glycan patterns of secreted and intracellular IgG produced in *Trichoplusia ni* cells. J. Biol. Chem. 272: 9062-9070.

Jarvis, D. L. 1993. Foreign gene expression in insect cells. In: Insect cell culture engineering. Goosen M. F. A., Daugulis, A. J., and Faulkner, P. eds. Marcel Dekker, Inc. New York, Basel, Hong Kong. pp. 195-219.

Jarvis, D.L., and Garcia, A., Jr. 1994. Biosynthesis and processing of the *Autographa californica* nuclear polyhedrosis virus gp64 protein. Virology. 205: 300-313.

Jarvis, D.L., Fleming, J.G.W., Kovacs, G.R., Summers, M.D., and Guarino, L.A. 1990. Use of early baculovirus promoters for continuous expression and efficient processing of foreign gene products in stably-transformed lepidopteran cells. Biotechnology. 8: 950-955.

Jarvis, D.L., and Finn, E.E. 1996. Modifying the insect cell N-glycosylation pathway with immediate early baculovirus expression vectors. Nat. Biotechnol. 14: 1288-1292.

Jarvis, D.L., Weinkauf, C., and Guarino, L.A. 1996. Immediate-early baculovirus vectors for foreign gene expression in transformed or infected insect cells. Protein Expr. Purif. 8: 191-203.

Jarvis, D.L., Wills, L., Burow, G., and Bohlmeyer, D.A. 1998. Mutational analysis of the N-linked glycans on *Autographa californica* nucleopolyhedrovirus gp64. J. Virol. 72: 9459-9469.

Joshi, L., Shuler, M.L., and Wood, H.A. 2001. Production of a sialylated N-linked glycoprotein in insect cells. Biotechnol. Prog.17: 822-827.

Kalman, V.K., Erdman, R.A., Maltese, W.A., and Robishaw, J.D. 1995. Regions outside of the CAAX motif influence the specificity of prenylation of G protein gamma subunits. J. Biol. Chem. 270: 14835-14841.

Karacali, S.E.A. 1997. Presence of sialic acid in prothoracic glands of *Galleria mellonella* (Lepidoptera). Tissue Cell. 29: 315-321.

Kato, K., Der, C.J., and Buss, J.E. 1992. Prenoids and palmitate: lipids that control the biological activity of Ras proteins. Semin. Cancer Biol. 3: 179-188.

Keith, M.B., Farrell, P.J., Iatrou, K., and Behie, L.A. 1999. Screening of transformed insect cell lines for recombinant protein production. Biotechnol. Prog. 15: 1046-52.

Kisselev, O.G., Ermolaeva, M.V., and Gautam, N. 1994. A farnesylated domain in the G protein gamma subunit is a specific determinant of receptor coupling. J. Biol. Chem. 269: 21399-21402.

Kitts, P.A., Ayres, M.D., and Possee,R.D. 1990. Linearization of baculovirus DNA enhances the recovery of recombinant virus expression vectors. Nucleic Acids Res. 18:5667-5672

Kitts, P.A., and Possee, R.D. 1993. A method for producing recombinant baculovirus expression vectors at high frequency. Biotechniques. 14: 810-817.

Kramer, S.F., Kostov, Y., Rao, G., and Bentley, W.E. 2003. Ex vivo monitoring of protein production in baculovirus-infected *Trichoplusia ni* larvae with a GFP-specific optical probe. Biotechnol. Bioeng. 83: 241-247.

Kulakosky, P.C., Shuler, M.L., and Wood, H.A. 1998. N-glycosylation of a baculovirus-expressed recombinant glycoprotein in three insect cell lines. *In vitro* Cell. Dev. Biol. Anim. 34: 101-108.

Lawrence, S.M., Huddleston, K.A., Tomiya, N., Nguyen, N., Lee, Y.C., Vann, W.F., Coleman, T.A., and Betenbaugh, M.J. 2001. Cloning and expression of human sialic acid pathway genes to generate CMP-sialic acids in insect cells. Glycoconjugate J. 18: 205-213.

Lee, D.F., Chen, C.C., Hsu, T.A., and Juang, J.L. 2000. A baculovirus superinfection system: efficient vehicle for gene transfer into *Drosophila* S2 cells. J. Virol. 74: 11873-11880.

Lee, J.C., Chen, H.H., and Chao, Y.C. 1998. Persistent baculovirus infection results from deletion of the apoptotic suppressor gene p35. J. Virol. 72: 9157-9165.

Lehtolainen, P., Tyynela, K., Kannasto, J., Airenne, K.J., and Yla-Herttuala, S. 2002. Baculoviruses exhibit restricted cell type specificity in rat brain: a comparison of baculovirus- and adenovirus-mediated intracerebral gene transfer *in vivo*. Gene Ther. 9: 1693-1699.

Lerch, R.A., and Friesen, P.D. 1993. The 35-kilodalton protein gene (p35) of *Autographa californica* nuclear polyhedrosis virus and the neomycin resistance gene provide dominant selection of recombinant baculoviruses. Nucleic Acids Res. 21:1753-1760.

Licari, P.J., Jarvis, D.L., and Bailey, J.E. 1993. Insect cell hosts for baculovirus expression vectors contain endogenous exoglycosidase activity. Biotechnol. Prog. 9: 146-152.

Lindorfer, M.A., Sherman, N.E., Woodfork, K.A., Fletcher, J.E., Hunt, D.F., and Garrison, J.C. 1996. G protein gamma subunits with altered prenylation sequences are properly modified when expressed in Sf9 cells. J. Biol. Chem. 271: 18582-18587.

Lo, H.R., Chou, C.C., Wu, T.Y.J., Yuen, P.Y., and Chao, Y.C. 2002. Novel baculovirus DNA elements strongly stimulate activities of exogenous and endogenous promoters. J. Biol. Chem. 277: 5256-5264.

Lopez-Ferber, M., Sisk, W.P., and Possee, R.D. 1995. Baculovirus transfer vector Methods Mol. Biol. 39: 25-63.

Lu, M., Farrell, P.J., Johnson, R., and Iatrou, K. 1997. A baculovirus (*Bombyx mori* nuclear polyhedrosis virus) repeat element functions as a powerful constitutive enhancer in transfected insect cells. J. Biol. Chem. 272: 30724-30728.

Luckow, V.A., Lee, S.C., Barry, G.F., and Olins, P.O. 1993. Efficient generation of infectious recombinant baculoviruses by site-specific transposon-mediated insertion of foreign genes into a baculovirus genome propagated in Escherichia coli. J. Virol. 67: 4566-4579.

Luckow, V.A., and Summers, M.D. 1988. Signals important for high-level expression of foreign genes in *Autographa californica* nuclear polyhedrosis virus expression vectors. Virology. 167: 56-71.

Ma, L., Tamarina, N., Wang, Y., Kuznetsov, A., Patel, N., Kending, C., Hering, B.J., and Philipson, L.H. 2000.Baculovirus-mediated gene transfer into pancreatic islet cells. Diabetes. 49: 1986-1991.

McIntosh, A.H., Ignoffo, C.M., and Andrews, P.L. 1985. *In vitro* host range of five baculoviruses in lepidopteran cell lines. Intervirology. 23: 150-156.

Merrihew, R.V., Clay, W.C., Condreay, J.P., Witherspoon, S.M., Dallas, W.S., and Kost, T.A. 2001. Chromosomal integration of transduced recombinant baculovirus DNA in mammalian cells. J. Virol. 75: 903-909.

Monsma, S.A., Oomens, A.G.P., and Blissard, G.W. 1996. The gp64 envelope fusion protein is an essential baculovirus protein required for cell-to-cell transmission of infection. J. Virol. 70: 4607-4616.

Mottershead, D., van der Linden, I., von Bonsdorff, C.H., Keinanen, K., and Oker-Blom, C. 1997. Baculoviral display of the green fluorescent protein and rubella virus envelope proteins. Biochem. Biophys. Res. Commun. 238: 717-722.

Murakami, K., Uchiyama, A., Kokuho, T., Mori, Y., Sentsui, H., Yada, T., Tanigawa, M., Kuwano, A., Nagaya, H., Ishiyama, S., Kaki, H., Yokomizo, Y., and Inumaru, S. 2001. Production of biologically active recombinant bovine interferon-gamma by two different baculovirus gene expression systems using insect cells and silkworm larvae. Cytokine. 13: 18-24.

Ojala, K., Mottershead, D.G., Suokko, A., and Oker-Blom, C. 2001. Specific binding of baculoviruses displaying gp64 fusion proteins to mammalian cells. Biochem. Biophys. Res. Commun. 284: 777-784.

Okano, F., Satoh, M., Ido, T., Okamoto, N., and Yamada, K. 2000. Production of canine IFN-gamma in silkworm by recombinant baculovirus and characterization of the product. J. Interferon Cytokine Res. 20: 1015-1022.

Oomens, A.G.P., Monsma, S.A., and Blissard, G.W. 1995. The baculovirus gp64 envelope fusion protein - synthesis, oligomerization, and processing. Virology. 209: 592-603.

O'Reilly, D.R., Miller L.K., and Luckow, V.A. 1992. Baculovirus Expression Vectors: A Laboratory Manual. W. H. Freeman and Company, New York.

Palomares, L.A., Joosten, C.E., Hughes, P.R., Granados, R.R., and Shuler, M.L. 2003. Novel insect cell line capable of complex N-glycosylation and sialylation of recombinant proteins. Biotechnol. Prog. 19:185-192.

Patterson, R.M., Selkirk, J.K., and Merrick, B.A. 1995. Baculovirus and insect cell gene expression: review of baculovirus biotechnology. Environ. Health Perspect. 103: 756-759.

Peakman, T.C., Harris, R.A., and Gewert, D.R. 1992. Highly efficient generation of recombinant baculoviruses by enzymatically medicated site-specific *in vitro* recombination. Nucleic Acids Res. 20:495-500.

Pennock, G.D., Shoemaker, C., and Miller, L.K. 1984. Strong and regulated expression of Escherichia coli beta-galactosidase in insect cells with a baculovirus vector. Mol Cell Biol. 4:399-406.

Pieroni, L., Maione, D., and La Monica, N. 2001. *In vivo* gene transfer in mouse skeletal muscle mediated by baculovirus vectors. Hum. Gene Ther. 12: 871-881.

Possee, R.D., and Howard, S.C. 1987. Analysis of the polyhedrin gene promoter of the *Autographa californica* nuclear polyhedrosis virus. Nucleic Acids Res. 15:10233-10248.

Possee, R.D. 1986. Cell-surface expression of influenza virus haemagglutinin in insect cells using a baculovirus vector. Virus Res. 5: 43-59.

Ramachandran, A., Jain, A., Arora, P., Bashyam, M.D., Chatterjee, U., Ghosh, S., Parnaik, V.K., and Hasnain, S.E. 2001. Novel Sp family-like transcription factors are present in adult insect cells and are involved in transcription from the polyhedrin gene initiator promoter. J. Biol. Chem. 276: 23440-23449.

Rankin, C., Ooi, B.G., and Miller, L.K. 1988. Eight base pairs encompassing the transcriptional start point are the major determinant for baculovirus polyhedrin gene expression. Gene 70: 39-49.

Ray, K., Kunsch, C., Bonner, L.M., and Robishaw, J.D. 1995. Isolation of cDNA clones encoding eight different human G protein gamma subunits, including three novel forms designated the gamma 4, gamma 10, and gamma 11 subunits. J. Biol. Chem. 270: 21765-21771.

Ren, J., Bretthauer, R.K., and Castellino, F.J. 1995. Purification and properties of a Golgi-derived (alpha 1,2)-mannosidase- I from baculovirus-infected lepidopteran insect cells (IPLB-SF21AE) with preferential activity toward mannose6-N-acetylglucosamine2. Biochemistry. 34: 2489-2495.

Richardson, C.D., Banville, M., Lalumiere, M., Vialard, J., and Meighen, E.A. 1992. Bacterial luciferase produced with rapid-screening baculovirus vectors is a sensitive reporter for infection of insect cells and larvae. Intervirology. 34: 213-27.

Sarkis, C., Serguera, C., Petres, S., Buchet, D., Ridet, J.L., Edelman, L., and Mallet, J. 2000. Efficient transduction of neural cells *in vitro* and *in vivo* by a baculovirus-derived vector. Proc. Natl. Acad. Sci. USA. 97: 14638-14643.

Seo, N.S., Hollister J.R., and Jarvis, D.L. 2001. Mammalian Glycosyltransferase Expression Allows Sialoglycoprotein Production by Baculovirus-Infected Insect Cells. Protein Expr. Purif. 22: 234-241.

Shoji, I., Aizaki, H., Tani, H., Ishii, K., Chiba, T., Saito, I., Miyamura, T. and Matsuura, Y. 1997. Efficient gene transfer into various mammalian cells, including non-hepatic cells, by baculovirus vectors. J. Gen. Virol. 78: 2657-2664.

Shuler, M.L., Hammer, D.A., Granados, R.R., and Wood, H.A. 1995. Overview of baculovirus-insect cell culture system. In: Baculovirus Expression Systems and Biopesticides. M.L. Shuler, H.A. Wood, R.R. Granados, and D.A. Hammer eds. Wiley-Liss, New York. Chapter 1, pp. 1-11.

Smith, G.E., Fraser, M.J., and Summers, M.D. 1983a. Molecular engineering of the *Autographa californica* nuclear polyhedrosis virus genome: Deletion mutations within the polyhedrin gene. J. Virol. 46: 584-593.

Smith, G.E., Summers, M.D., and Fraser, M.J. 1983b. Production of human beta interferon in insect cells infected with a baculovirus expression vector. Mol. Cell. Biol. 3: 2156-2165.

Tamura, T., Thibert, C., Royer, C., Kanda, T., Abraham, E., Kamba, M., Komoto, N., Thomas, J.L., Mauchamp, B., Chavancy, G., Shirk, P., Fraser, M., Prudhomme, J.C., Couble, P., Toshiki, T., Chantal, T., Corinne, R., Toshio, K., Eappen, A., Mari, K., Natuo, K., Jean-Luc, T., Bernard, M., Gerard, C., Paul, S., Malcolm, F., Jean-Claude, P., and Pierre, C. 2000. Germline transformation of the silkworm *Bombyx mori* L. using a piggyBac transposon-derived vector. Nat. Biotechnol. 18: 81-84.

Tanaka, T., Takeno, T., Watanabe, Y., Uchiyama, Y., Murakami, T., Yamashita, H., Suzuki, A., Aoi, R., Iwanari, H., Jiang, S.Y., Naito, M., Tachibana, K., Doi, T., Shulman, A.I., Mangelsdorf, D.J., Reiter, R., Auwerx, J., Hamakubo, T., and Kodama, T. 2002. The Generation of Monoclonal Antibodies against Human Peroxisome Proliferator-activated Receptors (PPARs). J. Atheroscler. Thromb. 9: 233-242.

Thomsen, D.R., Post, L.E., and Elhammer, A.P. 1990. Structure of O-glycosidically linked oligosaccharides synthesized by the insect cell line Sf9. J. Cell. Biochem. 43: 67-79.

Tomita, M., Munetsuna, H., Sato, T., Adachi, T., Hino, R., Hayashi, M., Shimizu, K., Nakamura, N., Tamura, T., and Yoshizato, K. 2003. Transgenic silkworms produce recombinant human type III procollagen in cocoons. Nat. Biotechnol. 21: 52-56.

Tomiya, N., Howe, D., Aumiller, J.J., Pathak, M., Park, J., Palter, K.B., Jarvis, D.L., Betenbaugh, M.J., and Lee, Y.C. 2003. Complex-type biantennary N-glycans of recombinant human transferrin from *Trichoplusia ni* insect cells expressing mammalian [beta]-1,4-galactosyltransferase and [beta]-1,2-N-acetylglucosaminyltransferase II. Glycobiology. 13: 23-34.

van Die, I., van Tetering, A., Bakker, H., van den Eijnden, D.H., and Joziasse, D.H. 1996. Glycosylation in lepidopteran insect cells: identification of a beta 1—>4-N-acetylgalactosaminyltransferase involved in the synthesis of complex-type oligosaccharide chains. Glycobiology. 6: 157-164.

Velardo, M.A., Bretthauer, R.K., Boutaud, A., Reinhold, B., Reinhold, V.N., and Castellino, F.J. 1993. The presence of UDP-N-acetylglucosamine:alpha-3-D-mannoside beta 1,2-N-acetylglucosaminyltransferase I activity in *Spodoptera frugiperda* cells (IPLB-SF-21AE) and its enhancement as a result of baculovirus infection. J. Biol. Chem. 268: 17902-17907.

Venter, J.C., Adams, M.D., Myers, E.W., Li, P.W., Mural, R.J., Sutton, G.G., Smith, H.O., Yandell, M., Evans, C.A., Holt, R.A., Gocayne, J.D., Amanatides, P., Ballew, R.M., Huson, D.H., Wortman, J.R., Zhang, Q., Kodira, C.D., Zheng, X.H., Chen, L., Skupski, M., Subramanian, G., Thomas, P.D., Zhang, J., Gabor Miklos, G.L., Nelson, C., Broder, S., Clark, A.G., Nadeau, J., McKusick, V.A., Zinder, N., Levine, A.J., Roberts, R.J., Simon, M., Slayman, C., Hunkapiller, M., Bolanos, R., Delcher, A., Dew, I., Fasulo, D., Flanigan, M., Florea, L., Halpern, A., Hannenhalli, S., Kravitz, S., Levy, S., Mobarry, C., Reinert, K., Remington, K., Abu-Threideh, J., Beasley, E., Biddick, K., Bonazzi, V., Brandon, R., Cargill, M., Chandramouliswaran, I., Charlab, R., Chaturvedi, K., Deng, Z., Di Francesco, V., Dunn, P., Eilbeck, K., Evangelista, C., Gabrielian, A.E., Gan, W., Ge, W., Gong, F., Gu, Z., Guan, P., Heiman, T.J., Higgins, M.E., Ji, R.R., Ke, Z., Ketchum, K.A., Lai, Z., Lei, Y., Li, Z., Li, J., Liang, Y., Lin, X., Lu, F., Merkulov, G.V., Milshina, N., Moore, H.M., Naik, A.K., Narayan, V.A., Neelam, B., Nusskern, D., Rusch, D.B.,

Salzberg, S., Shao, W., Shue, B., Sun, J., Wang, Z., Wang, A., Wang, X., Wang, J., Wei, M., Wides, R., Xiao, C., Yan, C., Yao, A., Ye, J., Zhan, M., Zhang, W., Zhang, H., Zhao, Q., Zheng, L., Zhong, F., Zhong, W., Zhu, S., Zhao, S., Gilbert, D., Baumhueter, S., Spier, G., Carter, C., Cravchik, A., Woodage, T., Ali, F., An, H., Awe, A., Baldwin, D., Baden, H., Barnstead, M., Barrow, I., Beeson, K., Busam, D., Carver, A., Center, A., Cheng, M.L., Curry, L., Danaher, S., Davenport, L., Desilets, R., Dietz, S., Dodson, K., Doup, L., Ferriera, S., Garg, N., Gluecksmann, A., Hart, B., Haynes, J., Haynes, C., Heiner, C., Hladun, S., Hostin, D., Houck, J., Howland, T., Ibegwam, C., Johnson, J., Kalush, F., Kline, L., Koduru, S., Love, A., Mann, F., May, D., McCawley, S., McIntosh, T., McMullen, I., Moy, M., Moy, L., Murphy, B., Nelson, K., Pfannkoch, C., Pratts, E., Puri, V., Qureshi, H., Reardon, M., Rodriguez, R., Rogers, Y.H., Romblad, D., Ruhfel, B., Scott, R., Sitter, C., Smallwood, M., Stewart, E., Strong, R., Suh, E., Thomas, R., Tint, N.N., Tse, S., Vech, C., Wang, G., Wetter, J., Williams, S., Williams, M., Windsor, S., Winn-Deen, E., Wolfe, K., Zaveri, J., Zaveri, K., Abril, J.F., Guigo, R., Campbell, M.J., Sjolander, K.V., Karlak, B., Kejariwal, A., Mi, H., Lazareva, B., Hatton, T., Narechania, A., Diemer, K., Muruganujan, A., Guo, N., Sato, S., Bafna, V., Istrail, S., Lippert, R., Schwartz, R., Walenz, B., Yooseph, S., Allen, D., Basu, A., Baxendale, J., Blick, L., Caminha, M., Carnes-Stine, J., Caulk, P., Chiang, Y.H., Coyne, M., Dahlke, C., Mays, A., Dombroski, M., Donnelly, M., Ely, D., Esparham, S., Fosler, C., Gire, H., Glanowski, S., Glasser, K., Glodek, A., Gorokhov, M., Graham, K., Gropman, B., Harris, M., Heil, J., Henderson, S., Hoover, J., Jennings, D., Jordan, C., Jordan, J., Kasha, J., Kagan, L., Kraft, C., Levitsky, A., Lewis, M., Liu, X., Lopez, J., Ma, D., Majoros, W., McDaniel, J., Murphy, S., Newman, M., Nguyen, T., Nguyen, N., Nodell, M., Pan, S., Peck, J., Peterson, M., Rowe, W., Sanders, R., Scott, J., Simpson, M., Smith, T., Sprague, A., Stockwell, T., Turner, R., Venter, E., Wang, M., Wen, M., Wu, D., Wu, M., Xia, A., Zandieh, A., and Zhu, X. 2001. The sequence of the human genome. Science. 291: 1304-1351.

Vlak, J.M., Schouten, A., Usmany, M., Belsham, G.J., Klinge-Roode, E.C., Maule, A.J., Van Lent, J.W.M., and Zuidema, D. 1990. Expression of cauliflower mosaic virus gene I using a baculovirus vector based upon the p10 gene using a novel selection method. Virology. 179: 312-320.

Wenzel-Seifert, K., Hurt, C.M., and Seifert, R. 1998. High constitutive activity of the human formyl peptide receptor. J. Biol. Chem. 273: 24181-24189.

Wilson, L.E., Wilkinson, N., Marlow, S.A., Possee, R.D., and King, L.A. 1997. Identification of recombinant baculoviruses using green fluorescent protein as a selectable marker. Biotechniques. 22: 674-676, 678-681.

Wittinghofer, A., and Waldmann, H. 2000. Ras- A Molecular Switch Involved in Tumor Formation. Agnew. Chem. Int. Ed. 39: 4192-4214.

Yang, G.Z., Chen, Z.Z., Cui, D.F., Li, B.L., and Wu, X.F. 2002. Production of recombinant human calcitonin from silkworm (B. mori) larvae infected by baculovirus. Protein Pept. Lett. 9: 323-329.

Yang, S., and Miller, L.K. 1998. An efficient way to introduce unique restriction endonuclease sites into a baculovirus genome. J. Virol. Methods 76: 51-58.

Zhang, F.L., and Casey, P.J. 1996. Protein prenylation: molecular mechanisms and functional consequences. Annu. Rev. Biochem. 65: 241-269.

9

Heterologous Expression of Genes in Chinese Hamster Ovary Cells

Gyun Min Lee and Sun Ok Hwang

Abstract

For the production of therapeutic proteins, recombinant Chinese hamster ovary (rCHO) cells with dihydrofolate-reductase (DHFR)-mediated gene amplification have been most widely used in industry. The popularity of rCHO cells is likely to persist as the demand for therapeutic proteins continues to increase. In this chapter, we review the basic features and the strategies of constructing rCHO cell lines suitable for large-scale production of therapeutic proteins.

Introduction

Polypeptides which are glycosylated in their native state require glycosylation for full *in vivo* efficacy and their therapeutic profile is profoundly influenced by glycosylation (Boyd *et al.*, 1995; Lifely *et al.*, 1995; Savage, 1997; Kukuruzinska and Lennon, 1998; Hills *et al.*, 2001). Thus, human therapeutic proteins, the majority of which are glycoproteins, should be designed to achieve correct glycosylation. Carbohydrate content varies from about 3% (as in the case of immunoglobulin [Ig] G (Peppard *et al.*, 1989)) to more than 50% (as in the case of thrombopoietin [TPO] (Kato *et al.*, 1998)). To date, cDNAs encoding a number of therapeutically important glycoproteins such as erythropoietin (EPO) and antibodies have been cloned and expressed in mammalian cells. The major advantage of using mammalian cells as an expression system for overproduction of recombinant proteins destined for therapeutic and diagnostic use is that these cells are able to perform complex post-translational modifications, including glycosylation, in an authentic manner.

For high-level expression of therapeutic proteins, one of the most widely used mammalian expression systems is the gene amplification procedure offered by the use of dihydrofolate reductase-deficient (DHFR⁻) Chinese hamster ovary (CHO) cells (Kaufman *et al.*, 1985; Lin *et al.*, 1985; Kaufman, 1990; Page and Sydenham, 1991; Wurm *et al.*, 1992; Trill *et al.*, 1995; Sinacore *et al.*, 1996; Wurm *et al.*, 1997; Kim *et al.*, 1998b). A variety of therapeutic proteins, including tissue plasminogen activator (tPA), EPO, Factor VIII, and antibodies, have been commercially produced using this system.

This chapter will focus on the development of recombinant CHO (rCHO) cells with DHFR-mediated gene amplification for the production of therapeutic proteins.

Popularity of CHO Cells as Mammalian Hosts for Therapeutic Protein

For the production of therapeutic proteins, CHO cells have been the most popular host in industry since the first product, human tPA, was approved for marketing in the mid 1980's. As of 2002, more than 50 recombinant proteins produced from CHO cells have been approved for marketing. The popularity of CHO cells as hosts for therapeutic protein production can be attributed to the following reasons.

First, since CHO cells have been demonstrated to be safe hosts, it may be easier to get approval for marketing therapeutic proteins from the regulatory agency such as the US Food and Drugs Administration (Wurm, 1997). Human pathogenic viruses like Polio, Herpes, Hepatitis B, HIV, Measles, Adenoviruses, Rubella and Influenza do not replicate in CHO cells (Wiebe *et al.*, 1989). Thus, the risk that a viral adventitious agent is carried along with the product of interest is very low (Wurm, 1997).

Second, one of the disadvantages of using mammalian cells for protein production is their low specific productivity (q). For CHO cells, however, a powerful amplification system – the DHFR-mediated gene amplification system – is available. Accordingly, q can be increased significantly by gene amplification. Pendse *et al.* (1992) reported that the q of rCHO cells increased with gene copy number. When subjected to successive rounds of selection in medium containing stepwise increments in methotrexate (MTX) levels, rCHO cells with specific antibody productivity (q_{Ab}) as high as $100 \ \mu g/10^6$ cells/day were obtained by Page and Sydenham (1991).

Third, CHO cells have the capability to carry out efficient post-translational processing of complex proteins and the glycosylation pattern of recombinant proteins produced in CHO cells is very similar to that of native proteins (Takeuchi *et al.*, 1988; Adamson, 1994; Jenkins *et al.*, 1996).

Fourth, CHO cells can be easily adapted to suspension growth which is preferred to monolayer culture for large-scale production. Since the *in vivo* dosage for therapeutic proteins is high, large-scale cultures are necessary to meet a reasonable market size. In the case of therapeutic antibodies, *in vivo* dosage ranges from 0.5 to more than 5 mg/kg (Aulitzky *et al.*, 1991). Currently, stirred tank bioreactors over 10,000L are used for serum-free suspension cultures of rCHO cells producing tPA and antibodies.

DHFR/MTX-Mediated Gene Amplification

This system is based on the DHFR gene encoding the DHFR enzyme that catalyzes the conversion of folate to tetrahydrofolate. Tetrahydrofolate is required for biosynthesis of glycine from serine, that of thymidine monophosphate from deoxyuridine monophosphate, and for purine biosynthesis (Kaufman, 1990). MTX, a folic acid analog, binds to and inhibits the DHFR enzyme, leading to cell death. However, DHFR⁻ CHO cells, which have taken up an expression vector containing the DHFR gene, can develop resistance to MTX by amplifying it. Since single-step high-level resistance to MTX may result in cells synthesizing a mutant, MTX-resistant DHFR, or in cells with altered MTX-transport properties, gene amplification is usually achieved by stepwise selection for resistance to increasing concentrations of MTX (Kaufman, 1990). Depending on the vector construct and the protocol employed, incremental increases in concentration (2-4-fold for each step) from a few nanomolar to several millimolar MTX can be utilized. Highly MTX-resistant cells may contain several thousand copies of the DHFR gene, expressing several thousand-fold elevated levels of DHFR enzyme. Because the amplification unit is much larger (100 to 300 kilobase [kb]) than the size of the DHFR gene, the specific gene of interest, which is co-linked to the DHFR gene in the same expression vector or adjacently resides in the host chromosome, is coamplified (Kaufman *et al.*, 1983).

Amplification of the gene of interest by MTX selection is frequently accompanied by increased production of the desired gene product, whereas rapid loss of the amplified gene during long-term culture in the absence of MTX is common and responsible for decreased production (Michel *et al.*, 1985; Stark *et al.*, 1989; Kim *et al.*, 1998a, 1998b).

DHFR⁻ CHO Cell Lines

The convenience of the DHFR selection system relies on the availability of DHFR⁻ CHO cell lines. These lines do not grow in the absence of nucleosides (thymidine, glycine, and hypoxanthine) unless they acquire a functional DHFR gene.

The two most frequently used DHFR⁻ CHO cell lines are DXB11 and DG44. The DXB11 cell line (also called DUK-XB11 or DUKX) was originally derived from CHO-K1 and is available from ATCC (catalog number CRL-9096). CHO-K1, a derivative of the CHO cell line that was established by Puck (Kao and Puck, 1968), is also available from ATCC (catalog number CCL-61). DXB11 contains a single point mutant allele, and the other DHFR allele has been deleted (Urlaub and Chasin, 1980; Graf and Chasin, 1982). DG44 was derived from CHOpro3⁻ cells (also referred to as CHO Toronto) in 1982. CHOpro3⁻ is itself a derivative of the CHO cell line that was established in the late 1950's. DG44, a double deletion mutant that contains no copies of the hamster DHFR gene, is most useful for documenting the status of an introduced DHFR minigene without any background interference (Flintoff *et al.*, 1976, 1982; Urlaub *et al.*, 1983, 1986).

rCHO cell lines could be developed more efficiently if serum-free, suspension-adapted phenotypes were incorporated into DHFR⁻ CHO cell lines (Sinacore *et al.*, 1996). In our experience, DXB11 is easier to grow in suspension than DG44, but we have observed reversion of DXB11 during adaptation to serum-free, suspension culture. By contrast, no revertants of DG44 have been seen.

Methods of Gene Transfer

A wide variety of methods are available for introducing foreign DNA into mammalian cells. One has a choice among physical (electroporation, laser poring, microinjection, and biolistics), chemical (calcium phosphate, DEAE-dextran, polybrene-DMSO, and lipofection), and biological (protoplast fusion, retroviruses, erythrocyte ghosts, and receptor-mediated transfer) gene transfer methods (Schlokat *et al.*, 1997). Originally, calcium phosphate co-precipitation was the most common method for non-viral gene transfer into mammalian cells, particularly in the case of rapidly growing adherent cells (Graham and van der Eb, 1973; Maitland and McDougall, 1977;

Wigler *et al*., 1977). A thorough investigation of calcium phosphate DNA precipitate formulation and its optimization have resulted in much higher transfection rates and greater reliability (Jordan *et al*., 1996). Polybrene-DMSO mediated gene transfer has been also reported to be successful in stably transforming CHO cells (Chaney *et al*., 1986). More recently, the development of a new generation of liposome forming agents represented by the Lipofectin reagent, composed of a mixture of DOTMA (N-(1-(2,3-dioleyloxy)propyl)-N,N,N-trimethylammonium) and DOPE (dioleoyl phosphatidyl ethanolamine), has greatly improved conventional liposome-mediated gene transfer (lipofection). A wide variety of Lipofectin-like reagents are commercially available in easy-to-use kits, allowing straightforward gene transfer to CHO cells.

Selection Vectors and Selection Strategies for Coamplification

Two classes of vectors for co-amplification – a single and a double vector system – have been described (Kaufman, 1990). In the single vector system, the gene of interest and the selection gene transcription units are contained in the same vector, whereas product gene transcript unit and selection gene transcription unit are encoded on separate plasmids in the double vector system. One advantage of the latter is that the ratio of product to selection gene can be varied in favor of the product gene. On the other hand, the efficiency of double vector systems depends on the ability of the two DNAs to be ligated in the cell.

Circular plasmid DNA is usually used for transfection. Within hours, endonuclease activity, which is abundant in nucleus, transforms circular DNA into linear DNA with free 5'- and 3' ends at random location. This step is essential for the integration of the linear backbone within a host chromosome (Finn *et al*., 1989). Linearization of circular DNA by the restriction enzyme digestion prior to transfection has been recommended for improving transfection efficiencies (Wurm, 1997).

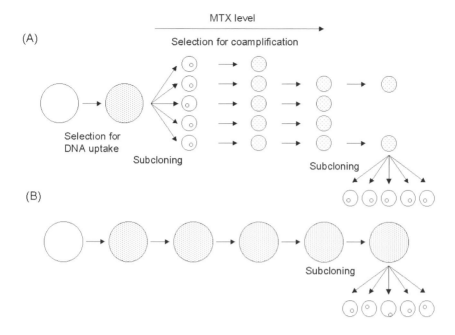

Figure 1. Strategies for the establishment of high producer clones. A: selection based on individual parental clones. B: selection based on parental cell pool.

After transfection, parental clones that have taken up the DHFR gene can be isolated in selective medium lacking glycine, hypoxanthine and thymidine. To improve selection efficiency for high producing parental clones, antibiotics such as neomycin, hygromycin, zeocin and puromycin can be added to the medium in concert with plasmids that provide the corresponding resistance genes. The selective medium may also be supplemented with a low (1-4 nM) concentration of MTX.

To select clones with highly amplified heterologous genes, two strategies are commonly used (Figure 1). In scheme A, individual parental clones, usually isolated by limiting dilution method in 96 well culture plates, are independently grown in increasing concentrations of MTX for gene amplification. Although the final clones resistant to high levels of MTX are derived clonally, they become heterogeneous with respect to q (Kim et al., 1998b). Thus, subcloning for high-producing clones is necessary. The selection outlined in scheme A is very labor-intensive. In scheme B, a pool of parental clones is collected and grown in increasing concentrations of MTX. The final resistant cell pool is cloned by the limiting dilution method at the end of the selection process. The advantage of scheme B is that it requires less effort to amplify large numbers of individual parental clones within the pool, but it has the potential of losing high producing clones.

Since the isolation of clones with high expression level clearly depends on the number of clones that have been screened, screening procedures are time-consuming and tedious regardless of the selection strategy used. As a result, there have been significant efforts to simplify screening procedures (Walls and Grinnell, 1990; Schneider and Rusconi, 1996; Wirth, 1997; Gains and Wojchowski, 1999; Chung *et al.*, 2000). Automation of many processes involved in the culture of clones for testing has significant advantages. A small number of companies such as Cytogration, The Automation Partnership, RTS Life Science International, and CRS Biodiscovery manufacture fully automated cell culture systems (Kempner and Felder, 2002). The use of automated cell culture system will facilitate the selection of co-amplification.

Characterization of Antibody Producing rCHO Cells in the Course of DHFR-Mediated Gene Amplification and Stability in the Absence of Selective Pressure

When rCHO cells are subjected to successive rounds of selection in medium containing stepwise increments in MTX concentration, the gene of interest is supposed to be co-amplified with the DHFR gene, leading to increased q (Kaufman *et al.*, 1983). In reality, the q of rCHO cells does not always increase linearly with the MTX concentration and the enhancement in q varies significantly among clones (Kim *et al.*, 1998b). To develop a strategy for making high-producing rCHO cell lines for therapeutic proteins, we systematically characterized therapeutic antibody producing rCHO cells during DHFR-mediated gene amplification and assessed their stability during long-term culture

(Kim *et al.*, 1998a, 1998b; Kim and Lee, 1999). The following paragraphs summarize our results.

Throughout this chapter, we have used the term "clone" for consistency. However, immortalized mammalian cells must be considered as heterogeneous even after very short period of cultivation. Therefore, the term "clonally derived population" would be more appropriate. For the expression of genetically engineered Ig in CHO cells, a two-vector system with each Ig chain on a separate plasmid (Wood *et al.*, 1990; Page and Sydenham, 1991; Fouser *et al.*, 1992; Tada *et al.*, 1994), as well as a one-vector system with both Ig chains on a single plasmid (Page and Sydenham, 1991; Newman *et al.*, 1992; Kim *et al*, 2001) have been utilized. Equivalent expression level and stability of the resulting cell lines were observed, and additional data will be needed to determine which is better.

We selected the two-vector system because of the ease of inserting the heavy chain (HC) and light chain (LC) coding sequences into individual plasmids. The LC and HC cDNAs encoding a chimeric antibody against the S surface antigen of HBV were constructed and separately placed under the control of the hCMV promoter in expression vectors. The DHFR gene linked to acrippled SV40 early promoter was included in the LC expression vector because of the ease of selecting producer cells. Furthermore, because most IgG-producing mouse myeloma tumors and cell lines are known to synthesize more LC than HC, we attempted to express more LC than HC by including the DHFR gene in the LC expression vector. The resulting plasmids were co-transfected into DHFR⁻ CHO cells using Lipofectin. To select the rCHO cell with the highest number of amplified heterologous genes, we followed the scheme A of Figure 1.

Briefly, drug selection was carried out by seeding 10^4 cells/well in 96 well plates containing α-MEM supplemented with 10% dialyzed fetal bovine serum (dFBS) and drugs. The culture supernatant of drug-resistant clones isolated from 96-well plates was tested for assembled antibody production by ELISA. Twenty parental clones with high productivity were exposed to increasing levels of MTX. Among them, the CS13 parental clone showed significantly elevated antibody expression levels following incremental increases in MTX concentration.

To isolate the high producer (HP) subclone at each MTX level, cells collected at various MTX levels were plated at 0.3-0.8 cells/well in

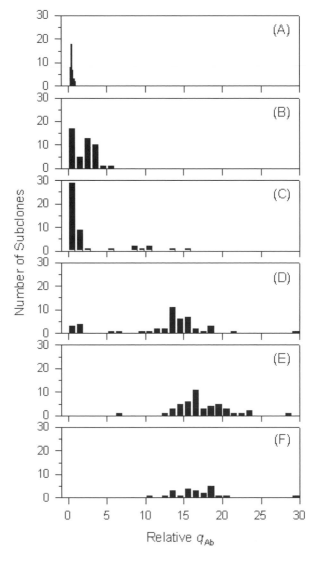

Figure 2. Clonal analysis of amplified cells at various MTX levels (Kim *et al.*, 1998b). A: CS13 parental cells. B: CS13-0.02 cells. C: CS13-0.08 cells. D: CS13-0.32 cells. E: CS13-1.0 cells. F: CS13-4.0 cells.

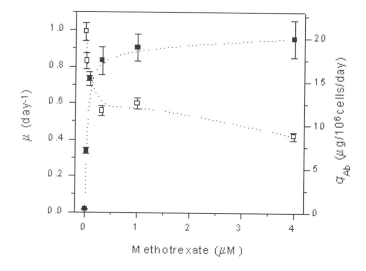

Figure 3. Specific growth rate (μ, □) and antibody production rate (q_{Ab},■) of HP subclones selected at various MTX levels (Kim *et al.*, 1998b).

96-well plates containing α-MEM supplemented with 10% dFBS and the corresponding level of MTX. Higher MTX concentrations resulted in lower cloning efficiency, indicating that MTX inhibits cell growth. The cell population, although derived from the CS13 parental clone, became heterogeneous with respect to q_{Ab} at elevated MTX levels (Figure 2). However, there was a general trend toward higher q in subclones isolated from cells selected at increasing MTX levels. The q_{Ab} of subclones reached a maximum at 1 μM MTX. The HP subclone was isolated at each MTX level (CS13*, CS13*-0.02, CS13*-0.08, CS13*-0.32, and CS13*-4.0, respectively) and μ and q_{Ab} were determined.

Figure 3 shows that the μ of HP subclones was inversely related to the MTX level. On the other hand, q_{Ab} rapidly increased as the MTX level went up to 0.08 μM and gradually increased thereafter to 20.0 ± 2.1 μg/10^6 cells/day at 4 μM MTX. Compared with the q_{Ab} of the parental subclone (CS13*), the q_{Ab} of CS13*-4.0 was enhanced approximately 61-fold. Southern blot analysis revealed that the higher q_{Ab} was related to the extent of Ig gene amplification (Figure 4) and that an up to 55-fold amplification resulted in a corresponding increase

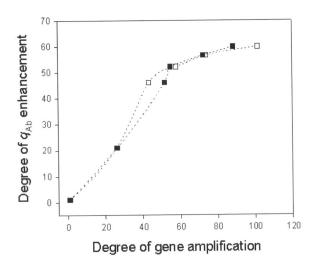

Figure 4. Relationship between the enhancement of q_{Ab} and the extent of LC (□) and HC (■) gene amplification (Kim *et al.*, 1998b). The q_{Ab}, LC and HC gene copies of amplified subclones were normalized by those of CS13* subclones, respectively.

in q_{Ab}. However, further gene amplification did not affect q_{Ab}, suggesting that there might be transcriptional and/or post-translational bottleneck in highly amplified subclones. Northern blots were carried out in order to determine the LC and HC mRNA contents in amplified HP subclones. When plotted as a function of the extent of gene amplification, the LC and HC mRNA content increased almost linearly with Ig gene amplification eliminating the possibility of transcriptional limitation in highly amplified subclones. However, for economical antibody production, it may not be an optimal to operate cells with high number of cloned gene copies since their detrimental effect on cell growth rate may outweigh the benefits of enhanced q_{Ab}.

For large-scale production of antibodies, not only the expression level but the stability of cell lines is critical. The instabilities of murine hybridomas (Frame and Hu, 1990; Ozturk and Palsson, 1990; Chuck and Palsson, 1992; Coco-Martin *et al.*, 1992; Merritt and Palsson, 1993), quadromas (Salazar-Kish and Heath, 1993), and transfectomas (Bae *et al.*, 1995; Kim *et al.*, 1996) in long-term cultures have been extensively studied. Changes in rCHO populations after extended culture in the presence and absence of selective pressure have been reported (Kaufman et al., 1985; Zettlmeissl et al., 1987; Weidle et al., 1988; Sinacore et al., 1996). However, there are few reports assessing the quantitative effect of increasing number of cloned gene copies on the stability of CHO cells with respect to q (Michel et al., 1985).

To determine the stability of amplified subclones at various MTX levels during long-term cultivation, cells were grown in the absence or presence of the corresponding level of MTX. Cell growth and antibody characteristics were monitored over 100 days. Amplified subclones at higher MTX level, which have higher gene copy numbers, displayed lower growth rates. In addition, in most cases, cell growth rates were higher in the absence of MTX. The growth rates of HP subclones, particularly CS13*-0.08, improved during long-term culativation in MTX-free medium. This improvement in cell growth was due to loss of genes (see below).

In the presence of MTX, the q_{Ab} of amplified subclones decreased slightly during culture (Figure 5). At the 21st passage (approximately 60 generations), the q_{Ab} of all subclones (except CS13*-0.08) was still more than 90% of the initial q_{Ab}. In the absence of MTX, the decrease in q_{Ab} was significant. Furthermore, the relative extent of decrease in q_{Ab} varied among amplified subclones. The CS13*-1.0 subclone which was the most amplified clone among all subclones tested, was most stable and retained 59% of the initial q_{Ab} at the 21st passage. On the other hand, CS13*-0.08 was the most unstable and retained only 23% of the initial q_{Ab} at the 21st passage. This was unexpected since the clone with higher gene copies was expected to be under greater metabolic stress and hence to be more unstable in the absence of selective pressure.

To gain insights into the genetic basis for the different degrees of stability in HP subclones, changes in genomic organization and chromosomal distribution of amplified LC and HC cDNAs during long-term culture were characterized by Southern blot hybridization and fluorescence in situ hybridization (FISH). In addition, we investigated the possible role of $(TTAGGG)_n$ in the stabilization of amplified structures by co-localizing it with amplified sequences. It is

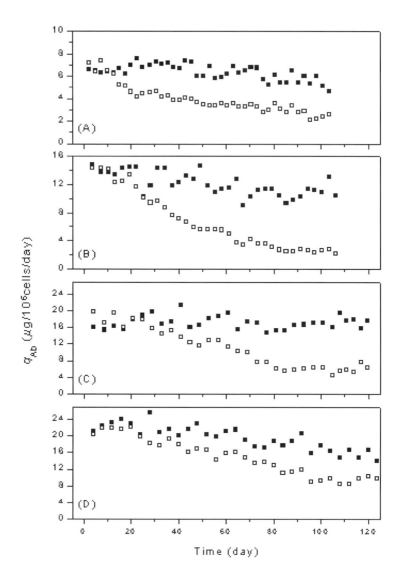

Figure 5. Antibody production characteristics of HP subclones during long-term culture in the presence (■) and absence (□) of MTX (Kim *et al.*, 1998b). A: CS13*-0.02 subclone, B: CS13*-0.08 subclone, C: CS13*-0.32 subclone, D: CS13*-1.0 subclone.

known that simple repetitive sequences like $(TTAGGG)_n$ occur at the ends of linear chromosomes in several species, including Chinese hamster, and that they are essential for chromosome integrity (Meyne *et al.*, 1989; Blackburn, 1991). Acquisition of $(TTAGGG)_n$ has also been suggested to be important in the stabilization of the chromosomal structures formed during gene amplification (Toledo *et al.*, 1992; Bertoni *et al.*, 1994).

Southern analysis of HP subclones revealed that LC and HC cDNAs were in close proximity (within 23 kb on an amplification unit), and that their configuration was not disrupted during long-term culture in the absence of MTX. However, when LC and HC genes were located on the metaphase chromosomes of HP subclones using FISH, amplified sequences were present as an extended array on diverse marker chromosomes. HP subclones exposed to high levels of MTX had more kinds of marker chromosomes: CS13*-0.02 isolated at 0.02 μM MTX had only one marker chromosome, whereas CS13*-1.0 isolated at 1 μM MTX had five different ones. Each marker chromosome exhibited a different fate during long-term culture in the absence of MTX, resulting in different degree of stability among HP subclones. Cells containing stable marker chromosomes constituted the dominant sub-population in CS13*-1.0, and this subclone became the most stable with regard to antibody production. Furthermore, dual-color FISH showed that the telomeric ends of amplified arrays on stable marker chromosomes were always surrounded by $(TTAGGG)_n$, indicating that these sequences are closely related to the stability and evolution of amplified sequences. Thus, assessment of the genotypic stability of amplified CHO cells is a prerequisite for understanding their production stability during long-term culture in the absence of selective pressure. Recently, a thorough review on the stability of protein production from recombinant mammalian cells has been published (Barnes *et al.*, 2003).

Although they were derived clonally, rCHO cells became quite heterogeneous with respect to q_{Ab} at elevated MTX levels (Figure 2). Likewise, their stability during long-term culture may also be heterogeneous. If so, clonal analysis of the entire cell population may allow one to screen for clones exhibiting better stability. However, this requires extensive effort. If the initial properties of a clone such as q_{Ab}, μ, and gene copy numbers help predict stability, much effort toward screening a stable clone will be saved. To test this possibility, the stability of 20 subclones randomly selected from the amplified CHO cell population (CS13-1.0) shown in Figure 2E, was examined during long-term culture in the absence of selective pressure to: (i) determine clonal heterogeneity; (ii) screen for stable subclones; and (iii) determine whether subclone stability can be predicted from their initial properties.

During 8 weeks of cultivation in the absence of selective pressure, the μ of most subclones did not change significantly. On the other hand, their q_{Ab} decreased from 30% to 80% in the various subclones. Among the initial properties examined (q_{Ab}, μ, and gene copy numbers), only q_{Ab} was useful to predict stability. Subclones with high q_{Ab} were relatively stable with regard to antibody production during long-term culture in the absence of selective pressure ($P<0.005$, ANOVA). Taken together, the clonal heterogeneity in an amplified CHO cell population necessitates clonal analysis for screening stable clones with high q_{Ab}.

In the above section, we highlighted a strategy for making high producing rCHO cell lines for therapeutic proteins by following scheme B of Figure 1. However, it is hard to conclude which strategy is better and, depending on the situation (time, labor, final purpose of the cell line, etc), a different strategy for developing a rCHO cell line may be selected.

Key Determinants in the Occurrence of Clonal Variation

As mentioned earlier, single-step high-level resistance to MTX may result from mechanisms other than gene amplification. However, when we performed gene amplification by stepwise selection for resistance to increasing concentrations of MTX, the q of rCHO cells did not increase with the drug concentration and the enhancement in q varied significantly among clones (Kim *et al.*, 1998b). The occurrence of clonal variations in regard to q enhancement during gene amplification had not been clearly elucidated. Only fragmented explanations such as MTX transport defects and altered affinity of DHFR for MTX have been proposed. Assaraf and Schimke (1987) reported that among 22 MTX-resistant clones, only 5 showed DHFR gene amplification while the others did not exhibit any increase in either DHFR gene copy numbers or steady-state mRNA levels. These authors suggested that mutations in the MTX transport system of the cell membrane might be a major factor for acquiring resistance to the drug. As observed in clonal isolates of MTX-resistant CHO cells by Flintoff and Essani (1980), enhanced MTX-resistance which is not related of gene amplification events may also be attributed in part to DHFR variants with altered affinity for MTX.

To investigate the reasons for the occurrence of clonal variations with respect to antibody expression during gene amplification, we examined 23 representative amplified clones using competition assay with an antifolate agent, fluorescent MTX (F-MTX), and Southern blot analysis (Kim et $al.$, 2001). When subjected to stepwise MTX selection, all 23 parental clones survived but only one (hu17) displayed enhanced q_{Ab} with increasing MTX level up to 0.32 µM. Hu17-0.32 did not experience severe genetic rearrangement during gene amplification. Moreover, it had only one 49-kb amplification unit including the LC and HC cDNAs and an intact MTX transport system. Clones that did not display enhanced q_{Ab} at 0.32 µM MTX developed resistance through other mechanisms such as an impaired MTX transport system. Thus, although stepwise selection for resistance to increasing concentrations of MTX was applied, most clones did not undergo gene amplification. This result was rather surprising because we gradually increased MTX levels (0, 0.02, 0.08 and 0.32 µM) over a period of 4 months to avoid selecting for MTX resistance mechanisms other than DHFR gene amplification which often occur when larger selection steps are employed (Kaufman, 1990).

Importance of Clone Selection for Manufacturing

In many cases, a cell line is established based on clones exhibiting high productivity, and process development for manufacturing is carried out using that cell line. However, if a specific cell culture process is in mind, clones suitable for that process should be selected because of clonal variability. In this section, we review two examples of clonal variability indicative of the importance of clone selection for manufacturing.

Temperature is one of the key parameters influencing cell growth and recombinant protein production. Most mammalian cells, including CHO cells, are cultivated at 37°C to simulate the body environment. Although lowering the culture temperature below 37 °C decreases μ, a number of studies have demonstrated its beneficial effects. These include maintaining high viability for longer periods (Furukawa and Ohsuye, 1998; Reuveny *et al.*, 1986), reducing glucose and glutamine consumption rates (Barnabe and Butler, 1994; Weidemann *et al.*, 1994), reducing the specific oxygen uptake rate and protease activity (Chuppa *et al.*, 1997), and improving tolerance against shear stress (Ludwig *et al.*, 1992).

Unlike its influence on μ, the effects of lowering the culture temperature on q cannot be generalized. A few reports suggest that the q of CHO cells increases at low temperature (Furukawa and Ohsuye, 1998, 1999; Hendrick *et al.*, 2001; Yun *et al.*, 2003). When the culture temperature was lowered from 37°C to 33°C, a more than 2.5-fold increase in maximum EPO concentration was achieved (Yun *et al.*, 2003). This did not only result from extended culture longevity and a slower release of proteolytic enzymes from dead cells, but mainly because of 3.9-fold increase in q_{EPO}. The quality of EPO produced at 33°C with respect to isoform pattern, sialic acid content and *in vivo* biological activity was comparable to, or better than that of EPO synthesized at 37°C. These results suggested that low temperature cultivation has potential for the commercial production of EPO in rCHO cells. On the other hand, we and others observed that the q of CHO cells was not enhanced by lowering culture temperature (Ryll *et al.*, 2000). Thus, the beneficial effect of low temperatures on q appears to depend on cell type, target protein and, possibly, on the integration site of the foreign gene. To examine the latter possibility, we cultivated 10 antibody-producing parental CHO clones with different integration sites at 32°C and 37°C. We observed enhanced q at 32°C in 6 cases while there was no change or a decrease in q in the other 4 clones. These results demonstrate that the beneficial effect of lowering the culture temperature on q depends on the integration site of a foreign gene. Thus, clonal selection is a key factor to the success of the application of low culture temperature to enhanced foreign protein production in CHO cells.

Hyperosmotic pressure, which can be induced by adding cheap salts to the medium, has been recognized as an economical solution to increase q in mammalian cells. However, since cell growth is depressed at elevated osmolality, the use of hyperosmolar media in batch cultures does not increase maximum product titers substantially (Ozturk and Palsson, 1991; Oh et al, 1993, 1995; Øyaas et al., 1994a, 1994b; Reddy and Miller, 1994; Lee and Park, 1995; Park and Lee, 1995; Ryu and Lee, 1997; Ryu et al., 2000). To overcome this drawback, several strategies such as adaptation to hyperosmotic pressure (Oh et al, 1993, 1995), use of osmoprotectants in media (Øyaas et al., 1994a, 1994b), and two-stage culture (Park and Lee, 1995) have been used to improve production. Among these, the use of strong osmoprotectants such as glycine betaine (Oh et al., 1995), is probably most feasible because this chemical can be easily added to the culture medium. Indeed, addition of glycine betaine was found to improve cell growth of rCHO cell line (4B3) at elevated osmolality without affecting q (Chen et al., 1998). Contrary to this result, we observed that although glycine betaine improved cell growth of other rCHO cell lines at elevated osmolality, q reduced significantly (Ryu et al., 2000). Thus, the efficacy of simultaneously using hyperosmotic pressure and glycine betaine as a means to improve foreign protein production appears to vary among different rCHO cell lines.

To investigate the influence of clonal variations on the response of rCHO cells to hyperosmotic pressure and glycine betaine, hyperosmolar batch cultures of 23 rCHO clones producing TPO were carried out in the absence and presence of 15 mM glycine betaine (Kim et al., 2000). Glycine betaine was found to have a strong osmoprotective effect on all 23 clones and enabled 22 of them to grow at 542 mOsm/kg, but at a cost of a reduced q_{TPO}. Furthermore, the relative decrease in q_{TPO} varied significantly among clones. Overall, 6 clones out of 23 displayed more than a 40% increase in maximum TPO titer in hyperosmolar medium containing glycine betaine, compared to standard medium with physiological osmolality.

The above examples highlight the fact that integration of target genes at different sites can affect the q of rCHO cell lines subjected to low temperature cultivation or growth in hyperosmolar medium supplemented with glycine betaine and emphasize the importance of clone selection for manufacturing.

Recently, there has been growing interest on site-specific integration of foreign genes into host chromosomes using Cre/loxP or Flp/frt-assisted homologous recombination (Baer and Bod, 2001; Koduri *et al.*, 2001; Wilson and Kola, 2001). The rationale is that integration of a foreign gene within a high transcription rate locus will result in high q. In selecting optimal integration sites for rCHO cells, a number of factors such as transcription rate, easy amplification and other characteristics suitable for large-scale cultivation should be considered. This will require that the genome of CHO cells be more fully understood.

CHO Cell Engineering

To improve the characteristics of rCHO cells in regard to cell growth and foreign protein production, overexpression of anti-apoptotic genes, chaperones, or glycosyltransferases have been attempted. These examples of CHO cell engineering are summarized in the following section.

Apoptotic cell death induced in mild suboptimal conditions is an active, genetically controlled process of cell suicide mediated by the activation of a series of caspases. Therefore, apoptotic cell death can be regulated to some extent by genetic modification such as the expression of the family of bcl-2 survival proteins (Simpson *et al.*, 1997; Perani *et al.*, 1998; Kim and Lee, 2001). In contrast, necrotic cell death induced at high stress level is a passive, genetically uncontrolled death. Thus, overexpression of antiapoptotic gene for extending culture longevity is effective on the condition that cells die by apoptosis, not necrosis.

CHO cells in serum-free batch culture were found to die by apoptosis, which is a physiological and developmental mode of death in various cell types including CHO (Singh *et al.*, 1994; Moore *et al.*, 1995; Goswami *et al.*, 1999). Apoptotic cell death can be triggered by several adverse conditions such as unavailability of nutrients, serum, growth

Table 1. Examples of CHO cell engineering for suppressing apoptosis.

Gene expressed	Apoptosis-inducing condition	Observation	References
bcl-2	Insulin and transferrin withdrawal, glutamine depletion	Extended culture durations; did not assess productivity	Goswami et al. (1999); Sanfeliu and Stephanopoulos (1999)
bcl-2	Exposure to 2.5 mM ammonia, treatment with 4mM thymidine	Improved viability and extended culture duration, but no improvement in titer	Tey et al. (2000)
bcl-2, bcl-x_L	Alphaviral infection	Improved viability and improved titer	Mastrangelo et al. (2000a, 2000b)
bcl-2	Hyperosmotic stress up to 522 mOsm/kg, treatment with 5mM sodium butyrate	Extended culture duration and improved titer	Kim and Lee (2001, 2002a)
Antisense RNA of caspase 3	Treatment with 5mM sodium butyrate	Extended culture duration, but no improvement in titer	Kim and Lee (2002b)

factors and accumulation of toxic metabolites during cultivation. Sodium butyrate (NaBu), which is widely used in rCHO cell cultures to enhance foreign protein expression, can also induce apoptotic cell death (Chang *et al.*, 1999; Kim and Lee, 2001). It has been reported that overexpression of survival genes can suppress such apoptotic cell death (Table 1)(Zanghi *et al.*, 1999; Mastrangelo *et al.*, 2000b; Tey *et al.*, 2000). For instance, overexpression of bcl-2 in rCHO cells producing a humanized antibody considerably suppressed NaBu-induced apoptosis by inhibiting caspase 3 activity and extended culture longevity. This resulted in a 2-fold increase in final antibody concentration, compared to rCHO cells that did not overexpress bcl-2 (Kim and Lee, 2001). Recently, we developed an apoptosis-resistant DHFR⁻ CHO host cell line by introducing the bcl-2 gene into the DHFR⁻ CHO cell line (DUKX-B11) and subsequently selecting clones that stably overexpress bcl-2 in the absence of selective pressure (Lee and Lee, 2003). The availability of this cell line should expedite the development of apoptosis-resistant rCHO cells for the production of therapeutic proteins.

Post-translational modifications of a protein include various rate-limiting interactions with numerous chaperones and enzymes in the secretory pathway. The efficient secretion of recombinant proteins from eukaryotic cells requires a simple way of targeting the proteins to endoplasmic reticulum (ER) and directing their translocation to the lumenal compartment of the ER. The lumen of the ER contains a number of molecular chaperones that assist in the late stages of protein biosynthesis and folding. Proteins are secreted from the lumen of the ER after conformational quality control is complete. Protein disulfide isomerase (PDI) also functions as a molecular chaperone and has been found to be associated with misfolded proteins in the ER. A number of studies have shown specific interactions between newly synthesized glycoproteins and putative chaperones such as calnexin and calreticulin (Zapun *et al*, 1998; Oliver *et al.*, 1999). Like calnexin and calreticulin, ERp57, a member of the PDI family, also modulates glycoprotein folding (Gething and Sambrook, 1992).

The effects of modulating the intracellular concentration of ER chaperones on the secretion of foreign proteins have been studied in

CHO cells. Overexpression of PDI decreased the secretion of TNFR:Fc and led to its retention in the ER (Davis *et al.*, 2000). Reduction of endogenous GRP78 levels by introduction of antisense GRP78 genes improved secretion of a tissue plasminogen activator variant (Dorner *et al.*, 1988), while overexpression of GRP78 blocked secretion of von Willebrand factor and a mutant form of factor VIII but did not affect the secretion of M-CSF. Thus, increased expression level of GRP78 may selectively lower the secretion efficiency of proteins that are normally associated with this chaperone (Dorner *et al.*, 1992). Recently, we reported that doxycyclin-regulated ERp57 expression in rCHO cells increased q_{TPO} (Hwang *et al.*, 2003). Overall, the effects of overexpression of ER chaperones on the secretion of foreign proteins in CHO cells appear to vary depending on the combination of chaperone and foreign protein (Dorner *et al.*, 1988, 1992; Davis *et al.*, 2000; Hwang *et al.*, 2003).

Glycosylation is one of key factors in determining the efficacy of many therapeutic proteins (Hansen *et al.*, 1988; Szkudlinski *et al.*, 1993). Improved glycoproteins have been obtained by genetic manipulation of the glycosylation pathway, which can be achieved by introducing glycosyltransferase and glycosidase genes into host cells (Umaña *et al.*, 1999a, 1999b) or by antisense inhibition of endogenous glycosyltransferase activity (Prati *et al.*, 1998).

Conclusion

CHO cells have been the most popular host in industry for the production of therapeutic proteins. Although the molecular mechanisms of gene transfer and amplification still need to be fully elucidated, the popularity of this host is likely to persist as the demand for antibodies and other therapeutic proteins continues to increase.

References

Adamson, R. 1994. Design and operation of a recombinant mammalian cell manufacturing process for rFVIII. Ann. Hematol. 68: S9-S14.

Assaraf, Y.G. and Schimke, R.T. 1987. Identification of methotrexate transport deficiency in mammalian cells using fluoresceinated methotrexate and flow cytometry. Proc. Natl. Acad. Sci. USA. 84: 7154-7158.

Aulitzky, W.E., Schulz, T.E., Tilg, H., Niederwieser, D., Larcher, K., Ostberg, L., Scriba, M., Martindale, J., Stern, A.C., Grass, P., Mach, M., Dierich, M., and Huber, C. Human monoclonal antibodies neutralizing cytomegalovirus (CMV) for prophylaxis of CMV disease: Report of a Phase I trial in bone marrow transplant recipients. 1991. J. Infect. Dis. 163: 1344-1347.

Bae, S.W., Hong. H.J., and Lee, G.M. 1995. Stability of transfectomas producing chimeric antibody against the pre-S2 surface antigen of hepatitis B virus during a long-term culture. Biotechnol. Bioeng., 47: 243-251.

Baer, A. and Bode, J. 2001. Coping with kinetic and thermodynamic barriers: RMCE, an efficient strategy for the targeted integration of transgenes. Curr. Opin. Biotechnol. 12: 473-480.

Barnabe, N. and Butler, M. 1994. Effect of temperature on nucleotide pools and monoclonal antibody production in a mouse hybridoma. Biotechnol. Bioeng. 44: 1235-1245.

Barnes, L.M., Bentley, C.M, and Dickson, A.J. 2003. Stability of protein production from recombinant mammalian cells. Biotechnol. Bioeng. 81: 631-639.

Bertoni, L., Attolini, C., Tessera, L., Mucciolo, E., and Giulotto, E. 1994. Telomeric and nontelomeric (TTAGGG)$_n$ sequences in gene amplification and chromosome stability. Genomics. 24: 53-62.

Blackburn, E.H. 1991. Structure and function of telomeres. Nature. 350: 569-573.

Boyd, P.N., Lines, A.C., and Patel, A.K. 1995. The effect of the removal of sialic acid, galactose and total carbohydrate on the functional activity of Campath-1H. Mol. Immunol. 32: 1311–1318.

Chaney, W.G., Howard, D.R., Pollard, J.W., Sallustio, S., and Stanley, P. 1986. High-frequency transfection of CHO cells using polybrene. Somat. Cell Mol. Genet. 12: 237-244.

Chang, K.H., Kim, K.S., and Kim, J.H. 1999. N-acetylcysteine increases the biosynthesis of recombinant EPO in apoptotic Chinese hamster ovary cells. Free Radic. Res. 30: 85-91.

Chen, Z., Liu, H., and Wu, B. 1998. Hyperosmolality leads to an increase in tissue-type plasminogen activator production by a Chinese hamster ovary cell line. Biotechnol. Tech. 12: 207-209.

Chuck, A.S. and Palsson, B.O. 1992. Population balance between producing and nonproducing hybridoma clones is very sensitive to serum level, state of inoculum, and medium composition. Biotechnol. Bioeng. 39: 354-360.

Chung, J.Y., Kim, T.K., and Lee, G.M. 2000. Morphological selection of parental Chinese hamster ovary cell clones exhibiting high-level expression of recombinant protein. Biotechniques. 29: 768-774.

Chuppa, S., Tsai, Y.-S., Yoon, S., Shackleford, S., Rozales, C., Bhat, R., Tsay, G., Matanguihan, C., Konstantinov, K., and Naveh, D. 1997. Fermentor temperature as a tool for control of high-density perfusion cultures of mammalian cells. Biotechnol. Bioeng. 55: 328-338.

Coco-Martin, J.M., Oberink, J.W., Brunink, F., van der Velden-de Groot, T.A.M., and Beuvery, E.C. 1992. Instability of a hybridoma cell line in a homogeneous continuous perfusion culture system. Hybridoma. 11: 653-665.

Davis, R., Shooley, K., Rasmussen, B., Thomas, J., and Reddy, P. 2000. Effect of PDI overexpression on recombinant protein secretion in CHO cells. Biotechnol. Prog. 16: 736-743.

Dorner, A. J., Krane, M. G., and Kaufman, R. J. 1988. Reduction of endogenous GRP78 levels improves secretion of a heterologous protein in CHO cells. Mol. Cell. Biol. 8: 4063-4070.

Dorner, A.J., Wasley, L.C., and Kaufman, R.J. 1992. Overexpression of GRP78 mitigates stress induction of glucose regulated proteins and blocks secretion of selective proteins in Chinese hamster ovary cells. EMBO J. 11: 1563-1571.

Finn, G.K., Kurz, B.W., Cheng, R.Z., and Shmookler Reis, R.J. 1989. Homologous plasmid recombination is elevated in immortally transformed cells. Mol. Cell. Biol. 9: 4009-4017.

Flintoff, W.F., Davidson, S.V., and Siminovitch, L. 1976. Isolation and partial characterization of three methotrexate-resistant phenotypes from Chinese hamster ovary cells. Somatic Cell Genet. 2: 245-261.

Flintoff, W.F. and Essani, K. 1980. Methotrexate-resistant Chinese hamster ovary cells contain a dihydrofolate reductase with an altered affinity for methotrexate. Biochemistry. 19: 4321-4327.

Flintoff, W.F., Weber, M.K., Nagainis, C.R., Essani, A.K., Robertson, D., and Salser, W. 1982. Overproduction of dihydrofolate reductase and gene amplification in methotrexate-resistant Chinese hamster ovary cells. Mol. Cell. Biol. 2: 275-285.

Fouser, L.A., Swanberg, S.L., Lin, B.Y., Benedict, M., Kelleher, K., Cumming, D.A., and Riedel, G.E. 1992. High level expression of a chimeric anti-ganglioside GD2 antibody: Genomic κ sequences improve expression in COS and CHO cells. Bio/Technology. 10: 1121-1127.

Frame, K.K. and Hu, W.S. 1990. The loss of antibody productivity in continuous culture of hybridoma cells. Biotechnol. Bioeng. 35: 469-476.

Furukawa, K. and Ohsuye, K. 1998. Effect of culture temperature on a recombinant CHO cell line producing a C-terminal α-amidating enzyme. Cytotechnology. 26: 153-164.

Furukawa, K. and Ohsuye, K. 1999. Enhancement of productivity of recombinant α-amidating enzyme by low temperature culture. Cytotechnology. 31: 85-94.

Gains, P. and Wojchowski, D.M. 1999. pIRES-CD4t, a dicistronic expression vector for MACS- or FACS-based selection of transfected cells. Biotechniques 26: 683-688.

Gething, M.J. and Sambrook, J. 1992. Protein folding in the cell. Nature. 355: 33-45.

Goswami, J., Sinskey, A.J., Steller, H., Stephanopoulos, G.N., Wang, D.I.C. 1999. Apoptosis in batch cultures of Chinese hamster ovary cells. Biotechnol. Bioeng. 62: 632-640.

Graf, L.H. and Chasin, L.A. 1982. Direct demonstration of genetic alterations at the dihydrofolate reductase locus after gamma irradiation. Mol. Cell. Biol. 2: 93-96.

Graham, F.L. and van der Eb, A.J. 1973. A new technique for the assay of infectivity of human adenovirus 5 DNA. Virology. 52: 456-467.

Hansen, L., Blue, Y., Barone, K., Collen, D., and Larsen, D.R. 1988. Functional effects of asparagine-linked oligosaccharide on natural and variant human tissue-type plasminogen activator. J. Biol. Chem. 263: 15713-15719.

Hendrick, V., Winnepenninckx, P., Abdelkafi, C., Vandeputte, O., Cherlet, M., Marique, T., Renemann, G., Loa, A., Kretzmer, G., and Werenne, J. 2001. Increased productivity of recombinant tissular plasminogen activator (t-PA) by butyrate and shift of temperature: A cell cycle phase analysis. Cytotechnology. 36: 71-83.

Hills, A.E., Patel, A., Boyd, P., and James, D.C. 2001. Metabolic control of recombinant monoclonal antibody N-glycosylation in GS-NS0 cells. Biotechnol. Bioeng. 75: 239-251.

Hwang, S.O., Chung, J.Y., and Lee, G.M. 2003. Effect of doxycyclin-regulated ERp57 expression on specific thrombopoietin productivity of recombinant CHO cells. Biotechnol. Prog. 19: 179-184.

Jenkins, N., Parekh, R.B., and James, D.C. 1996. Getting the glycosylation right: Implications for the biotechnology industry. Nat. Biotechnol. 14: 975-981.

Jordan, M., Schallhorn, A., and Wurm, F.M. 1996. Transfecting mammalian cells: Optimization of critical parameters affecting calcium-phosphate precipitate formation. Nucleic Acids Res. 15: 596-601.

Kao, F.T. and Puck, T.T. 1968. Genetics of somatic mammalian cells, VII. Induction and isolation of nutritionally mutants in Chinese hamster cells. Proc. Natl. Acad. Sci. USA. 60: 1275-1281.

Kato, T., Matsumoto, A., Ogami, K., Tahara, T., and Miyazaki, H. 1998. Native thrombopoietin: Structure and function. Stem Cells.16: 322-328.

Kaufman, R.J. 1990. Selection and coamplification of hetrologous genes in mammalian cells. Methods Enzymol. 185: 537-566.

Kaufman, R.J., Sharp, P.A., and Latt, S.A. 1983. Evolution of chromosomal regions containing transfected and amplified dihydrofolate reductase sequences. Mol. Cell. Biol. 3: 699-711.

Kaufman, R.J., Wasley, L.C., Spiliotes, A.J., Gossels, S.D., Latt, S.A., Larsen, G.R., and Kay, R.M. 1985. Coamplification and coexpression of human tissue type plasminogen activator and murine dihydrofolate reductase sequences in Chinese hamster ovary cells. Mol. Cell. Biol. 5: 1750-1759.

Kempner, M.E. and Felder, R.A. 2002. A review of cell culture automation. JALA. 7: 56-62.

Kim, J.H. Bae, S.W., Hong, H.J., and Lee, G.M. 1996. Decreased chimeric antibody productivity of KR12H-1 transfectoma during long-term culture results from decreased antibody gene copy number. Biotechnol. Bioeng., 51: 479-487.

Kim, N.S., Byun, T.H., and Lee, G.M. 2001. Key determinants in the occurrence of clonal variation in humanized antibody expression of

CHO cells during dihydrofolate reductase mediated gene amplification. Biotechnol. Prog. 17: 69-75.

Kim, N.S., Kim, S.J., and Lee, G.M. 1998a. Clonal variability within dihydrofolate reductase-mediated gene amplified Chinese hamster ovary cells: Stability in the absence of selective pressure. Biotechnol. Bioeng. 60: 679-688.

Kim, N.S. and Lee, G.M. 2001. Overexpression of bcl-2 inhibits sodium butyrate-induced apoptosis in Chinese hamster ovary cells resulting in enhanced humanized antibody production. Biotechnol. Bioeng. 71: 184-193.

Kim, N.S. and Lee, G.M. 2002a. Response of recombinant Chinese hamster ovary cells to hyperosmotic pressure: effect of Bcl-2 overexpression. J. Biotechnol. 95: 237-248.

Kim, N.S. and Lee, G.M. 2002b. Inhibition of sodium butyrate-induced apoptosis in recombinant Chinese hamster ovary cells by constitutively expressing antisense RNA of capase-3. Biotechnol. Bioeng. 78: 217-228.

Kim, S.J., Kim, N.S., Ryu, C.J., Hong, H.J., and Lee, G.M. 1998b. Characterization of chimeric antibody producing CHO cells in the course of dihydrofolate reductase-mediated gene amplification and their stability in the absence of selective pressure. Biotechnol. Bioeng. 58: 73-84.

Kim, S.J. and Lee, G.M. 1999. Cytogenetic analysis of chimeric antibody producing CHO cells in the course of dihydrofolate reductase-mediated gene amplification and their stability in the absence of selective pressure.1999. Biotechnol. Bioeng. 64: 741-749.

Kim, T.K., Ryu, J.S., Chung, J.Y., Kim, M.S., and Lee, G.M. 2000. Osmoprotective effect of glycine betaine on thrombopoietin production in hyperosmotic Chinese hamster ovary cell culture: Clonal variations. Biotechnol. Prog. 16: 775-781.

Koduri, R.K., Miller, T.J. and Thammana, P. 2001. An efficient homologous recombination pTV(1) contains a hot spot for increased recombinant protein expression in Chinese hamster ovary cells. Gene. 280: 87-95.

Kukuruzinska, M.A. and Lennon, K. 1998. Protein N-glycosylation: Molecular genetics and functional significance. Crit. Rev. Oral. Biol. Med. 9: 415-448.

Lee, S.K. and Lee, G.M. 2003. Development of apoptosis-resistant dihydrofolate reductase-deficient Chinese hamster ovary cell line. Biotechnol. Bioeng. 82: 872-876.

Lee, G.M. and Park, S.Y. 1995. Enhanced specific antibody productivity of hybridomas resulting from hyperosmotic stress is cell line-specific. Biotechnol. Lett. 17: 145-150.

Lifely, M.R, Hale, C., Boyce, S., Keen, M.J., and Phillips, J. 1995. Glycosylation and biological activity of CAMPATH-1H expressed in different cell lines and grown under different culture conditions. Glycobiology. 5: 813–822.

Lin, F.K., Suggs, S., Lin, C.H., Browne, J.K., Smalling, R., Ergie, J.C., Chen, K.K., Fox, G.M., Martin, F., Stabinsky, Z., Badrawi, S.M., Lai, P.M., and Goldwasser, E. 1985. Cloning and expression of the human erythropoietin gene. Proc. Natl. Acad. Sci. USA. 82: 7580-7584.

Ludwig, A., Tomeczkowski, J., and Kretzmer, G. 1992. Influence of the temperature on the shear stress sensitivity of adherent BHK 21 cells. Appl. Microbiol. Biotechnol. 38: 323-327.

Maitland, N.J. and McDougall, J.K. 1977. Biochemical transformation of mouse cells by fragments of herpes simplex virus DNA. Cell. 11: 233-241.

Mastrangelo, A.J., Hardwick, J.M., Zou, S., and Betenbaugh, M.J. 2000a. Part I. Bcl-2 and Bcl-x_L limit apoptosis upon infection with alphavirus vectors. Biotechnol. Bioeng. 67: 544-554.

Mastrangelo, A.J., Hardwick, J.M., Zou, S., and Betenbaugh, M.J. 2000b. Part II. Overexpression of bcl-2 family members enhances survival of mammalian cells in response to various culture insults. Biotechnol. Bioeng. 67: 555-564.

Merritt, S.E. and Palsson, B.O. 1993. Loss of antibody productivity is highly reproducible in multiple hybridoma subclones. Biotechnol. Bioeng. 42: 247-250.

Meyne, J. Ratliff, R.L., and Moyzis, R.K. 1989. Conservation of the human telomere sequence $(TTAGGG)_n$ among vertebrates. Proc. Natl. Acad. Sci. USA. 86: 7049-7053.

Michel, M.L., Sobczak, E., Malpiéce, Y., Tiollais, P., and Streeck, R.E. 1985. Expression of amplified hepatitis B virus surface antigen genes in Chinese hamster ovary cells. Bio/Technology. 3: 561-566.

Moore, A., Donahue, C.J., Hooley, J., Stocks, D.L., Bauer, K.D., and Mather, J.P. 1995. Apoptosis in CHO cell batch cultures: Examination by flow cytometry. Cytotechnology. 17: 1-11.

Newman, R., Alberts, J., Anderson, D., Carner, K., Heard, C., Norton, F., Raab, R., Reff, M., Shuey, S., and Hanna, N. 1992. Primatization of recombinant antibodies for immunotherapy of human diseases: A macaque/human chimeric antibody against human CD4. Bio/Technology. 10: 1455-1460.

Oh, S.K.W., Vig, P., Chua, F. Teo, W.K., and Yap, M.G.S. 1993. Substantial overproduction of antibodies by applying osmotic pressure and sodium butyrate. Biotechnol. Bioeng. 42: 601-610.

Oh, S.K.W., Chua, F.K.F., and Choo, A.B.H. 1995. Intracellular responses of productive hybridomas subjected to high osmotic pressure. Biotechnol. Bioeng. 46: 525-535.

Oliver, J.D., Roderick, H.L., Llewellyn, D.H., and High, S. 1999. ERp57 functions as a subunit of specific complexes formed with the ER lectins calreticulin and calnexin. Mol. Biol. Cell. 10: 2573-2582.

Øyaas, K., Ellingsen, T.E., Dyrset, N. and Levine, D.W. 1994a. Utilization of osmoprotective compounds by hybridoma cells exposed to hyperosmotic stress. Biotechnol. Bioeng. 43: 77-89.

Øyaas, K., Ellingsen, T.E., Dyrset, N., and Levine, D.W. 1994b. Hyperosmotic hybridoma cell cultures: Increased monoclonal antibody production with addition of glycine betaine. Biotechnol. Bioeng. 44: 991-998.

Ozturk, S. and Palsson, B.O. 1990. Loss of antibody productivity during long-term cultivation of a hybridoma cell line in low serum and serum-free media. Hybridoma. 9: 167-175.

Ozturk, S. and Palsson, B.O. 1991. Effect of medium osmolarity on hybridoma growth, metabolism, and antibody production. Biotechnol. Bioeng. 37: 989-993.

Page, M.J. and Sydenham, M.A. 1991. High level expression of the humanized monoclonal antibody Campath-1H in Chinese hamster ovary cells. Bio/Technology. 9: 64-68.

Park, S.Y. and Lee, G. M. 1995. Enhancement of monoclonal antibody production by immobilized hybridoma cell culture with hyperosmolar medium. Biotechnol. Bioeng. 48: 699-705.

Pendse, G.E., Karkare, S., and Bailey, J.E. 1992. Effect of cloned gene dosage on cell growth and hepatitis B surface antigen synthesis and secretion in recombinant CHO cells. Biotechnol. Bioeng. 40: 119-129.

Peppard, J.V., Hobbs, S.M., and Jackson, L.E. 1989. Role of

carbohydrate in binding of IgG to the Fc receptor of neonatal rat enterocytes. Mol. Immunol. 26: 495-500.

Perani, A., Singh, R.P., Chauhan, R., and Al-Rubeai., M. 1998. Variable functions of bcl-2 in mediating bioreactor stress induced apoptosis in hybridoma cells. Cytotechnology. 28: 177-188.

Prati, E.G.P., Scheidegger, P., Sburlati, A.R., and Bailey, J.E. 1998. Antisense strategies for glycosylation engineering of Chinese hamster ovary cells. Biotechnol. Bioeng. 59: 445-450.

Reddy, S. and Miller, W.M. 1994. Effects of abrupt and gradual osmotic stress on antibody production and content in hybridoma cells that differ in production kinetics. Biotechnol. Prog. 10: 165-173.

Reuveny, S., Velez, D., Macmillan, J.D., and Miller, L. 1986. Factors affecting cell growth and monoclonal antibody production in stirred reactors. J. Immunol. Methods. 86: 53-59.

Ryll, T., Dutina, G., Reyes, A., Gunson, J., Krummen, L., and Etcheverry, T. 2000. Performance of small-scale CHO perfusion cultures using an acoustic cell filtration device for cell retention: Characterization of separation efficiency and impact of perfusion on product quality. Biotechnol. Bioeng. 69: 440-449.

Ryu, J.S., Kim, T.K., Chung, J.Y., and Lee, G.M. 2000. Osmoprotective effect of glycine betaine on foreign protein production in hyperosmotic recombinant Chinese hamster ovary cell cultures differs among cell lines. Biotechnol. Bioeng. 70: 167-175.

Ryu, J.S. and Lee, G.M. 1997. Influence of hyperosmolar basal media on hybridoma cell growth and antibody production. Bioprocess. Engr. 16: 305-310.

Salazar-Kish, J.M. and Heath, C.A. 1993. Comparison of a quadroma and its parent hybridomas in fed batch culture. J. Biotechnol. 30: 351-365.

Sanfeliu, A. and Stephanopoulos, G. 1999. Effect of glutamine limitation on the death of attached Chinese hamster ovary cells. Biotechnol. Bioeng. 64: 46-53.

Savage, A. 1997. Glycosylation: A post-translational modification. In: Mammalian Cell Biotechnology in Protein Production. H. Hauser and R. Wagner, eds. Walter de Gruyter, Berlin, p. 233-276.

Schlokat, U., Himmelspach, M., Falkner, F.G., and Dorner, F. 1997. Permanent gene expression in mammalian cells: Gene transfer and selection. In: Mammalian Cell Biotechnology in Protein Production.

H. Hauser and R. Wagner, eds. Walter de Gruyter, Berlin, p. 33-63.

Schneider, S. and Rusconi, R. 1996. Magnetic selection of transiently transfected cells. Biotechniques. 21: 876-880.

Simpson, N.H., Milner, A.E., and Al-Rubeai, M. 1997. Prevention of hybridoma cell death by bcl-2 during suboptimal culture conditions. Biotechnol. Bioeng. 54: 1-16.

Sinacore, M.S., Charlebois, T.S., Harrison, S., Brennan, S., Richards, T., Hamilton, M., Scott, S., Brodeur, S., Oakes, P., Leonard, M., Switzer, M., Anagnostopoulos, A., Foster, B., Harris, A., Jankowski, M., Bond, M., Martin, S., and Adamson, S.R. 1996. CHO DUKX cell lineages preadapted to growth in serum-free suspension culture enable rapid development of cell culture processes for the manufacture of recombinant proteins. Biotechnol. Bioeng. 52: 518-528.

Singh, R.P., Al-Rubeai, M., Gregory, C.D., and Emery, A.N. 1994. Cell death in bioreactors: A role for apoptosis. Biotechnol. Bioeng. 44: 720-726.

Stark, G.R., Debatisse, M., Giulotto, E., and Wahl, G.M. 1989. Recent progress in understanding mechanisms of mammalian DNA amplification. Cell. 57: 901-908.

Szkudlinski, M.W., Thotakura, N.R., Bucci, I., Joshi, L.R., Tsai, A., East-Palmer, J., Shiloach, J., and Weintraub, B.D. 1993. Purification and characterization of recombinant human thyrotropin (TSH) isoforms produced by Chinese hamster ovary cells: The role of sialylation and sulfation in TSH activity. Endocrinology. 133: 1490-1503.

Tada, H., Kurokawa, T., Seita, T., Watanabe, T., and Iwasa, S. 1994. Expression and characterization of a chimeric bispecific antibody against fibrin and against urokinase-type plasminogen activator. J. Biotechnol. 33: 157-174.

Takeuchi, M., Takasaki, S., Miyazaki, H., Kato, T. Hoshi, S., Kochibe, N., and Kobata, A. 1988. Comparative study of the asparagine-linked sugar chains of human erythropoietin purified from urine and the culture medium of recombinant Chinese hamster ovary cells. J. Biol. Chem. 263: 3657-3663.

Tey, B.T., Singh, R.P., Piredda, L., Piacentini, M., and Al-Rubeai, M. 2000. Influence of bcl-2 on cell death during the cultivation of a Chinese hamster ovary cell line expressing a chimeric antibody. Biotechnol. Bioeng. 68: 31-43.

Toledo, F., Le Roscouet, D., Buttin, G., and Debatisse, M. 1992. Coamplified markers alternate in megabase long chromosomal inverted repeats and cluster independently in interphase nuclei at early steps of mammalian gene amplification. EMBO J. 11: 2665-2673.

Trill, J.J., Shatzman, A.R., and Ganguly, S. 1995. Production of monoclonal antibodies in COS and CHO cells. Curr. Opin. Biotechnol. 6: 553-560.

Umaña, P., Jean-Mairet, J., and Bailey, J.E. 1999a. Tetracycline-regulated overexpression of glycosyltransferases in Chinese hamster ovary cells. Biotechnol. Bioeng. 65: 542-549.

Umaña, P., Jean-Mairet, J., Moudry, R., Amstutz, H., and Bailey, J.E. 1999b. Engineered glycoforms of an antineuroblastoma IgG1 with optimized antibody-dependent cellular cytotoxicity. Nat. Biotechnol. 17: 176-180.

Urlaub, G. and Chasin, L.A. Isolation of Chinese hamster cell mutants deficient in dihydrofolate reductase activity. 1980. Proc. Natl. Acad. Sci. USA. 77: 4216-4220.

Urlaub, G., Kas, E., Carothers, A.M., and Chasin, L.A. 1983. Deletion of the diploid dihydrofolate reductase locus from cultured mammalian cells. Cell. 33: 405-412.

Urlaub, G., Mitchell, P.J., Kas, E., Chasin, L.A., Funanage, V.L., Myoda, T.T., and Hamlin, J. 1986. Effect of gamma rays at the dihydrofolate reductase locus: Deletions and inversions. Somat. Cell Mol. Genet. 12: 555-566.

Walls, J.D. and Grinnell, B.W. 1990. A rapid and versatile method for the detection and isolation of mammalian cell lines secreting recombinant proteins. Biotechniques. 8: 138-142.

Weidemann, R., Ludwig, A., and Kretzmer, G. 1994. Low temperature cultivation – A step towards process optimization. Cytotechnology. 15: 111-116.

Weidle, U.H., Buckel, P., and Wienberg, J. 1988. Amplified expression constructs for human tissue-type plasminogen activator in Chinese hamster ovary cells: Instability in the absence of selective pressure. Gene. 66: 193-203.

Wiebe, M.E., Becker, F., Lazar, R., May, L., Casto, B., Semense, M., Fautz, C., Garnick, R., Miller, C., Masover, G., Bergman, D., and Lubiniecki., A.S. 1989. A multifaceted approach to assure that recombinant tPA is free of adventitious virus. In: Advances in Animal Cell Biology and Technology for Bioprocesses. R.E. Spier, J.B. Griffiths, J. Stephenne, and P.J. Crooy, eds. and R. Wagner, eds.

Butterworth, London, p. 68-71.

Wigler, M., Silverstein, S., Lee, L.S., Pellicer, A., Cheng, Y., and Axel, R. 1977. Transfer of purified herpes virus thymidine kinase gene to cultured mouse cells. Cell. 11: 223-232.

Wilson, T.J. and Kola, I. 2001. The LoxP/CRE system and genome modification. Methods Mol. Biol. 158: 83-94.

Wirth, M. 1997. Isolation of recombinant cell clones exhibiting high-level expression of the introduced gene. In: Mammalian Cell Biotechnology in Protein Production. H. Hauser and R. Wagner, eds. Walter de Gruyter, Berlin, p. 121-137.

Wood, C.R., Dorner, A.J., Morris, G.E., Alderman, E.M., Wilson, D., O'Hara, R.M. Jr., and Kaufman, R.J. 1990. High level synthesis of immunoglobulins in Chinese hamster ovary cells. J. Immunol. 145: 3011-3016.

Wurm, F.M., Petropoulos, C.J., and O'Connor, J.V. 1992. Manufacture of proteins based on recombinant Chinese hamster ovary cells: Assessment of genetic issues assurance of consistency and quality. In: Transgenic Organisms and Biosafety. E.R. Schmidt and T. Hankeln, eds. Springer-Verlag, Berlin, p. 283-301.

Wurm, F.M. 1997. Aspects of gene transfer and gene amplification in recombinant mammalian cells. In: Mammalian Cell Biotechnology in Protein Production. H. Hauser and R. Wagner, eds. Walter de Gruyter, Berlin, p. 87-120.

Yun, S.K., Song, J.Y., and Lee, G.M. 2003. Effect of low culture temperature on specific productivity, transcription level, and heterogeneity of erythropoietin in Chinese hamster ovary cells. Biotechnol. Bioeng. 82: 289-298.

Zanghi, J.A., Fussenegger, M., and Bailey, J.E. 1999. Serum protects protein-free competent Chinese hamster ovary cells against apoptosis induced by nutrient deprivation in batch culture. Biotechnol. Bioeng. 64: 108-119.

Zapun, A., Darby, N.J., Tessier, D.C., Michalak, M., Bergeron, J.J., and Thomas, D.Y. 1998. Enhanced catalysis of ribonuclease B folding by the interaction of calnexin or calreticulin with ERp57. J. Biol. Chem. 273: 6009-6012.

Zettlmeissl, G., Ragg, H., and Karges, H. 1987. Expression of biologically active human antithrombin III in Chinese hamster ovary cells. Biotechnology (N.Y). 5: 720-725.

10

Heterologous Protein Expression in Methylotrophic Bacteria

Kelly A. Fitzgerald and Mary E. Lidstrom

Abstract

Methylotrophic bacteria are attractive hosts for recombinant protein production because of the low cost of the single carbon substrates that sustain them. This cost advantage is particularly notable in the case of labeled substrates for generating labeled proteins. Although a large body of genetic and biochemical work has been performed on the methylotroph *Methylobacterium extorquens* AM1, the genome has only recently been sequenced and little is known about regulatory mechanisms in this organism. In order to develop strains for commercial production of heterologous proteins, an efficient expression system is necessary. Work in this area is ongoing, and this chapter represents the current state of the field.

Introduction

Heterologous protein expression is routinely carried out in several strains of bacteria as well as in yeast and a variety of other eukaryotic cell systems. These systems have been in use for decades, and many tools are available for the expression of proteins in these organisms. However, protein expression is still problematic in many cases, and a broader range of expression hosts is desirable to increase the diversity of potential substrates for protein production.

One group of bacteria of interest for heterologous protein expression is methylotrophic bacteria, which grow on single carbon compounds such as methanol (Anthony, 1982). These single carbon compounds are relatively inexpensive, making their use attractive for bulk production. In addition, the ability of these organisms to grow on single carbon compounds is particularly useful in the labeling of proteins for ^{13}C NMR. Uniform ^{2}H, ^{13}C, and ^{15}N labeling of proteins is essential if large polypeptide structures are to be solved. Because methanol can be synthesized quantitatively in a single step from ^{2}H and ^{13}CO, it is the most economical substrate available for protein labeling.

In addition to protein labeling, methylotrophs could be used as hosts for commercial protein production or in the metabolic engineering of strains for chemical production or bioremediation. Engineering strains for chemical production or bioremediation can involve the expression of multiple heterologous genes. Because methylotrophic bacteria use different metabolic pathways than traditional bacterial expression hosts, some chemicals and biodegradative capabilities may be easier to engineer in methylotrophs.

Work has been undertaken by a few laboratories to optimize *Methylobacterium* strains for large-scale heterologous protein expression, and this chapter will review the current status of this work as well as the inherent properties of these organisms that make them interesting targets for the optimization of expression methods. One of the best studied groups of methylotrophs is the pink-pigmented facultative methylotrophs that have been classified as *Methylobacterium* species (Lidstrom, 1991; Green, 1992). These bacteria use the serine cycle for formaldehyde assimilation and class with α-2 proteobacteria (Lidstrom, 1991; Green, 1992). Most *Methylobacterium* strains can utilize substrates such as methanol, methylamine, and methylated glycines, C_2 compounds such as ethanol and ethylamine, and a variety of C_3 and C_4 compounds such as pyruvate and succinate. The first *Methylobacterium* strain to be studied in any detail was *Methylobacterium extorquens* AM1 (Peel *et al.*, 1961). This strain was originally called *Pseudomonas* AM1, but was later reclassified (Green, 1992).

M. extorquens AM1 has served as a model system for the study of methylotrophic metabolism and methylotrophic enzymes, and consequently a large body of biochemical and genetic information has been generated. In addition, several new genetic tools have recently been developed for use in *Methylobacterium extorquens* AM1. This chapter will focus on *Methylobacterium* strains as hosts for recombinant proteins. However, in an attempt to broaden the utility of this chapter, generalizations to other methylotrophic bacteria will be made wherever possible, and methods for adapting specific *Methylobacterium* tools to other strains will be mentioned.

Metabolism

The defining characteristic of methylotrophic bacteria is their ability to grow on single carbon substrates and to use these compounds as their sole source of carbon and energy. This group of substrates encompasses methane, methanol, methylated amines, halogenated methanes, and methylated sulfur species (Anthony, 1982). Methylotrophic bacteria are able to synthesize all of the carbon-carbon bonds necessary for building cell components. Aerobic methylotrophic bacteria are found in diverse aquatic and terrestrial environments and class into several branches of the proteobacteria as well as the high and low GC Gram-positive bacteria (Lidstrom, 2001). Three routes of carbon assimilation are available to methylotrophs. Some organisms grow autotrophically by converting methane or methanol to carbon dioxide and thus assimilate carbon via the Calvin-Benson-Bassham cycle. Two other assimilatory pathways operate at the level of formaldehyde: the ribulose monophosphate (RuMP) cycle or the serine cycle. The following discussion on methylotrophic assimilation will focus on the latter two pathways and is a brief review of a large body of work on the biochemistry of methylotrophic bacteria, which is described in more detail in other reviews (Anthony, 1982).

Methylobacterium extorquens AM1 utilizes the serine cycle for the assimilation of single carbon units. The serine cycle contains carboxylic acid and amino acid intermediates, and phosphoglycerate is synthesized from 2 molecules of formaldehyde and 1 molecule of carbon dioxide. Carbon flows through malyl-CoA to acetyl-CoA, which is then oxidized to glyoxylate to finish the cycle. At this point two variants exist based on how acetyl-CoA is converted to glyoxylate: one less common variant involving isocitrate lyase (faculatative methylotrophs unable to use methane or methanol, *Pseudomonas* MA, *Pseudomonas aminovorans* (Anthony, 1991)), and one more common *icl⁻* variant (Type II methanotrophs, *Methylobacterium*, *Hyphomicrobium* (Anthony, 1991))involving a novel pathway via hydroxybutyrate, butyrate, and propionyl CoA (Korotkova *et al.*, 2001; Korotkova *et al.*, 2002).

The RuMP cycle consists of carbohydrate intermediates rather than amino acids and carboxylic acids. One molecule of pyruvate or dihydroxyacetone is synthesized from three molecules of formaldehyde. In this cycle, formaldehyde is condensed with ribulose 5-phosphate yielding hexulose phosphate, and is followed by isomerization to fructose 6-phosphate (FMP). Two variants of this pathway are known. In one, a molecule of FMP is converted to 2-keto 3-deoxy 6-phosphogluconate (KDPG) and in another, it is converted to the bisphosphate (FBP). Ribulose 5-phosphate is regenerated in the rearrangement part of the pathway via one of two sequences: the sedoheptulose bisphosphatase or the transaldolase variant. The most common variant is the KDPG aldolase/transaldolase and can be found in type I methanotrophs (*e.g. Methylococcus capsulatus*) and obligate methylotrophs unable to use methane (*e.g. Methylophilus methylotrophus*)(Anthony, 1991).The less common variant is the FBP aldolase/SBP phosphatase, which is found in facultative methylotrophs unable to use methane (*e.g. Arthrobacter* P1, *Bacillus* sp., *Acetobacter methanolicus* MB58)(Anthony, 1991).

Methylotrophic species that use the ribulose monophosphate cycle tend to have both a higher growth rate and a higher growth yield than those that rely on the serine cycle for assimilation. This is due to the fact that the RuMP cycle is more efficient with respect to ATP and NADH requirements (Goldberg *et al.*, 1991). In early work on this subject, four strains – two RuMP cycle bacteria and two serine cycle bacteria – were grown in continuous culture at steady state. Maximum yield was found to average 19.1 g cell dry wt/mol substrate utilized in methanol-fed RuMP cycle strains and 13.5 g cell dry wt/mol substrate utilized in methanol-fed serine cycle strains (Rokem *et al.*, 1978). RuMP cycle organisms are attractive as expression hosts because of their higher growth yield, but higher cell yield does not always translate to a higher protein production rate. The expression system must be optimized based on the yield of the heterologous protein of interest, and this may depend upon other factors than growth yield.

Numerous single-carbon compounds have been found to serve as substrates for methylotrophic bacteria. These compounds differ in the amount of energy that can be gained from their oxidation based on their degree of reduction. In addition, the yields of different bacterial species per gram of substrate may differ according to how efficiently the organism oxidizes the substrate, although the yield per carbon atom will depend on both metabolic efficiency and the substrate used for growth. In order to compare growth yields on different substrates, yield per carbon must be calculated. Erickson has described a method of relating yields to substrates based on elemental analysis of substrate, biomass, and product (Erickson, 1987). With this approach, the yields of different organisms on varied substrates can be compared, and an in-depth analysis has been carried out (Erickson *et al.*, 1991). In order to use the equations developed in the above work, true growth yield data must be obtained. A concise compilation of growth data for methylotrophs has been reported (Erickson *et al.*, 1991) and is briefly summarized here. Most notably, the substrate yields of methylotrophs are highest (0.5 to 1) on methane. Yields on methanol range from 0.3 to 0.54, and yields on formaldehyde and formate are 0.24 to 0.32 and 0.073 to 0.150, respectively (Erickson *et al.*, 1991). Substrate yields on dimethyl sulfide, dimethyl disulfide, and chloromethane are comparable to yields on methanol.

465

The choice of a substrate for an expressions system depends on other practical considerations in addition to growth yield. The possible set of methylotrophic substrates ranges from gaseous to liquid substrates, highly toxic to relatively mild compounds, and compounds that are abundant in nature to those that are more rare. If the goal of heterologous expression is to produce the largest amount of protein for the lowest cost, the choice of substrate will likely be an abundant, low-cost liquid with low toxicity. In the case of heterologous expression of biodegradative enzymes to produce a bioremediation strain, the substrate may be fixed and the choice of organism will depend on its ability to degrade the specified compound.

Methanol as a Substrate

Methane is the most reduced of the single carbon substrates, and it is the main component of natural gas. Although methane is a possible substrate for methylotrophic protein production, the potential to form explosive mixtures with oxygen, its low solubility in water, and a high heat transfer requirement make it less than ideal for large-scale processes (Goldberg *et al.*, 1991). Methanol, on the other hand, can be made efficiently from methane, and is more compatible with large-scale bioprocesses than methane.

Methanol is a colorless, volatile liquid with a vapor pressure of 16.94 kPa at 25° C. It is also a polar, acid-base neutral, and water-miscible organic solvent capable of dissolving a wide range of inorganic salts, and it is generally considered to be non-corrosive (Cheng *et al.*, 1994). Although methanol is toxic to humans, it is not considered a great environmental threat as it is biodegradable in soils and wastewater (Cheng *et al.*, 1994). Industrial methanol production is typically carried out either by indirectly converting natural gas to syngas, or by directly oxidizing methane to methanol. The indirect syngas route is currently the more economical choice and is also advantageous because remote natural gas can be converted to methanol at the source and then shipped to more desirable locations (Cheng *et al.*, 1994). A detailed collection of physical and chemical properties of methanol as well as its production methods can be found in *Methanol*

Production and Use edited by W. Cheng and H. H. Kung (Cheng *et al.*, 1994). Methanol is an abundant fuel that can be readily dissolved in aqueous bacterial culture medium.

Because methanol is the most plentiful and least toxic of the liquid single carbon substrates, a methylotroph that can grow on methanol is likely to be the host of choice for most applications. However, methylamine has a specific advantage in the generation of labeled proteins, which is the potential for a double label in both C and N.

Genome

The genome of *Methylobacterium extorquens* AM1 has been sequenced by the University of Washington Genome Center and Integrated Genomics Inc. in collaboration with the Lidstrom laboratory at the University of Washington. Although the finished sequence is not available at the time of this writing, the available contigs can be downloaded at the following public website:
http://www.integratedgenomics.com/genomereleases.html#list6.
The genome is 7 MB with a G+C content of 67.47% and approximately 7000 open reading frames.

Promoters

Development of robust expression systems requires an understanding of promoters and regulatory sequences. This is an area of active study in *M. extorquens* AM1, but information on promoters is confined mainly to those involved in methylotrophy. Currently, no consensus for a promoter sequence has been identified, but several promoters involved in methanol oxidation have been mapped with regard to transcriptional start site and region of promoter activity. Twenty-six genes of *M. extorquens* AM1 that are known to be involved in methanol oxidation are organized in six gene clusters: *mxaFJGIRSACKLDEHB*, *mxaW*, *mxbDM*, *mxcQE*, *pqqABC/DE* and *pqqFG* (Zhang *et al.*, 2003). Promoters have been studied upstream of *mxaF*, *mxaW*, *mxbD*, *pqqA*, and *pqqF* (Zhang *et al.*, 2003). These promoters have a sequence similar to the -35 region of *Escherichia coli* σ^{70}-dependent promoters but the -10 regions are more divergent, being more noticeably GC-rich. The only strong consensus among promoter regions is a hexanucleotide sequence, AAGAAA, found upstream of the -35 regions of several methylotrophy genes in *Methylobacterium extorquens* AM1 (Zhang *et al.*, 2003). These promoter regions are located 46 to 225 bp upstream of the respective translational start sites, and the complete promoter regions encompass on the order of 100 bp.

All of the known promoters involved in methanol oxidation are expressed at 3-6 fold higher levels in cells grown on methanol compared to succinate (Chistoserdov *et al.*, 1994; Zhang *et al.*, 2003), suggesting that some regulatory circuit must exist for this system. Five regulatory genes, *mxaB*, *mxbDM*, and *mxcQE*, are known. *mxaB*, *mxbM*, and *mxcE* are predicted to encode response regulators while *mxbD* and *mxcQ* are predicted to encode sensor kinases. All five genes are required for detectable transcription of the *mxaF* promoter (Springer *et al.*, 1997) but only *mxbDM* are required for transcription of other methanol oxidation promoters. The promoter upstream of *mxbDM* has been shown to be regulated by *mxcQE* (Springer *et al.*, 1997).

Foreign promoters have been shown to be functional in other methylotrophs. The *tac* promoter was used to successfully express a cDNA encoding human interferon αF in *M. flagellatum* at levels two to three-fold higher than those obtained in *E. coli* (Chistoserdov *et al.*, 1987). Additionally, both *tac* and *trp* promoters have been shown to function in *M. methylotrophus* (Byrom, 1984). However, no foreign promoters have been found that provide high expression in *Methylobacterium extorquens* AM1.

Genetic Tools

A sophisticated group of genetic tools has been developed for *Methylobacterium* strains. Broad-host range vectors, conjugation, and electroporation have been used successfully to carry out studies of genes and gene expression. Directed insertional mutagenesis using allelic exchange techniques is routine (Holloway *et al.*, 1987; Lidstrom *et al.*, 1990; Chistoserdova, 1996). Small, completely sequenced IncP vectors are now available for cloning, promoter identification and analysis, chromosomal insertion, and for generating unmarked chromosomal deletion mutations (Marx *et al.*, 2001, 2002). The ability to generate unmarked directed mutations and targeted insertions is especially important, as it allows sequential manipulation of chromosomal genes without excluding subsequent addition of genes to be expressed (see below).

P_{mxaF} is the promoter that drives the expression of the large subunit of methanol dehydrogenase, the most highly expressed protein in *Methylobacterium extorquens* AM1. It has been used for expression in the small IncP expression vectors designed by Marx (Marx *et al.*, 2001). Three vectors were developed for high-level expression in *Methylobacterium extorquens* AM1, pCM80, pCM160, and pCM110. Two promoter probe vectors have also been developed to examine relative strengths of promoters in *M. extorquens* AM1 (Table 1). These latter two vectors are attractive tools for identification and testing of promoters in a variety of gram-negative bacteria.

Antibiotic Markers and Tools for Generating Mutants

Few antibiotic markers are available for selection in *Methylobacterium extorquens* AM1. A rifamycin-resistant strain of *M. extorquens* AM1 is used routinely for genetic studies (Nunn *et al.*, 1986), and both

Table 1. Expression vectors for *Methylobacterium* using P_{mxaF} and related promoter cloning vectors.

Plasmid	Resistance	Promoter	Reporter	Replicon
pCM80	Tc	P_{lac}-P_{mxaF}	none	ColE1-IncP
pCM160	Km	P_{lac}-P_{mxaF}	none	ColE1-IncP
pCM110	Tc	P_{mxaF}	none	IncP
pCM130	Tc	none	XylE	ColE1-IncP
pCM132	Km	none	LacZα	ColE1-IncP

tetracycline and kanamycin have been used successfully for selection. Several attempts were made to develop other antibiotics such as erythromycin and gentamycin for use in *Methylobacterium extorquens* AM1 but they have not been successful (Lidstrom unpublished). Because of this paucity of markers, a broad-host-range system for antibiotic marker recycling has been created using *cre-lox* methodology (Marx *et al.*, 2002). This approach uses Cre recombinase, a site-specific recombinase from phage P1, to excise DNA regions flanked by co-directional *loxP* recognition sites (Palermos *et al.*, 2000). The system consists of two parts: an allelic exchange vector with a *loxP*-flanked antibiotic resistance cassette, and an IncP plasmid expressing the Cre recombinase. Recycling antibiotic resistance markers allows for multiple genetic manipulations within a single strain, and this capability has opened the door to more sophisticated genetic experiments. Although this system is described here as a tool for *M. extorquens* AM1, it was also successfully used in *Burkholderia fungorum* LB400 (Marx *et al.*, 2002). Because the system requires only the ability of a strain to uptake IncP plasmids and that Cre recombinase be sufficiently expressed, it should be applicable to a wide variety of other gram-negative bacteria. This tool makes genetic manipulations possible in strains in which few genetic markers are available.

Another tool that is particularly helpful in the engineering of overproduction strains is a system for generating chromosomal insertions with selection for double crossovers and counter-selection for single crossovers. The insertions can then be unmarked using the previously described *cre-lox* methodology (Marx and Lidstrom unpublished).

Heterologous Protein Expression in *M. extorquens* AM1

As noted above, a number of studies have reported on the expression of antibiotic markers and reporter genes in *Methylobacterium*. However, in these cases, the heterologous proteins were genetic tools, not tests of heterologous expression.

Several heterologous proteins have been expressed in methylotrophic bacteria (Lidstrom, 1992). The glutamate dehydrogenase gene from *E. coli* and two eukaryotic cDNA's encoding chicken ovalbumin and mouse dihydrofolate reductase were cloned and expressed in *M. methylotrophus* (Windlass *et al.*, 1980; Hennam *et al.*, 1982). In these cases, vector promoters drove expression at lower levels than in *E. coli*. The *E. coli* pyruvate dehydrogenase complex was also expressed at low levels in *Hyphomicrobium* X, but the promoters were not determined (Dijkhuizen *et al.*, 1984).

The first account of attempts to optimize heterologous protein expression in a *Methylobacterium extorquens* strain involved expression of green fluorescent protein (GFP) from the jellyfish *Aequorea victoria* in *Methylobacterium extroquens* ATCC 55366 (Figueira *et al.*, 2000). The gene encoding GFP was cloned downstream of two different promoters into several expression vectors ranging in size from 7 to 26 kb. The highest level of expression (850-1000 µg of GFP/g biomass) was achieved in a strain expressing GFP under control of the *Escherichia coli lacZ* promoter in plasmid pRK310 (Figueira *et al.*, 2000).

Although expression of a heterologous protein was achieved using the *lacZ* promoter, several native *Methylobacterium extroquens* AM1 promoters have much higher activity and were therefore attractive targets for expression tools. As described above, a set of broad-host-range vectors has been designed for protein expression in *M. extorquens* AM1(Marx *et al.*, 2001). Using one of these, pCM80, two heterologous proteins, GFP and XylE (1,2-catechol dioxygenase) could be overexpressed at levels detectable by SDS-PAGE (Marx *et al.*, 2001) and XylE was found to be active in the degradation of catechol. A P_{mxaF}-*xylE* construct was shown to have 40 times the activity of a P_{lac}-*xylE* construct in *M. extorquens* AM1 (Marx *et al.*, 2001). Therefore, this promoter has potential as a general tool for high level expression of heterologous proteins in *M. extorquens* AM1.

The pCM80 vector containing P_{mxaF} has been tested for expression of other proteins in order to determine how applicable this vector may be for heterologous protein expression. The enzyme haloalkane dehalogenase (DhlA) from *Xanthobacter autotrophicus* GJ10 was chosen as an example of a biodegradative enzyme of commercial interest. Although DhlA was expressed at detectable levels using pCM80, it was synthesized at much lower concentrations than MxaF. Attempts were made to optimize expression, but the levels of DhlA did not increase after changing the Shine-Dalgarno sequence to that of *mxaF*, or adding the MxaF signal sequence which directs MxaF to the periplasm. It was not possible to test for DhlA in the insoluble fraction, as the poly-β-hydroxybutyrate (PHB) granules normally produced by the cell interfered with the analysis. Therefore, expression of DhlA was tested in a mutant deficient in the production of PHB. Surprisingly, this mutant was capable of producing large amounts of active DhlA, comparable to the wild-type production of MxaF (FitzGerald *et al.*, 2003). The reason for this difference is not clear, as neither GFP nor LacZ were expressed at higher levels in the mutant compared to the wild-type strain, and it was not possible to detect DhlA adsorbed to PHB granules in wild-type cells. However, the PHB-negative mutant provides another valuable tool for heterologous protein expression in this methylotroph.

Future Directions and Tool development

Although expression vectors have been developed for *Methylobacterium extorquens* and have been shown to be useful for the overexpression of heterologous proteins, products are limited to those that are not toxic to the host. The *mxaF* promoter is especially problematic in this regard, as it is induced at high levels in the absence of methanol, and its activity increases about 3-fold in the presence of methanol (Zhang *et al.*, 2003). A tightly-regulated, inducible expression system is crucial to expand the classes of proteins that can be produced in *M. extorquens* AM1. The only known strong, tightly-regulated promoter is that driving the expression of methylamine dehydrogenase, but it is also strongly repressed by methanol and other substrates. This characteristic makes the transition to full expression logistically difficult and rules out the use of methanol as a feedstock (Lidstrom and Chistoserdov unpublished).

The ideal inducer would be an inexpensive and non-toxic compound that could be added easily to a fermentor. An inducible expression system would be useful at the laboratory scale as well as in large-scale production of proteins, and toxicity and cost would not be as important for small cultures. Optimally, several expression systems should be developed using an array of promoters suitable for different applications.

Transcriptional control elements are not the only important elements of a robust expression system. Modifying translational regulatory elements such as Shine-Dalgarno sequences may also boost production. Preliminary investigation of SD sequences shows that changing this sequence to that of the most highly expressed protein in *Methylobacterium extorquens* AM1 did not enhance heterologous protein expression (FitzGerald *et al.*, 2003). However, other SD sequences have inhibited expression of heterologous proteins in *M. extorquens* AM1 (Strovas, FitzGerald, and Lidstrom unpublished). A systematic examination of SD sequences would be useful to identify optimal tools for expression.

Still another avenue of exploration that would likely improve protein yield is in the stage of protein folding. For example, in *Escherichia coli*, co-overexpression of GroES/L and other chaperones has been shown to enhance the proper folding and expression of heterologous proteins (Baneyx, 1999). These types of modifications to the strain may be necessary for optimal protein production in methylotrophs.

Yet another area for research is in large scale culturing techniques. Methods for obtaining continuous cultures with high cell densities will be useful for producing large amounts of heterologous proteins. Investigating different reactor designs and media as well as determining optimal induction times will be necessary in the future.

Heterologous protein expression in methylotrophic bacteria is a new and growing field. The minimal necessary genetic tools have been developed, and several heterologous proteins have been expressed at high levels in *Methylobacterium extorquens* strains. The tools that have been developed for *M. extorquens* can be easily adapted for use in other methylotrophs. The next step toward a viable expression host is the development of an inducible expression system. The ability to control the expression of heterologous proteins will allow for a wider range of proteins to be expressed. Finally, strain optimization much like that carried out in *Escherichia coli* should allow for optimal protein expression.

References

Anthony, C. 1982. The biochemistry of methylotrophs. Associated Press, London.

Anthony, C. 1991. Assimilation of carbon by methylotrophs. Biology of Methylotrophs. Goldberg, I. and Rokem, J.S. Butterworth-Heinemann, Boston: 79-109.

Baneyx, F. 1999. Recombinant protein expression in *Escherichia coli*. Curr. Opin. Biotechno. 10: 411-421.

Byrom, D. 1984. Host-vector systems for *Methylophilus methylotrophus*. Microbial Growth on C_1 compounds. Crawford, R.L. and Hanson, R.S. ASM Press, Washington D.C.: 221-223.

Cheng, W.-H. and Kung, H.H. 1994. Methanol production and use. Marcel Dekker, Inc., New York.

Chistoserdov, A.Y., Chistoserdova, L.V., McIntire, W.S. and Lidstrom, M.E. 1994. Genetic organization of the *mau* gene cluster in *Methylobacterium extorquens* AM1: the complete nucleotide sequence and generation and characteristics of *mau* mutants. J. Bacteriol. 176: 4052-4065.

Chistoserdov, A.Y., Eremashvili, M., Mashko, S., Lapidus, A., Skvortsova, M. and Sterkin, V. 1987. Expression of human interferon F gene in obligate methylotroph *Methylobacillus flagellatum* KT and *Pseudomonas putida* (in Russian). Mol. Genet. Microbiol. Virol. 8: 36-42.

Chistoserdova, L. 1996. The utilization of formaldehyde in *Methylobacterium extorquens* AM1: metabolic and genetic characterization. Microbial Growth on C_1 compounds. Kluwer Academic Publishers, Dordrecht.

Dijkhuizen, L., Harder, W., DeBoer, L., Van Boven, A., Clement, W., Bron, S. and Venema, G. 1984. Genetic manipulation of the restricted facultative methylotroph *Hyphomicrobium* X by the R-plasmid mediated introduction of the *Escherichia coli pdh* genes. Arch. Microbiol. 139: 311-318.

Erickson, L.E. 1987. Thermal and energetic studies of cellular biological systems. James, A.M. Wright, Bristol, England: 14-33.

Erickson, L.E. and Tuitemwong, P. 1991. Growth yields, productivities, and maintenance energies of methylotrophs. Biology of Methylotrophs. Goldberg, I. and Rokem, J.S. Butterworth - Heinemann, Boston: 149-169.

Figueira, M.M., Laramée, L., Murrell, J.C., Groleau, D. and Miguez, C.B. 2000. Production of green fluorescent protein by the methylotrophic bacterium *Methylobacterium extorquens*. FEMS Microbiol. Lett. 193: 195-200.

FitzGerald, K.F. and Lidstrom, M.E. 2003. Overexpression of a heterologous protein, haloalkane dehalogenase, in a poly-β-hydroxybutyrate-deficient strain of the facultative methylotroph *Methylobacterium extorquens* AM1. Biotechnol. Bioeng. 81: 263-268.

Goldberg, I. and Rokem, J.S., Eds. 1991. Biology of methylotrophs. Biotechnology.Butterworth-Heinemann, Paris, France.

Green, P. 1992. The genus methylobacterium. The Prokaryotes II. Balows, A. *et al*. Springer-Verlag, New York.

Hennam, J.F., Cunningham, A.E., Sharpe, G.S. and Atherton, K. 1982. Expression of eukaryotic coding sequences in *Methylophilus methylotrophus*. Nature. 297: 80-82.

Holloway, B.W., Kearney, P.P. and Lyon, B.R. 1987. The molecular genetics of C1 utilizing microorganisms. An overview. Ant. van Leeuw. 53: 47-53.

Korotkova, N., Chistoserdova, L., Kuksa, V. and Lidstrom, M.E. 2002. Glyoxylate regeneration pathway in the methylotroph *Methylobacterium extorquens* AM1. J. Bacteriol. 184: 1750-1758.

Korotkova, N. and Lidstrom, M.E. 2001. Connection between poly-β-hydroxybutyrate biosynthesis and growth on C_1 and C_2 compounds in the methylotroph *Methylobacterium extorquens* AM1. J. Bacteriol. 183: 1038-1046.

Lidstrom, M.E. 1991. The methylotrophic bacteria. The Prokaryotes II. Balows, A. et. al. Springer-Verlag, New York.

Lidstrom, M.E. 1992. Genetics and molecular biology of methanol -

utilizing bacteria. Methane and Methanol Utilizers. Murrell, C.J. and Dalton, H. Plenum Press, New York: 183 - 206.

Lidstrom, M.E. 2001. The methylotrophic bacteria. The Prokaryotes. Dworkin, M. Springer-Verlag, New York.

Lidstrom, M.E. and Stirling, D.I. 1990. Methylotrophs: Genetics and commercial applications. Annu. Rev. Microbiol. 44: 27-58.

Marx, C.J. and Lidstrom, M.E. 2001. Development of improved versatile broad-host-range vectors for use in methylotrophs and other Gram-negative bacteria. Microbiology 147: 2065-2075.

Marx, C.J. and Lidstrom, M.E. 2002. A broad-host-range *cre-lox* system for antibiotic marker recycling in Gram-negative bacteria. Biotechniques 33: 1062-1067.

Nunn, D. and Lidstrom, M.E. 1986. Isolation and complementation analysis of 10 methanol oxidation mutant classes and identification of the methanol dehydrogenase structural gene of *Methylobacterium* sp. Strain AM1. J. Bacteriol. 166: 581-590.

Palermos, B., Wild, J., Szybalski, W., Le Borgne, S., Hernandez-Chavez, G., Gosset, G., Valle, F. and Bolivar, F. 2000. A family of removable cassettes designed to obtain antibiotic-resistance-free genomic modifications of *Escherichia coli* and other bacteria. Gene 247: 255-264.

Peel, D. and Quayle, J.R. 1961. Microbial growth on C1 compounds. 1. Isolation and characterization of *Pseudomonas* AM1. Biochem. J. 81: 465-469.

Rokem, J.S., Goldberg, I. and Mateles, R.I. 1978. Maintenance requirements for bacteria growing on C_1-compounds. Biotechnol. Bioeng. 20: 1557-1564.

Springer, A.L., Morris, C.J. and Lidstrom, M.E. 1997. Molecular analysis of *mxbD* and *mxbM*, a putative sensor-regulator pair required for oxidation of methanol in *Methylobacterium extorquens* AM1. Microbiology 143: 1737-1744.

Windlass, J., Worsey, M., Pioli, E., Pioli, D., Barth, P., Atherton, K., Dart, E., Byrom, D., Powell, K. and Senior, P. 1980. Improved conversion of methanol to single-cell protein by *Methylophilus methylotrophus*. Nature. 287: 396-401.

Zhang, M. and Lidstrom, M.E. 2003. Promoters and transcripts for genes involved in methanol oxidation in *Methylobacterium extorquens* AM1. Microbiology. 149:1033-1040.

11

Secretion or Presentation of Recombinant Proteins and Peptides Mediated by the S-layer of *Caulobacter crescentus*

John F. Nomellini, Michael C. Toporowski, and John Smit

Abstract

The production of heterologous proteins is a challenge in biotechnology because it is difficult to identify expression systems that are appropriate or capable of expressing proteins of interest. Secretion of heterologous polypeptides is a way to incorporate a significant purification step at the earliest stages of production and there is wide interest in finding bacterial systems capable of doing so in an efficient and faithful manner. The S-layer secretion system of the Gram-negative bacterium *Caulobacter crescentus* has been adapted for heterologous protein secretion. It is a Type I secretion mechanism that is naturally adapted for high level expression and the C-terminal secretion signal appears to mediate the export of a wide variety of non-native proteins through a large hydrophilic channel that traverses both membranes. The S-layer is a self-assembling, regularly structured surface layer that is present at about 40,000 copies per cell and a candidate for high density display of peptides for a variety of uses ranging from whole cell vaccines to library display. Display in other bacterial systems is often limited by the size of foreign peptides tolerated or the copy number per cell. The S-layer gene of *C. crescentus* has been adapted for displaying proteins 10-200 amino acids in length at high density and with an unexpectedly high degree of success.

Introduction

Most gene expression systems have a convoluted history of development, involving multiple research groups. This chapter describes the development of an expression system that, up to now, has largely occurred in a single research laboratory, based on a single gene and its cognate transporter. Caulobacter expression must be described as in its infancy, compared to virtually all the other technologies described in this volume. Yet, it shows great promise for protein production at both research and commercial levels and as a means to present proteins and peptides on the surface of the bacterium at high densities. The marriage of these two capabilities should prove very useful when it is desirable to quickly move from discovery to production. The potential of a new gene expression system is best evaluated in the context of the characteristics of others that are already in use. The first sections of this chapter attempt to do so and set standards for comparison.

Gene Expression and the Case for Protein Secretion

Recombinant or heterologous gene expression technology is in many ways "coming of age". Early on, the production of recombinant hormones, cytokines, vaccines, antibodies and other proteins of commercial or therapeutic interest was done largely by adapting

Escherichia coli for the task. This was largely driven by the large number of people experienced in *E. coli* genetics, and the extensive number of developed "tools" (specialized strains, mutant genes, regulated promoters, and high copy number plasmid vectors to name a few). In many ways, that momentum continues, especially in such areas as high throughput screening methodologies. Indeed, the term "soluble proteins" (without reference to the protein's origins) has entered the lexicon of structure analysis experts as an undefined reference to the fraction of *E. coli*-produced heterologous proteins that, for incompletely understood reasons, do not form inclusion bodies when overexpressed inside the bacterium. The problem in this case is the difficulty in purifying active proteins from inclusion bodies and its extent is significant since the fraction of "soluble" proteins has been estimated to be only about 30% of the proteins expressed from large, random clone libraries.

The matter of soluble versus insoluble recombinant proteins in part defines the paradox of heterologous gene overexpression in *E. coli*. Aside from its many advantages, this bacterium is not ideal for heterologous protein production. The formation of inclusion bodies stems from the fact that expression generally does not involve secretion. Thus, high level synthesis in the confined space of the cell interior encourages internal protein crystalloids to form. With some notable exceptions, such as methods that take advantage of secretion to the periplasm of this Gram-negative bacterium (and later encourage leakage of the protein across the outer membrane), or of the Type I secretion mechanism used for α-hemolysin secretion by some strains (see below), *E. coli* does not secrete proteins. From the standpoint of protein purification, this means that recombinant proteins must be separated from a large number of other resident *E. coli* proteins, as well as nucleic acids, membrane lipids and other biomolecules. This impacts not only the ease and cost of purifying proteins but also introduces the challenge of producing a protein that is "active" (*i.e.* properly folded) so that it exhibits native function.

There are additional challenges in purifying proteins from a complex milieu of other proteins, nucleic acids, oligosaccharides and other biomolecules at the commercial scale. Generally, multiple chromatography steps are required, including reverse phase chromatography. Each step has significant impact on so-called "downstream processing" costs and purification techniques such as reverse phase chromatography involve organic solvents, thereby generating a waste stream, the handling of which incurs additional costs and hazards.

Another challenge in the purification of protein from bacteria is the presence of endotoxin. For Gram-negative bacteria, endotoxin is synonymous with lipopolysaccharide (LPS), the ubiquitous amphipathic molecule located in the outer membrane. In the case of enteric Gram-negative bacteria, the endotoxic component capable of inducing sepsis, septic shock and pyrogenic responses, is localized in the Lipid A portion of LPS. The specific structure of Lipid A is recognized by the LPS binding protein, CD14 and Tol-like receptor 4 as part of the mammalian innate immune response (Schletter *et al.*, 1995). Gram-positive bacteria, such as *Bacillus* sp., can also produce endotoxic effects that have ascribed to the peptidoglycan or other cell wall fragments (Heumann *et al.*, 1998). In any event, *E. coli* produces a potent endotoxin and large-scale protein purification strategies must include careful monitoring and elimination of endotoxin contamination of target proteins. Interestingly, many of the current expression targets are proteins of potential human therapeutic use, broadly categorized as cytokines. These proteins have specific effects on immune responses or other cell signaling mechanisms. In such cases, rigorous removal of endotoxin is required before functional assays are performed. Thus, even at research scale, the presence of endotoxin in the bacterium producing recombinant protein frequently persists as a problem.

The ability to grow *E. coli* cultures to very high densities (approaching an effective value of 200 units when measured by optical density determination at 600 nm) in large-scale fermentation, has been an important advantage in the use of this bacterium and has essentially set the bar high when considering alternative organisms. A more recent and important challenge to the industry of large-scale recombinant protein production has been restrictions on the use of animal-derived proteins as a component of the growth medium used in fermentation. This was a response to concerns about the introduction of prions, such as that responsible for variant Creutzfeldt-Jakob disease, into therapeutic products. While some microorganisms can rely on plant-based protein hydrolysates (often derived from soybeans), this is done at relatively high cost and the ideal medium will be composed of simple sugars, micronutrient salts and mineral sources of nitrogen and phosphorus. High density *E. coli* fermentation typically can be accomplished with such medium restrictions, setting another high bar for comparison to alternative expression organisms.

The point of the foregoing is neither to be critical nor supportive of efforts to produce recombinant proteins in *E. coli*, but to emphasize that there are numerous reasons to consider other microorganisms for recombinant protein expression, both at the research and commercial scales. Ideally the alternative organisms will be appropriate at both levels of scale. More importantly, it is our view that the next generation of microbes used for commercial level expression will be those that secrete heterologous proteins. Secretion of target proteins into the culture medium nearly always means that subsequent purification steps will be fewer, simpler and involve less chromatography media and solvent waste. As a result, it is more likely that the purification process will bypass protein denaturation steps, which often means a higher chance of maintaining function of the expressed protein.

Protein secretion presents a considerable biological "hurdle": the transport of protein across membranes. This usually requires that the protein be competent for transport, which often involves its insertion into a membrane in an unfolded or alternatively folded state in a pathway-specific (or organism-specific) manner. Many secreted proteins may have evolved to match their mode of secretion and this does not bode well for the research community who expect an expression system to have as general a utility as possible. Moreover, in eukaryotic systems, the secretion pathway is complex and involves transport through the endoplasmic reticulum and the Golgi, where glycosylation and other forms of postranslational modification occur. Often, such modifications are cell type-specific and may be essential to accomplish ultimate export and proper folding. In many cases, however, it is not desirable to have the target protein modified by glycosylation and when postranslational modifications are needed, they are likely to be very specific and something that no general secretion-competent microorganism can readily achieve. For Gram-negative, Gram-positive and Archaeabacteria, and with some notable exceptions including those described below, secretion beyond the cytoplasmic membrane is most often accomplished via the General Secretory (or Sec-dependent) Pathway (GSP) (Van Wely *et al.*, 2000). This pathway involves an N-terminal signal or leader peptide that initiates secretion by insertion into a transport complex located within the cell membrane. The secreted protein, often held in an unfolded state by cytoplasmic molecular chaperones, follows and either becomes integrated within the membrane or emerges on its outer face. These events are followed by cleavage of the signal leader peptide by an appropriate peptidase.

A final factor worth mentioning when considering new generations of gene expression organisms is the desirability of aerobic fermentation (*i.e.* that they require oxygen for growth, using respiratory pathways). Oxygen-mediated respiration offers large efficiency advantages over anaerobic fermentative pathways. For nearly all applications, such efficiencies are necessary for the cost-effective conversion of feedstock to target protein. *Obligate* respiratory behavior is even more desirable. For example, *E. coli* is capable of aerobic respiration or of mixed acid anaerobic fermentation and will switch from aerobic to fermentative mode when oxygen is limited. The result is rapid utilization of feedstock, acidification of the culture medium and an early, unproductive end to the fermentation run. To prevent this from occurring requires control of either dissolved oxygen levels or feedstock levels. In practice, the solution has been to maintain careful control of feedstock addition in the presence of adequate oxygen levels while maintaining good growth rates. This is a complicated task, especially in high-density fermentations, requiring skilled operators and fermentors with considerable process control. As will be discussed below, microbes with obligate aerobic metabolism have the potential to considerably simplify these needs.

The foregoing is not intended to be a comprehensive analysis of the issues surrounding the needs of recombinant protein expression and some important characteristics have been left out (such as translational codon usage bias, the necessity to form intramolecular disulfide bridges, or the need for the protein to assemble as multimers). However, as a summary, the ideal microbe for heterologous protein expression should have the following characteristics. The microbe should be an obligate aerobe, capable of growth to high densities in completely defined mineral salts and simple sugars media. It should contain a secretion mechanism capable of exporting a wide variety of heterologous protein varying in properties (*e.g.* hydrophobic, hydrophilic, charged or neutral) and sizes. The secretion mechanism should not require postranslational glycosylation of the protein or require it to pass through a hydrophobic membrane interior in an unfolded state. Ideally, the organism would secrete little else other than the recombinant protein and particularly not proteases that could degrade the target product. The microbe should produce no endotoxin or other toxic or immune-stimulating substances. It should be equally appropriate for research and commercial scale production. Finally, it should exhibit many of the current *E. coli* advantages, such as rapid and simple transformation (normally using plasmid vectors) and adaptability for high-throughput methods (in order to screen large numbers of uncharacterized proteins and convert genome sequence to proteome functional information). This means rapid growth as colonies on solid media.

Although the list of requirements is not complete, it is already apparent that no available microbe fulfils all of the above needs. Nevertheless, many expression hosts in use today possess a large number of these characteristics and are reviewed in detail in other chapters of this book. One of the first microbes to be exploited was the yeast *Saccharomyces cerevisiae*. Like *E. coli*, this was due to a long history of genetics development, a large number of experienced researchers and the discovery that elements of the mating system could be exploited for protein secretion. *Bacillus subtilis* strains received attention because they were well known for native protein secretion and because fermentation expertise was well developed based on long history of production of subtilisin variants, primarily for the detergent industry. The fact that the bacterium secretes numerous native proteases was an impediment for heterologous protein secretion until protease-deficient strains were developed (Wong, 1995; Wu *et al.*, 1998, 2002). Yeasts that use methanol as a food source (*Pichia pastoris,*

P. methanolica and *Hansenula polymorpha*) have been used because of the low cost of feedstock, their secretion capabilities, and their ability to grow to very high densities (Cereghino and Cregg, 2000; Gellissen, 2000). Filamentous fungi were also chosen because of their protein secretion capability and established expertise in fermentation. The latter expression hosts suffer from less developed genetics, but strides are being made. For example, completion of the *P. pastoris* genome sequence will undoubtedly assist strain development. All of the organisms listed above can be fermented in animal protein-free media and do not produce an LPS-type endotoxin.

Such is the background to adding another expression system to this arsenal of possibilities. This chapter will indicate that heterologous gene expression using the bacterium *Caulobacter crescentus* and its S-layer secretion mechanism fulfils a large number of the requirements mentioned above. As a preliminary summary, Caulobacters secrete recombinant proteins using the Type I secretion system of the S-layer, which does not rely on the potentially limiting General Secretory Pathway (GSP). The bacteria are obligate aerobes and grow to high densities in defined minimal media with minimal requirements for process control. They do not secrete other proteins, have an LPS that is unexpectedly low in endotoxin potential, are easily manipulated due to a long history of genetics development (including a sequenced genome) and are readily grown as colonies on solid media. In addition, the secretion mechanism provides opportunities for rapid initial purification at the research scale via production of recombinant fusion proteins as hydrated aggregates, while producing soluble proteins in large-scale fermentation. Yields can approach the gram per liter scale even in laboratory shake flask culture, although the 100 mg/liter range is more typical.

Presentation of Heterologous Peptides and Proteins: Size Matters

The ability to present heterologous peptides or entire proteins on the surfaces of viruses or bacteria ("display") has been explored by many laboratories. There are numerous applications to this technology, ranging from vaccine development to toxin absorbents and affinity purification agents, to the preparation of large random libraries of peptides or gene segments. The latter can be used to screen for binding activities useful in biotechnology or to analyze the interactions of biomolecules in cells.

Peptide display has been accomplished in many species of both Gram-positive and Gram-negative bacteria (for a recent review, see Samuelson *et al.*, 2002). Most studies in Gram-negative bacteria have focussed on the use of outer membrane proteins and rely on the fact that nearly all of them have a β-barrel structure in which membrane-spanning antiparallel β-strands are interspersed with surface exposed loops suitable for the display of heterologous peptides. Porins or surface receptors (*e.g.* OmpF, LamB, FhuA) as well as autotransporter proteins (now called Type V secretion) and outer membrane lipoproteins have all been used for this purpose (Etz *et al.*, 2001; Samuelson *et al.*, 2002). C-terminal display of peptides is possible in cases such as OmpA or the IgA protease autotransporter. The general result has been that small peptides, ranging in size up to 60 amino acids, are often successfully displayed. However, with larger peptides, successes are often accompanied by a reduction in the amount of protein displayed or by the degradation of the recombinant protein. This is presumably because the extraneous insertion has caused some degree of impediment in transport or folding processes, allowing an opportunity for proteolytic cleavage. It is tempting to speculate that much of this occurs by exposure of at least a portion of the protein to the periplasm, where a variety of proteases act, perhaps because the proteins are not able to properly interact with periplasmic folding factors (Bulieris *et al.*, 2003; Voulhoux *et al.*, 2003). The concern then is the degree of "editing" that occurs with such display vehicles when used to produce libraries of combinatorial peptides or for gene fragment display.

Display using flagellar and fimbrial filament proteins has also been demonstrated for a variety of different genes (Klemm and Schembri, 2000; Lu *et al.*, 1995). In the case of flagellar display, there is an opportunity to bypass the GSP pathway and the need to insert an unfolded target protein into the Sec transport apparatus, since transport of filament proteins occurs by a mechanism analogous to Type III secretion and through a pore created by the developing flagellar motor apparatus (Plano *et al.*, 2001). But this is apparently balanced by constraints in the folding pathway required to produce a flagellum and size is restricted to about 25 amino acids. Fimbrial (pilus) display seems to have a limit of 10-20 amino acids due to analogous folding constraints. This limit is often counterbalanced by the large number of monomers produced by some fimbrial types, which may be an advantage for the screening of certain types of libraries or for whole cell vaccine applications.

Display in Gram-positive bacteria has often relied on the adaptation of the sortase pathway. Here, proteins are secreted through the cytoplasmic membrane in a GSP-dependent manner, whereupon they are cleaved in a multiprotein complex at a conserved motif (LPXTG) and covalently attached to the cell wall at the site of cleavage (Mazmanian *et al.*, 2001; Wernerus *et al.*, 2002). Relatively large (≈ 400 residues long) proteins have been successfully displayed in this manner using Staphylococcal or lactic acid bacteria strains. This mode of display allows N-terminal exposure of the library (as opposed to display within exposed loops) which may be essential for the proper folding of certain proteins, but presumably has limits. In some cases, relatively high copy numbers of displayed protein (1000 to 10,000 copies/cell) have been achieved. The use of Gram-positive bacterial display for use as vaccines or oral therapeutics is predicated upon the reasonable belief that the food grade bacteria chosen will be safe for use in humans (Kruger *et al.*, 2003; Lee *et al.* 1999; Pouwells *et al.*, 2001; Wernerus *et al.*, 2002).

Other motifs derived from proteins that bind cell wall components non-covalently have also been exploited for Gram-positive bacteria display. Most notable are those derived from the peptidoglycan binding motif discovered in the lactococcal cell wall hydrolase AcmA (Leenhouts *et al.*, 1999). Similarly, the so-called S-layer like homology (SLH) regions, first defined in an S-layer from a Gram-positive organism (Lupas *et al.*, 1994) provide anchoring to the Gram-positive cell wall. These non-covalent attachment regions have allowed display of the N- or C-terminus of the desired heterologous protein. In both cases three tandem repeats are found in the native gene and elimination of one or two of these leads to less efficient cell wall attachment (Leenhouts *et al.*, 1999; Mesnage *et al.*, 1999).

Display using bacterial viruses for the purpose of producing libraries of variants has been an active area of research. Most efforts have been directed towards the *E. coli* filamentous phage M13, where there are two major approaches for display (Irving *et al.*, 2001). Combinatorial peptide libraries have been displayed at the N-terminus of the high copy number gene VIII protein, where packing constraints limit the size of displayed peptides to perhaps less than 20 amino acids (although work is under way to improve this aspect). Libraries of much larger proteins, such as antibodies (in the form of single chain Fv [ScFv] or Fab' segments) have been displayed on the low copy number gene III protein, usually at less than one recombinant protein per virus particle. This occurs because the system involves the use of unmodified genes to produce "helper" proteins that ensure the formation of intact phage particles. Phage display has been largely used for the discovery of useful binding activities and to improve binding through mutation ("maturation") technologies rather than for producing an actual "end product". More recently, there have been significant advances in adapting recombinant phages for applications such as vaccines and diagnostics (Irving *et al.*, 2001).

The use of phages to display peptides and proteins has technological limitations. The small size of phage particles means that the use of flow cytometry (most commonly under the form of a fluorescence activated cell sorter [FACS]) is impractical. This means that binding phages must be selected by attaching the target ligand to a solid surface (usually glass or plastic), applying the library, and eluting bound phages. Several cycles of enrichment are generally required, making selection a relatively slow process compared to FACS-based screening. There is also a growing concern that the tightest – and often most desirable – binding clones cannot be retrieved without destroying the phage particle. Some have approached the latter problem by incorporating enzyme cleavage sites between the random peptide and the remainder of the phage coat protein. In addition, proteins that form filamentous phage coats must pass through the cell membranes and there is concern that, in this process, some filtering of useful clones is inevitable. The use of the lytic phage T7 for library presentation was adopted in part to address this issue (Danner and Balasco, 2001).

Saccharomyces cerevisiae has also been adapted for display, either by covalent linkage of the C-terminus to the cell wall or by use of the Aga2p binding domain of **a** agglutinin mating receptor, which in turn forms disulfide bonds to the Aga1p cell wall protein (Boder and Wittrup, 2000; Wittrup, 2001). Among other things, this allows cleavage or release of the target protein, if there is a concern about missing tight binders. It has also been argued that this eukaryotic host, with its attendant secretion pathway, may be much more appropriate than phage or bacteria for the display of certain proteins of eukaryote origin, such as T cell receptors. As with bacterial display systems, yeasts are large enough to be amenable for rapid screening by FACS using fluorescent ligands, suggesting the possibility of high-speed screening and retention of tight binders in the pool of positive clones.

Once again, the foregoing is not a complete analysis of the issues surrounding peptide display, its uses (for example, we have made little mention of genome fragment display in library applications or the value of certain bacteria as potential vaccine delivery agents), or even a complete list of the organisms adapted for this purpose. However, an important emphasis is that no display system has ideal characteristics. In our view, a significant limitation is that the GSP-dependency of most bacterial and phage display systems may be responsible for editing out a fraction of clones from the library because they cannot be synthesized and pass through the membrane in an unfolded state. Furthermore, the maximum size tolerated is often not adequate for many applications, especially for vaccines. Finally, the size of the displaying entity and the number of copies displayed per cell determines whether flow cytometry can be applied, which offers many advantages in screening. The latter factors may also have significant impact on the success of this technology for vaccination, toxin sorbents or agents for diagnosis.

Perhaps one way of summarizing would be that a bacterium useful for display in the broadest number of applications should be able to secrete a wide variety of peptides by a pathway that does not involve the

GSP, and be able to display large peptides, without degradation and at high surface density. The organism should be safe to use (including oral ingestion by humans and animals), yet be sufficiently stimulating to the immune system to help elicit immunity. Display of peptides with either a free C- or N-terminal should be possible. Library display applications also require considerable capabilities for genetic manipulation (including high copy number plasmids capable of being shuttled to *E. coli* for certain methods such as rapid DNA sequencing), in addition to the usual requirements of ready growth in simple media and growth as colonies on solid media. Of course, it will be suggested that the Caulobacter S-layer display capabilities meet many of these criteria. Caulobacter S-layer display, however, is a newcomer to both the display and heterologous gene expression fields. While certain aspects are promising, more direct evaluation in particular areas of application are still needed. One key advantage may be that the S-layer display protein is secreted by the Type I secretion mechanism in a manner that is somehow optimized for high level expression.

Type I Secretion in Gram Negative Bacteria

So far, five main secretion pathways (*i.e.* mechanisms that transport proteins from the cell interior through both membranes to the external milieu) have been characterized to a greater or lesser extent in Gram-negative bacteria (Thanassi and Hultgren, 2000). In the context of this chapter, Types II, IV and V secretion use the GSP as the means to traverse the cytoplasmic membrane. Type III secretion is used largely as a means for pathogenic bacteria to insert products into host cells. Type I secretion is decidedly different from the others, in terms of simplicity (number of transporter components), size of the aqueous pore created for transport (relatively large as predicted from crystal structure, Koronakis *et al.*, 2000, 2001), use of an uncleaved C-terminal secretion signal, and finally, because it involves a member of the large family of so-called ABC (ATP binding cassette) transporters. This family has members engaged in the export of many types of drugs, toxins and proteins and sometimes in protein import (Holland and Blight, 1999).

The best characterized Type I secreted protein is the *E. coli* α-hemolysin protein (Blight and Holland, 1994; Gentschev *et al.*, 1996, 2002; Mollenkopf *et al.*, 1996). A number of other proteases (often involved in pathogenesis) and occasionally an S-layer protein are also secreted by the Type I pathway. Early on, the understanding that such proteins were part of a large family was slowed by the observation that the C-terminal secretion signals of individual Type I secreted proteins exhibited little sequence homology. It was not until the transporter genes from a number of different Type I proteins were functionally identified that the family became clarified. This is because although secretion signals diverge, the transporter components of various Type I systems share a much higher degree of homology. Based on these results and the observation that secretion of one Type I protein by the transport apparatus of another was often possible, it was concluded that C-terminal secretion signals share a similar secondary structure (*i.e.* an overall shape).

The minimal secretion signals can usually be assigned to the last 60 amino acids or so of the protein (Binet *et al.*, 1997). However, upstream of the C-terminus of all Type I secreted proteins are tandemly repeated motifs that presumably bind calcium. Called repeats in toxin (RTX) region, these motifs are somehow involved in either the efficient secretion of the protein or activity of the protein being secreted. In any event, most or all of the Type I proteins require calcium for proper functioning and although there is not complete clarity in this area, it is often assumed that the RTX region, with bound calcium, is needed for efficient folding, perhaps for both secretion and subsequent function. Proteins secreted by Type I mechanism presumably are completely translated before secretion begins and so may have achieved at least partial folding into their functional form before transport across the membranes begins.

The transport apparatus is composed of three components. The ATP binding cassette (ABC) transporter spans the cytoplasmic membrane, recognizes the C-terminal secretion signal and hydrolyzes ATP during transport. The membrane fusion protein (MFP), which is anchored in the cytoplasmic membrane, and an outer membrane protein (OMP) that interacts with the MFP and ABC transporter to create a gated

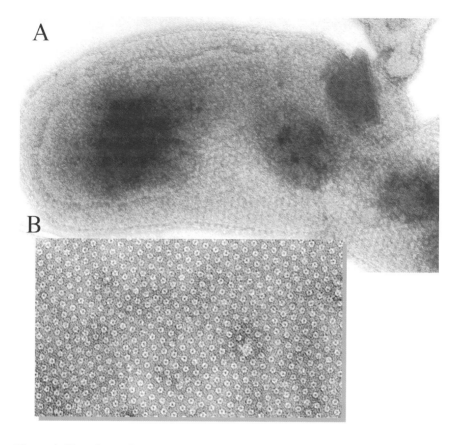

Figure 1. Negative stain electron microscopy of: A: A lysed cell demonstrating the S-layer on the cell surface and, B: Isolated membrane fragment showing the structural detail of the S-layer assembly more clearly.

pore that spans both membranes complete the system (see Figure 2). In *E. coli* the OMP is TolC, a protein that functions in other transport systems. Most other Type I systems have a dedicated OMP. In terms of genome organization, for most Type I systems, the ABC transporter and MFP are adjacent to the gene for the secreted protein on the 3' side, whereas the OMP may be immediately downstream of the MFP or elsewhere in the genome. Crystallographic studies (Koronakis *et al.*, 2000; Andersen *et al.*, 2001) have shown that TolC forms a trimer when engaged to the other two components, spanning the outer membrane with anti-parallel β-barrel sheet motifs and traversing the periplasm with α-helical structure to link with the other two transporter proteins. The formation of a trimeric assembly accounts for the large internal aqueous cavity predicted by structural studies. The large size of this pore may allow transport of partially folded proteins. This feature could explain the success of Type I systems for the secretion of heterologous proteins.

The Type I mechanism for hemolysin secretion in *E. coli* has been adapted for the export of heterologous proteins (Mollenkopf *et al.*, 1996; Tzschaschel *et al.*, 1996). At least two groups have described its general utility for secretion of proteins such as vaccines or enzymes (Blight and Holland, 1994; Gentschev *et al.*, 2002), while others have subsequently used the transporter for secretion of ScFvs (Fernandez *et al.*, 2000). The hemolysin secretion system seems to have the capabilities we will suggest for the Caulobacter S-layer Type I system. It can secrete hydrophilic proteins at relatively high yields, often in an active conformation and via a GSP-independent pathway. The system must be induced for maximal expression, which can be 3-5% of cell protein. Because the mechanism bypasses exposure to the periplasm, there is no opportunity for intramolecular disulfide bond formation to be catalyzed by the DsbA and DsbC proteins (Collet and Bardwell, 2002). Thus, one might expect that the folding of some heterologous proteins would be affected. Nevertheless, in the case of certain ScFvs, the disulfide bonds needed for functionality were formed, apparently in the cytoplasm (Fernandez and de Lorenzo, 2001). Because hemolysin secretion occurs in enteric bacteria, LPS endotoxin contamination of secreted proteins is an issue. On the other hand, this system shows promise as a vehicle for vaccines produced in attenuated strains of Salmonella (Tzschaschel *et al.*, 1996), whose cell internalization capabilities may be important for vaccine development. Because the secretion mechanism relies on an uncleaved C-terminal signal, removal of the secretion signal (by incorporation of an enzyme cleavage site and subsequent enzyme treatment) is required in many applications.

Caulobacter crescentus, the S-layer and its Type I Secretion Mechanism

C. crescentus is a non-pathogenic, Gram-negative freshwater bacterium that has been the subject of molecular genetics-based laboratory experimentation for more than 30 years, largely because of its use as a model for cell differentiation (England and Gober, 2001; Jenal *et al.*, 1995; Wheeler *et al.*, 1998). Toward this end, many genetic methods that are routinely used in *E. coli* have been developed for *C. crescentus*. These including electroporation of plasmids (Gilchrist and Smit, 1991; Smit *et al.*, 2000a), transposon mutagenesis (Ely, 1991), allelic exchange, and the selection of classical point mutations. The genome of strain CB15 has been sequenced (Nierman *et al.*, 2001), adding another useful tool for rapid strain development.

C. crescentus strains grow rapidly in simple glucose and salts media to high densities and grow especially well in media containing glutamate in place of some or all of the ammonia nitrogen (first noted by Poindexter, 1998). Growth is optimal at 30°C with a generation time that can be as short as 2 hours. They are obligate aerobes and thus are easy and inexpensive to cultivate. We and others have grown Caulobacter in standard fermentors up to 25-30 OD units (at 600 nm) using simple batch feeding methods, where the limitation on growth is the rate of oxygen addition. This is unlikely to be an upper limit. Thus, in sharp contrast to fermentation of *E. coli* (a facultative aerobic bacteria), in the presence of widely varying concentrations of organic carbon, no pH adjustment is required using a growth medium that is virtually unbuffered (J. Smit, unpublished). One wag has commented that one needs neither a Ph.D. nor a pH meter to accomplish fermentation of the bacterium.

C. crescentus secretes large amounts of a protein (98-kDa in size), which assembles into a regular hexagonally packed pattern on the cell surface (Figure 1) (Smit *et al.*, 1981; Smit and Agabian, 1984; Gilchrist *et al.*, 1992). The crystalline surface (S)-layer that results is composed of ring-like structures spaced at 22 nm intervals and there are approximately 40,000 monomers on an average cell (Beveridge *et al.*, 1997; Bingle *et al.*, 1993b; Smit *et al.*, 1992). The S-layer attaches to the surface of *C. crescentus* using a variant of the outer membrane LPS with a long side chain (Walker *et al.*, 1994; Awram and Smit, 2001). The protein is hydrophilic (25% of the residues are serine or threonine) and lacks cysteine residues (as do nearly all S-layers). S-layers (*i.e.* geometrically arranged protein layers on the outermost surface of a bacterium) are common in many genera of microorganisms, including Gram-positive, Gram-negative and Archaebacteria (Boot and Pouwells, 1996; Sleytr and Beveridge, 1999; Smit, 1986). Most are secreted by the GSP pathway, using Type II secretion in Gram-negative bacteria to traverse the outer membrane. However, the S-layer proteins of *C. crescentus* (Awram and Smit, 1998) and those of *Campylobacter fetus* (a pathogen of sheep and cattle causing spontaneous abortions) are secreted by the Type I mechanism (Thompson *et al.*, 1998).

A key factor for the success in adapting the S-layer for heterologous protein secretion and presentation is likely due to the flexibility engendered by the Type I pathway, especially in the case of hydrophilic proteins. In contrast to hemolysin secretion or that of other Type I secreted proteases, the secretion of the S-layer monomer has been in some way adjusted for high level expression. We have estimated the S-layer monomer (RsaA) accounts for 10-12% of the cell protein. RsaA synthesis occurs without need for induction, and the protein is produced continuously throughout the cell cycle (Fisher *et al.*, 1988; Smit and Agabian 1982b). A growing body of information suggests that export may be limited by some sort of feedback mechanism that slows protein secretion when the surface is "full". In particular, we have noted that truncated forms of RsaA (where the truncated S-layer monomer is no longer surface attached or crystallized) can be synthesized to about 30% of the total cell protein. One factor that may account for the high level of secretion is that two OMPs are involved in the process (M. C. Toporowski, J.F. Nomellini and J. Smit, unpublished; illustrated in Figure 2). Interrupting the OMP gene (*rsa*F973) that is adjacent to the ABC transporter and MFP genes (*rsa*D and *rsa*E) reduces S-layer monomer (RsaA) secretion by about 50%. Interruption of the other known OMP gene (*rsa*F1984) reduces S-layer secretion by only about 25%. However, disruption of both genes abolishes all RsaA secretion suggesting that the RsaF1984 OMP paralog assists the main OMP

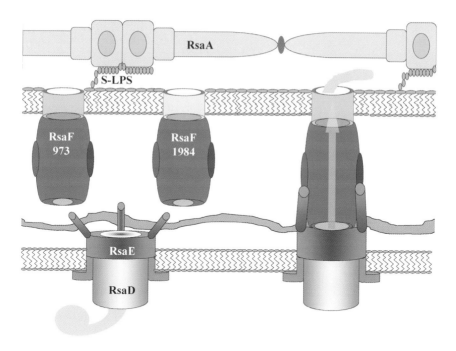

Figure 2. RsaA Type I transporter. The role of each of the RsaF genes in the assembled transporter is purposely ambiguous; we presume that they form trimers but do not yet know if they are hetero- or homotrimers.

transporter to achieve high level secretion. High level secretion may be driven by a need in the natural environment to cover the surface with S-layer at all times, since it is presumed that a common function of S-layers is to provide a measure of protection for the cell from attacking agents such as proteases, viruses and parasitic bacteria (Sara and Sleytr, 2000; Sleytr and Beveridge, 1999). It is known, for example, that a *C. crescentus* S-layer proficient strain is resistant to a Bdellovibrio-like organism, while an S-layer negative variant is sensitive to attack (Koval and Hynes, 1991).

Development of a Gene Expression System with the Caulobacter S-Layer

Our early efforts in defining the secretion signal of the *C. crescentus* S-layer were carried out without knowing that it was secreted by a

Type I mechanism. Thus, initial efforts were directed at determining whether the somewhat hydrophobic N-terminus (Fisher *et al.*, 1988) was a secretion signal for a GSP-like pathway. Curiously, results suggested a periplasmic targeting capability that we now believe was cryptic (Bingle *et al.*, 1993a; Bingle and Smit, 1994). These results led to a more general inspection of the *rsaA* gene using a linker mutagenesis procedure to define functional regions and learn whether there were sites that would tolerate the addition of heterologous sequence (Bingle *et al.*, 1997b). We indeed found numerous positions (largely in the central 60-70% of the gene) that tolerated the addition of four to six heterologous amino acids and later learned that many of these sites would accept the addition of an epitope of about 15 amino acids with no loss of functionality (Bingle *et al.*, 1997a). That is, the protein was still secreted, surface attached, assembled into a crystalline structure and the epitope was accessible to binding by a monoclonal antibody. This of course has evolved into the presentation aspects of the S-layer that are described below.

We also learned that those sites that were tolerant for presentation were also sites at which the gene could be truncated without affecting the secretion of the C-terminal remainder of the protein. Moreover, we discovered that this was also true when a "passenger protein" was incorporated at the N-terminus. The minimum amount of C-terminal sequence required to accomplish secretion is not more than 82 amino acids (Bingle *et al.*, 2000) and by sequence comparison to other Type I secretion signals is about 65 amino acids. However, increasing the length of the C-terminus in the constructs to up to about 240 amino acids resulted in significantly greater amounts of secreted protein, more than could be accounted for by the larger monomer sizes, assuming a fixed number of transport "events". Thus, although the extreme C-terminus provides an essential secretion signal, additional amounts of the C-terminal region are needed for optimal expression levels. We presume that this is due to an optimal shape of the partially folded protein adjacent to the secretion signal. One important region for proper secretion may well be the tandemly repeated 9 amino acid RTX motif adjacent to the minimal secretion signal.

Aggregate Formation: A Built-In Chromatography Tool

Of particular interest in the development of a heterologous protein secretion system was the observation that the secreted S-layer truncates or the fusion proteins produced by adding a heterologous partner would form loose aggregates of gel-like protein in the culture medium. These aggregates can be readily filtered away from the cells with a mesh filter and further purified with a simple rinse with water (or repeated cycles of low speed centrifugation to pellet the gel, but not the bacteria). The typical result is a recombinant protein that is at least 90% pure and is often more than 99% uniform, as illustrated in Figure 3. The gel-like material is 95-98% water, yet remains insoluble during multiple weeks of storage. Therefore, Type I secretion has not only the advantages summarized above but provides a built in "chromatography" capability that enables rapid concentration and separation from other organic components or cell debris present in the culture medium.

The aggregated fusion proteins produced in this manner can be prepared in a soluble form. Urea is the least disruptive agent and most of our experiments involve solubilization with 4M urea, which is capable of almost completely dissolving the material. The aggregates do exhibit some degree of heterogeneity as evidenced by the fact that a fraction of the protein can be rendered soluble by using lower concentrations of urea (even 0.5 M urea releases useful amounts of protein from aggregates). The latter property may be important in instances where no denaturation of a protein is desired. Indeed, it is believed by some that urea can be used up to a 1M concentration without denaturing effects on most proteins. Following urea solubilization and removal by dialysis results in soluble fusion proteins if concentrations are maintained at about 2 mg/ml or lower. Careful solubilization and high speed centrifugation to remove micro-aggregates allows maintenance at still higher concentrations (about 5 mg/ml, although our experience is limited).

More recently, we have used disruption of aggregates by sonication as a means to prepare soluble protein. Operationally, the recombinant proteins are indeed soluble, enzyme cleavage sites are more accessible, and a large fraction of the aggregates can be converted to the soluble form. Since this method circumvents the use of urea, some recombinant proteins are more likely to maintain or adopt a properly folded conformation. The stability of such preparations against reaggregation over time is still being evaluated.

The exact reason why truncated S-layer monomers (or shed full-length protein) form highly hydrated aggregates is not clear. It does not seem related to the RTX (presumably calcium-binding) regions and the calcium available in the medium. Treatment with EGTA does not disrupt the aggregates, calcium has only a minimal effect on encouraging their formation and some aggregates still form when the shortest truncated variants (those that do not contain the RTX region) are expressed. The best explanation at the moment is that the high number of hydroxyls in the abundant serine and threonine residues of the S-layer protein order water molecules and encourage hydrated aggregate formation.

The formation of aggregates can greatly speed initial purification of recombinant fusion proteins, not only because of the ease of collection, but the fact that contaminating proteins, polysaccharides or cell debris can be removed by a simple flushing with water or buffer. There is, however, a certain consistency required in such factors as shaker speed and ratio of medium volume to flask size to obtain the maximum amount of (or in some cases, any) aggregate material. Essentially excessive shear force (caused by too little medium in the flask or too fast a shaker speed) results in poor aggregate formation or aggregates that are fine and difficult to collect. On the other hand, too much medium in the flask or too slow a shaking speed results in little or no aggregate formation. This is in part due to a reduction in the amount of oxygen flux and the resulting decrease in cell growth. For research level production, we currently use Erlenmeyer flasks 250 ml in size or greater (or Fernbach flasks for the larger volumes) with a 40-45% volume of liquid medium. For 250 ml flasks, agitation at120 rpm is suitable for most shakers. Larger flasks must be shaken at slower speeds (60-80 rpm).

Culture Media and the Growth of an Obligate Aerobe

Caulobacters are perhaps best known by microbiologists as an example of an oligotrophic bacterium, one that prefers or requires low organic carbon fluxes. On that basis, they would not be immediate candidates for use as an "industrial" bacterium capable of growth to high densities in a robust manner. This may well be true for the majority of Caulobacter strains isolated from natural sources (MacRae and Smit, 1991) which must survive extended periods of starvation and do not grow in defined mineral salts medium or to high densities in complex media. Those with S-layers in this category are generally grouped under the species *Caulobacter vibrioides* (Abraham *et al.*, 1999; Walker *et al.*, 1992). Much less commonly found are freshwater Caulobacters that do grow readily in defined mineral salts media (and also produce S-layers). These are strains of *C. crescentus*, despite the fact that they do not differ from *C. vibrioides* in 16S rDNA sequence and other characters (Abraham *et al.*, 1999; Poindexter, unpublished). The S-layer expression system described here is present in strains CB2 and CB15. These strains are able to grow in defined media containing glucose as carbon source and ammonium chloride as nitrogen source. They grow especially well and to high density in defined media where at least some of the nitrogen requirement is met using glutamate (Smit *et al.*, 1981). As mentioned above they can be grown to at least 30 OD units (at 600 nm) in media that do not contain animal proteins, an important consideration for industrial production.

It should also be noted that freshwater Caulobacters do not grow well at salt levels greater than about 50 mM and do not grow at all above about 120 mM (Anast and Smit, 1988; Smit, unpublished). This is a useful feature since, as a consequence, *C. crescentus* does not grow in tissue culture media. This issue is becoming increasingly important as it is common for companies to maintain several recombinant protein expression systems and it is difficult and expensive to keep bacterial expression systems separated from tissue culture facilities in order to prevent contamination of mammalian cell cultures. Presumably this would not be an issue in Caulobacter fermentation.

Growth in fermentors is an especially uncomplicated matter, largely due to the obligately aerobic metabolism. Our normal fermentation method involves batch feeding of glucose and glutamate. Growth is ultimately controlled by the rate of oxygen addition, which is nearly always limiting. *C. crescentus* apparently tolerates oxygen stress well, and there has no tendency to form glycolytic byproducts (such as lactic or formic acid) during high density growth. No pH adjustment apparatus is necessary. In fermentation runs to 30 ODs variations of no more than 0.1 pH unit from (from pH 7) are typical in a medium that contains only modest amounts of imidazole (5 mM) and phosphate (initially 2 mM) as pH buffering agents. We have not yet attempted continuous fermentation, but there is no apparent reason why this would not be possible.

Despite the ease of growing *C. crescentus* under fermentation conditions, we have so far been unable to find reproducible conditions for aggregate formation in a standard fermentor. This is undoubtedly due to the shear forces produced by stirring and air and oxygen addition, especially at higher cell densities. On the other hand, we and others have learned that recombinant fusions proteins are nevertheless produced and are in a soluble form. That is, the proteins do not pellet with the cells under typical centrifugation speeds and can be readily identified in a clarified supernatant. Thus, while aggregate formation is often viewed as a convenience at the research scale, it is not especially helpful at production scale and happily, can be easily avoided. Thus the fusion proteins can be treated as secreted proteins in a medium that is largely devoid of organic molecules by the end of a fermentation run. Moreover, aside from the production of a polar flagellum (which is discarded by the cell during the development cycle, but readily pellets with the cells) and small amounts of polar pili (Smit and Agabian 1982a), *C. crescentus* secretes no other proteins and, in particular, no native proteases.

Plasmid Vector Development

Like nearly all non-enteric bacteria, Caulobacter strains do not replicate the commonly used colE1 replicon type plasmid vectors (such as pUC vectors). Initial studies with S-layer gene expression system was accomplished using broad host range plasmid vectors (*e.g.* Bingle and Smit, 1990). Although many of the known broad host range plasmid complementation groups work in Caulobacter (Anast and Smit, 1988), we have mainly used vectors based on the IncQ replicon. These vectors are relatively large and have low copy numbers in both *E. coli* and Caulobacter. As a result, they are not amenable to a variety of methods, including ease of plasmid purification in large amounts for rapid cloning, DNA sequence confirmation and the like. We therefore developed a generation of *E. coli/C. crescentus* shuttle vectors, based on the colE1 vector pUC8, to which we added the vegetative origin of the IncQ plasmid RSF1010 (Umelo *et al.*, 2001). This increased the vector size by about 400 base pairs. The *rep*BAC genes, required for recognition and replication of this replicon, were installed into the chromosome of selected *C. crescentus* strains, enabling them to replicate the plasmid. The result was a typical high copy number plasmid that performed as well as other pUC- or pTZ-type plasmids during manipulations in *E. coli*. Unexpectedly, we noted a relatively high copy number in *C. crescentus* strains as well (approximately 22 copies/cell). This feature now enables us to retrieve plasmids from Caulobacter strains for re-introduction into *E. coli*. To date, we have been unable to cure Caulobacter strains of these plasmids, even after numerous cycles of cultivation without antibiotic selection. Plasmid stability is presumably related to the fact that, at higher copy numbers, no cell is free of at least one copy of the plasmid following division. The same vector is used for peptide/protein display within the complete S-layer gene, although the high copy number required modifications of the *rsa*A 5' untranslated region (see below).

Expression of Recombinant Fusion Proteins

At the time of this writing, we have attempted expression of approximately 50 different peptide or protein types in our laboratory. In addition, in a number of cases, we have combined several types of peptides together in tandem. This is important in projects related to vaccine development, where a typical strategy to optimize formulation is to use B- and T-cell antigens in different combinations, orders and tandem repeats (*e.g.* Umelo *et al.*, 2000). We have also generated successful fusions with portions of the C-terminus of the S-layer (most often at amino acids 690 or 784, although amino acids 723 or 622 have also proven useful).

A number of the proteins for which we attempted expression were part of evaluation projects with commercial entities and often a complete description of the targets is not possible. In our laboratory we have produced fusion proteins where the recombinant target protein has been as small as 6 and as large as 450 amino acids. We have not systematically explored the upper limits of size, but presumably some recombinant segments could be secreted that are as large as the region that was deleted to produce the secretion signal vector. They may even be larger since, in some instances of S-layer display (see below), we are exporting chimeric S-layer proteins that are more than 600 amino acids larger than the native RsaA monomer.

In general, the highest degree of success has been obtained with hydrophilic proteins and protein segments. For example, viral surface or coat protein segments (often evaluated as vaccine candidates) are often successfully produced as fusion proteins (*e.g.* Simon *et al.*, 2001; Umelo *et al.*, 2000). We have only attempted a small number of proteins with multiple transmembrane segments and these have not proven capable of secretion. However, production of segments with a single predicted transmembrane region (such as from Type I viral glycoproteins) have been successful, as has a bacterial outer membrane protein predicted to be a β-barrel type of integral membrane protein (characterized by anti-parallel β-sheets). As another example, we have successfully secreted the entire FimH fimbrial tip monomer from Type I *E. coli* fimbriae (Figure 3), despite the property of being very insoluble if not properly assembled into the pilus structure (Choudhury *et al.*, 1999). This latter S-layer protein fusion produced hydrated aggregates that were readily separated from the cells and appeared similar to those produced with other fusion partners. Constructs which contained an enzyme cleavage site between FimH and the S-layer secretion signal were also prepared. Once cleaved, the FimH monomer became highly insoluble and resistant to solubilization, much like the natively produced protein (J. Smit, unpublished).

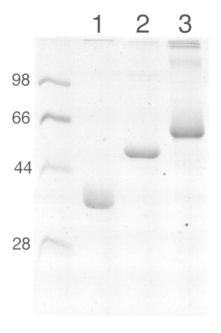

Figure 3. Expression of FimH fimbrial tip monomer from Type I *E. coli* fimbriae as a fusion with the C-terminal 336 amino acid of RsaA. SDS-PAGE with Coomassie Blue staining of the hydrated aggregates produced by the expression system with no additional purification steps. Lane 1- 336 amino acid C-terminal secretion signal only; Lane 2-Fusion of the N-terminal 164 amino acids of FimH with the C-terminal signal; Lane 3-Fusion of the entire 278 amino acids of FimH with the secretion signal. Size standards are shown at left with values in kilodaltons.

Proteins containing regions of high positive charge density may also not be suitable for this system. For example, we installed a typical furin protease cleavage site (recognition sequence arginine-lysine-lysine-arginine) in several fusion protein constructs (including one with just these four amino acids as fusion partner). We also inserted this sequence at several places in the complete RsaA monomer. In all cases, the result was a loss of protein secretion (J.F. Nomellini and J. Smit, unpublished). We have had similar results with certain short sequences of candidate cationic antibiotic peptides with high levels of positively charged amino acids. It is likely these charged amino acids are surface exposed in the passenger protein and we assume that interaction with charged residues lining the interior of the Type I transporter may cause the problem. We are addressing the matter with strategies to alter negatively charged residues in the predicted interior of the RsaA OMP genes, but for the moment avoidance of a high density of positive charges is the recommended solution.

In terms of other practical issues in gene expression, we have noted few problems in the stability of tandem repeats of genes, despite the fact that we have indeed made a number of such constructs in the course of investigations in subunit vaccine development as well as protein polymers. The current expression hosts have the plasmid replication genes integrated into the chromosome within the *rec*A gene (Umelo *et al.*, 2001), which presumably contributes to stability against deletion of tandem repeats of gene sequence. We have had occasional difficulties in aberrant translation starts from internal methionines (I. Dorocicz, J.F. Nomellini, J. Smit, unpublished; see Simon *et al.*, 2001 and Umelo-Njaka *et al.*, 2002 for an example). These are readily corrected by change of the affected methionine. Another opportunity to minimize the occurrence of such aberrant starts is to synthesize the gene and adjust the DNA sequence to minimize accidental Shine-Delgarno-like sequences ahead of the ATG codons.

Endotoxin Issues

As briefly mentioned above, we have investigated, in collaboration with Nilo Qureshi (University of Missouri at Kansas City), the Lipid A structure of the Caulobacter lipopolysaccharide to determine if we are likely to encounter the same endotoxin problems associated with *E. coli* expression. Fortuitously, we learned that *C. crescentus* Lipid A has a distinctly different structure (N. Qureshi and J. Smit, unpublished). The typical enteric Lipid A is composed of a disaccharide of N-acetyl glucosamine, bounded with phosphate residues and derivatized with multiple fatty acids, including those with 3-hydroxy acyl chains. Removal of fatty acids or the phosphate residues causes significant reduction in endotoxin-inducing effects. Examination of the *C. crescentus* Lipid A has revealed significant differences: a disaccharide of diaminoglucose replaces the N-acetylglucosamine disaccharide and the phosphate residues are replaced with residues of galacturonic acid. The effect, when assayed by a tumor necrosis factor induction assay for endotoxin has been somewhat variable but is centered about a reduction of about 1,000 fold, as compared with the *E. coli* Re LPS response. This obviously bodes well for a heterologous protein expression system. The reduced endotoxin effects led us to consider Caulobacter LPS as an adjuvant for vaccine studies. This is addressed in the next section.

Codon Usage

The GC content of *C. crescentus* is more than 65% and, as a consequence, it will exhibit a biased codon usage compared to many other organisms and often a heterologous gene may contain a high proportion of codons that are rarely used by Caulobacter. This is especially true when expressing genes derived from low GC content bacteria such as strains of Bacillus, Clostridia or Lactococcus. We define rare as those codon used less than 5 times per thousand codons. The actual codon usage can be found at http://www.kazusa.or.jp/codon/. In practice, we have expressed gene segments with as high as 35% rare codons, but have not systematically addressed the problem. Larger proteins, the extent of tandem rare codons, and whether rare codons occur very near the N-terminus of the heterologous protein are all negative factors that must be considered when assessing the effect on translation efficiency. In any case, codon usage is a factor when heterologous genes are used directly by subcloning or, more commonly, by PCR amplification of a target sequence. With the reduction in the cost of gene synthesis in the last several years, it is often cost-effective to simply have the desired gene sequence synthesized using the Caulobacter codon usage table as a guide.

Enzyme Cleavage of Heterologous Proteins from the RsaA Secretion Signal

When using the S-layer expression system as a means to produce proteins for applications such as raising antiserum or for monoclonal antibody generation, purification of the intact fusion protein is often sufficient. For other applications, such as human vaccine candidates, active enzymes or cytokines, it may be desirable to remove the S-layer secretion signal. This is most logically done by incorporating a cleavage enzyme recognition site at the junction between the target protein and the secretion signal. Most of our efforts in engineering a cleavage site have involved the use of a Clostridial collagenase recognition site (PL|GP) because it is inexpensive and leaves only two non-native amino acids at the C-terminus of the target protein. To maximize opportunity for cleavage, this sequence is sometimes tandemly repeated to produce two or three consecutive cleavage sites. Enzyme cleavage works most reliably on fusion proteins that have been solubilized with urea, followed by removal of urea by dialysis. Alternatively our recent success in cleavage using sonication of aggregates (as described earlier) merits additional development. Finally we are sometimes able to treat aggregates directly with enzyme and get significant levels of cleavage. In this case, the target protein is released into a liquid phase after cleavage and the aggregates are removed from the reaction by centrifugation or filtering. We are working toward determining an ideal spacer segment between the cleavage site and the S-layer secretion signal to maximize direct cleavage of aggregates, allowing significant purification of the target protein with no chromatography and minimal handling of the sample.

Development of a Peptide or Protein Display System with the Caulobacter S-Layer

As described earlier, initial studies aimed at defining functional regions of the S-layer protein also revealed the presence of sites that could be used to display small peptides (about 20 amino acids in length, including the residues specified by the linker sequences). In recent years we have learned that a number of these sites are actually tolerant of much larger sequences (commonly in the 70-200 amino acid range and more than 650 amino acids in one case so far; see Table 1). Tolerance in this instance is defined as: (i) insertion of the foreign sequence while maintaining minimal ability to secrete (*e.g.* 10-20% of normal levels although most often secretion levels approach wild type expression), and (ii) surface attachment and assembly of the monomer into a two-dimensional array, while still displaying the recombinant insertion to labeling tools such as antibodies. An example is shown in Figure 4. With a few exceptions, the successful display of these size ranges, in high copy number and without degradation, exceeds the capabilities of other Gram-negative display systems. This makes possible a variety of potential applications, ranging from peptide and gene fragment display libraries to serodiagnostic assay tools to whole cell vaccines for animals and humans.

Sites of Display

Sites have been prepared for insertion of heterologous material at gene positions corresponding to amino acids 266, 622, 690, 723, 784, 860 and 944 of the complete RsaA protein and heterologous sequences has been displayed at all of these positions. For various reasons, most current work has gravitated to sites 690, 723 and 944 and we are accumulating examples where the chimaeric RsaA protein appears to be more efficiently or fully secreted when a peptide is presented at one position compared to others. For example, we have displayed a 73 amino acid immunoglobulin binding domain derived from Protein G at several of the above positions and have found that display at position 944 result in ≈ 50% of the normal secretion levels while the amount displayed is lower at other positions (I. Dorocicz, J.F. Nomellini and J. Smit, unpublished). Recently, we have constructed a plasmid vector containing a display site at amino acid 2, as part of an effort to determine whether we can display heterologous material at the extreme N-terminus of RsaA (only the initial methionine of RsaA is removed during maturation, Gilchrist *et al*., 1992). We have been successful at inserting small peptides without adverse effects (notably shedding) and are learning whether this position is tolerant of larger protein fusions (J.F. Nomellini and J. Smit, unpublished). If this is the case, it may be possible to display libraries that have employed N-terminal display in other systems such as combinatorial peptide or ScFv libraries in M13 phage display.

Initial efforts in display were carried out using a low copy number, broad host range plasmid vectors (Bingle and Smit 1990; Bingle *et al*., 1997a). More recently, we adapted the high copy number

Table 1. Examples of presentation of peptides and proteins in the *C. crescentus* S-layer

Peptide presented	Size (AA*)	Site(s) presented (AA position in RsaA)	Approximate expression level or surface coverage	Notes	Ref. for peptide
Small peptides (10-50 AA)					
•c-myc epitope	10	690, 723, 944	WT¶		Rabbitts *et al.* 1983
•*Pseudomonas aeruginosa* type IV pilus "adhesintope	14	69, 277, 353, 450, 467 551, 574, 622, 690, 723, 944	WT		Wong *et al.* 1992
•Cadmium binding peptide	18	69, 277, 353, 450, 467 551, 574, 622, 690, 72, 944	WT		Ghadiri *et al.* 1990
•MUC1 conserved repeat sequence of mucins	28, 42	723	WT		Lan *et al.* 1990
Medium sized peptides (50-100 AA)					
•Tandem repeats of P. aeruginosa adhesintope.	3X=60, 6X=129	690, 723	WT		Wong *et al.* 1992
•Tandem repeats of MUC1 conserved repeat of mucins	4X= 91	723	WT		Lan *et al.* 1990
•Intimin binding domain of the Tir protein of *E. coli* O157:H7 and O127:H7	68	690, 723	70-100% of WT		de Grado *et al.* 1999
•Immunoglobulin Fc binding domain of Streptococcal Protein G	73	690, 723, 944	20-50% of WT		Perna *et al.* 2001 Fahnestock *et al.* 1986

Large peptides (100-200 AA)

•Intimin domain 3 and 4 from *E. coli* O157:H7 (AA 754-934)	181	690, 723	10-20% of WT		Perna *et al.* 2001
•Tandem repeats of MUC1 conserved repeat of mucins	8X= 175	723	WT		Lan *et al.* 1990
•IPNV salmonid fish virus glycoprotein VP2: AA145 to 257.	112	690, 723	WT		Engelking *et al.* 1989
•IPNV salmonid fish virus glycoprotein VP2: AA145 to 257 plus two two T-cell epitopes P2=17, MVF= 17	165	690, 723	50% of WT		Engelking *et al.* 1989; Panina-Bordignon *et al.* 1989 Valmori *et al.* 1992
•IHNV salmonid fish virus polyprotein AA336 - 444.	109	69, 277, 450	50 - 80% of WT		Morzunov *et al.* 1995
•IHNV salmonid fish virus polyprotein AA270 - 453	184	69, 353, 690, 723, 944	50 - 80% of WT		Morzunov *et al.* 1995
•Domain IV of the Anthrax Protective Antigen	144	690, 723	WT		Petosa *et al.* 1997
•Intimin binding domain of the Tir protein of *E. coli* O157:H7 flanked with copies of the *P. aeruginosa* type IVpilus "adhesintope	122	723	80% of WT		de Grado *et al.* 1999 Wong *et al.* 1992

Very large peptides (>200 AA)

•24-32 tandem repeats of MUC1 conserved repeat of mucins	16X= 343 24X= 511 32X= 679	723 723 723	WT	Considerable shedding of unassembled protein	Lan *et al.* 1990

*AA: amino acids
†WT: Wild type

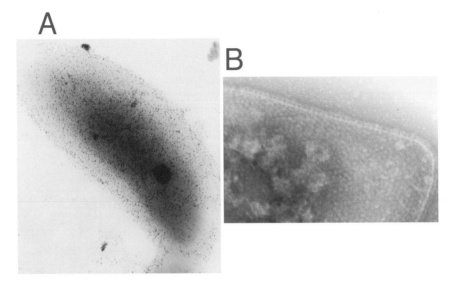

Figure 4. Caulobacter cells presenting IPNV salmonid fish virus glycoprotein VP2, amino acids 145 to 257 plus two T-cell epitopes within the S-layer (see Table 1). A: Immunolabel with IPNV-specific antisera and a protein A/colloidal gold conjugate; B: Negative stain electron microscopy of a lysed cell of the same construct demonstrating that S-layer export, attachment and assembly is still occurring.

ColE1/IncQ dual origin vectors described earlier for display purposes, using a complete copy of *rsaA*, since many of the anticipated uses for display require small, high copy number vectors with a minimum of extraneous restriction sites. The initial result was apparent toxicity. No clones could be obtained by electroporation when either the native *rsa*A promoter or several variants of the β-galactosidase (*lac*) promoter (used for fusion protein expression) were employed. We achieved success by modifying the spacing between the putative Shine-Dalgarno sequence and the ATG start codon of the native 5' untranslated region (J.F. Nomellini and J. Smit, unpublished). Presumably, this lowered the efficiency of translation starts such that the circa 20 copies of the *rsa*A gene per cell were tolerated. With vectors modified in this manner, we see no indication of either reduced expression or toxicity.

The S-Layer Associated Protease

In the initial report of display of small peptides, we noted that insertion in many places of the *rsa*A gene, while allowing for successful presentation (*i.e.* while maintaining S-layer attachment and assembly), also resulted in degradation of a notable fraction of the protein produced (Bingle *et al.*, 1997a). Unlike degradation in other Gram-negative display systems, which results from non-specific proteolytic action and produces a "smeared band" on SDS-gels, the proteolysis noted is typically the result of a single cleavage event. With the exception of a site at about 200 amino acids from the N-terminus, which seems especially prone to cleavage, the single site of hydrolysis was always within the recombinant insertion. The cleavage is sufficiently precise that unambiguous N-terminal amino acid sequence of the truncated segments can be obtained. Curiously, the proteolytic activity was only able to affect recombinant insertions in the middle two thirds of the gene. Insertions in the extreme N- or C-termini were unaffected. Since the N-terminal region is involved with attachment to the smooth LPS, many of these N-terminal insertion also resulted in S-layer monomer "shedding" (Bingle *et al.*, 1997a). Analogously, there are few available sites at the C-terminus of RsaA that do not disrupt the secretion signal. Initial expression of larger heterologous insertions also showed the degradation effect and it was often pronounced. We devised assays, using a combination of classical chemical and transposon insertion mutagenesis, to identify the protease involved. We discovered that a single gene (*sap*) was responsible (Umelo-Njaka *et al.*, 2002). The Sap protease appears to be the result of a gene rearrangement in which a Type I secreted serine protease has lost its C-terminal secretion signal and this region has been replaced by a sequence including a domain sharing moderate homology with an N-terminal portion of the RsaA monomer. Our working hypothesis is that this region enables Sap to remain associated with the RsaA monomer while still inside the cell (perhaps with a region derived from RsaA that is required for subunit-subunit interactions outside the cell for S-layer assembly). This would allow Sap to cleave non-native sequences in the middle of the RsaA monomer but be unable to "reach" either end of the protein. If correct, such a hypothesis could also explain why we have not seen comparable degradation of heterologous fusion proteins (even for those containing passenger proteins that were degraded when cloned in display vectors), since these do not have an N-terminal region to interact with Sap. Sap is present in the cell but is not able to cleave fusion proteins efficiently. A number of approaches toward clarifying the exact role of Sap are underway, but for the purposes of heterologous display, we were able to delete the chromosomal *sap* by standard gene knockout methods (Umelo-Njaka *et al.*, 2002), as well as to produce classical point mutants. With this single protease eliminated, we have seen no proteolytic activity in any of our subsequent display efforts. This observation is especially important when considering production of libraries of combinatorial peptides or random gene fragments, where minimization of any filtering process is important to maintain as large and as unbiased a library as possible.

Library Display: Use with Fluorescence Activated Cell Sorters (FACS)

Indications so far are that the S-layer presentation system should be capable of presenting libraries of heterologous segments and those segment sizes could be in the range of 10 to 200 amino acids. A bacterium displaying peptides at high density allows consideration of FACS as a means to rapidly screen libraries. In addition to speed, the ability to use soluble ligands (albeit derivatized to be fluorescent) may overcome the concerns related to attachment to solid supports (*e.g.* cost, high level of false positives and loss of best binders). We have carried out preliminary experimentation with Caulobacter cultures containing recombinant cells displaying the c-myc epitope at abundances as low as one in ten million. We have been able to readily detect and sort cells displaying the c-myc epitope at the lowest abundance (Andrew Johnson and J. Smit, unpublished). The process of screening and sorting into wells of three billion cells of such a "library" can be accomplished in about one hour. Individual cells then expand in the microtiter dish wells to a stationary culture in 3 to 4 days, growing at 30°C. Because Caulobacters produce an adhesive holdfast at the end of their stalk (Merker and Smit, 1988) which results in growth of a monolayer of cells on the walls of the microtiter well (Smit *et al.*, 2000b), many methods are possible for high throughput characterization of hits. These include direct staining of wells with specific reagents, and PCR-based sequence analysis of hits without centrifugation steps. We are presently engaged in several efforts to produce and display combinatorial peptide, gene fragment and antibody libraries to validate this display method for use in FACS.

Use in Whole Cell Vaccines or Anti-Infectives

We are also exploring use of the Caulobacters displaying candidate vaccine antigens to determine if this way to present antigens is superior to simple subunit vaccine proteins. For example, in farmed fish the ability to vaccinate by immersion (where antigens of a particulate nature are taken up by circulating macrophage in the gills) offers significant cost savings as well as other advantages in terms of being able to vaccinate fish that are too small for injection-based vaccination. Similarly, addition of cells displaying key viral or bacterial antigens in the feed may stimulate immunity or help prevent transmission of viral or bacterial infections by competing for binding sites. This may be true not only for fish but also for large farmed animal such as cows and pigs. Bacterial or particulate antigens may be important for stimulating strong T cell responses, which is a minimum requirement in the quest for anti-cancer vaccines. We are also engaged in projects to develop such an anti-cancer vaccine strategy.

An important feature of Caulobacters in these efforts is that the LPS of the bacterium does not stimulate a significant endotoxin response and in turn, presumably, no sepsis response or localized reaction to bacterial injection. In collaboration with others we have also determined that in at least some instances, the LPS can function as an adjuvant when used in combination with subunit vaccine proteins (Smit et al., 2002). This combination of features bodes well for use of Caulobacters as whole cell vaccines.

Development of Diagnostic Particles

Caulobacters displaying peptides at high density could supplant dipstick-based serodiagnostic assays for assessing the presence of specific antibodies. Often, colored latex beads chemically coupled to antibody-binding peptides form the basis of the assay. However, to do this it would be ideal if the bacteria were highly colored, like the latex beads, thereby enabling simple readouts. We have done preliminary experimentation with S-layer negative Caulobacter cells that have been chemically stabilized with glutaraldehyde and stained with Crystal Violet. These cell-based particles are still capable of serving as an attachment surface for the recrystallization of recombinant S-layer protein (J. Smit, unpublished), using methods developed for RsaA protein extraction and recrystallization (Nomellini *et al.*, 1997). Thus it appears that producing colored "particles" displaying peptides useful for serodiagnosis is possible.

Concluding Remarks

Both the gene expression and display aspects of the Caulobacter S-layer system are still in active development in numerous areas to better understand the limits of the system and to create better expression strains by modification or amplification of the transporter complex. In addition, we are presently developing vectors for auxotrophic selection, to eliminate the need for antibiotic selection or the presence of antibiotic resistance genes. We are also designing vectors for co-expression of two different recombinant S-layers, using tandemly repeated *rsaA* gene copies and ways to quickly convert "hits" detected by the library display system to vectors producing the target as a fusion protein. Evaluation of enzyme cleavage sites for removal of the secretion signal of fusion proteins and N-terminal display vectors are areas of development that were mentioned earlier. For the past two years, a version of the Caulobacter gene expression system has been marketed by Invitrogen (Carlsbad, CA) under the name PurePro™, as part of a licensing agreement with our university. That relationship ended recently and we anticipate that both the gene expression and display aspects of Caulobacter will be commercially available during the next year from another source. In the meantime, interested readers are encouraged to contact the authors for access to the system.

Acknowledgements

The studies reported here were supported by grants to J. S. from the Natural Sciences and Engineering Research Council of Canada, as well as support from a number of individual corporate sponsors. In particular, we acknowledge the cooperation and assistance of Dr. Sol Langermann and Medimmune, Inc and Dr. William W. Kay and Microtek International for Figures 3 and 4, respectively. We also acknowledge the contributions of former laboratory members, especially those of Wade H. Bingle.

References

Abraham, W.-R., Strömpl, C., Meyer, H., Lindholst, S., Moore, E.R.B., Christ, R., Vancanneyt, M., Tindall, B.J., Bennasar, A., Smit, J., and Tesar, M. 1999. Phylogeny and polyphasic taxonomy of Caulobacter species. Proposal of *Maricaulis* gen. nov. with *Maricaulis maris* (Poindexter) comb. nov. as the type species, and emended description of the genera *Brevundimonas* and *Caulobacter*. Int'l J. System. Bacteriol. 49:1053-1073.

Anast, N., and Smit, J. 1988. Isolation and characterization of marine Caulobacters and assessment of their potential for genetic experimentation. Appl. Environ. Micro. 54: 809-817.

Andersen, C., Hughes, C., and Koronakis, V. 2001. Protein export and drug efflux through bacterial channel-tunnels. Curr. Opin. Cell Biol. 13(4):412-6.

Awram, P., and Smit, J. 1998. The *Caulobacter crescentus* paracrystalline S-layer protein is secreted by a ABC transporter (type I) secretion apparatus. J. Bacteriol. 180: 3062-3069.

Awram, P., and Smit, J. 2001. Identification of lipopolysaccharide O-antigen synthesis genes required for attachment of the S-layer of *Caulobacter crescentus*. Microbiol. 147:1451-1460.

Beveridge, T.J., Pouwels, P.H., Sara, M., Kotiranta, A., Lounatmaa, K., Kari, K., Kerosuo E., Haapasalo, M., Egelseer E.M., Schocher, I., Sleytr, U.B., Morelli, L., Callegari, M.L., Nomellini, J.F. Bingle, W.H. Smit, J., Leibovitz, E., Lemaire M. Miras, I., Salamitou, S.,

Beguin, P., Ohayon, H., Gounon, P., Matuschek, M., and Koval, S.F. 1997. Functions of S-layers. 6. Form, function and utility studies with the *Caulobacter crescentus* S-layer protein. FEMS Microbiol. Rev. 20: 121-127.

Binet, R., Letoffe, S., Ghigo, J.M., Delepelaire, P., and Wandersman, C. 1997. Protein secretion by Gram-negative bacterial ABC exporters—a review. Gene. 192: 7-11.

Bingle, W.H., Kurtz, Jr.,H.D, and Smit, J. 1993a. An "all-purpose" cellulase reporter for gene fusion studies and application to the paracrystalline surface (S)-layer protein of *Caulobacter crescentus*. Can. J. Microbiol. 39: 70-80.

Bingle, W.H., and Smit, J. 1994. Alkaline phosphatase and a cellulase reporter protein are not exported from the cytoplasm when fused to large N-terminal portions of the *Caulobacter crescentus* surface (S)-layer protein. Can. J. Microbiol. 40: 777-782.

Bingle,W.H., Nomellini, J.F. and Smit, J. 1997a. Cell surface display of a *Pseudomonas aeruginosa* PAK pilin peptide within the paracrystalline S-layer of *Caulobacter crescentus*. Molec. Microbiol. 26: 277-288.

Bingle, W.H., Nomellini, J.F., and Smit, J. 1997b. Linker mutagenesis of the *Caulobacter crescentus* S-layer protein: Toward a definition of an N-terminal anchoring region and a C-terminal secretion signal and the potential for heterologous protein secretion. J. Bacteriol. 179: 601-611.

Bingle, W.H., Nomellini, J.F, and Smit, J. 2000. Secretion of the *Caulobacter crescentus* S-layer Protein: Further localization of the C-terminal secretion signal and its use for secretion of recombinant proteins. J. Bacteriol. 182: 3298-3301.

Bingle, W.H.. and Smit, J. 1990. High level plasmid expression vectors for *Caulobacter crescentus* incorporating the transcription and transcription-translation initiation regions of the paracrystalline surface layer protein gene. Plasmid. 24: 143-148.

Bingle, W.H., Walker, S.G., and Smit, J. 1993b. Definition of form and function for the S-layer of *Caulobacter crescentus*. Advances in Bacterial Paracrystalline Surface Layers, T.J. Beveridge, S.F. Koval, eds., Plenum Press, pp. 181-192.

Blight, M.A.. and Holland, I.B. 1994. Heterologous protein secretion and the versatile Escherichia coli haemolysin translocator. Trends in Biotechnol. 12: 450-455.

Boder, E.T., and Wittrup K.D. 2000. Yeast surface display for directed evolution of protein expression, affinity, and stability. Methods in Enzymol. 328: 430-44.

Boot, H.J., and Pouwels, P.H. 1996. Expression, secretion and antigenic variation of bacterial S-layer proteins. Molec. Microbiol. 21: 1117-1123.

Bulieris, P.V., Behrens, S., Holst, O., and Kleinschmidt. J.H. 2003. Folding and insertion of the outer membrane protein OmpA is assisted by the chaperone Skp and by lipopolysaccharide. J. Biol. Chem. 278: 9092-9099.

Cereghino, J.L., and Cregg, J.M. 2000. Heterologous protein expression in the methylotrophic yeast *Pichia pastoris*. FEMS Microbiol. Rev. 24: 45-66.

Choudhury, D., Thompson, A., Stojanoff, V., Langermann, S., Pinkner, J., Hultgren, S.J., and Knight, S.D. 1999. X-ray structure of the FimC-FimH chaperone-adhesin complex from uropathogenic *Escherichia coli*. Science. 285: 1061-1066.

Collet, J.F., and Bardwell, J.C. 2002. Oxidative protein folding in bacteria. Mol. Microbiol. 44: 1-8.

Danner, S., and Belasco, J.G. 2001. T7 phage display: a novel genetic selection system for cloning RNA-binding proteins from cDNA libraries. Proc. Natl. Acad. Sci. USA 98: 12954-12959.

de Grado, M., Abe, A., Gauthier, A., Steele-Mortimer, O. and Finlay, B.B. 1999. Identification of the intimin binding domain of Tir of enteropathogenic *E.coli* (EPEC) Cell. Micro. 1: 7-17.

Ely, B. 1991. Genetics of *Caulobacter crescentus*. Methods Enzymol. 204: 372-384.

England, J.C., and Gober, J.W. 2001. Cell cycle control of cell morphogenesis in Caulobacter. Curr. Opin. Microbiol. 4: 674-680.

Engelking, H.M., and Leong, J.C. 1989. The glycoprotein of infectious hematopoietic necrosis virus elicits neutralizing antibody and protective responses. Virus Res. 13: 213-230.

Etz, H., Minh, D.B., Schellack, C., Nagy, E., and Meinke, A. 2001. Bacterial phage receptors, versatile tools for display of polypeptides on the cell surface. J. Bacteriol. 183: 6924-6935.

Fahnestock, S.R., Alexander, P., Nagle, J., and Filpula, D. 1986. Gene for an immunoglobulin-binding protein from a group G streptococcus. J.Bacteriol. 167: 870-880.

Fernandez, L.A., and de Lorenzo, V. 2001. Formation of disulphide bonds during secretion of proteins through the periplasmic-independent type I pathway. Mol. Microbiol. 40:332-346.

Fernandez, L.A. Sola, I., Enjuanes, L., and de Lorenzo, V. 2000. Specific secretion of active single-chain Fv antibodies into the supernatants of *Escherichia coli* cultures by use of the hemolysin system. Appl. Environ. Microbiol. 66: 5024-5029.

Fisher, J., Smit, J., and Agabian, N. 1988. Transcriptional analysis of the major surface array gene of *Caulobacter crescentus*. J. Bacteriol. 170: 4706-4713.

Ghadiri, M.R., and Choi, C. 1990. Secondary structure nucleation in peptides. Transition metal ion stabilized alpha-helices. J.Am. Chem Soc.112: 1630-1632

Gellissen, G. 2000 . Heterologous protein production in methylotrophic yeasts. Appl. Microbiol. Biotechnol. 54: 741-750.

Gentschev, I., Dietrich, G., and Goebel, W. 2002. The E. coli alpha-hemolysin secretion system and its use in vaccine development. Trends Microbiol. 10: 39-45.

Gentschev, I., Maier, G., Kranig, A., and Goebel, W. 1996 . Mini-TnhlyAs: a new tool for the construction of secreted fusion proteins. Mol. Gen. Genet.. 252: 266-74.

Gilchrist, A., Fisher, J.A., and Smit, J. 1992. Nucleotide sequence analysis of the gene encoding the *Caulobacter crescentus* paracrystalline surface layer protein. Can. J. Microbiol. 38: 193-202.

Gilchrist, A., and Smit, J. 1991. Transformation of freshwater and marine Caulobacters by electroporation. J. Bacteriol. 173: 921-925.

Heumann, D., Glauser, M.P., and Calandra, T. 1998. Molecular basis of host-pathogen interaction in septic shock. Curr. Opin. Microbiol. 1: 49-55.

Holland, I.B., and Blight, M.A. 1999. ABC-ATPases, adaptable energy generators fuelling transmembrane movement of a variety of molecules in organisms from bacteria to humans. J. Mol. Biol. 293: 381-399.

Irving, M.B., Pan, O. and Scott, J.K. 2001. Random-peptide libraries and antigen-fragment libraries for epitope mapping and the development of vaccines and diagnostics. Curr.Opin. Chem. Biol. 5: 314-324.

Jenal, U. Stephens, C. Shapiro, L. 1995. Regulation of asymmetry and polarity during the Caulobacter cell cycle. Advances in Enzymology and Related Areas of Molecular Biology. 71: 1-39.

Klemm, P., and Schembri, M.A. 2000. Fimbrial surface display systems in bacteria: from vaccines to random libraries. Microbiol. 146: 3025-3032.

Koronakis, V. Andersen, C. and Hughes, C. 2001. Channel-tunnels. Curr. Opin. Struct. Biol. 11: 403-407.

Koronakis, V., Sharff, A., Koronakis, E,. Luisi, B., and Hughes, C. 2000. Crystal structure of the bacterial membrane protein TolC central to multidrug efflux and protein export. Nature. 405: 914-919.

Koval, S.F., and Hynes, S.H. 1991. Effect of paracrystalline protein surface layers on predation by Bdellovibrio bacteriovorus. J. Bacteriol. 173: 2244-2249.

Kruger, C. Hu, Y., Pan, Q., Marcotte, H., Hultberg, A., Delwar, D., van Dalen, P.J., Pouwels, P.H., Leer, R.J., Kelly, C.G., van Dollenweerd, C., Ma, J.K., and Hammarstrom, L. 2002. *In situ* delivery of passive immunity by lactobacilli producing single-chain antibodies. Nat. Biotechnol. 20: 702-706.

Lan, M.S., Batra, S.K., Qi, W.N., Metzgar, R.S., and Holingsworth, M.A. 1990. Cloning and sequencing of a human pancreatic tumor mucin cDNA. J. Biol Chem. 265: 15294-15299.

Lee, S.F., March, R.J., Halperin, S.A., Faulkner, G., and Gao, L. 1999. Surface expression of a protective recombinant pertussis toxin S1 subunit fragment in *Streptococcus gordonii*. Infect. Immun. 67: 1511-1516.

Leenhouts, K., Buist, G., and Kok, J. 1999. Anchoring of proteins to lactic acid bacteria. Antonie van Leeuwenhoek. 76: 367-376.

Lu, Z, Murray, K.S., Van Cleave, V., LaVallie, E.R., Stahl, M.L., and McCoy, J.M. 1995. Expression of thioredoxin random peptide libraries on the *Escherichia coli* cell surface as functional fusions to flagellin: a system designed for exploring protein-protein interactions. Biotechnol. (N.Y.) 13: 366-372.

Lupas, A., Engelhardt, H., Peters, J., Santarius, U., Volker, S., and Baumeister, W. 1994. Domain structure of the *Acetogenium kivui* surface layer revealed by electron crystallography and sequence analysis. J. Bacteriol. 176: 1224-1233.

MacRae, J.D., and Smit, J. 1991. Characterization of Caulobacters isolated from wastewater treatment systems. Appl. Environ. Microbiol 57: 751-758.

Mazmanian, S.K., Ton-That, H., and Schneewind, O. 2001. Sortase-catalysed anchoring of surface proteins to the cell wall of *Staphylococcus aureus*. Molec. Microbiol. 40(5):1049-57.

Merker, R.I., and Smit, J. 1988. Analysis of the adhesive holdfast of marine and freshwater Caulobacters. Appl. Environ. Micro. 54: 2078-2085.

Mesnage, S., Tosi-Couture, E,. Mock, M,. and Fouet, A. 1999. The S-layer homology domain as a means for anchoring heterologous proteins on the cell surface of *Bacillus anthracis*. J. Appl. Microbiol. 87: 256-260.

Mollenkopf, H.J., Gentschev, I., and Goebel, W. 1996. Conversion of bacterial gene products to secretion-competent fusion proteins. Biotechniques. 21: 854, 856-60.

Morzunov, S.P., Winton, J.R., and Nichol, S.T. 1995. The complete genome structure and phylogenetic relationship of infectious hematopoietic necrosis virus. Virus Res.38: 175-192.

Nierman, W.C., Feldblyum, T.V., Laub, M.T., Paulsen, I.T., Nelson, K.E., Eisen, J., Heidelberg, J.F., Alley, M.R.K., Ohta, N., Maddock, J.R., Potocka, I., Nelson, W.C., Newton, A., Stephens, C., Phadke, N. D., Ely, B., DeBoy, R.T., Dodson, R.J., Durkin, A.S., Gwinn, M.L., Haft, D.H., Kolonay, J.F, Smit, J., Craven, M.B., Khouri, H., Shetty, J., Berry, K., Utterback, T., Tran, K., Wolf, A., Vamathevan, J., Ermolaeva, M., White, O., Salzberg, S.L., Venter, J.C., Shapiro, L., and Fraser, C.M. 2001. Complete genome sequence of *Caulobacter crescentus* strain CB15. Proc. Natl. Acad. Sci. USA. 98: 4136-4141.

Nomellini, J.F., Kupcu, S., Sleytr, U.B., and Smit, J. 1997. Factors controlling *in vitro* recrystallization of the *Caulobacter crescentus* paracrystalline S-layer. J. Bacteriol. 179: 6349-6354.

Panina-Bordignon, P., Tan, A., Termitijtelen, A., Demotz, S., Corradin, G. and Lanzavecchia, A. 1989. Universally imunogenic T cell epitopes; Promiscuous binding to human MHC classII and promiscuois recognition by T cells. Eur. J. Immunol. 19: 2237-2242.

Perna N.T., Plunkett G. 3rd., Burland V., Mau B., Glasner J.D., Rose D.J., Mayhew G.F., Evans P.S., Gregor J., Kirkpatrick H.A., Posfai G., Hackett J., Klink S.., Boutin A., Shao Y., Miller L., Grotbeck

E.J., Davis N.W., Lim A., Dimalanta E.T., Potamousis K.D., Apodaca J., Anantharaman T.S, Lin J., Yen G., Schwartz D.C., Welch R.A., Blattner F.R. 2001. Genome sequence of Enterohaemorrhagic *Escherichia coli*: O157:H7. Nature 409: 529-533.

Petosa, C., Collier, R.J., Kimpel, K.R., Leppla, S.H. and Liddington, R.C. 1997. Crystal structure of the anthrax toxin protective antigen. Nature 385: 833-838.

Plano G.V., Day J.B., Ferracci, F. 2001. Type III export: new uses for an old pathway. Molec. Microbiol. 40: 284-293.

Poindexter, J.S. 1978. Selection for nonbuoyant morphological mutants of *Caulobacter crescentus*. J. Bacteriol. 135:1141-1145.

Pouwels, P.H., Vriesemam A., Martinez, B., Tielen, F.J., Seegers, J.F., Leer, R.J., Jore, J., and Smit, E. 2001. Lactobacilli as vehicles for targeting antigens to mucosal tissues by surface exposition of foreign antigens. Methods Enzymol. 336: 369-89.

Rabbitts, T.H., Hamlyn, P.H., and Baer, R. 1983. Altered nucleotide sequences of a translocated C-myc gene in Burkitts lymphoma. Nature 306: 760-765.

Samuelson, P., Gunneriusson, E., Nygren, P.A., and Stahl, S. 2002. Display of proteins on bacteria. J. Biotechnol. 96: 129-154.

Sara, M., and Sleytr, U.B. 2000. S-Layer proteins. J. Bacteriol. 182: 859-868.

Schletter J., Heine H., Ulmer A.J., Rietschel, E.T. 1995. Molecular mechanisms of endotoxin activity. Arch. Microbiol. 164: 383-389.

Simon, B., Nomellini, J.F., Chiou, P., Bingle, W.H., Thornton, J., Smit, J., and Leong, J-A. 2001. Recombinant vaccines against Infectious Hematopoietic Necrosis Virus: Production by the *Caulobacter crescentus* S-layer protein secretion system and evaluation in laboratory trials. Diseases of Aquatic Organisms. 44: 17-27.

Sleytr, U.B., and Beveridge, T.J. 1999. Bacterial S-layers. Trends in Microbiol. 7(6):253-60.

Smit, J., 1986. "Protein surface layers of bacteria", Chapter 13 in Outer Membranes as Model Systems, M. Inouye, editor, J. Wiley and Sons, pp. 343-376.

Smit, J., and Agabian, N. 1982a. *Caulobacter crescentus* pili: analysis of production during development. Dev. Biol. 89: 237-247.

Smit, J., and Agabian, N. 1982b. Cell surface patterning and morphogenesis: biogenesis of a periodic surface array during Caulobacter development. J. Cell Biol. 95: 41-49.

Smit, J., and Agabian, N. 1984. Cloning the major protein of the *Caulobacter crescentus* periodic surface layer: detection and characterization of the cloned peptide by protein expression assays. J. Bacteriol. 160: 1137-1145.

Smit, J., Engelhardt, H., Volker, S., Smith, S.H., and Baumeister, W. 1992. The S-layer of *Caulobacter crescentus*: Three-dimensional image reconstruction and structure analysis by electron microscopy. J. Bacteriol. 174: 6527-6538.

Smit, J., Grano, D.A., Glaeser, R.M., and Agabian, N. 1981. A periodic surface array in *Caulobacter crescentus*: Chemical and fine structure analysis. J. Bacteriol. 146: 1135-1150.

Smit, J., Hermodson, M., and Agabian, N. 1981. *Caulobacter crescentus* pilin: purification, chemical characterization and amino-terminal amino acid sequence of a structural protein regulated during development. J. Biol. Chem. 256: 3092-3097.

Smit, J., Langermann, S., Koenig, S., and Qureshi, N. 2002. Caulobacter LPS Immunoadjuvant. U.S. Patent Office #6,368,599.

Smit, J., Nomellini, J.F., and Bingle, W.H. 2000a. "Electroporation of plasmids into freshwater and marine strains of Caulobacter", in Electrotransformation of Bacteria, N. Eynard and J. Teissie, ed., Springer-Verlag; Chapt. 33, pp 271-280.

Smit, J., Sherwood, C., and Turner, R.F.B. 2000b. Characterization of high density monolayers of the biofilm bacterium *Caulobacter crescentus*: Evaluating prospects for developing immobilized cell bioreactors. Can. J. Microbiol. 46: 339-349.

Thanassi, D.G., and Hultgren, S.J. 2000. Multiple pathways allow protein secretion across the bacterial outer membrane. Curr Opin Cell Biol. 12(4):420-30.

Thompson, S.A., Shedd, O.L., Ray, K.C., Beins, M.H., Jorgensen, J.P., and Blaser, M.J. 1998. *Campylobacter fetus* surface layer proteins are transported by a type I secretion system. J. Bacteriol. 180: 6450-6458.

Tzschaschel, B.D., Guzman, C.A., Timmis, K.N., and de Lorenzo, V. 1996. An *Escherichia coli* hemolysin transport system-based vector for the export of polypeptides: export of Shiga-like toxin IIeB subunit by *Salmonella typhimurium* aroA. Nature Biotechnol. 14: 765-769.

Umelo-Njaka, E., Bingle,W.H., Borchani, F., Le, K.D., Awram, P., Blake, T., Nomellini, J.F., and Smit, J. 2002. *Caulobacter crescentus* synthesizes an S-layer editing metalloprotease possessing a domain sharing sequence similarity with its paracrystalline S-layer protein. J. Bacteriol. 184: 2709-2718.

Umelo, E., Nomellini, J.F., Bingle, W.H., Glasier, L., Irvin, R.T., and Smit, J. 2000. Expression and testing of *Pseudomonas aeruginosa* vaccine candidate proteins prepared with the *Caulobacter crescentus* S-layer protein expression system. Vaccine 19: 1406-1415.

Umelo-Njaka, E., Nomellini, J.F., Yim, H., and Smit, J. 2001. Development of small high copy number plasmid vectors for gene expression in *Caulobacter crescentus*. Plasmid 46: 37-46.

Valmori, D., Pessi, A., Bianchi, E., and Corradin, G. 1992. Use of human universally antigenic tetanus toxin T cell epitopes as carriers for human vacination. J.Immunol. 149: 717-721.

Van Wely, K. H.M., Swaving., J., Freudl, J. Driessen, A. J.M. 2000. Translocation of proteins across the cell envelope of Gram-positive bacteria. FEMS Microbiol. Rev. 25: 437-454.

Voulhoux, R., Bos, M.P., Geurtsen, J., Mols, M., Tommassen, J. 2003. Role of a highly conserved bacterial protein in outer membrane protein assembly. Science 299: 262-265.

Walker, S.G., Kurunaratne, D.N., Ravenscroft, N., and Smit, J. 1994. Characterization of mutants of *Caulobacter crescentus* defective in surface attachment of the paracrystalline surface layer. J. Bacteriol. 176: 6312-6323.

Walker, S.G., Smith, S.H., and Smit, J. 1992. Isolation and comparison of the paracrystalline surface layer proteins of freshwater Caulobacters. J. Bacteriol. 174: 1783-1792.

Weber, V. Weigert, S. Sara, M., Sleytr, U.B., and Falkenhagen, D. 2001. Development of affinity microparticles for extracorporeal blood purification based on crystalline bacterial cell surface proteins. Therapeutic Apheresis. 5: 433-438.

Wernerus, H., Lehtio, J., Samuelson, P., and Stahl, S. 2002. Engineering of staphylococcal surfaces for biotechnological applications. J. Biotechnol. 96: 67-78.

Wheeler, R.T., Gober, J.W., and Shapiro, L. 1998. Protein localization during the *Caulobacter crescentus* cell cycle. Curr. Opin. Microbiol. 1: 636-642.

Wittrup, K.D. 2001. Protein engineering by cell-surface display. Curr. Opin. Biotechnol. 12: 395-399.

Wong, S.L. 1995. Advances in the use of *Bacillus subtilis* for the expression and secretion of heterologous proteins. Curr. Opin. Biotechnol. 6: 517-522.

Wong, W.Y., Irvin, R.T., Paranchych, W., and Hodges, R.S. 1992. Antigen-antibody interactions: elucidation of the epitope and strain-specificity of a monoclonal antibody directed against the pilin protein adherence binding domain of *Pseudomonas aeruginosa* strain K. Protein Sci. 1: 1308-1318.

Wu, S.C., Ye, R., Wu, X.C., Ng, S.C., and Wong, S.L. 1998. Enhanced secretory production of a single-chain antibody fragment from *Bacillus subtilis* by coproduction of molecular chaperones. J. Bacteriol. 180: 2830-2835.

Wu, S.C., Yeung, J.C., Duan, Y., Ye, R., Szarka, S.J., Habibi, H.R., and Wong, S.L. 2002. Functional production and characterization of a fibrin-specific single-chain antibody fragment from *Bacillus subtilis*: effects of molecular chaperones and a wall-bound protease on antibody fragment production. App. Envir. Microbiol. 68: 3261-3269.

Index

oxidoreductases 101-105, 117-120
peptidyl-prolyl *cis-trans* isomerases
98-99, 116-117
plasmids 87-89, 158-162
promoters 89-92, 163-169
proteases 105-109, 120-125
RNA processing and decay 169-173
secretion 110-115
sigma factors 89
translational control 179

Fab 114, 116
Farnesylation 412-413
Fermentation
batch 316
continuous 321
fed-batch 316-320, 493, 500
feed rate 317-318, 328-329
high density 480, 482, 500
mixed feed 319-320, 329, 331-332
solid-state 359-361
viscosity 355-357
FfH 114, 214
Filamentous fungi
advantages and limitations 346-347,
484
chaperones 370-371
fermentation 354-361
genomics 372-374
morphology 355-357
promoters 351-352
recombinant protein expression
361-369
regulation of protein expression
352-353
selectable markers 350-351
transformation 348-350
FimC 116
FkpA 116-117
FLP recombinase 154, 158, 446
FRT sites 154, 158, 446
FtsH 107, 109
FtsY 114, 214
Fusion protein 264, 362-363, 364-366, 496,
501-502

GC content 26, 45-46, 468, 505
General secretory pathway *See* Sec
pathway

Genome
acidophiles 6
alkaliphiles 6
analysis 278-279
Bacillus subtilis 235-236
Caulobacter crescentus 493
Fungi 372
halophiles 6
insect cells 391
insects 391
Methylobacterium extorquens AM1
467
minimization 236-238
psychrophiles 6
shuffling 181-182
thermophiles 5-6
GFP *See* Green fluorescent protein
Glutaredoxin pathway 103-104
Glycosylation
engineering 312, 369, 410-411, 449
hyperglycosylation 271, 312
in CHO cells 449
in fungi 368-369
in human proteins 427-428
in insect cells 407-411
in *Pichia* 311
in *Saccharomyces cerevisiae* 269-271
N-linked 269-270, 311-312, 368,
407-410
O-linked 269-270, 311-312, 407, 411
GmMNPV 406
gor 103-105
gp120 402
gp64 401-402
GPCR *See* G-protein coupled receptors
G-protein coupled receptors 24, 275-277,
282, 412
GRAS 37, 200, 347, 429
Green fluorescent protein
as a reporter of expression 405-406,
471-472
as a reporter of folding 47
as a reporter of mRNA stability 172
as a reporter of protein localization
278-279
as a reporter of secretion 113
for recombinant baculoviruses
detection 400
GroEL 93-95, 99-101, 110, 204, 215, 473.
See also Hsp60